PATTERNS IN PREHISTORY

PATTERNS IN

Humankind's First
Three Million Years

PREHISTORY

ROBERT J. WENKE

THIRD EDITION

New York   Oxford
OXFORD UNIVERSITY PRESS
1990

For Chris and Ann Wenke

Oxford University Press

Oxford   New York   Toronto
Delhi   Bombay   Calcutta   Madras   Karachi
Petaling Jaya   Singapore   Hong Kong   Tokyo
Nairobi   Dar es Salaam   Cape Town
Melbourne   Auckland

and associated companies in
Berlin   Ibadan

Published by Oxford University Press, Inc.,
200 Madison Avenue, New York, New York 10016

Library of Congress Cataloging-in-Publication Data
Wenke, Robert J.
Patterns in prehistory : humankind's first three million years /
Robert J. Wenke. — 3rd ed.
p.   cm.   Includes bibliographical references.
ISBN 0-19-506848-3
ISBN 0-19-505522-5 (pbk)
1. Man, Prehistoric.   2. Archaeology.   I. Title.
GN740.W46   1990
573.3—dc20   89-78441   CIP

2 4 6 8 9 7 5 3

Printed in the United States of America
on acid-free paper

Preface

My objective in writing this book was to review the first three million years of the human past. To express the difficulties of compressing such a subject into a single yet still portable volume, I prefaced a previous edition by quoting Woody Allen, who said, "I took a course in speed reading . . . and I was able to read *War and Peace* in twenty minutes. It's about Russia." This kind of superficiality remains an obvious danger for a book like *Patterns,* especially given the exponential increase in recent years of archaeological research, but my objective has always been more to convey a *perspective* on the past than to recount the details of our antiquity. In Douglas Adams' delightful *The Restaurant at the End of the Universe,* a scientist's family, frustrated by his obsession with astronomy and cosmology, kept nagging him to give up his endless study of the universe: "Have some sense of proportion," they would say, as often as thirty-eight times in a single day. In response the scientist built the "Total Perspective Vortex": by peering into this machine, one could see one's self in relation to the size of the entire universe—a disproportion so vast and awesome that it instantly destroyed his wife's mind—proving that the one thing people "cannot afford to have is a sense of proportion."

This, the third edition of *Patterns,* like the first two, is an attempt to provide the reader with a sense of proportion and perspective only slightly less grand than that offered by the Total Perspective Vortex. Specifically, I hope to assist readers in seeing themselves in relation to the great evolutionary transformations of our past: (1) the evolution of "culture" itself, the initial use of tools and human social forms by our ape-like ancestors, who survived the bloody struggle of life on the African savannas for several million years; (2) the first appearance of "us," *Homo sapiens,* out of the mix of intrepid Ice Age hominid hunter-foragers, and our penetration of the world's frontiers, from Australia to Argentina; (3) the evolution of "agriculture," in the form of the small-time farmers of a few thousand years ago, who brought the village way of life to most of the world; and (4) the first appearances of cultural and social "complexity," in the form of the earliest and greatest civilizations of antiquity.

For reasons discussed at length in Chapter 1, I remain convinced that to see oneself in relation to these transformations and the great antiquity of humankind is an indispensable part of any liberal education, especially for college students, whether they be potential engineers, captains of industry, mortuary scientists, or theologians—perhaps especially if they are engineers or theologians.

All academicians feel this way about their subjects, of course, and I have no illusions that the study of world prehistory will eradicate poverty or lead to world peace. In fact, the reader will find that throughout this book I pose the great questions of prehistory and then conclude that we don't really know what "caused" bipedalism, tool-use, human brain expansion, agriculture, writing, etc. In describing yet another loss, baseball great Yogi Berra observed that watching the game was like "déjà vu all over again," and by the end of this book the reader may feel the same way about archaeology, as I recount how time after time archaeologists have gone up against the great problems of prehistory, and in the end have concluded, mainly, that "more research is needed."

Yet I believe fundamentally in the Major Premise of Archaeology, which is that much of what we can hope to know about ourselves and our future can come only from a study of the past. Philosophy teaches us how to think effectively, physics instructs us in the mechanics of our universe, and psychology informs us about the nature of human cognition. But only archaeology and the study of the past have the potential to tell us where we came from, and by what processes. And thus archaeology can set ourselves and our own world into the context of all that has gone before.

If archaeology were simply a matter of collecting interesting objects and speculating about what they might have been used for, no ponderous theoretical discussions would be required, and this book would be much shorter. But all over the world archaeologists are trying to understand what we can hope to learn about the past and what the past "means," and anyone who hopes to understand modern archaeology must know something of these debates. My own theoretical perspective in the first two editions of this text was a rather straightforward, simple, cultural ecological evolutionism. This general perspective has been widely questioned in contemporary archaeology. And yet, although I am not blind to its limitations, I've retained much the same perspective taken in the first two editions. That is, I've tried to illustrate the importance of basic economic, technological, demographic, and environmental factors in determining our past. Many nonarchaeologists think that archaeology is primarily concerned with lavish tombs and lovely temples. There is still some truth to that, and I've devoted a lot of this book to the more spectacular remains of antiquity, but I've also tried to explain, for example, why for every ancient temple there were 200 fly-blown villages, and why it is the remains of the villages that, in many ways, are the most important remnants of our past.

At the same time, I've also tried to explain why there has been something of a loss of faith in techno-environmental deterministic interpretations of history and the goal of making archaeology an empirical, fundamentally economic, quantitatively sophisticated science of history—and why not every archaeologist now thinks that multivariate statistics will reveal the laws of the cultural universe. Many archaeologists believe that archaeology is not the science of what happened in the past so much as it is the study of a past we ourselves create, in that we necessarily read into ancient bones and stones elements of our own beliefs, values, and other sociopolitical identity. I appreciate the importance of this point, and I have provided several examples of alternative "post-processual" archaeological analyses, so that the reader can at least see what some of the issues are (and why to some "archaeological theory" is a classic oxymoron). But I persist in the belief that there is much of interest about world prehistory that can be analyzed in the format of an empirical science, and this book reflects that bias.

There are, of course, other biases as well. Much of this edition is devoted to the problem of explaining the origins of complex societies because I remain convinced that this is one of the most important transitions in human history, and because I have spent many years pursuing archaeological evidence of this transition in Egypt, Iran, Turkey, Italy, Mexico, and even darkest Missouri.

To laypeople archaeology sounds as if it is the most exciting work one could do, and sometimes it is. But it's also usually a lot of bloody hard labor as well, and I've tried to acknowledge as many as I could of the thousands of scholars who have gone out and done the fieldwork on which surveys such as this are based. I've also tried to give them a chance to speak for themselves, by quoting them at length.

Professors David Webster and Julie Stein gave me many useful suggestions about this third edition, and I appreciate their help. I am particularly grateful to Valerie Aubry, my editor at Oxford, for her assistance, encouragement, and exactly the right amount of patience. Niko Pfund, assistant editor at Oxford, expended much of that firm's profits on this book in telephone charges and FAXs to get illustrations permissions, and I very much appreciate his assuming this difficult task. Rosemary Wellner did an excellent job of text editing, and with commendable speed as well.

For fifteen years, students in a variety of courses at the University of Washington have listened to me lecture on the topics in this book, and I appreciate not only their patience and interest but also their many questions and comments that have helped shape *Patterns*.

The staff of the Humanities and Arts Computing Center at the University of Washington magically and magnanimously put the second edition of this book on computer discs with an optical scanner, saving me many months of work. I would also like to thank Maurice and Lois Schwartz for their gracious help in supporting my initial fieldwork in Egypt, the National Science Foundation for allowing me to spend much of the last ten years doing field archaeology in Egypt, and Janet Long, Richard Redding, Paul Buck, Emilia Zartman, Michal Kobusiewicz, Karla Kroeper, Lech Krzyzaniak, Douglas Brewer, Maria Casini, and Hisham al-Hegazy for their continued efforts on the Kom el-Hisn Project. I thank Dr. Mary Ellen Lane, Executive Director of the Council of American Overseas Research Centers, and Mr. Colin Davies, founder and Past President of the Middle East and North African branch of the Buddy Holly Appreciation Society, for educating and entertaining me during the past decade, in association with my goddaughter, Julia Lane Davies. Shireen al-Fostati generously assisted in all phases of manuscript preparation.

Most of all I am grateful to Dr. Nanette Marie Pyne, who has helped prepare all three editions of this book and who collaborated in much of the fieldwork on which it is based. *Albo lapillo noto omniam diem.*

Seattle   R. J. W.
January 1990

Contents

1. Prehistory, History, and Archaeology 3

2. Fundamentals of Archaeology 39

3. The Origins of Culture 75

4. The Origins of *Homo sapiens sapiens* 136

5. The First Americans 196

6. The Origins of Agriculture 225

7. The Evolution of
   Socially Complex Cultures 277

8. The Origins of Complex
   Societies in Southwest Asia 318

9. The Origins of Egyptian Civilization 370

10. Indus Valley Civilization 407

11. Early Chinese Civilization 428

12. Secondary Old World States 452

13. The Evolution
    of Mesoamerican Civilization 476

14. Andean Civilizations 529

15. Early Cultural Complexity
    in North America 557

16. Prehistory in Perspective 594

Credits 606

Index 609

PATTERNS IN PREHISTORY

1

Prehistory, History, and Archaeology

History is philosophy teaching by examples.
Dionysius of Halicarnassus (c. 40 B.C.)

History is bunk.
Henry Ford (A.D. 1919)

The archaeologist-adventurers of film and fiction never seem to have any doubts about what they are looking for or why they want it. "It" may be the Biblical Ark of the Covenant, or a curse-protected pharaoh's tomb treasures, or the Golden City of the Incas, but whatever these heroes are after, it is always something intrinsically valuable and extraordinarily interesting.

In its early years, archaeology *was*, in fact, somewhat like this romantic conception of it. In the early 1800s, for example, Giovanni Belzoni, an Italian machinery salesman, looted dozens of ancient Egyptian tombs in Thebes (modern Luxor). Belzoni crawled through miles of tunnels in the stinking, dusty air of these crypts, smashing hundreds of mummies into powder as he went:

> [Al]though, fortunately I am destitute of the sense of smelling, I could taste that the mummies were rather unpleasant to swallow. After the exertion of entering into such a place, through a passage of . . . perhaps six hundred yards, nearly overcome, I sought a resting place. . . . but when my weight bore on the body of an Egyptian, it crushed like a bandbox. . . . I sank altogether among the broken mummies, with a crash of bones, rags, and wooden cases . . . every step I took I crushed a mummy . . . I could not pass without putting my face in contact with that of some decayed Egyptian; but as the passage inclined downwards, my own weight helped me on: however, I could not avoid being covered with bones, legs,

3

1.1   This sixteenth-century engraving of Florida Indians exemplifies ethnocentrism, or interpreting other cultures and early cultures in one's own terms. The four women on the right are highly reminiscent of Italian Renaissance renderings of Venus and the Three Muses. Many European scholars of the last three centuries were ethnocentric in their analyses of ancient and non-European cultures—most of which they assumed had incompletely evolved to the high level of European civilization.

> arms, and heads rolling from above. . . . The purpose of my researches was to rob the Egyptians of their papyri.[1]

Even the more scholarly of the early archaeologists had a clear simple vision of what it was they were after and why. In 1876, for example, the German aristocrat Heinrich Schliemann, his imagination fired by his readings of Homer, resolved to find the remains of the ancient city of Troy. Based on his studies of *The Odyssey,* he ravaged the archaeological site of Hissarlik, in western Turkey, looking for the home of the heroes of the Trojan War. Schliemann found gold masks affixed to the decayed bodies of what he took to be Trojan warriors and other treasures. When he died many years later he was happily ignorant of the fact that what he thought were the remains of Troy were actually of a much earlier period, and that he had hacked right through the settlement occupied during the presumed period of the Trojan War.

The Meaning of the Past:
What Are Archaeologists Looking for?

Archaeology has changed a great deal since the days of Belzoni and Schliemann, in both methods and objectives and perhaps in "romance." If a professional archaeologist were to reexcavate the parts of Troy that Schliemann left, he or she would be expected to use a lot of high-technology equipment (discussed in Chapter 2) to locate the buried walls of the city, to estimate its age, and to analyze the composition and probable origin of any gold masks found.

But the most profound difference between the archaeology of Schliemann's day and that of the present is that most contemporary archaeologists look beyond the objects they find and seek some more profound understanding of the past. Most archaeologists spend years gathering small bits of stone, bone, and pottery that would not arrest the attention of a museum-goer for more than a few seconds; they then spend much of the rest of their lives in Hamlet-like self-examination, asking themselves, "What does all this stuff mean?" "What is the point of investigating the past?" "Are there causal factors that we can identify that explain the course and nature of human history?" "Does the past have any relevance for our own lives?"

This concern with the meaning of the past is one of the most difficult concepts for the nonarchaeologist to understand about the discipline, yet one cannot comprehend what contemporary archaeology—or this book— is all about, until one understands this search for patterns and meaning in the past. Most nonarchaeologists think that archaeologists are simply trying to answer specific questions about what our hominid ancestors first used stone tools for, or why the Neandertals disappeared, or how ancient Egyptians built pyramids. In every archaeological excavation and study specific questions are at issue, but the context of every archaeological excavation and study involves the more abstract goals of trying to understand in some sense the "meaning" of the past and to formulate some kind of explanation of why history has turned out as it has.

It may seem odd that archaeologists think of something so apparently chaotic as human history as having some discernible patterns, but anyone who has spent tedious years—as most archaeologists have—looking at stones and bones and pots, inevitably begins to see patterns in the past and wonder what it all means.

It is not as if we do not have a lot of evidence to consider these questions. All over the world museum shelves groan under the weight of ceramic pots, stone tools, and the broken skulls of our ancestors. Archaeologists have excavated everything from the first known human camps of two million years ago to this year's refuse in the municipal dump of Tucson, Arizona. And every year, archaeologists in the thousands spread out across the globe to do research, which they then inflict on their colleagues and the world in the form of hundreds of thousands of books, articles, and lectures.

In the rest of this book, I will review hundreds of these studies, testing the reader's patience with seemingly innumerable names for archaeological artifacts, sites (i.e., concentrations of artifacts), cultures, periods, and the like. But before wading into this sea of facts and figures, the reader should join archaeologists in musing on what is the point of all this sifting through the garbage of the past.

In general, archaeologists have tried both to *describe* the past and to *explain* it. Specifically, the traditional objectives of archaeology have been to describe the past through (1) the reconstruction of the lifeways of an-

1.2   Excavating a typical mound site, at Chogha Sefid, Iran. Note the outlines of mudbricks.

cient peoples and (2) the compilation of culture histories; and to explain it through (3), the analysis of cultural processes.[2]

This does not mean that there are competing groups of archaeologists who concentrate on one or another of these goals; most archaeologists do analyses that would fit—loosely, perhaps—in each of these categories. And many contemporary archaeologists no longer conceive of what they are doing as described by these traditional goals.[3] But these goals nonetheless still serve as convenient descriptions of archaeology as it has been practiced during the past several decades.

Reconstructing Past Lifeways

Much of the last century of archaeology has been devoted to trying to reconstruct as completely as possible the diet, technology, residences, burial practices, seasonal movements—in short, the lifeways—of ancient peoples. Archaeologists have usually done this by excavating ancient remains and interpreting their discoveries with the help of analogy, inference, and liberal amounts of imagination.

The reconstruction of past lifeways is a form of cerebral time-traveling. Many archaeologists, given one and only one such time-travel opportunity, would probably have a difficult time choosing among going to Africa's Olduvai Gorge about 1.7 million years ago to see how our ancestors made a living (a problem to which much of Chapter 3 is devoted); or to Palestine of about 90,000 years ago, to see if Neandertals and people physically like ourselves really did live at the same time (see Chapter 4); or to Pakistan's Indus Valley of 4000 years ago to learn what is contained in the thousands of Harappan written documents—one of the world's undeciphered languages (Chapter 10); or perhaps to London of the later nineteenth century, just to chat with Charles Darwin or Karl Marx or another of the founding fathers of modern historical sciences (I would probably be unable to resist the temptation to go to Kom el-Hisn, in the Egyptian Delta, on April 15, 2478, B.C., to solve some cosmically unimportant problems of stratigraphic archaeology that have puzzled me for two field seasons).

In a sense, cultural reconstruction is what any perceptive person does while walking through an archaeological museum: one looks at ancient pots, stone tools, etc. and is irresistibly drawn into musing on how they were used and the nature of the people who used them. Most museums help the viewer in these cultural reconstructions by displaying dioramas, scale replicas of Indian villages, for example, and other reconstructions of scenes of daily life in extinct societies.

A classic example of cultural reconstruction is Henri de Lumley's analysis of Lazaret, the 300,000-year-old remains of human occupation of a cave in France.[4] By measuring the location of every artifact and bit of bone that remained, de Lumley was able to show quite convincingly that the cave's occupants had piled up kelp (inferred from the remains of snails that live only on kelp) and covered it over with bearskins (inferred from the position of the claws) to form a comfortable bed in a shelter in this cave. By studying the kinds of stone tools and other debris, de Lumley could estimate the number of people in the cave, their diet, the seasons of the year in which they visited the cave, and so on.

In some cases cultural reconstructions are particularly compelling. On August 24, A.D. 74, a group of people in an Italian coastal town concluded a funeral service for a friend by sitting down to a final ceremonial meal. We can only wonder about what was said at the meal, but we know what food they ate, how they were dressed, and where they sat because this town was Pompeii, and at some point in the banquet, poison gases from the eruption of the volcano of Vesuvius killed them all (Figure 1.3). They and their fellow citizens, and their neighbors in Herculaneum, were preserved in the midst of their daily activities. A half-cooked suckling pig, bread, and other foods were found in ovens; money was left near half-eaten meals in restaurants; wax tablets and papyri lay on library tables; bodies of townspeople were found curled over the children they were trying to protect; and in one case a woman's skeleton was found scattered over the floor of room filled with volcanic ash, near the intact skeleton of a dog chained to a stake—strongly suggesting that the dog slowly starved and eventually ate the body of its mistress.[5]

Written language aids greatly in "fleshing out" cultural reconstructions. Pompeii's and Herculaneum's daily life were described in surviving documents; we even have graffiti on toilet walls that tells us of the appeal of specific prostitutes and other information common to such venues. Pompeii's and Herculaneum's destruction were in fact recorded by people who took boats out to sea to watch the inferno.

Thus, just as an ethnographer describes the daily life of the people he or she lives with and is studying, an archaeologist, substituting analogy and inference for direct observation, uses ancient objects to describe daily life in extinct cultures.

Compared to the riot of full life evident to the ethnographer, even analyses of towns like Pompeii may seem a pale and cold investigation into the dry stalks and stems of dead cultures. But the great advantage that archaeologists claim over ethnographers is that whereas the cultural anthropologist is limited to the study of societies as they exist today or in the recent past, archaeologists can study change over the long term, over the three million years of our history as a genus.

The last several decades have seen major advances in the techniques of cultural reconstructionism (reviewed in Chapter 2). An impressive array

1.3 The excavations of Pompeii in the mid-eighteenth century stimulated worldwide interest in antiquity and archaeology.

of electron microscopes, satellite imagery, chemical analyses, and other techniques are available to help the archaeologist make ever more plausible inferences about the past. Elaborate mathematical models and computers are now routinely used to reconstruct the past. But despite these many improvements, cultural reconstructions often include a large measure of speculation, particularly when dealing with extremely ancient societies, whose artifacts and lifeways may have no historical analogs. And the possibilities for cultural reconstructions are essentially infinite: one can always excavate another site and fill in one more tiny bit of knowledge. Once the thrill of discovery has worn off, most archaeologists begin to wonder just how important it is to know, for example, whether this or that group ate more salmon than reindeer, or vice versa.

And of course there is really no way ever to confirm—to be certain— about one's reconstructions. The reconstruction of the dog at Herculaneum eating the body of his mistress may seem compelling, but one can imagine alternative interpretations, and reconstructions of less well-preserved and earlier sites may be much less persuasive than those concocted about the living dead of Pompeii and Herculaneum. In short, you can make up extremely convincing stories about what happened at some ancient place, or what some ancient artifacts were used for, but you'll never be able to confirm conclusively your inferences—at least not in the way an astronomer's understanding of celestial mechanics is confirmed by exact prediction of eclipses.

Consider, for example, our crocodiles. Excavating in Egypt in 1981, we found that one site of the late fourth millennium B.C. had many crocodile bones but no crocodile heads or teeth. Our geologist convinced us that this particular site was located near the edge of the ancient Fayyum Lake, and our other finds suggested that it was probably a temporary camp, occupied in spring and fall, where people came to fish and hunt migratory birds. Fueled by one, perhaps two, Stella beers, we evolved the plausible notion that they were hunting these things for God knows what reason and, to aid in transporting their game back to camp, they gutted the animals, cut off their heads and other inedible parts, and left them to rot on the shore.

Now, we could go back and "test" this hypothesis by excavating other sites to see if the same pattern is found, or by checking the crocodile body bones carefully for cut marks, or in any number of other plausible ways. We could also generalize this specific idea and test some hypotheses about general cultural processes, such as the notion that hunters and gatherers tended to transport the shortest distance those animal parts with the least meat on them. Or we could generalize still further and test the notion that people tend to avoid work as much as possible.

One problem here, however, is that one can imagine many different alternative explanations for our finding few crocodile head parts at this site, and in the end all that one could hope to do is show that one hypothesis is more probable than the others. And such hypotheses would really only be particularly interesting in the context of some theory about ancient cultures—some set of generalized principles—which at present we do not have—about the factors that determine the course of cultures in general, including our rag-tag bunch of ancient Fayyum crocodile hunters.

Currently archaeologists argue heatedly about what is and what is not cultural reconstruction and what the uses of cultural reconstructions might be, and some have rejected some forms of reconstruction as worthy objectives of archaeologists.[6] All archaeologists *do* cultural reconstructions, in one sense or another, but there is a widespread sense that archaeologists can be more than just speculative ethnographers of long-dead people.

Compiling Culture Histories

> Very few things happen at the right time and the rest do not happen at all. The conscientious historian will correct these defects. (Mark Twain)

Archaeologists, like most people, are fascinated with "firsts." Some of the most intense debates now going on in the profession concern problems of time, such as when the first *Homo sapiens* appeared, when the first people arrived in the New World, and when agriculture was first practiced.

In a complex way, much of Western art, philosophy, religion, and other cultural ideas have been directly related to how old humanity and the world were thought to be. The recognition—or at least the widespread acceptance—of the idea that we and our world are millions of years old is only about a century old. Until the nineteenth century, most Western people accepted the Biblical implication that the world was about 6000 years old and that almost from the very beginning our ancestors had lived in towns, engaged in agriculture, and organized themselves in great states and empires. Many scholars pondered the Scriptures for clues to the date of Creation. James Ussher (1581–1656) computed the ages of Old Testament figures and their children to date the Great Flood to 2349 B.C. and the Creation of the world at 4004 B.C. Dr. John Lightfoot (1828–1889), Vice Chancellor of Cambridge University, dated Creation to 3928 B.C. Lightfoot concluded that the universe had been created on September 12, at 9 o'clock in the morning, and that God had rested the following Saturday (Sabbath) morning.

But even by Dr. Lightfoot's time many people had concluded that the world must be hundreds of thousands of years old or older. Nineteenth-century excavations deep below the ground levels of many European cities revealed the bones of animals long extinct, such as mammoths and cave bears. Amidst many of these bones were stone tools. By the early twentieth century the great antiquity of humankind was evident, and today no knowledgeable scientist doubts that our hominid, tool-using ancestors go back millions of years.

But specifically where and when did our ancestors live? Eventually these questions coalesced into the goal of locating every prehistoric culture in time and space and arranging all of them "in a way which accurately reveals their generic affinities."[7]

Culture histories are fairly easy and accurate when one deals with, say, the Roman Empire, where there are written records and the evidence of thousands of cities. But how do you write culture histories of prehistoric people who left no written records and of whom we have little more than the occasional skeleton and crude flint tools?

The main method of culture history is to make large collections of artifacts (the stone tools, pots, and everything else made by people) from each "site" (the houses, graves, or workplaces of ancient people) and then make a lot of brave inferences about the cultural relationships among the people who made these pots, stone tools, or whatever.

Modern chemical dating techniques have helped arrange collections of artifacts in time, but a lot of culture history has been done on simple comparisons of artifacts. My own adventures in culture history are in this pattern. I and my associates once walked the plains of southwestern Iran

1.4 Archaeology in the early years of this century was often more like looting than scientific research. Here, workmen at a site near Les Eyzies, France, destroy a Paleolithic site while looking for nicely fashioned stone tools.

for five months, collecting bags of broken pottery from the thousands of ancient villages and towns to be found there. Most of these old settlements are now just mounds (called a *tell* in Arabic, *teppeh* in Persian) of decomposed mudbricks and garbage and are littered with thousands of scraps of broken pottery and stone. Through dating methods, we knew the styles of pottery that were in use in different ancient periods in this part of the world—to the experienced analyst a fragment of a clay pot of the Sasanian period (c. A.D. 225–640) is utterly different from a Late Uruk (c. 3200 B.C.) sherd (though both are impressively ugly). Each day when we finished collecting our pottery samples, we emptied out each bag and counted up the numbers of different types of pottery styles from each site. Sites that had very similar kinds of pottery were assumed to have been occupied at about the same time and to have interacted socially or economically; those with markedly different pottery styles were assumed to have been occupied at different times. On this basis, I constructed settlement maps for several successive periods of occupation in this area, and then reconstructed a 3000-year culture history, complete with estimates of changes in population density, irrigation systems, possible wars, probable economic collapses—in other words, a rich, even emotionally moving, history of people I knew almost exclusively from 4134 bags of pottery bits.[8] I (and the ten or twelve other people in the world who really cared about this period of southwest Iranian history) still believe in most of my culture history, although some aspects of it are probably wrong.

Such culture history "works" in the sense that through using these methods, we now know roughly what kinds of cultures inhabited most of the world during most of the past. When we are dealing with the relatively slow change of styles in early stone tools, even with the help of modern chemical dating techniques we usually can form only rough categorizations of time into periods of hundreds of thousands of years long. Nonetheless, most of the world's past cultures have been arranged in an approximate temporal sequence.

It is unfortunate but probably inescapable that culture histories were often done in such a way that all the world's cultures were arranged like a giant ladder, leading to the pinnacle of Western societies. Because American and Western European cultures of the nineteenth and early twentieth centuries were technologically the most "advanced," the archaeological record has often been viewed in terms of the way in which modern Western humans evolved from the first culture-bearing animals. Thus, cultural historical "explanations" of, for example, the appearance of agriculture and urban communities have tended to assume that these developments were the "natural" and inevitable products of prehistoric peoples who, like many Westerners, were constantly trying to improve their standard of living.

Most archaeologists recognize the non-explanatory nature of culture histories: simply arranging archaeological sites in time and sorting them into categories does not explain how they came to be and their interrelationships. By combining culture histories and cultural reconstructions, archaeologists have provided descriptions of much of antiquity, but they assume that they need principles to make sense of the world's culture histories.

Explaining Cultural Processes

As long as archaeology was mainly cultural reconstruction and culture history, there were comparatively few nasty professional quarrels about what archaeology was all about.

All this changed in the 1960s, when bitter arguments broke out about the goals of archaeology. Some of the nicer terms archaeologists have recently used in print to describe their colleagues' research are "pseudoscience" and "naive." Many archaeologists genuinely believe that the research of other archaeologists is positively damaging, in that they see these projects as exhausting an irreplaceable part of the past without adequate analysis and interpretation of it.

How could a discipline as musty and vague as archaeology provoke such intense feelings? It has to do mainly with the problem of explaining cultural processes and with the search for the meaning of the past.

The reader anxious to get to the blood and sex of prehistory may at this point recoil from a plunge into the philosophy of science and history, but it is simply impossible to understand what contemporary archaeologists are doing and why without considering carefully the intellectual history of archaeology.

The problem is that the only word whose meaning archaeologists can agree on in the proposition that a major goal of archaeology is "to explain cultural processes" is the "to."

The common denominator in these philosophic positions, however, is the idea that we can do more than just describe the past: we can understand it in terms of general principles, even if these general principles are not yet formulated.

Take, for example, something as apparently simple as the origins of agriculture. After three million years as a hunter-forager, about 10,000 years ago our ancestors began to domesticate plants and cultivate them. Farming did not appear in one place and then spread to the rest of the world; it developed in many different places at about the same time, both in the Old World and the New, and involved thousands of species, from yaks to cats, and wheat to yams. Surely, one would think, with patient archaeological research one should be able to find the *causes* of agricultural origins—perhaps a combination of, for example, changed population densities, climate, and technology that caused agriculture to occur—that explain in effect why agriculture appeared in Southwest Asia, for example about 10,000 years ago, and not 20,000 or 5000 years ago, and why it was practiced first in Southwest Asia, and only much later in Southwest North America. And if we were to explain agricultural origins in this fashion, it seems not implausible that we might formulate similar explanations for the origins of writing, cities, warfare, state religions, etc. And having constructed such explanations, one would think, we might be able to bring them together in a set of general propositions about people and history.

In fact, however, as we will see in detail, there is nothing simple, nothing obvious about such attempts to explain the past, and archaeology is currently in the midst of a great identity crisis concerning such issues as whether or not we will ever be able to explain the past in these terms, what cultural processes, in fact, are in essence, and what "explanation" is. The philosophical origins of these questions go back thousands of years.

A Short History of Attempts to
Understand the Past

It's about love, death, torture, infinity. It's a comedy.
(Woody Allen)

The world's literature is littered with attempts to understand our place in the cosmos and make sense of our past. Ancient Middle Eastern literature, especially as reflected in the comparatively late form of the Old Testament, envisioned a static, created world in which great changes came about through divine intercession, and where the ultimate explanation of events was God's will—which we see only "as in a glass, darkly." For the devout Christian, Muslim, and many others as well, even if we could see the divine design of history more clearly, we would see that it is predestined, beyond the realm of human initiative. As an Oxford student named M. E. Hare once put it,

> There once was a man who said, "Damn!
> It is borne in upon me I am
> An engine that moves
> In predestinate grooves,
> I'm not even a bus I'm a tram."

"We are all Greeks," it has been said, because so much of modern Western thought is ultimately traceable to ancient Greece. Greeks like Thucydides and Herodotus were the first, as far as we know, to travel widely and compile extensive descriptions of the people, cultures, and places of their world. In his late 4th-century B.C. history of the Peloponnesian War, Thucydides tried to explain how the struggle began. He described the personalities involved, the strategies of the warring powers, and the economic realities of the time. In short, by arranging the events and circumstances preceding the war in what he thought was a causal chain, he did what any modern historian would do in trying to explain the events of the past. Archaeologically based culture histories are founded precisely on this logic. It is a mark of how Greek we are that what Thucydides did seems so normal and natural to us as hardly to deserve comment. But in truth this sort of real-world explanation of the course of history seems to have been first widely used by the Greeks and became common in many other cultures only recently.

Another central idea of modern archaeology that has Greek roots is evolution. Evolutionary theory was first clearly stated by Charles Darwin and some of his nineteenth-century contemporaries, but already by the fifth century B.C. the Athenian scholar Empedocles had described the basic principle of natural selection.[9]

1.5 The ancient Greek view of the world was that it reflected the "perfection" of the gods in its symmetry and wholeness—attributes reflected in their art, as in this bronze figure from a Greek shipwreck of about 460 B.C.

But this idea of natural selection was in contrast to a more powerful ancient Greek view of the world, the concept of the "Great Chain of Being" or the *Scala Naturae*, which is founded on Greek ideas about the nature of God and "perfection."[10] Greek philosophers found it inconceivable that the world they knew could have arisen by chance, because there seemed to be such a precise design in its every part. The intricate interdependences between plants and animals, the regularity of the seasons—the whole natural world exhibited to them proof of the existence of a Supreme Intelligence; they therefore defined God as the perfect being who created and controls the world. The Greeks' conception of perfection, however, had a somewhat different connotation than it does for us, for they understood it to be in essence wholeness, or completeness (a concept vividly evident, for example, in classical Greek statues).

Aristotle, in particular, formulated the idea of the world as a Great Chain of Being, perfect in its completeness. He concluded that the natural world was rationally ordered according to what he charmingly called "powers of soul," representing different levels within the perfectly whole universe. Thus, a horse is higher than a sunflower because a horse can think, after a fashion, and a man is higher than a horse because he can reason and apprehend God. Early European scholars were greatly influenced by the Great Chain of Being and found it impossible that there should be any "missing links" in the chain, or that any parts of the chain should cease to exist. God, being perfect, could not create an imperfect, that is, incomplete, universe; nor could his sustaining powers allow a whole level of this perfection to vanish:

> Vast chain of being! which from God began,
> Natures aethereal, human, angel, man,
> Beast, bird, fish, insect, what no eye can see,
> No glass can reach; from Infinite to thee,
> From thee to nothing.—On superior pow'rs
> Were we to press, inferior might on ours;
> Or in the full creation leave a void,
> Where, one step broken, the great scale's destroy'd-
> From Nature's chain whatever link you strike,
> Tenth, or ten thousandth, breaks the chain alike.
> (Alexander Pope)

The idea of this Great Chain of Being pervades literature and science well into our own time, and, obviously, grave difficulties arose in reconciling this concept with evolutionary theories and the discoveries of modern archaeology. For if God had created and sustained every bacterium, every sparrow, every Neandertal, how could it happen that thousands of species had come into existence, flourished, and vanished? How could humans possibly have evolved from lower primate species and lived for millions of years as "subhuman" races now long since extinct?

Yet it is easy to understand the attractiveness of the Great Chain. It placed humanity, the masterwork of the Creator, "but little lower than the angels." Furthermore, it explained why we were here and why the world seems so marvelously intricate and designed: God designed the world in every detail for His purposes.

In many ways, the last two hundred years of Western philosophy and archaeology have been mainly a long struggle to come to terms with the realization that the "argument from design," the idea that the intricacies and perfections of the world *prove* that humanity and history are under divine control, may be wrong.

The Enlightenment

It was not until the "Enlightenment" of the eighteenth century that the intellectual foundations of modern archaeology were securely established. In the eighteenth and nineteenth centuries the *scientific method* was combined with *determinism, materialism,* and *evolutionism* to try to explain the world and history.

There can be no absolute definition of the scientific method, for rather than a method it is a state of mind along with a few assumptions. The key ideas are that most, if not all, things and events can be understood in terms of identifiable, measurable, physical forces, and that the best way to identify and measure these forces is to conceive ideas and then expose them to rejection through scientific experimentation.

Although this sounds to most people like nothing more than common sense, the documents of history show that it is a rather late and rare perception of the world. For most ancients and for many moderns, the world swarms with phenomena and forces that can never be understood by science—to them, in fact, it is precisely those aspects of human existence that cannot be scientifically understood or explained that are most important.

The demise of that view of the world in the Western world began in the Enlightenment. Building on Greek ideas, seventeenth- and eighteenth-century scholars observed natural, social, and historical phenomena, devised hypotheses about their causes, and then tested these ideas by dropping balls from the tops of buildings (Galileo), flying kites in lightning storms (Franklin), and so on. What was scientifically essential was that the ideas be put into operation, or expressed in terms that could be measured; that they be exposed to falsification or contradiction by some sort of experimentation; and that, above all, the explanation of something be considered not absolute, eternal truth but just the best current hypothesis, subject to correction in the light of new research.

The materialist and determinist elements in this kind of science were the assumptions that the phenomena of this world—including historical and cultural phenomena—had some ultimate causation (that is, were in some way determined by) knowable, measurable, material factors like population growth, technological change, genetic mutation, infections, or gravity. It was in the deterministic laws and processes governing the attributes of phenomena that one should look for explanations, not in human decisions or divine agency. The evolutionary component was the notion that over time there had been and would continue to be an increase in complexity in the biological, historical, and political world.

Few of the scholars of the Enlightenment were atheists. Benjamin Franklin was largely serious when he said that wine was a continuing proof that God loves us very much and wants us to be happy. These Enlightenment scholars made brilliant advances in determining the physical mechanics of the universe, but most found no conflict between their science and their belief that God created these mechanics and set them into operation.

The Enlightenment was more than the crucible of modern science, it was in many ways the period in which the social perspective of the West first formed, particularly with regard to the notion of progress and conceptions of material culture.[11]

By the end of the eighteenth century, science had shown that the natural world was understandable in terms of the elegant (i.e., comprehensive yet reduced to the simplest possible terms) ideas of mathematics and physics. One result was that scholars everywhere began to apply the scientific method to the understanding of human history and the problem of cultural origins. For example, the Marquis de Condorcet (1743–1794), a French philosopher, proposed a series of universal laws he thought governed the history of human social organizations, and he went so far as to use his analysis to try to predict the future of the world.

Such direct applications of the physical science model to history might seem strange, but one must understand the eighteenth-century mind. "Common sense" tells us that history and culture are far too complex to explain in terms of simple mathematical laws, but common sense also tells us that we walk about on a flat earth, around which revolves the sun. The scholars of the Enlightenment had only recently been shown how treacherous was common sense and how the mysteries of the universe were being reduced to the commonplaces of science.[12] And all around them great advances in the biological and historical sciences were being made.

In France, Georges Cuvier (1769–1832) undertook an extensive analysis of fossilized bones and concluded that hundreds of animal species had become extinct and that there seemed to be an evolutionary trajectory to the biological world. The French naturalist Jean Lamarck (1744–1829) published various arguments that the world was much older than the 6000 years described in the Bible, and he arranged the biological world in a sequence, from human beings to the smallest invertebrates, in a way similar to later evolutionary schemes.

Scientific archaeology based on evolutionary and materialist assumptions was also emerging in the late eighteenth century, when P. F. Suhm published in 1776 a *History of Denmark, Norway, and Holstein,* based on the recognition that in many parts of Europe ancient people first made tools of stone, then of bronze, and finally of iron.[13] This Three-Age System was first clearly stated and then developed by the Danish archaeologist Christian Thomsen in 1836 and by J. Worsae (1821–1885).[14] But its origins are much earlier. Lucretius (c. 98–55 B.C.), the great Roman scholar, wrote that "The earliest weapons were the hands, nails, and teeth; then came stone and clubs. These were followed by iron and bronze, but bronze came first, the use of iron not being known until later." And as K. C. Chang noted, a near-contemporary of Lucretius, the Chinese scholar Yuan K'ang, stated essentially the same idea.[15]

Nineteenth-Century Evolutionism, Materialism, and Determinism

Excavations for the London Underground Railroad in the mid-1800s produced many curious finds, among them the bones of animals that sorely troubled the scientists of the time. Great elephant-like animals, standing 13 feet at the shoulder, had left their skeletons amidst those of gigantic cave-bears and many other animals that no longer lived in Britain—or anywhere else in the world. "Animals dead before the Noachian deluge," some concluded. But others sought scientific explanations. In England in the 1830s, William "Strata" Smith and Charles Lyell, among others, attempted to show that the earth was formed through the action of slow geological processes—processes still in effect. Lyell's contributions were particularly important because the dawning realization of the earth's great age had led some scientists and clergy to a belief in a series of "catastrophes," the last of which was Noah's Flood. Adherents of this position saw the fossil animal bones deep in the earth's strata as evidence that God had "destroyed" the world at various times with floods.

In 1848 John Stuart Mill published an evolutionary analysis of history involving a sequence of six stages: (1) hunting; (2) pastoralism; (3) Asiatic (by which he meant the great irrigation civilizations of China and the Near East); (4) Greco-Roman; (5) feudal; and (6) capitalist. He complemented this classification with an extensive analysis of the economic factors determining these stages.

At about the same time Mill was writing, another Englishman, Herbert Spencer, applied the concepts of "natural selection" to human societies some years before Darwin connected them to the biological world, though his constructs were quite different from those eventually arising out of Darwinian thought.

Spencer was much influenced by Thomas Malthus, who in 1798 had noted that human societies—and indeed all biological species—tended to reproduce in numbers far faster than they increased the available food supply. For human groups this meant a life of struggle in which many were on the edge of starvation and more "primitive" societies lost out in the struggle for survival to the more "advanced" cultures. Spencer believed that eventually natural selection would produce a perfect society:

> Progress therefore, is not an accident, but a necessity. Instead of civilization being artifact, it is part of nature; all of a piece with the development of the embryo or the unfolding of a flower. The modifications mankind have undergone, and are still undergoing, result from a law underlying the whole organic creation; and provided the human race continues, and the constitution of things remains the same, those modifications must end in completeness. . . . So surely must the things we call evil and immorality disappear; so surely must man become perfect.[16]

In some ways Spencer has been more influential in modern archaeology than Charles Darwin, and some scholars (many of whom have not read him) consider Spencer to have been a racist whose notions of progress have misled a century of social science. But Spencer was a brilliant analyst, and he was only expressing the basic eighteenth- and nineteenth-century assumptions that history is subject to natural laws, that we can know these laws, and that—as Spencer's whole life experience showed him—applications of science to human affairs could only lead to progress.[17]

CHARLES DARWIN

God is really only another artist. He invented the giraffe,
the elephant and the cat. He has no real style. He just goes
on trying other things. (Pablo Picasso)

Spencer's ideas permeate archaeological analyses of early civilizations and social forms, but Charles Darwin's writings are the basis for analyzing the physical evolution of human beings, the nature of plant and animal domestication, and many other issues central to contemporary archaeology.

On a warm Saturday afternoon in June 1860, about a thousand people gathered in Oxford, England, to witness a debate on Charles Darwin's theory of biological evolution. For years Darwin had studied the animals and plants of South America, and he had formulated ideas about "descent with modification." But for various reasons he was reluctant to publish his views. Only when he knew that others were about to publish similar evolutionary analyses did he advance his opinion that for centuries the biological sciences had been in error concerning the origins and nature of biological species. Before Darwin most scholars had assumed that all varieties of plants and animals were the direct product of God's creative might; humankind itself was viewed as a special act of creation.

But Darwin's research shook his religious faith in a designed universe. He had been particularly impressed by the great diversity of plant and animal life in the Galapagos Archipelago, near Ecuador. There he found islands geologically similar and within sight of one another, but nevertheless inhabited by significantly different species of plants and animals. Why should there be such diversity in such a small area?

It was evident (after some reflection) that such facts as these could only be explained on the supposition that species gradually become modified; and the subject haunted me. But it was equally evident that neither the action of the surrounding conditions, nor the will of the organism . . . could account for the innumerable cases in which organisms of every kind are beautifully adapted to their habits of life—for instance . . . a tree-frog to climb trees, or a seed for dispersal by hooks or plumes. I had always been much struck by such adaptations, and until these could be explained it seemed to me almost useless to endeavor to prove by indirect evidence that species had been modified.[18]

1.6 Charles Darwin (1809–1882) altered forever human conceptions of the dynamics of the physical world and the nature of history.

Darwin knew of course that for millennia farmers had used selective breeding to improve their animals in specific ways, such as milk production in cows. But these changes were the result of purposeful intervention in these animals' breeding patterns. How could such selection come about in the natural world?

Darwin was influenced by Malthus's idea of population "pressure" and Adam Smith's concepts of economic competition, and he was much impressed with the importance of competition in all spheres of life:

> Being well prepared to appreciate the struggle for existence which everywhere goes on from long-continued observation of the habits of animals and plants, it at once struck me that under these circumstances favorable variations would tend to be preserved and unfavorable ones to be destroyed.[19]

With these observations and simple conclusions, Darwin provided the world with answers to a whole range of perplexing questions. Why was there such variety in the biological world? Because many different environments could be inhabited, and natural selection was constantly shaping biological populations to fit into any newly created environments. Why did animals and plants change over time? Because their environments had changed and those best adapted to these new environments survived to pass on their personal characteristics.

Darwin knew nothing about the genetic mechanisms we now recognize as the agencies through which biological diversity arises and natural selection operates, and he believed that characteristics acquired by an organism in its lifetime could be passed on to its offspring. We now know this to be a misconception. We also know that Darwin was intellectually in debt to other scientists.[20] But all this does not detract from his great contribution.

Darwin put in motion an intellectual revolution that has continued to the present and has battered the very foundations of Western ideas about the nature of God, humanity, and history. Through logic and evidence, Darwin and his proponents showed there had been hundreds of millions of years in which the world had been dominated by reptiles, eons in which there were no people. The Victorians had to consider how God could be glorified by countless generations of snakes and lizards and dinosaurs breeding, fighting, and dying in primeval swamps, and why they should consider humanity a special act of creation if people, too, developed from earlier, simpler forms, from ancestors who were no more imaginative, intelligent, creative, or religious than any other animal.

There was no *necessary* conflict of these ideas with Christianity or other religions; indeed, many founders of Western science remained true believers until their deaths. They simply assumed that God used the natural processes of the world to work out His Divine and Unknowable Plan. Thus the Austrian Augustinian monk Gregor Mendel (1822–1884) could serenely work out the genetic basis of biological evolution and still die untroubled in his faith. And Darwin thought that through evolutionary processes people would eventually become altruistic and profoundly civilized.[21]

But the great mass of people were shocked. Evolutionary biology is now the only generally accepted theory of biology, and few scientists doubt its essential validity. But in that room at Oxford in 1860, Darwin and his advocate, Thomas Huxley, were reviled and ridiculed. This hostility characterized reaction to Darwinian ideas well into our own times, and it is not at all surprising that this should be so.

> Until Darwin, what was stressed . . . was precisely the harmonious cooperative working of organic nature, how the plant kingdom supplies animals with nourishment and oxygen, and the animals supply plants with manure, ammonia, and carbonic acid. Hardly was Darwin recognized before these same people saw everywhere nothing but struggle.[22]

Even humankind's ethical sense, which some felt was the most important distinction between us and other animals, could be seen in Darwinian terms as just an extension of the social behavior of wolves, chimpanzees, and other creatures—simply another evolutionary strategy that has evolved in some species as a part of its entirely natural adaptation.

Man Found only in a Fossil State—Reappearance of Ichthyosauri

A **LECTURE:** "You will at once perceive," continued Professor Ichthyosaurus, "that the skull before us belonged to some of the lower order of animals; the teeth are very insignificant, the power of the jaws trifling, and altogether it seems wonderful how the creature could have procured food."

1.7   Many nineteenth-century cartoonists were amused by the idea of biological evolution.

In a sense, Darwin completed Galileo's revolution. A critic of Galileo "proved" finally and utterly that Galileo was wrong because Galileo said there were moons orbiting Jupiter. As this critic pointed out, if these moons were too small to be seen, they were too small to affect the earth, and since God designed the earth as the focal point of the whole universe, Jupiter could, therefore, have no moons. Galileo showed that the earth was one among an inconceivable number of celestial bodies, without apparent special claim to centrality; Darwin showed that today's human being is one of many related life forms and is always a "transitional form," constantly changing.

The profound effects of Darwin's ideas permeated all of Western arts and sciences. In "Dover Beach," Matthew Arnold likened the Christian faith and view of the world to a comforting tide that surrounded the world as a shining sea, but which was withdrawing under the onslaught of revolutionary ideas of the nineteenth century including Darwinism:

The Sea of Faith
Was once, too, at the full, and round earth's shore
Lay like the folds of a bright girdle furl'd.
But now I only hear
Its melancholy, long, withdrawing roar,
Retreating, to the breadth
Of the night-wind, down the vast edges drear
And naked shingles of the world.

Ah, love, let us be true
To one another! for the world, which seems
To lie before us like a land of dreams,
So various, so beautiful, so new,
Hath really neither joy, nor love, nor light,
Nor certitude, nor peace, nor help for pain;
And we are here as on a darkling plain
Swept with confused alarms of struggle and flight,
Where ignorant armies clash by night.

Similarly, Leo Tolstoy, in a letter written in 1910 on his deathbed to his son and daughter advised them:

The views you have acquired about Darwinism, evolution, and struggle for existence won't explain to you the meaning of your life and won't give you guidance in your actions, and a life without an explanation of its meaning and importance, and without the unfailing guidance that stems from it is a pitiful existence. Think about it. I say it, probably on the eve of my death, because I love you.[23]

As Stephen Jay Gould points out, Tolstoy's lament is somewhat unfair. Evolutionary theory makes no pretense of explaining the ultimate nature or purpose of human existence—if it is humanly possible, at all, ever to answer such questions. Evolutionary theory simply supplies powerful explanations of change over time in human biology and society—leaving the problem of ultimate causation to some other forms of investigation.

It is important to note in this context that evolutionary theory cannot *prove* that the "argument from design" is wrong. It simply provides an alternative explanation of how the biological and cultural history of earth might have happened. And even in scientific circles the argument from design has had lasting traces. One of novelist Gabriel García Marquez's characters says, "I don't believe in God but I'm afraid of Him," and one finds vestiges of the same thought among cosmologists.

One reason for this is that from some perspectives the evolution of life is so extraordinarily improbable. If the orbit of the earth had been 5% closer to the sun, for example, the earth would have been far too hot for life to evolve; if the earth were 1% farther out, all the water on the planet would be locked in glacial ice.[24] Some scientists accept that the complex chemicals constituting the self-replicating compounds necessary for biological life *could* have evolved by random processes but consider the few billion year history of organic earth life to be too short for this to be a probable occurrence. Hence, some hypothesize that the earth passed through a space cloud of organic molecules, from which we are all descended.

But such subtle points were unknown to people of Darwin's age. Darwin and other evolutionists had posed cosmic questions and had provided compelling answers. Their arguments and research showed that not just biological species but cities, ships, pyramids, farms, religions, parliaments—all things cultural had evolved out of earlier, simpler forms; they implied that we have no special claim to centrality or exemption from the processes of the universe. And, perhaps most important, Darwin's ideas made it reasonable to ask whether or not there were principles with which to understand our cultural as well as our physical evolution.

KARL MARX

An early anthropological expression of evolutionary and materialist ideas that had considerable impact on studies of the past was the work of Lewis Henry Morgan (1818–1881), who divided history into a series of stages on the basis of initial uses of fire, bow and arrow, pottery, domesticated animals, writing, etc.[25] One of the people influenced by Morgan was Karl Marx (1818–1883), and the full impact of nineteenth-century evolutionism was achieved only when it was combined with Marx's ideas about materialist determinism. Even archaeologists who consider Marxism a bankrupt political ideology are in debt to Marxian thought in many of its elements of materialist determinism. Marxism, or at least materialist notions, are a "natural" for archaeology, because archaeologists spend their lives amidst heaps of materials—the houses, stone tools, storage bins, pots, weapons, irrigation canals, and other items that constitute the "means of production" of ancient peoples.

Certainly one of the most diverse—not to say solipsistic—associations of all time is that group of individuals who have tried to explain what Marx meant. People have killed each other in disputes over Marxian interpretations, and neo-Marxist variants on the basic Marxian gospel are so diverse as to defy summarization. The student with a sense of humor and a German dictionary is invited to read Marx's original description of the all-important "modes of production" and then to follow the exegesis of this term into the contemporary era.[26] Scholars with every inclination of Platonic Idealists now claim to be Marxist materialist-determinists, all on the strength of their interpretation of modes of production. Little wonder that some scholars see more of Groucho than Karl in Marxist anthropology.

But what has all this got to do with archaeology? Precisely this: for some

1.8 Karl Marx (1818–1883) profoundly influenced anthropology and archaeology by revealing the links between the economic basis of societies and their political institutions, social structures, and other cultural characteristics.

of the most influential American and European archaeologists, the only way to explain the past is with the concepts derived from Marx and his intellectual descendants.

As we will see in Chapter 7, Marx argued that much of human history can be understood if we examine how a society produces and distributes its wealth. He attempted to show that everything—wars, social classes, poverty, parliaments, religion, art—could be explained if one examined the technology, economy, and environment of a given society. In recent years there have been numerous reworkings of Marxian theory, and the most recent archaeological expressions of these idea stress the social relations that people enter into in producing and consuming goods.[27] These, rather than just the blunt forces of climate, crops, and technology, are seen by Marxist archaeologists as the determinant factors of history.

We will consider contemporary expressions of Marxist ideas in archaeology more fully in Chapter 7. It is sufficient here to note that materialist determinism is strongly embedded in modern archaeology and that many archaeologists draw directly on the works of Marx in trying to understand and explain the past.

Early Twentieth-Century Archaeology

From the mid-nineteenth century and the beginnings of the twentieth, archaeology entered a remarkable era of "Discovery and Decipherment."[28] This was the period in which Egyptian hieroglyphs and Mesopotamian cuneiform were first deciphered and major archaeological excavations were undertaken everywhere. In 1922 Lord Carnarvon and Howard Carter opened the tomb of King Tutankhamen in Egypt, in 1926 Sir Leonard Woolley discovered the Royal Tombs of Ur, in Mesopotamia, in 1922 Sir John Marshall began excavating the great Harappan civilization in the Indus Valley, and all over the world at this time marvelous discoveries were being made.

Archaeology made great progress in this era, but it was also a time in which archaeology was dominated by Europeans and Americans in an ethnocentric and imperialist way that even today gives the discipline a negative image in many parts of the world. The French and British in particular, but all Western nations to some extent, looted the antiquities of other lands, especially Greece, Egypt, and Persia (Iran). European governments collaborated with archaeologists to extract antiquities concessions from weak governments. When the British and French controlled Egypt, they always made one of their own citizens the director of the Egyptian Museum. Archaeologists, in turn, often acted as spies for their governments.

World War I, with its horrible carnage, the Depression of the 1930s, and the other dismal events of this era, disillusioned many intellectuals and caused them to doubt any notion of human social evolution—at least in the sense of a world growing more rational and moral, of *progress* in any sense. This was the age of existentialism and in many senses a rejection of rationalism. Darwinian notions of struggle and godlessness, the irrationality of world wars, all these destroyed for many people the last vestiges of the "argument from design," and resulted in a profound sense of cosmic isolation.

> For the listener, who listens in the snow,
> And, nothing himself, beholds,
> Nothing that is not there and the Nothing that is.
> (Wallace Stevens, *The Snow Man*)

The philosophical currents of the early twentieth century produced in archaeology a largely nontheoretical discipline in which most scholars contented themselves with patient accumulation of artifacts and a minimum of interpretation. This period was the "Golden Age" of cultural reconstruction and culture history. It was widely assumed that progress in explaining prehistoric cultural developments would be made only when much more archaeological evidence had been accumulated and the "facts be allowed to speak for themselves."

In the United States the federal government invested considerable sums in archaeological investigations, in part as a way to employ people during the Depression. Even by the 1930s, however, there was a growing frustration with the idea of an archaeology limited just to an endless series of inferences about ancient houses and diets.

One of the earliest and most influential attempts to move archaeology beyond simple data collection and description was Julian Steward's 1949 classic, "Cultural Causality and Law; a Trial Formulation of Early Civilization." Steward tried to relate worldwide similarities in the evolution of cities, writing, warfare, urbanism, etc., to basic determinants of ecology, technology, and demography.

Archaeology 1960–1990

A hundred thousand lemmings can't be wrong.
(Anonymous)

Between 1960 and 1990 hundreds of archaeologists have struggled to formulate a discipline of archaeology that was fundamentally *scientific*.[29] Anyone unfamiliar with archaeology might wonder in what sense archaeologists ever hoped to create a truly scientific discipline of archaeology, and the answer, alas, is truly difficult to explain. In fact, almost every archaeologist had individual ideas about what kind of a science archaeology might become. For many of them, though, physics, chemistry, and even biology were models that—if they could not be equaled by a science of archaeology—should at least be imitated as closely as possible. Why would physics and these other sciences be so attractive as models for archaeologists? Primarily because they offer powerful *explanations*. Natural scientists use a limited number of principles, such as the principle of relativity, thermodynamics, etc., and powerful mathematical formulations to explain much of the workings of the physical universe. We *know* what causes eclipses, thermonuclear explosions, malaria, and other natural processes to the extent that we can control these forces and make nuclear reactors, televisions, and genetically engineered plants and animals.

Many archaeologists of the past several decades have sought to establish a discipline of archaeology that was similarly powerful, mathematical, and fundamentally *explanatory*. Archaeologists have wanted to know what *caused* the first vaguely human animals to use stone tools, what factors drove these early hominids into the northern temperate latitudes, why people changed from hunting-gathering to agriculture, what forces impelled people to establish cities, develop written languages, wage international wars—in short, why history has turned out the way it has.

Archaeologists recognize that human history is in some ways fundamentally different from some of the material phenomena of the world, such as atoms. For one thing, human history is the product of *natural selection* in a way that does not seem true of the history of galaxies and sub-atomic particles (this point is discussed in later chapters). The human past by definition is a sequence in time and through a particular space— the planet earth—and the principles of physics and chemistry are timeless and spaceless at a fundamental level. That the volume of a gas is determined by its pressure and temperature is true now and forever, here and in the farthest reaches of the universe.

But biology, population genetics, ecology—these and other life sciences are powerful mathematical explanatory disciplines: why can't a science of the human past be like them? For many archaeologists these life sciences not only still offer models of sciences that archaeology may become, they are also useful for analyzing human history. Other scholars continue to pursue the idea that modifications of Marxian theory will provide an explanation of the human past, still others are investigating cognitive psychology, computer-based artificial intelligence, and many other diverse subjects—all in the hope that in some way a deeper, more profound understanding of the past is possible.

One of the most influential scholars in the attempt to make archaeology a powerful science has been Lewis Binford. While a student at the University of Michigan, Binford was greatly impressed by anthropologist Leslie White's evolutionary, materialistic vision of anthropology.[30] White focused on the material culture of past and present societies and related cultural evolution, from the first hominids to modern societies, to the evolution in ways of harnessing energy.

In an influential and programatic series of papers Binford argued that archaeologists should turn their attention from endless excavations and attempts to reconstruct ancient cultures and culture histories, and concentrate instead on the study of cultural processes and the formulation of cultural laws. Binford particularly stressed the importance of problem orientation and testing hypotheses. He argued, for example, that the origins of agriculture were caused by, among other factors, climate changes, population growth, and various cultural adaptations of peoples in certain environments. He also has argued that systematic big-game hunting was not a common human practice until after about 30,000 years ago and that, therefore, the many ideas about human evolution based on the idea of "man the hunter" are suspect. In all of his analyses, Binford has tried to show how certain kinds of factors, such as climate, technological innovations, population growth, human social systems, etc., were causally and mutually related in ways that explain how ancient cultures functioned, and why and how they changed over time throughout the world.

Beginning in the 1960s and continuing into the present, many other archaeologists have sought to establish a science of history, and although many of them disagree as to how this might be accomplished, there are several common ideas.[31]

First, many archaeologists believe that *mathematics* can play a powerful role in archaeological analyses. Archaeologists have admired the kinds of explanations provided by physics and have entertained the hope that archaeology could in some ways produce explanations of history like those physicists use to explain the natural world. Martin Wobst, for example, used a computer to simulate the births, deaths, movements, and various social interactions of sixty-one imaginary hunting-gathering societies in ice age Europe, in an effort to understand how these social interactions and demographic characteristics might determine the kinds of stone tools and other archaeological remains we might find.[32] And Susan Gregg[33] and Arthur Keene[34] have both devised complex computer models that simulate the diets and economies hunter-foragers would need in temperate environments under varying conditions.

Second, almost all attempts to make archaeology a science have assumed that this would be an *evolutionary* science—at least in the simple sense that Bruce Trigger described as the assumption that "all significant differences among cultures can be regarded as differing states of development from simple to complex."[35] The application of Darwinian ideas of evolution to human societies has in many ways just begun, and some archaeologists see great potential in this approach.[36]

Third, the New Archaeology also has stressed *cultural ecology,* in the sense that archaeologists look for the causes of major cultural changes primarily in factors like climate changes, variability in the agricultural productivity of environments, technological changes, and demographic factors. Karl Butzer and Fekri Hassan, for example, have shown strong statistical connections between ancient Egyptian settlements and subsistence practices and the basic determinants of agriculture in the Nile Valley.[37]

A fourth important notion of the New Archaeology was the assumption that cultures are complex "*systems*"—that is, the patterns of causality are complexly interrelated. Thus, if one were to consider why people in ancient Iraq first began cultivating plants, one would expect that the answer would involve many different factors, such as changes in population densities, climate fluctuations, evolving technologies, the genetic properties of certain plant genera, etc. Human societies are assumed to have regulating mechanisms that promote the group's adaptation. Most human societies, for example, have developed means of birth control to regulate their population to balance resources; the intensity with which birth control is practiced can change as the resources change. To some extent, the systems-view of culture is that societies are self-regulating entities, open to influences from their environments. Explanations based on this model are often *functional* explanations: just as a biologist might account for the evolution of the mammalian heart as a way to circulate blood, a Marxist archaeologist might account for the rise of state religions in all early civilizations by the need to organize and control the mass of the population for military and economic purposes.

The goal of reformulating archaeology as a mathematical, evolutionary, ecological science of complex systems has spread to much of the world, well beyond Western European and American circles.[38] But it should also be noted that in recent years several archaeologists have argued that archaeology can never be a neutral, value-free science of artifacts. T. Cuyler Young, for example, expresses this difference of opinion as, on the one hand (the New Archaeology side), the view that "history is what *happened* in the past," as opposed to his belief that history "is what a living society *does* with the past."[39]

This may seem an obvious point, that archaeologists' interpretations of the past are in part a function of their own sociocultural context. One might think that archaeologists would be less prone to this than, say, ethnologists, but archaeologists regularly report temples, social stratification, intensified storage, social class, warfare, states, and other "things" that do not, technically, exist. Our own culture determines to some extent what we make of the past. We can dig up people, sort through their feces, measure their bodies, sift their garbage for their food remains, translate their writings, measure their buildings, and do all kinds of scientific things to them, but in the end what we make of them will have a lot to do with our own lives and personalities.

An archaeologist, for example, working in any one of many different areas of the world, from Peru to China, might find the remains of what he thinks was a "market." But for most Westerners "market" implies an area in which goods and services are exchanged by people using money who are "rational consumers" in that they are trying to satisfy their needs and wants at the best prices, with their long-term economic self-interest very much in their minds.

But we have no reason to believe that ancient peoples perceived markets in this way, or that the term "market" is really applicable. From what we know of modern peasant societies, what we call markets can be imbued with social, civil, and even religious motives and connotations that differ radically from markets in Western industrial societies. In any case, what have we done when we excavate something and call it a "market"? We simply define some part of the past in terms of our own experience. For some archaeologists the solution to this difficulty is to discard the term market or to define it in ways that can be measured, such as architecture, plant remains, etc. For other archaeologists the solution is to try to understand the significance of "markets" in the terms of the ancient society itself, as that society is known from the whole corpus of its remains.

Indeed, just as some ethnographers have concluded that human life and society have "only meanings rather than causes,"[40] some archaeologists think that the emphasis of archaeology should be on assessing the meaning of the past for ourselves, rather than on futile attempts to understand why history has happened the way it has. These ideas are discussed in Chapter 16.

The Goals of Archaeology: Conclusions and Summary

We must now return to the two fundamental questions that started this chapter: What can we know about the past and how should we go about knowing it?

As we have seen, answers have ranged from the notion that history is God's plan and ultimately unknowable to the idea that history can be the subject of scientific methods of analysis and can ultimately be expressed in eternal, universal laws. Contemporary archaeologists include those who are working on complex computer models designed to extract historical processes, those who are trying to produce highly detailed culture histories, and those who have given up all pretense of doing anything other than finding interesting items (the "gold-bowls-and-mummies" school of archaeology).

In summary, the layperson could not be blamed for thinking that any philosopher of archaeology who sees it as a key to the nature of human existence is on a slippery slope indeed. In the following chapters, I will try to illustrate some of the more promising approaches to the problem of analyzing cultural processes. None of these is so powerful at present that it is worth exploring in painful detail; most are understandable only in terms of specific examples. The reader should bear in mind four basic points: (1) most archaeologists continue to search for more powerful ways to understand cultural processes and the nature of history; (2) recent attempts to make archaeology a "science" so that we can comprehend cultural processes have been logically flawed and not terribly impressive in their results; (3) the diversity of theoretical approaches to archaeology suggests that as a discipline, archaeology has entered a period of loss of confidence in its theoretical structure—the kind of period, as it happens, that in other disciplines has often presaged great advances in method and theory; and (4) history may seem to teach few clear and unambiguous lessons, but it does seem evident that one should not underestimate the ingenuity and power of the human intellect. Contemporary arts and sciences would seem absolutely miraculous to anyone who lived more than a few decades ago, and there is no reason to suspect that we are even approaching the limits of our potential knowledge of life, death, the universe, and everything. It is estimated that 85% of all the scientists who ever lived are alive today,[41] and so the increase in human knowledge—and of the dynamics of history and human society—in the near future can be expected to be rapid.

Notes

1. Quoted in Daniel, *The Origins and Growth of Archaeology*, pp. 48–49.
2. Binford, "Archaeological Perspectives"; Willey and Phillips, *Method and Theory in American Archaeology*.
3. See, e.g., Shanks and Tilley, *Re-Constructing Archaeology*.
4. de Lumley, *Le paléolithique inférieur et moyen du Midi Mediterranéen dans son cadre géologique*.
5. Ceram, *Gods, Graves, and Scholars*, pp. 8–9.
6. See, e.g., Dunnell, "Methodological Issues in Americanist Artifact Classification"; Binford, "Reply to J. F. Thackeray 'Further Comment on Fauna from Klasies River Mouth.'"; Gould and Watson, "A Dialogue on the Meaning and Use of Analogy in Ethnoarchaeological Reasoning"; Wylie, "An Analogy By Any Other Name Is Just as Analogical."
7. Binford, "Archaeological Perspectives," p. 8.
8. Wenke, "Western Iran in the Partho-Sasanian Period."
9. This issue is discussed in Kirk and Raven, *The Presocratic Philosophers*, p. 338.
10. Lovejoy, *The Great Chain of Being: A Study of the History of an Idea.*
11. See, respectively, Nisbet, *History of the Idea of Progress*, and Miller, *Material Culture and Mass Consumption*.
12. Langer et. al., *Western Civilization*, pp. 120–79.
13. Daniel, *The Origins and Growth of Archaeology*, p. 90.
14. Reviewed in Graslund, *The Birth of Prehistoric Chronology*, p. 18.
15. Chang, *The Archaeology of Ancient China*, p. 5.
16. Spencer, *Social Statics*, p. 80.
17. See Turner, *Herbert Spencer. A Renewed Appreciation.*
18. Darwin, quoted in Adams, *Eternal Quest*, p. 334.
19. Quoted in Adams, *Eternal Quest*, p. 335.
20. Eiseley, *Darwin and the Mysterious Mr. X.*
21. Richards, *Darwin and the Emergence of Evolutionary Theories of Mind and Behavior.*
22. Friedrich Engels, quoted in Meek, *Marx and Engels on Malthus*, p. 186.
23. Quoted in Gould, "Kropotkin Was No Crackpot," p. 12.
24. Michael Hart, quoted in Easterbrook, "Are We Alone?"
25. Morgan, *Ancient Society.*
26. Harris, *Cultural Materialism.*
27. See, e.g., Friedman and Rowlands, *The Evolution of Social Systems;* Kohl, "Materialist Approaches in Prehistory"; Spriggs, ed., *Marxist Perspectives in Archaeology;* Patterson and Gailey, *Power Relations and State Formation.*
28. Daniel, *The Origins and Growth of Archaeology.*
29. One of the most important books on this subject is Watson et al., *Explanation in*

Archaeology; also see Watson, et al. *Archaeological Explanation;* Bamforth and Spaulding, "Human Behavior, Explanation, Archaeology, History, and Science"; Schiffer,
"Some Issues in the Philosophy of Archaeology."
30. White, *The Science of Culture.*
31. Trigger, "Archaeology at the Crossroads: What's New?"; also see Bamforth and
    Spaulding, "Human Behavior, Explanation, Archaeology, History, and Science";
    Salmon, *Philosophy and Archaeology.*
32. Wobst, "Boundary Conditions for Paleolithic Social Systems."
33. Gregg, *Foragers and Farmers.*
34. Keene, *Prehistoric Foraging in a Temperate Forest.*
35. Trigger, "Archaeology at the Crossroads: What's New?" p. 278.
36. Dunnell, "Evolutionary Theory and Archaeology"; Dunnell, "Science, Social Sci-
    ence, and Common Sense: The Agonizing Dilemma of Modern Archaeology";
    Dunnell, "Methodological Issues in Americanist Artifact Classification"; Wenke,
    "Explaining the Evolution of Cultural Complexity: A Review;" Kirch and Green,
    "History, Phylogeny, and Evolution in Polynesia."
37. Butzer, "Geological and Ecological Perspectives on the Middle Pleistocene"; and
    Hassan, "Desert Environment and Origins of Agriculture in Egypt."
38. See, e.g., Paddayya, "Theoretical Archaeology—A Review."
39. Young, "Since Herodotus, Has History Been a Valid Concept?" p. 7.
40. O'Meara, "Anthropology as Empirical Science."
41. This estimate is from *Harper's,* August 1989, p. 11.

Bibliography

Adams, A. B. 1969. *Eternal Quest.* New York: Putnams.
Ayer, A. J. 1946. *Language, Truth, and Logic.* New York: Dover.
Bamforth, D. B. and A. C. Spaulding. 1982. "Human Behavior, Explanation, Archaeol-
    ogy, History, and Science." *Journal of Anthropological Archaeology* 1(2):179–195.
Binford, L. R. 1968. "Archaeological Perspectives." In *New Perspectives in Archaeology,*
    eds. Sally R. Binford and Lewis R. Binford. Chicago: Aldine.
————. 1981. *Bones: Ancient Men and Modern Myths.* New York: Academic Press.
————. 1986. "Reply to J. F. Thackeray 'Further Comment on Fauna from Klasies
    River Mouth.' " *Current Anthropology* 27(1):57–62.
Bintliff, J. 1986. "Archaeology at the Interface: An Historical Perspective." In *Archaeol-
    ogy at the Interface,* eds. J. L. Bintliff and C. F. Gaffney. Oxford: BAR Interna-
    tional Series 300.
Butzer, K. W. 1975. "Geological and Ecological Perspectives on the Middle Pleisto-
    cene." In *After the Australopithecines,* eds. K. W. Butzer and G. Isaac. The Hague:
    Mouton.
————. 1982. *Archaeology as Human Ecology.* Cambridge, England: Cambridge Univer-
    sity Press.
Caldwell, J. 1958. *Trend and Tradition in the Prehistory of the Eastern United States.* Men-
    asha: Memorial Series of the American Anthropological Association 88.
Ceram, C. W. 1967. *Gods, Graves, and Scholars.* New York: Knopf.
Chang, K. C. 1988. *The Archaeology of Ancient China.* New Haven: Yale University Press.
Cohen, G. A. 1978. *Karl Marx's Theory of History. A Defense.* Princeton: Princeton Uni-
    versity Press.
Daniel, G. 1967. *The Origins and Growth of Archaeology.* Baltimore: Penguin.
Dunnell, R. C. 1980. "Evolution Theory and Archaeology." In *Advances in Archaeological
    Method and Theory,* Vol. 3, ed. M. B. Schiffer. New York: Academic Press.
————. 1982. "Science, Social Science, and Common Sense: The Agonizing Dilemma
    of Modern Archaeology." *Journal of Anthropological Research* 38:1–25.

————. 1986. "Methodological Issues in Americanist Artifact Classification." *Advances in Archaeological Method and Theory* 9:147–207.

Earle, T. K. and R. W. Preucel. 1987. "Processual Archaeology and the Radical Critique." *Current Anthropology* 28(4):501–38.

Easterbrook, G. 1988. "Are We Alone?" *The Atlantic,* August 1988, pp. 25–38.

Eiseley, L. 1946. *The Immense Journey.* New York: Time, Inc.

————. 1979. *Darwin and the Mysterious Mr. X.* New York: Harcourt Brace Jovanovich.

Fildes, V. A. 1987. *Breasts, Bodies, and Babies: A History of Infant Feeding.* Edinburgh: Edinburgh University Press.

Flannery, K. V. 1973. "Archaeology with a Capital 'S.' " In *Research and Theory in Current Archaeology,* ed. Charles L. Redman. New York: Wiley.

Friedman, J. and M. J. Rowlands. 1977. *The Evolution of Social Systems.* Pittsburgh: University of Pittsburgh Press.

Gregg, S. A. 1988. *Foragers and Farmers.* Chicago: University of Chicago Press.

Gould, R. A., ed. 1978. *Explorations in Ethno-Archaeology.* Albuquerque: University of New Mexico Press.

Gould, R. A. and P. J. Watson. 1982. "A Dialogue on the Meaning and Use of Analogy in Ethnoarchaeological Reasoning." *Journal of Anthropological Archaeology* 1:355–381.

Gould, S. J. 1977. *Ever Since Darwin.* New York: Norton.

————. 1988. "Kropotkin Was No Crackpot." *Natural History* 97(7):12–21.

Graslund, B. 1987. *The Birth of Prehistoric Chronology.* Cambridge, England: Cambridge University Press.

Grayson, D. K. 1983. *The Establishment of Human Antiquity.* New York: Academic Press.

Harris, M. 1968. *The Rise of Anthropological Theory.* New York: Crowell.

————. 1979. *Cultural Materialism.* New York: Random House.

Harrold, F. B. and R. A. Eve. 1988. *Cult Archaeology and Creationism.* Iowa City: University of Iowa Press.

Hassan, F. A. 1986. "Desert Environment and Origins of Agriculture in Egypt." *Norwegian Archaeological Review* 19:63–76.

Hempel, C. B. 1966. *Philosophy of Natural Science.* Englewood Cliffs, N. J.: Prentice-Hall.

Janzen, D. 1976. "Why Bamboos Wait So Long to Flower," *Annual Review of Ecology and Systematics* 7:347–91.

Johnson, G. A. 1987. "Comment on Earle, T. K. and R. W. Preucel. 1987. *Processual Archaeology and the Radical Critique.*" *Current Anthropology* 28(4):517–18.

Kaplan, A. 1984. "Philosophy of Science in Anthropology." *Annual Review of Anthropology* 13:25–39.

Keene, A. S. 1981. *Prehistoric Foraging in a Temperate Forest.* New York: Academic Press.

Kirch, P. V. and R. C. Green. 1987. "History, Phylogeny, and Evolution in Polynesia." *Current Anthropology* 28(4):431–56.

Kirk, G. S. and J. E. Raven. 1966. *The Presocratic Philosophers.* Cambridge, England: Cambridge University Press.

Kohl, P. L. 1981. "Materialist Approaches in Prehistory." *Annual Review of Anthropology* 10:89–118.

Krebs, J. and N. B. Davies. 1981. *Introduction to Behavioral Ecology.* Oxford: Blackwell.

Langer, W. L., Gen. Ed., P. MacKendrick, D. Geanakoplos, J. H. Hexter, and R. Pipes. 1968. *Western Civilization.* New York: Harper & Row.

Leach, E. R. 1984. "Glimpses of the Unmentionable in the History of British Social Anthropology." *Annual Review of Anthropology* 13:1–23.

Leone, M., P. B. Potter, Jr., and P. A. Schackel. 1987. "Toward a Critical Archaeology." *Current Anthropology* 28(3):283–302.

Lloyd, M. and H. S. Dybus. 1966. "The Periodical Cicada Problem." *Evolution* 20:133–49.

Lovejoy, A. O. 1960. *The Great Chain of Being: A Study of the History of an Idea.* New York: Harper & Row.

Lumley, H. de. 1969. *Le paléolithique inférieur et moyen du Midi Mediterranéan dans son cadre géologique:* Vol. 1: *Ligurie Provence.* Paris: Éditions du Centre National de la Recherche Scientifique.

Marx, K. 1904. *The Critique of Political Economy,* trans. I. N. Stone. Chicago: International Library Publication Co.

Meek, R. L. 1953. *Marx and Engels on Malthus.* London: Lawrence and Wishart.

Miller, D. 1987. *Material Culture and Mass Consumption.* Oxford: Basil Blackwell.

Morgan, L. H. 1877. *Ancient Society.* Reprint ed. 1964, Ed. and with an introduction by L. A. White. Cambridge: Harvard University Press.

Nisbet, R. J. 1980. *History of the Idea of Progress.* New York: Basic Books.

O'Meara, J. T. 1989. "Anthropology as Empirical Science." *American Anthropologist* 91(2):354–369.

Orton, C. 1980. *Mathematics in Archaeology.* London: Collins.

Paddayya, K. 1985. "Theoretical Archaeology—A Review." In *Recent Advances in Indian Archaeology,* eds. S. B. Deo and K. Paddayya. Poona: Deccan College of Post–Graduate and Research Institute.

Patterson, T. C. and C. W. Gailey, eds. 1987. *Power Relations and State Formation.* Washington, D.C.: American Anthropological Association.

Richards, R. J. 1987. *Darwin and the Emergence of Evolutionary Theories of Mind and Behavior.* Chicago: University of Chicago Press.

Ross, E. B., ed. 1980. *Beyond the Myths of Culture. Essays on Cultural Materialism.* New York: Academic Press.

Sahlins, M. D. and E. R. Service, eds. 1960. *Evolution and Culture.* Ann Arbor: University of Michigan Press.

Salmon, M. H. 1982. *Philosophy and Archaeology.* New York: Academic Press.

Salmon, M. H. and W. C. Salmon. 1979. "Alternative Models of Scientific Explanation." *American Anthropologist* 81:61–74.

Salt, G. W., ed. 1984. *Ecology and Evolutionary Biology.* Chicago: University of Chicago Press.

Schiffer, M. B. 1981. "Some Issues in the Philosophy of Archaeology." *American Antiquity* 46:899–908.

———. 1976. *Behavioral Archaeology.* New York: Academic Press.

———. 1987. *Formation Processes of the Archaeological Record.* Albuquerque: University of New Mexico.

Shanks, M. and C. Tilley. 1987. *Re-Constructing Archaeology.* Cambridge, England: Cambridge University Press.

Spaulding, A. C. 1973. "Archeology in the Active Voice: The New Anthropology." In *Research and Theory in Current Archaeology,* ed. Charles L. Redman. New York: Wiley.

Spencer, H. 1883. *Social Statics.* New York: Appleton.

Spriggs, M., ed. 1984. *Marxist Perspectives in Archaeology.* Cambridge, England: Cambridge University Press.

Spuhler, J. N. 1985. "Anthropology, Evolution, and 'Scientific Creationism.'" *Annual Review of Anthropology* 14:103–33.

Strong, D. E., ed. 1973. *Archaeological Theory and Practice.* London: Seminar Press.

Trigger, B. G. 1984. "Archaeology at the Crossroads: What's New?" *Annual Review of Anthropology* 13:275–300.

Turner, H. 1985. *Herbert Spencer. A Renewed Appreciation.* Beverly Hills: Sage.

Watson, P. J., S. A. LeBlanc, and C. L. Redman. 1971. *Explanation in Archeology.* New York: Columbia University Press.

———. 1984. *Archeological Explanation.* New York: Columbia University Press.

Wells, P. S. 1980. *Culture Contact and Culture Change.* Cambridge, England: Cambridge University Press.

Wenke, R. J. 1975–1976. "Imperial Investments and Agricultural Developments in Parthian and Sasanian Khuzestan: 150 B.C. to A.D. 640." *Mesopotamia* 10–11:31–217.

———. 1981. "Explaining the Evolution of Cultural Complexity: A Review." In *Advances in Archaeological Method and Theory*, Vol. 4, ed. M. B. Schiffer. New York: Academic Press.

———. 1987. "Western Iran in the Partho-Sasanian Period: The Imperial Transformation." In *The Archaeology of Western Iran*, ed. F. Hole. Washington, D.C.: Smithsonian Institution Press.

White, L. A. 1949. *The Science of Culture.* New York: Grove Press.

———. 1959. "The Concept of Culture." *American Anthropologist* 61:227–51.

Willey, G. R. and P. Phillips. 1958. *Method and Theory in American Archaeology.* Chicago: University of Chicago Press.

Wobst, H. M. 1974. "Boundary Conditions for Paleolithic Social Systems: A Simulation Approach." *American Antiquity* 39(2):147–179.

Wylie, A. 1982. "An Analogy By Any Other Name Is Just as Analogical." *Journal of Anthropological Archaeology* 1:382–401.

Young, T. C., Jr. 1988. "Since Herodotus, Has History Been a Valid Concept?" *American Antiquity* 53(1):7–12.

2

Fundamentals of Archaeology

Though nothing can bring back the hour
Of splendour in the grass, of glory in the flower;
We will grieve not, rather find
Strength in what remains behind;
William Wordsworth (1770–1850)

People are messy animals. More than two million years ago, our ancestors began littering the African plains with stone tools and smashed animal bones, and ever since we have been sinking deeper and deeper into our own garbage. Strictly speaking, all this junk, from two-million-year-old quartz tools to today's eternal aluminum beer cans, is the *archaeological record*.

As indicated in the previous chapter, archaeologists see cosmic significance in the trash that is the archaeological record. The major premise of archaeology, in fact, is that much of what we will ever know about our origins and destiny is inherent in these layers of debris; that we can see in the contents, spatial arrangements, and history of the world's ancient garbage reflections of the factors that have shaped our physical and cultural evolution.

Archaeological Terminology

All scientific disciplines have their own jargon, and archaeology is no exception. Archaeologists analyze the archaeological record primarily in terms of *artifacts*,[1] which are things that owe any of their physical characteristics or their place in space to human activity. Thus, a beautifully shaped stone spear-point from a 20,000-year-old campsite in France is an artifact, but so are some undistinguished rocks pitched out of a Mississippi corn field a thousand years ago by a weary farmer. Even the faintest traces of hu-

39

man activity are considered artifacts, such as the footprints left several million years ago when a few of our ancestors strolled across a volcanic plain at Laetoli, in Tanzania (Figure 2.1).

Nor do things cease to be artifacts because of their recent origins. Since the early 1970s, archaeologist William Rathje and numerous archaeology students at the University of Arizona have been sorting through the artifacts added each day to the Tucson municipal dump and littered around about city roads, trying to discern how things are thrown away and what they say about the community that created the trash. (They learned among other things that the average Tucson resident wastes astounding amounts of meat and other food.)[2]

Another common archaeological term is *feature,* used to refer to nonportable clusters of artifacts, such as hearths, latrines, and burials, which are presumed (often on slight evidence) to reflect specific, often recurrent activities.

Perhaps the most common archaeological term is *site,* an imprecise term generally used to refer to relatively dense concentrations of artifacts and features. The ancient city of Babylon in Iraq, which 4000 years ago was the capital of a great empire, is a site, but so is any one of the many areas in Olduvai Gorge, Tanzania, where a few score stone tools and animal bones mark the spot where 1.8 million years ago a few of our ancestors killed or scavenged a gazelle.

Ancient village and town sites are hard to miss because they are marked by walls and massive quantities of pottery and other debris. It is convenient to think of the archaeological record in this case as composed of many discrete sites representing different settlements, but, in truth, the whole world is littered with artifacts and features.

Thousands of other terms are used in archaeology, most of them quite imprecise. *Prehistory* refers to the period before written records, but this time differs, of course, in different parts of the world. *Paleolithic, neolithic,* and *mesolithic,* refer to an old, middle, and new stone age, but these too, have no definite definition or absolute time span.

The Formation and Preservation of the Archaeological Record

The artifacts, features, and sites constituting the archaeological record vary widely in their contents and ages, but all must be understood to have been formed by a complex interplay of not only the activities of the people who created them but also the natural forces of decay.[3]

The laws of thermodynamics assert that matter is never destroyed or lost in the universe, but this is little consolation to the archaeologist looking, for example, at the smear of color that is all that is left of an ancient

2.1 Hominid footprints at Laetoli. A trail of hominid footprints was found preserved in volcanic ash at Laetoli, Tanzania. These prints demonstrate that hominids were fully bipedal by 3.6 million years ago.

corpse buried in the warm, wet soils of the Egyptian Delta. Even in drier, better preserved contexts, a depressing number of things can destroy archaeological remains. Floods wash them away, glacial ice sheets grind them to fragments, rodents go out of their way to burrow through them, and rivers and winds bury them under blankets of soil. Even the humble earthworm, as Julie Stein has shown, greatly alters the remains of the past.[4]

The greatest destruction, however, is caused by people—and not just in our own industrial age. The builders of the Egyptian pyramids used rubble from old villages as part of the mortar that binds the blocks of stone together. All over the ancient world, in fact, successive settlements were built on—and of—the remnants of earlier occupations. Still, our own generation is perhaps the worst despoiler of antiquities. In some areas, almost any new construction destroys antiquities; in London, for example, almost every time the foundations are dug for a new building, remnants of Roman Londinium are destroyed.

Industrialization at least has some apparent benefits, but the same cannot be said for the other great destroyer of the past, looting. In many countries, farmers loot sites and sell their finds to antiquities dealers, who sell them to brokers in America, Japan, and elsewhere. In the United States, vandals use all-terrain vehicles to raid even remote sites.

Looting destroys the only hope we have of analyzing cultural processes in the archaeological record because it obliterates the *context* of artifacts and features. Thus, for example, to study the origins of the first civilizations of Mexico, it is crucial to excavate sites in such a way that the goods people were buried with and the contents of their houses are meticulously recorded, so that the distribution of wealth in the community can be estimated. But once a looter has ripped through house floors to loot graves of their contents, the anthropological significance of the site is forever lost. Some (at least one) archaeologists are trying to have capital punishment instituted for archaeological looting, but even the mild fines and other penalties currently in force are only occasionally applied—although in 1988 someone was convicted and sentenced to jail for looting Indian burials in Colorado.

Natural processes of decay affect sites too, but if a site is not looted the effects of these natural processes can be discerned and taken into account in interpretations. Stone tools are almost indestructible, but bones, hides, wood, plants, and people rot. Organic decay is a chemical reaction, and the best preservation of organic remains is where there is not enough water or oxygen for the chemistry of decay to occur, such as in dry caves, permafrost, or deep, dark, cold water. Entire mammoths have been retrieved from frozen pits in Siberia, and well-preserved human corpses thousands of years old have been recovered from peat bogs and swamps in Europe (Figure 2.2). Eight-thousand-year-old corpses found in waterlogged levels of the Windover site in Florida still had enough brain tissue preserved that it may be possible to compare their genes to those of present-day Native Americans and even determine what language family they belonged to and how long their ancestors had been in North America.[5]

No one knows how powerful the analytical equipment of the future will be, so it is important that archaeologists preserve as much as possible—

2.2 "Tollund man," a 2,000-year-old hanging victim from the peat bogs of Denmark, illustrates the "pickling" properties of weakly acidic environments.

even deciding not to dig some extremely important sites in the hope that someday we will have equipment and techniques of vastly greater sophistication.

Archaeologists and the Practice of Archaeology

Before considering in more detail how archaeology is done, let us consider who actually does it. When trapped in airplanes or up against the wall at a party, archaeologists who reveal their occupation often are told "When I was growing up I wanted to be either a fireman or an archaeologist," or "It must be exciting to be an archaeologist!" But few nonarchaeologists realize that most archaeologists have to spend years in academic preparation, or that most spend at least two-thirds of their careers in universities or public agencies, and only a third or less actually digging.

Most archaeologists practicing in the United States have doctoral degrees, the dubious prize of an average of six years of post-graduate study. "Those who can, do; those who can't, teach" does not apply to archaeology, where almost every major professional holds a teaching position. Competition for university positions is ferocious. Most beginning professorships pay about $30,000 a year, and literally hundreds of scholars compete for every job. Many have spent twenty-two years in school preparing for a professional life in archaeology, only to discover that they cannot find an academic position.

Archaeology in North America is done primarily by two distinct kinds of professionals: anthropologists, who are the majority, and a minority made up mainly of those trained primarily as language scholars, such as Egyptologists, Assyriologists, Classicists (e.g., Greek), and Biblical scholars. Most anthropological archaeologists consider their discipline a science, with as much in common with the natural sciences as with the social sciences. Most graduate programs in anthropological archaeology require some mastery of mathematics, geology, and biology. In contrast, most language scholars consider themselves primarily historians and humanists. Language scholars tend to be particularly interested in relating archaeological remains to ancient written documents, such as the Bible, Greek and Roman texts, Egyptian hieroglyphic inscriptions, etc.

In Europe, Asia, and Africa, archaeology is often a separate university department—not connected with sociocultural anthropology or physical anthropology, as it is in the United States.

Archaeology is unusual among academic disciplines in that women make up a high proportion of practicing professionals and include some of the most prominent scholars in the field.

The Structure of Archaeological Research

No matter what an archaeologist's academic orientation, anyone who actually dons the pith helmet and digs holes in the ground has to do at least two things: choose where to dig and classify the items found into some analytical categories (e.g., sheep bones, Predynastic Red-slipped bowl, etc.). Choosing the place to dig is usually not so unstructured and speculative a procedure as imagined by nonarchaeologists—who frequently ask "How do you know where to look?" In modern archaeology, one rarely sets out on expeditions to remote places on the Micawberish assumption that something interesting will turn up, although some sites are still found by accident or unsystematic exploration. The construction of a subway system in Mexico City in the 1960s, for example, turned up parts of an Aztec temple. But increasingly archaeological remains are identified through a process of systematic survey. It does not take a trained archaeologist to find the pyramids of Egypt or Mexico, but most archaeological remains are less evident and accessible, such as those covered by drifting sand or alluvial soils, buried beneath contemporary settlements, or located in remote, unsurveyed areas.

Archaeological surveys and excavations are usually done within the context of a specific problem. If one were interested in the origin of maize agriculture in ancient Mexico, for example, one would read the many reports written on this subject, define an area where maize is likely to have been first cultivated, hypothesize some possible causes of the transition to maize agriculture, and then design a program of surveys and/or excavations to study this problem.

One usually incorporates these preliminaries in a written proposal for research funds (to, for example, the United States National Science Foundation), explaining precisely what kinds of evidence one is after, why it is important, and how to acquire it. This proposal will be judged by a group of one's peers, and if it is successful (in recent years only about 15% of National Science Foundation archaeology proposals were funded), one might begin to conduct archaeological surveys to locate relevant sites within this region.

Actually locating such sites might involve walking surveys, where five or ten archaeologists, working from maps or aerial photographs, simply line up and walk over a selected area, recording sites as they are found. Aerial photographs and other photogrammetric techniques can often be used to reveal ancient agricultural fields, roads, and other features not visible from the ground (Figure 2.3). However they are located, archaeological sites can either be simply mapped and recorded, or can be excavated—depending on the project's resources and objectives.

The methods used to excavate archaeological sites depend on the kind of remains involved and the objectives of the archaeologist. Normally the first step is to make a careful map of the site so that objects and features found can be given precise three-dimensional coordinates, the "provenience." Then the site is gridded into, say, five-by-five-meter blocks, and a sample of these blocks is selected for excavation. Actual digging is done with dental tools, paint brushes, trowels, shovels, bulldozers, or dynamite—depending, again, on the objectives and context.

Like every other profession, archaeology has its variants of Murphy's Laws: veteran field workers know that the most important find will be made on the last day of the season when there is no time or money to continue the excavations, and that particularly important finds are always located in the most inaccessible places. It's also a lot of hard work, usually. Anyone who has dug a backyard trench for a sewer pipe on a hot August day has already experienced many of the thrills of field archaeology.

The simple mechanics of excavation are within the range of abilities of almost any adult, except those who have difficulty envisioning strata in three dimensions. The best field archaeologists tend to be those who have

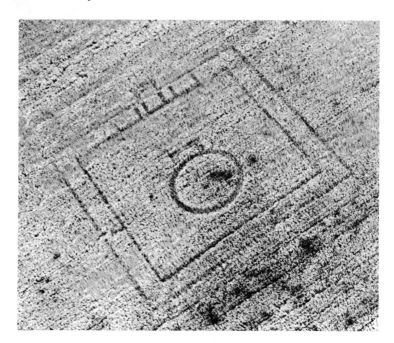

2.3 Aerial photographs often reveal archaeological remains that are not directly visible from ground level, as in the outline of this Roman temple on Hayling Island, Hampshire, England. Stone walls just below ground surface caused parching of grain just above them during the 1976 drought, revealing the outline of the temple walls.

a good sense of spatial relationships and enormous patience. "God is in the details," said a great architect, and the same is true of archaeology. One usually tries to excavate according to the cultural stratigraphy (Figure 2.4) of the site, so that the different layers of debris are removed in the layers in which they were deposited—as opposed to simply digging the site by arbitrary levels. In Tabun Cave in Palestine, for example, Neandertals came each year for a few months and built fires, made tools, butchered animals, and generally lived their unremarkable lives. Rocks falling from the ceiling and animals bringing their prey back to the cave when people were not there added to the layers of debris. Thus the excavators,[6] who were interested in subtle changes in diet and tool manufacture over the whole history of the cave's occupation, had to tease apart layer after layer of debris, trying to separate layers that were the result of short time intervals and single events.

Aaron Copland described listening to one of Ralph Vaughn William's symphonies as being like staring at a cow for 45 minutes, and studying archaeological strata can be similarly uneventful, but it is one of the most important activities in archaeology. In cave sediments, for example, one must try to discern faint traces of burrowing animals that may have tunnelled in from the surface and whose burrows were subsequently filled with charcoal, ash, and artifacts that date to periods long after their stratigraphic position would suggest.

Excavation also rewards imagination. British archaeologist Sir Leonard Woolley, while excavating Ur, in Mesopotamia, removed some debris and saw two holes in the ground, where something had apparently rotted away. He poured them full of plaster and when the plaster had hardened Woolley unearthed an almost complete cast of a wooden ancient musical instrument that had long since disintegrated. And Woolley was a clumsy hacker by the standards of Sir Mortimer Wheeler, who excavated some of the great cities of the Indus Valley with techniques still used today.[7]

2.4   This profile drawing shows the depositional history of a community at Tepe Sabz, near Deh Luran, Iran, from about 5500 to 3500 B.C. Skill in field archaeology is largely the ability to discern and interpret such cultural layers in the confusing jumble of mudbrick, stones, ash, and other debris.

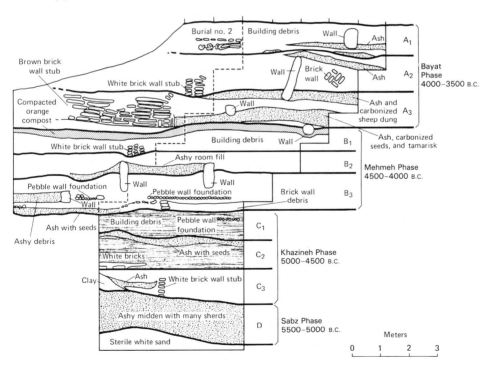

Artifact Analyses

Every archaeological site is unique and nonrenewable, and to make the most of the evidence, excavation staffs usually include geologists, botanists, palynologists (experts on plant pollen), architects, faunal experts, and other specialists.

Conserving finds once they are made has become a highly technical specialty, requiring advanced training in chemistry and other subjects.[8]

Aside from ancient buildings, in sheer bulk the largest part of the archaeological record is made up of stone tools and pottery fragments (sherds). Stone tools are the earliest known artifacts, having been first used more than two million years ago, and they have remained in use to the present day. In analyzing ancient stone tools, archaeologists have studied closely the mechanics of the process by doing it themselves. When a chunk of fine-grain stone is struck with another rock or a length of wood or bone at the proper angle and the correct force, a shock wave will pass through the stone and detach a flake of the desired size and shape (Figure 2.5). Classrooms all over the world are bloodied each year as instructors attempt to demonstrate this process, but with a little experience most become quite skilled. Few things are sharper than a fragment struck from fine-grain flint or glass (obsidian). Obsidian is so fine grained that flakes struck of it can have edges only about 20 molecules thick—hundreds of times thinner than steel tools. Archaeologist Richard Daughtery convinced his doctor to use obsidian tools as well as standard surgical scalpels during his recent heart surgery and claimed that the incisions made with obsidian healed faster.

Through experimentation, though, some archaeologists are able to produce copies of almost every stone tool type used in antiquity. A common research strategy is to make flint tools, use them to cut up animals, saw wood, clean hides, bore holes, etc., and then to compare the resulting wear traces with the marks found on ancient artifacts. Sometimes electron-scanning microscopes are used to study minute variations in these use-marks. Some rough correspondence can be found between the types of uses of a lithic and the characteristics of wear marks, but there are many ambiguities. Patrick Vaughn, Lawrence Keeley, and other archaeologists have shown how subtle and ambiguous the marks produced on stone tools by different uses can be.[9]

Ethnographic data from people who still use lithics, like Brian Hayden's study of use of stone in the Mexican highlands and Polly Weisner's study of how the !Kung hunter-gatherers use styles of stone spear-points to identify their social groupings,[10] indicate that even crude-looking stone tools may reflect a great amount of social life and economic forces.

Ceramics were in use much later than the first stone tools (appearing

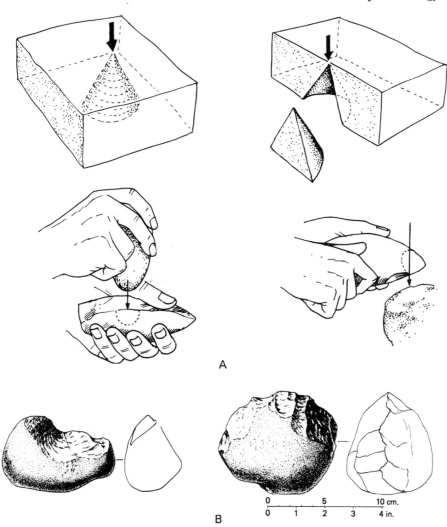

2.5 How stone tools are made. Fine grain stones will shatter in predictable ways when struck at the correct angle. Ancient stone tool makers could shape stone tools very precisely, but the earliest stone tools are just rocks with chips struck off one edge.

after about 10,000 years ago), but they were used in such massive quantities in antiquity that, for many archaeologists, life is mainly the slow sorting and analyzing of potsherds. Ceramic pots were first made by hand and dried in the sun or low-temperature kilns, but in many areas of the Old World, the invention of the potter's wheel and high-temperature kilns produced pottery that is a form of glass and therefore all but indestructible (Figure 2.6).

2.6 This "Susa A" style jar, from early fourth millennium B.C. Iran, exemplifies the hand-painted, highly decorated pottery styles that were widely distributed in Southwest Asia just before initial cultural complexity.

Ceramics form such a large part of archaeologists' lives because ceramics express so much about the people who made them.[11] Pots are direct indicators of function, in that they show how diets and economies changed over time. David Braun, for example, has documented how pottery in the American southeast changed in prehistoric times as a form of agriculture developed in which people boiled seeds of various native plants.[12]

Ceramics are almost always analyzed on the basis of their *style*. This idea of style is hard to define, but—as discussed below—changing styles are the basis on which archaeologists date much of the archaeological record. For many archaeologists, ceramics styles are more than just convenient devices for dating. Some even see ceramic decorations as "condensed symbols that express the cosmological assumptions that underpin social action."[13]

Although stone tools and ceramics make up much of the archaeological record, artifacts of wood, animal hides, metals, minerals, and almost

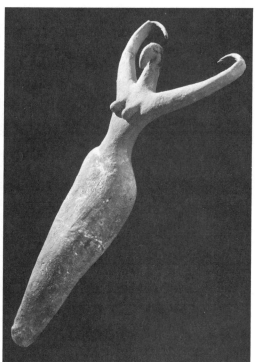

2.7 A great part of the world's archaeological record is composed of stone, wood, and clay artifacts. The flint knife depicted here dates from about 4000 B.C., from Egypt. Its ivory handle is carved with scores of delicate animal figures. The ceramic pot and figurine are also from Egypt, from about 3100 B.C.

everything else have been in use for thousands, even millions of years (Figure 2.7).

The study of animal remains, or faunal analysis, to use the common term, is a complex discipline in which in most cases the archaeologist is trying to reconstruct human diet and local environments. Faunal analysts usually tally the numbers and kinds of animals represented by the remains they find, and then use statistical methods to estimate food values, the ages and sexes of the animals involved, and changes in diets and the physical characteristics of the animals being exploited.[14] One of the most prolonged and heated arguments in contemporary archaeology now involves analyses of marks left by humans by cutting up animals with stone tools: for reasons discussed in Chapters 3 and 4, it is important to study butchered animal bones to try to distinguish between cases in which people butchered animals they had killed and those where they butchered animals they scavenged from kills of other animals, such as lions and hyenas.[15]

Throughout the history of our genus, plants have been the main source of food for most humans, and so analyses of floral remains are an extremely important part of archaeology, particularly in studies of how domesticated plants and animals and agricultural economies evolved.[16] Carbon is chemically quite stable, so charred plants and seeds preserve well. Carbonized plant remains can be retrieved by *water-flotation:* excavated sediments are mixed with water and the charred plant fragments floated to the surface where they can be skimmed off and identified.

A rapidly growing technical speciality within archaeology is *geoarchaeology,* the combination of archaeological and geological analyses.[17] People alter the surfaces they live on in many ways. In very ancient sites, the floors of caves and open-air camps often contain small particles of bones, plants, debris from making stone tools, and other remnants. Spilled food, human wastes, the manure of domestic animals—all these and many other factors associated with human life change the chemistry, texture, and contents of the surfaces people lived on. Geoarchaeologists rely on careful stratigraphic examination, chemical analyses, and other studies to recreate the forces producing the sediments that form the matrix of the archaeological record.

Classifying and Typing Artifacts

Here is an ancient Chinese classification of animals:

> Animals are divided into (a) those that belong to the Emperor, (b) embalmed ones, (c) those that are trained, (d) suckling pigs, (e) mermaids, (f) fabulous ones, (g) stray dogs, (h) those that are included in this classification, (i) those that tremble as if they were mad, (j) innumerable ones,

(k) those drawn with a very fine camel's hair brush, (l) others, (m) those that have just broken a flower vase, and (n) those that resemble flies from a distance.[18]

A zoologist working with the above classification of animals might develop exquisite sensibilities, but he or she would have a difficult time using this system to study comparative anatomy. For a fundamental procedure of science, or any form of analysis, is to construct classifications that facilitate certain kinds of research objectives. To understand how the world operates, we have to break it up into groups of similar things and discover the relationships between these groups. Modern chemistry or physics, for example, would be inconceivable were it not for classes such as electrons, atoms, and molecules, and the laws of thermodynamics. In the same way, evolutionary biology is possible only because of concepts of chromosomes, cells, and species, and the principles of population genetics. These notions about classification and analysis are quite straightforward and simple, but when we consider the kinds of data that archaeologists work with, we find that archaeological classifications and analyses have differed somewhat from those of other disciplines. The archaeologists' broken pottery, house foundations, and stone tools have not been organized in classifications in the same ways the atom and the cell have. A potassium atom is exactly the same thing to a Japanese chemist and an American chemist; but when a French archaeologist describes stone tools from southern France as "handaxes," those artifacts differ in many respects from North Chinese "handaxes" as described by a Chinese archaeologist. Archaeological classifications have generally been constructed with much more limited purposes than the units of the natural sciences. It is theory, whether biological, quantum, or Marxian, that tells the researcher how to break up the world for analysis, and in archaeology the only theories are relatively weak behavioral generalizations.

Most current archaeological classifications and typologies are inward-looking, descriptive, and remind one of Antony trying to describe a crocodile to Lepidus in Shakespeare's *Antony and Cleopatra:*

> LEP. What manner o thing is your crocodile?
> ANT. It is shaped, sir, like itself; and it is as broad as it hath breadth; it is just so high as it is, and moves with its own organs: it lives by that which nourisheth it; and, the elements once out of it, it transmigrates.
> LEP. What colour is it of?
> ANT. Of its own colour too.
> LEP. 'Tis a strange serpent.
> ANT. Tis so. And the tears of it are wet.
> CAES. Will this description satisfy him?

One of the most common classifications in archaeology has been in terms of *functional* types. Archaeologists, for example, frequently categorize the

1.75-million-year-old tools from Olduvai Gorge as "cleavers," "scrapers," and "hand-axes." Such a classificatory system is based in part on ideas about how our earliest ancestors actually used these tools. Obviously, imagination plays a role in creating functional types, particularly when archaeologists are dealing with very old remains left by people very unlike known or existing cultures. The use of high-powered microscopes to study wear patterns on stone tools and other technical advances has given archaeologists more confidence in their ability to infer the functions of artifacts, but there will always be an element of inference and error in these speculations.

Another widely used archaeological classificatory approach employs chronological types. Chronological (or "historical") types are artifacts whose combination of attributes is known to be limited to particular time periods. We have already noted that stylistic elements such as pottery decorations and house architecture have limited distribution in time, and by sorting artifacts into groups based on their similarity of stylistic elements we can often devise relative chronologies of archaeological remains.

While depending on chronological and functional types in most analyses, archaeologists continue to search for more powerful systems of arrangement. In contemporary archaeology, the logic and mechanics of arranging and classifying artifacts into analytical units continues, with some stressing a statistical approach, others more formal methods.[19]

Once archaeologists have grouped the artifacts of the archaeological record into classes or types, they analyze the distribution of these classes and type through time and space. In a film scene, W. C. Fields, while dealing cards, was asked by a prospective player, "Is this a game of chance?" Fields—felonious eyes agleam—replied, "Not the way I play it!"

Modern archaeology, on the other hand, is in many crucial ways, a game of chance, in the sense that we must use probability theory and statistics to interpret what we find. Chance in this sense enters in directly in the formation of what archaeologists have to work with, the archaeological record. Some 1.7 million years ago, for example, an individual who from the neck down looked very much like ourselves made a light lunch on a cow-like animal (probably the leavings of some other animal) and tossed some of its bones into some lakeside sediments, where the bone was preserved—cut marks intact—until Louis Leakey dug them out in the 1950s. Doubtless this same individual of 1.7 million years ago munched on other bones that were thrown away in areas where they rotted or were totally fragmented by hyenas, and have thus disappeared. And chance enters into not only the preservation of objects but also into their discovery. Most of the major archaeological sites in European countries, for example, are within a short distance of major roads—a sign that there are probably many other sites that have not yet been discovered because no one has happened on them.

Chance—or, more precisely, probability statistics—is also part of the analytical methods of modern archaeology. The costs in time and money of archaeology are such that even well-known sites, like the ancient city of Susa in Iran, where the biblical Esther lived 3000 years after the city's founding, are so large that even a century of excavation has removed only a small fraction of the site. Even in Egypt, where centuries of excavations and reoccupation have destroyed many sites, hundreds of huge sites have been only partly excavated.

The only reasonable archaeological strategy in the face of such a massive archaeological record is to sample: to excavate some parts of some sites in the hope that these samples will accurately reflect the whole.

The essentials of statistical sampling are familiar to most people. Polling organizations regularly ask a few thousand people how they are going to vote in an election and use this information to make very reliable predictions about the voting behavior of the larger population (all individuals who actually vote). One of the reasons sampling works in elections is that pollsters stratify samples: they know from previous elections that people in the North vote differently from those in the South, and that certain occupational groups are far more likely to vote than others. Thus, they break up, or stratify, their samples so that these and other subpopulations are proportionately represented. Then, by using procedures of statistical inference, they are often able to estimate election results quite precisely.

Archaeologists also use sampling theory and procedures.[20] If they wish to know something relatively straightforward, such as the number and kinds of sites in a large region, they can divide the area up into subareas—perhaps stratifying it according to ecological zones—and then go out and record the number of sites in perhaps 10% of all the subareas. Excellent results are usually obtained from such procedures, if the objective is simply an estimate of site densities. One critical sampling problem derives from the great size and complexity of the archaeological record. Suppose, for example, that you have the idea that trade in items such as flint and obsidian was a key element in the rise of the first states of ancient Mexico. The only way to test your idea would be to determine if there had been a significant increase in the amount or kinds of these commodities at sites occupied just prior to or during the period when the first states appeared. To do this with statistical precision, you would have to excavate at least portions of a statistically valid sample of at least thirty or forty sites—something just not feasible in today's archaeology. The result is that archaeologists are not purists when it comes to using statistics and probability models. Because so much of the residues of the past has decayed, and because of the costs of gathering and analyzing archaeological data, archaeologists tend to misuse statistics and probability theory by making sweeping inferences on the basis of inadequate data. No Wall Street executive or even drunk riverboat gambler would bet on the odds that archaeologists do when testing their hypotheses; but archaeologists deal only in history and science, whereas gamblers deal in money.

Archaeologists have opted for the only realistic compromise: they use statistical sampling techniques, knowing that they often don't meet the theoretical requirements of optimal statistical inference, but believing that useful—if not perfect—results can be obtained. Fortunately, most statistical sampling techniques are very "robust" in that one can strain their assumptions badly and still get quite reasonable results.

To a large extent archaeological interest in sampling—and many other aspects of modern archaeology as well—are side effects of the invention and improvement of the modern computer. Even simple statistical description and inference are impossibly time-consuming to be broadly applied to archaeology without computers. Quantification in archaeology is not just a matter of sampling; it underlies most other methodological advances.[21]

Archaeometry

The basics of excavating artifacts and features, classifying them, and counting them are relatively straightforward problems common to many sciences. But like other disciplines, archaeology involves many specialized forms of *measuring* artifacts in many different ways. This field is generically referred to as archaeometry.[22]

Dating Methods in Archaeology

The primary importance in archaeology of dating methods is in analyzing cultural changes. To take an example, some have argued (Chapter 6) that the reason people first domesticated sheep and goats and began farming wheat and barley in the Middle East was that human population densities had risen to the point that people could no longer survive on hunting and gathering. Other people suggest that rising population densities had little to do directly with the origins of agriculture in this area.

Our only hope of resolving such disputes—of testing hypotheses—about the mechanics of major cultural transformations is to look at the archaeological record. If we do archaeological surveys in the area of the Middle East where agriculture first appeared and determine what sites were occupied during what periods and how large they were, we can estimate population densities before, during, and after the period at which agriculture first appeared—about 10,000 years ago. If we discover that there is no significant rise in population densities just before and during the period when we find the first domesticated plants and animals and agricultural implements, we might reject the idea that rising population densities were the important direct cause of this change. In short, our only hope of determining cause and effect in ancient cultures is to show correlations in time and space.

But how are we to date artifacts to show such correlations?

Archaeologists rely on two different kinds of dating methods. In some situations the objective is to obtain a *chronometric* date: that is, an age expressed in years, such as "that house was built 7200 years ago." In many situations chronometric dates may be difficult to obtain or simply unnecessary for the problem at issue, and for these situations archaeologists have devised several methods of *relative* dating, in which the objective is to arrange sites or artifacts in a sequence that reflects the order in which they were created—even though we may not know for certain the actual age of any of them.

CHRONOMETRIC DATING

Most archaeologists dream of a small pocket-sized device, stuffed with microchips and Star Trekian dilithium oxide crystals, which when pointed at an artifact will read out the object's exact date of manufacture. Fanciful as this may sound, modern physiochemical dating methods have been greatly improved in the last decade, and age estimates are becoming increasingly reliable.

Perhaps the most precise and yet technologically simple form of chronometric dating is dendrochronology—the study of tree rings.[23] Most trees add a single "ring" each year to their circumference; thus, if we count the number of rings, the age of a tree can be precisely established. Normally the tree grows faster in wet years than in dry ones; therefore, over the centuries there is a unique series of changes in ring widths, and precise dates can be inferred by comparing cross sections of trees that overlapped in time (Figure 2.8). By comparing beams, posts, and other artifacts to cross sections taken from trees that live for long periods, it is often possible to determine the exact year in which the tree used to make the artifact was cut. But here's the rub: in dry climates tree trunks tend to be used and reused for very long times, so that the date that the tree actually was cut may be centuries older than the period it was used as a rafter in some house.

Also, since local climates vary, dendrochronological records must be built up for each region, and at present detailed records are available only for the North American West and a few other places.

Douglas Brewer has devised an interesting parallel to dendrochronol-

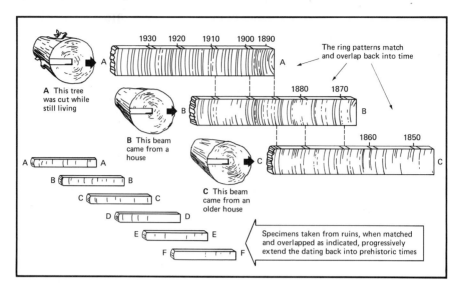

2.8   The most precise dates in archaeology are derived through dendrochronology. In many important areas of the world, however, a dendrochronological sequence has not been established, and in other areas, such as Mesopotamia, there are no native, long-lived species of trees.

ogical dating using growth rings in fish vertebrae.[24] Working with Nile catfish remains, he has computerized various patterns of growth and may eventually be able to get a complete sequence for some critical ancient periods.

The most widely used chronometric technique is ^{14}C, or carbon-14 dating, a theory first outlined in the 1940s by Nobel laureate Willard Libby.[25] When solar radiation strikes the upper atmosphere, it converts a small amount of atmospheric nitrogen into the radioactive isotope ^{14}C. Wind and other factors spread this ^{14}C throughout the atmosphere, and because all living organisms exchange gases with the atmosphere, the ratio of ^{14}C in their cells is equal to that in the atmosphere. When the organism dies, the ^{14}C trapped in its cells begins to revert to nitrogen. Because we know that approximately half of any given quantity of ^{14}C will disintegrate in about 5730 years, we can estimate the time an organism has been dead by measuring the amount of ^{14}C remaining in its cells. After about 50,000 years, too little persists to be measurable with standard laboratory methods, although with large samples and the most powerful equipment, reliable dates up to 100,000 years ago are theoretically possible.

Carbon-14 dating works best on wood and charcoal, but paper, leather, bone, skin, peat, and many other organic materials can also be dated by this method. Grains and grasses make excellent archaeological samples when charred by fire, because they preserve well and are short-lived compared to trees.

The ratio of ^{14}C in the atmosphere has not been constant over the last 50,000 years, and thus carbon-14 dates have had to be "corrected" by measuring the ratio of ^{14}C in tree rings dated through dendrochronology. Fortunately, some trees, such as the bristlecone pine of northern California, live thousands of years, and cores from their trunks can be dated through dendrochronology and then each ring can be radiocarbon dated to construct a "correction curve."[26] Logs found submerged in northern European bogs where they have been preserved for thousands of years have recently allowed the calculation of a radiocarbon correction curve extending back more than 7000 years (Figure 2.9). But samples dated by

2.9 Radiocarbon age estimates must be corrected for the fact that the amount of ^{14}C in the atmosphere has not been constant.

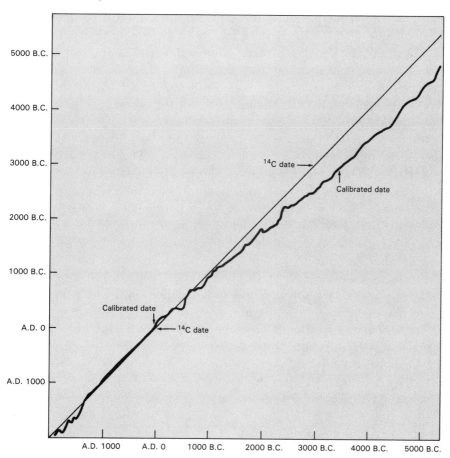

the ^{14}C method can still be contaminated with younger or older carbon sources, such as ground water or petroleum deposits.

A major advance in radiocarbon dating was made in the 1970s when various researchers used particle accelerators (the AMS method, or accelerator mass spectrometry) to date samples. This method allows reliable dates to be obtained from samples the size of a match-head, whereas older methods require about a handful of carbon. Accelerator dating has other advantages: samples can be more easily purified of contaminants, individual samples can be subdivided into very small amounts and tested for internal consistency, and older samples can be dated because problems involving background radiation have been obviated. Because accelerator dates can be done on such small samples, reliable dates can be obtained from cooking soot on pots, dung and other organic temper in pottery, slag, textiles, and many other materials.

In February 1989 an international team of twenty-one scientists reported the results of radiometric dating of the Shroud of Turin, a cloth that appears to bear the image of a man who has been whipped and crucified. Many people have believed that the Shroud was used to wrap the body of Jesus Christ. The scientists took three samples of cloth, each about 50 mg (about the size of a postage stamp), and sent them to three different laboratories, in England, Switzerland, and the United States. Using accelerator mass spectrometers, scientists at the three laboratories all concluded independently that the linen used for the Shroud was made about A.D. 1260–1390.

Interpretations of radiocarbon dates are rarely simple. In trying to date the Egyptian pyramids, for example, I, and my colleagues, Mark Lehner and Herbert Haas, seem to have encountered almost every problem intrinsic in the method.[27] We wanted to know when the Egyptian pyramids were built because we were engaged in a decade-long project to try to define the basic mechanics of ancient Egyptian cultural change. Construction of the enormous pyramids and other monuments in the Nile Valley was obviously a critical part since it must have required astounding investments of time and energy. But how do we know when they were built, and what their relation in time was to fluctuations in the Nile floods, political developments in neighboring areas, and other important events?

Not a single ancient text of the age when the pyramids were built describing their construction, or even referring to them, has ever been found. Egyptologists have dated the pyramids primarily on the basis of names on inscriptions in temples and tombs in areas near the pyramids. Ancient king-lists have been found and the reigns of specific kings are often given in inscriptions, so Egyptologists have been able to estimate the sequence of pharaohs and how long each ruled. Occasionally a text would record a specific astronomic event in the reign of a specific king, such as the rising of the star Sirius at a particular time and place on the horizon. Such events can be precisely dated, so we know the dates of some rulers with great accuracy. Unfortunately, these astronomical observations have not been found for the period when the pyramids appear to have been built.

Egyptologist Mark Lehner[28] convinced me that it would be worth trying to arrive at some date for the pyramids independent of ancient records. Lehner had observed that pieces of carbon could be found throughout the mortar used to bind the blocks of stone making up most of the pyramids, and we assumed that these bits of carbon were produced when gypsum was burned to create the powder that, when combined with water and other materials, constituted the mortar. We assumed, therefore, that if we could date these bits of mortar, we could date when the brush, trees, etc. had been cut to get the fuel to burn the limestone, and that from this we could estimate the age of the pyramids. We even hoped that if we took a lot of samples in sequence, from the base to the top of each pyramid, we might arrive at some estimate of how long it took to construct them and the sequence in which they were constructed.

After having obtained the necessary research funds and permissions, we started at the first course of the Great Pyramid of Khufu and began prising bits of carbon out of the mortar. Six months later we had just over a hundred samples from seventeen of the largest pyramids. Some samples were about the size of a pea, others constituted roughly a handful of carbon. We sent the larger samples to the Radiocarbon Laboratory at Southern Methodist University, for conventional radiocarbon dating, and we sent the smaller samples to a laboratory in Switzerland, to be dated with the recently developed (AMS) accelerator methods (see above). The majority of our dates came out almost four hundred years older than most Egyptologists would estimate as the ages for these various pyramids. We presented a paper on our results at a scientific conference and were informed by most Egyptologists and virtually everyone else that our radiocarbon dates had little or nothing to do with the ages of the pyramids. It was suggested that our dates came out too old because the ancient Egyptians used old wood to build the fires to produce the mortar, or because the carbon came from plants that naturally absorb relatively large amounts of radioactive carbon, or because the mortar itself had contaminated the carbon, or because the correlation curves we used were wrong.

All these factors may, in fact, have played a role in producing our dates, and even though we tried to control for as many of them as we could, we still are in no position to conclude that the traditional Old Kingdom chronology is wrong. More samples will be required, and we are currently testing modern vegetation for radiocarbon content. Eventually, by microscopic examinations of carbon samples, we also hope to determine if the material is from trees or reeds, etc. Reeds or some other short-lived plants would be preferable, since their time of death and the time they were burned would probably be roughly the same.

In any case, as the above indicates, radiocarbon dating can be very useful but interpretations of radiocarbon dates are usually difficult: dates that agree with one's suppositions tend to find a ready audience, dates that don't are often labeled "intrusive."

Another important form of archaeological dating is the potassium-argon method. Potassium-argon dating is based on the fact that a radioactive isotope of potassium (^{40}K), present in minute quantities in rocks and volcanic ash, decays into the gas argon (^{40}Ar) at a known rate (half of a given amount of ^{40}K will change into ^{40}Ar in about 1.3 billion years). Because ^{40}Ar is a gas, it escapes when rock is molten (as in lava), but when the rock cools, the ^{40}AR is trapped inside. By using sensitive instruments to measure the ratio of ^{40}K to ^{40}Ar, it is possible to estimate the time since the rock or ash cooled and solidified.

Because of the long half-life of ^{40}K (1.3 billion years), potassium-argon dating can be used to estimate dates of materials many millions of years old. The remains of our ancestors at Olduvai Gorge and other sites more than a million years old have been dated with the potassium-argon method.

Carbon-14 and potassium-argon dating remain the mainstays of chronometric dating, but there are now many other techniques involving chemical changes, most of which are subject to considerable error and many qualifications.[29] Thermoluminescence dating ("TL"), for example, has considerable potential. Its basis is that clays, silts, and other sediments contain naturally radioactive elements, so that when pottery is fired, or glass is melted—in general, when substances are heated to a certain point— the products of natural radiation are driven out of these materials, setting their radiological "clocks" at zero. Once these materials cool, their constituents continue the natural process of radioactive decay, but the pottery, glass, etc. now trap the electrons released by the natural radiation process. In theory, if one reheats these materials and measures very precisely how much of certain kinds of energy are given off in the process, one can estimate how long it has been since the material was last fired. This has the great advantage of dating the actual construction of the artifact, in contrast to the radiocarbon method, which dates the death of the organism from which an artifact was made and used—events that may be very far apart. Refinements of the TL method raise the possibility that campfires and even sun-baked surfaces can be dated with this method. But energy accumulation rates for each material and locale must be established, and currently the TL method is still in its early experimental stage.

RELATIVE DATING

As a novice graduate student, I was impressed as I watched two archaeologists wander over an ancient Mesopotamian town, occasionally stooping to pick up a potsherd and saying things like, "Obviously this area was occupied into the Late Uruk, whereas that part must have been abandoned by the Terminal Susa A Period."

This kind of relative dating involves the concept of style. Mesopotamian potters, like all other people, differed with the passage of time in the styles of pottery they produced. Therefore, the pottery in each succeeding century or two can be distinguished in form, color, and decoration.

The distribution of stylistic elements through time and space tends to follow certain patterns, whether the objects involved are skirt lengths, musical forms or stone tools. Styles originate in some small area, spread to adjacent ones, reach a peak in popularity, and then die out (Figure 2.10). To some extent, styles reflect rates of interaction and shared aesthetic preferences, and these are not always exact functions of time and distance. Dress styles, for example, in midtown Manhattan may be more similar to those on Rome's Via Condoti than to those in a small town in rural New Jersey, even though this pattern of stylistic similarity "reverses" their relative distances. And often a style dies out at its point of origin long before it reaches its ultimate dispersal.[30]

Relative dating, or seriation as it is usually called, is often used where many surface collections of artifacts have been made. Jeffrey Parsons and his students, for example, have surveyed most of the area around Mexico City, identifying thousands of settlements dating from the Spanish Conquest to 12,000 years ago.[31] Most of these sites are small mounds whose surfaces were littered with pottery sherds and obsidian tools. The differences in style between a Late Aztec Black-on-Orange dish (c. A.D. 900) and Middle Formative plainware jar (c. 550 B.C.) are so obvious that anyone can learn to date sites of these periods in a few days. On this basis, thousands of sites were dated without excavating them, simply by grouping them into a relative seriation of four or five major periods. Carbon-14 dating can be used to provide a few absolute dates to anchor this sequence, but pottery styles alone are all that is necessary to construct a seriation. Accurate relative seriations usually require massive quantities of data from artifacts of a highly decorated nature (like pottery) from a relatively small area, and they tend to be least precise when extended to largely undecorated objects such as early stone tools.

One of the most absorbing problems in archaeology has been to devise mathematical, computerized improvements on the graphical seriation technique illustrated in Figure 2.10. Dick Drennan and Stephen LeBlanc,

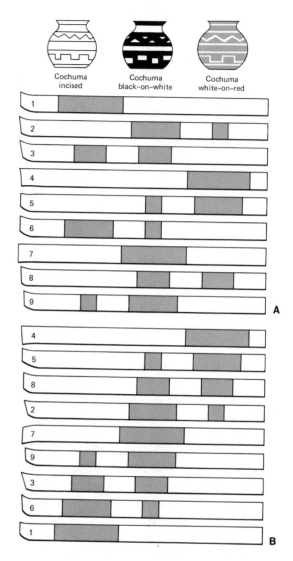

2.10 Relative seriation of nine archaeological sites on the basis of three pottery styles from the American Southwest. The percentage that each pottery style represents of the total pottery found at each site is shown by the width of the colored area on the strip of paper. Since most styles tend gradually to grow in popularity and then slowly die out, a seriation can be produced by arranging the paper strips in such a way that the three pottery styles have this "battleship shape" distribution through time. The inferred order of the nine sites is shown in B. Mathematical models and computer programs have been developed to sort scores of sites and pottery styles into these kinds of graphs. Such mathematical aids are often needed, because the number of possible unique orderings of 9 sites is 9!, or 362,880.

for example, have both devised methods using multidimensional scaling to produce seriations of archaeological data.[32]

Other Methods of Archaeological Analysis

An important field in current methodological research is *remote sensing.* Satellites in stationary orbits have been equipped with sensors that measure the reflectance of light from the earth's surface with such resolution that ancient roads, agricultural fields, and other features can be discerned where the observer on the ground sees little of interest (Figure 2.11).

Anyone can now buy the computer tape that has on it a digitized picture of a given area of the world, but the power of resolution of these Landsat images is such that the smallest objects that can be distinguished are about 80 meters on a side (French and Russian images are also for sale and have better resolution), and one needs about a million dollars worth of computer equipment to process the computer tape.

Satellites also can take photographs, as opposed to digitized Landsat records of reflectance properties. Newspaper reports of satellite photographs so clear that license plate numbers of cars in the Kremlin parking lot can be read have convinced many archaeologists that a lot of great archaeological data is in drawers in the basement of the Pentagon.

Various instruments for use on land surface have also been developed to try to identify buried buildings, canals, roads, etc. These instruments use measures of electrical resistance or the magnetic disturbances such features have on gravity.

Underwater archaeological survey methods have also greatly improved

2.11 A LANDSAT image of the Egyptian Delta. By analyzing the reflectance properties of such areas, the locations of ancient occupations, old channels of the Nile, and other archaeologically useful information can be recovered.

2.12 Underwater archaeology is expensive and time consuming but often rewarding. Here a diver examines a brick floor of what was part of the settlement at Port Royal, Jamaica, before the town was buried beneath the sea by an earthquake on June 7, 1692.

in recent years (Figure 2.12), but the cost of such surveys is high and the area covered usually very small.

The past decade has witnessed a great increase in the use of instruments and procedures such as chemical analyses to identify residues of food in pots, precise methods for separating and identifying pollen grains from sediments, and analyses of the chemical composition of finds. Some plants, such as maize, differ from others in the amount of the radioactive isotope ^{13}C they absorb; so by analyzing human or animal bones it is possible to determine if these people ate these plants—or ate the animals that ate these plants. Similarly, volcanic glass, obsidian, has a distinctive chemical composition determined by the volcano that was its source and, with neutron activation analysis, the origin of the raw material for the artifact can be determined—and trade patterns thereby inferred.

ETHNOARCHAEOLOGY

Many archaeologists have tried to interpret the past by studying contemporary societies. For example, the San, who live by hunting and gathering in the areas of South Africa in which we suspect some of our earliest human ancestors lived millions of years ago (Figure 7.2), have been carefully studied by several archaeologists.[33] By documenting how many San live together, what kinds of food they eat, where they throw their garbage (and what happens to it after it's discarded), and many other facets of their lives, archaeologists hope to infer something about early humans in this area.

Similarly, anyone who has excavated in the Middle East knows the irresistible temptation to use modern village life as a model of the community one is excavating: the remains of 7000-year-old houses in Iran, for example, greatly resemble those currently being occupied.[34]

The use of ethnographic data to interpret archaeological remains is a complex topic. We obviously cannot assume that contemporary societies perfectly reflect past ones, and, as discussed in Chapter 1, simple description of the past is not the only goal of archaeologists.

An Archaeological Example

Having considered various elements of modern archaeological methods and theory, it is perhaps useful to consider an example of a specific archaeological project. As one such example, I offer the Fayyum Archaeological Project, which I co-directed in Egypt in 1981.

The moving force behind this project, which was wholly conducted in Egypt, was in fact the Iranian Revolution. I had excavated and done regional archaeological surveys in Iran on several occasions in the early 1970s and was due to resume work there in 1979, on a day almost exactly between the Shah's departure from Iran and the first seizure of the American Embassy. I had received my first National Science Foundation grant and would have gone to Teheran despite the revolution, had the Iranians permitted. Through a series of events too baroque to recount here, I had the good fortune to be able to go instead to Egypt, where in 1980 I directed a small archaeological project.

I had long been interested in the origins of agricultural economies, and was much impressed by some new ideas about agricultural origins in Egypt[35] (see Chapter 6). So while in Egypt I started searching for an area in which to investigate the origins of Egyptian agriculture.

Many of the most important sites in Egypt have been excavated for decades and are already within the concession of another archaeologist; one cannot just decide to excavate this or that site. Professor Michael Hoffman of the University of South Carolina suggested I look at unsurveyed parts of the Fayyum Depression, where years of work by other archaeologists had shown evidence of early agriculture but which was currently not being explored. Archaeological projects in Egypt usually have on their staffs an Egyptologist—someone who reads ancient Egyptian writing—and I was fortunate enough to recruit a recent graduate from the Sorbonne, Dr. Mary Ellen Lane, as co-director and project bon vivant. With a representative of the Egyptian Antiquities Organization, we made several trips into the deserts of the southern Fayyum without finding much except a restaurant where I got deathly ill for only ninety piastres.

In the 1920s, the intrepid British archaeologist Gertrude Caton-Thompson had driven around the southern edge of the Fayyum Lake, noting here and there scatters of Neolithic-style stone tools. One day, driving in an area near where she had surveyed, we saw a large pile of bones. On inspection it proved to be the remains of a hippopotamus, and we were delighted to see that near it were stone projectile points ("arrowheads") of a Neolithic type. Within a few hours of survey, it was evident that we had found a dense scatter of hearths, pottery, stone tools, and animal bones, and the styles of artifacts suggested a date of about 4500 B.C.

Back in Cairo, our readings convinced us that what we had was significant, and that we should try for our first field season in the summer of 1981—a year later. All we needed was $200,000, a staff of at least twenty trained archaeologists, and permission from the Egyptian government.

Famed felon Willie Sutton, when asked why he robbed banks, patiently explained, "That's where the money is!" Archaeologists, too, must go where the money is, and in this era it is mainly to the government. After months of writing proposals, we received about $200,000 from the United States National Science Foundation and the United States Agency for International Development. We then recruited our staff of specialists in ancient plant remains, animal bones, and geology, and drafted eight graduate students to do the actual work.

On June 4, 1981, we left Cairo in several jeeps and trucks to make the four-hour trip to the Fayyum. We lived that summer and autumn in a large gray house that looked across a green palm grove and the blue of the Fayyum Lake to the white limestone cliff of the Jebel Qatrani. Our villa—the country home of a wealthy Cairo family—was a lovely International Style building with every convenience but three: water, electricity, and a sewage system. The provincial governor graciously arranged for a water-truck to visit us every three days, we bought a generator, and we devised an entertaining method of periodically napalming our open cesspool.

"Tell me what you eat, and I will tell you what you are," said the French gastronome Brilliant-Savarin. In our case, the remoteness of our field quarters meant that our diet was almost wholly composed of tuna fish, a vile processed cheese, rice, tomatoes, and several thousand chickens, who were executed on our kitchen steps and then converted into indescribable meals. "Fire-cracked Veal" and "Dreaded Veal Cutlet" were occasional holiday treats. We bored each other constantly with food fantasies.

The morbidity rates—physical and psychological—on archaeological projects are usually high, especially when, as in our case, water for cleaning was scarce and our cook did not believe in the germ theory of disease. We totaled at least five different strains of parasitical and bacterial infections among our crew, and lost many days to illness. There was also one emergency appendectomy (mine), performed in Cairo after a thought-provoking four-hour truck ride from the desert. In this adventure I was greatly assisted by Professor Fekri Hassan of Washington State University, whose truck and driver got me to Cairo in record time.

When we began our six months of fieldwork, we geared most of our efforts to reconstructing as precisely as possible the ways of life of the people who had lived in the Fayyum in the Qarunian period (c. 6500 B.C.), just before the appearance of domesticated plants and animals in this region, and in the succeeding Neolithic Fayyum A period (c. 5000 B.C.), when the first agriculturalists appeared here. We hoped to reconstruct the pattern of human settlement in the Fayyum between 7000 B.C. and A.D. 1500 and explain the changes in these settlement patterns over this long period.

We began by making a topological map of the area we intended to work in. We then devised a sampling program and collected every artifact in the sampling units defined, that is, in the hundreds of 5 × 5 meter squares in our study area. The average temperature during much of this work was over 40°C (104°F), and by mid-day the stone tools were often so hot we would have to juggle them as we bagged them. Afternoons were spent sorting, drawing, and photographing artifacts, drinking warm water, and drawing each other's attention to the heat. In some cases, "it's not the heat, it's the humidity" is not at all true. In September we began excavations, mainly of the hearths and pits that were the dominant feature of both the Qarunian and Fayyum A occupations. In most we found charred animal bones, some carbonized plant remains, and other debris.

To evaluate our "model" of how agriculture appeared in the Fayyum and why, we had to collect sufficient evidence to make statistical arguments about certain kinds of conditions and events in Fayyum prehistory. The details of these arguments are not relevant here, but it should be stressed that as in most archaeological projects, not all the information we had hoped would be there was actually found. But most was, and the preliminary analysis of this information required over five years, and is still continuing.[36] Hundreds of thousands of lines of numbers representing our measurements of artifacts, topographic elevations, animal bone types, and so on are still being subjected to statistical analyses.

In Cairo, after the season was over, we delivered the artifacts to the Egyptian Museum and made preparations to leave. It is traditional, after the privations of the field, to treat oneself to some rest and relaxation, and some project members agonized between such choices as the Club Med's Red Sea beaches or the "Il Buchetto" restaurant in Rome. I went to both.

Notes

1. For definitions of terms see Binford, "A Consideration of Archaeological Research Design"; Dunnell, *Systematics in Prehistory;* Hole and Heizer, *An Introduction to Prehistoric Archaeology;* Joukowsky, *A Complete Manual of Field Archaeology;* and Thomas, *Predicting the Past: An Introduction to Anthropological Archaeology.*
2. Rathje, "A Manifesto for Modern Material Culture Studies," p. 52.
3. Schiffer, *Formation Processes of the Archaeological Record.*
4. Stein, "Earthworm Activity: A Source of Potential Disturbance of Archaeological Sediments."
5. Zegura, "Blood Test."
6. Jelinek, "The Tabun Cave and Paleolithic Man in the Levant."
7. Wheeler, *Archaeology from the Earth.*
8. Sease, *A Conservation Manual for the Field Archaeologist.*
9. For example, Keeley, *Experimental Determination of Stone Tool Use: A Microwear Analysis;* Vaughn, *Use-wear Analysis of Flaked Stone Tools;* Jensen, "Functional Analysis of Prehistoric Flint Tools by High-Power Microscopy: A Review of West European Research"; Crabtree, *An Introduction to Flintworking.*
10. Hayden, *Lithic Studies Among the Contemporary Highland Maya;* Weisner, "Style and Social Information in Kalahari San Projectile Points"; Yellen, *Archaeological Approaches to the Present;* also see Lewenstein, *Stone Tool Use at Cerros. The Ethnoarchaeological and Use-wear Evidence;* Sackett, "Approaches to Style in Lithic Archaeology."
11. For an excellent review of the role of ceramics in archaeological analysis, see Rice, *Pottery Analysis.*
12. Braun, "Pots as Tools."
13. David, Sterner, and Gavua, "Why Pots Are Decorated."
14. Grayson, *Quantitative Zooarchaeology.*
15. Binford, "Human Ancestors: Changing Views of Their Behavior"; Bunn, "Archaeological Evidence for Meat-Eating by Plio-Pleistocene Hominids from Koobi Fora and Olduvai Gorge"; and Shipman, "Early Hominid Lifestyle: Hunting and Gathering or Foraging and Scavenging?"
16. Dimbleby, *The Palynology of Archaeological Sites;* Piperno, *Phytolith Analysis;* Renfrew, *Palaeoethnobotany;* Gilbert and Mielke, eds., *The Analysis of Prehistoric Diets;* and Bodner and Rowlett, "Separation of Bone, Charcoal, and Seeds by Chemical Flotation."
17. For examples see the journal *Geoarchaeology.*
18. Jorge Luis Borges, *Other Inquisitions: 1937–1952,* quoted in Aldenderfer and Blashfield, *Cluster Analysis,* p. 7.
19. See, e.g., Whallon and Brown, *Essays on Archaeological Typology;* Read, "The Substance of Archaeological Analysis and the Mold of Statistical Method: Enlightenment Out of Discordance?" pp. 45–86; Dunnell, "Methodological Issues in Americanist Artifact Classification"; Adams, "Archaeological Classification: Theory Versus Practice"; Deetz, *Invitation to Archaeology;* and Beck and Jones, "Bias and Archaeological Classification."
20. Reviewed in Shennan, *Quantifying Archaeology;* and Mueller, ed., *Sampling in Archaeology.*

21. Aldenderfer, ed., *Quantitative Research in Archaeology;* Aldenderfer and Blashfield, *Cluster Analysis;* Carr, ed., *For Concordance in Archaeological Analysis;* Reidhead, "Linear Programming Models in Archaeology"; Sabloff, ed., *Simulations in Archaeology;* Doran and Hodson, *Mathematics and Computers in Archaeology;* Leonard and Jones, eds., *Quantifying Diversity in Archaeology;* and Mueller, ed., *Sampling in Archaeology;* Kimes, et al., "A Method for the Identification of the Location of Regional Cultural Boundaries."

22. Reviewed in Leute, *Archaeometry.*

23. Baillie, *Tree-Ring Dating.*

24. Brewer, "Seasonality in the Prehistoric Faiyum Based on the Incremental Growth Structures of the Nile Catfish (Pisces: *Clarias*)."

25. Libby, *Radiocarbon Dating.*

26. Taylor, *Radiocarbon Dating. An Archaeological Perspective;* Browman, "Isotopic Discrimination and Correction Factors in Radiocarbon Dating." Until recently, radiocarbon dates were often published as a certain number of years before A.D. 1950, the benchmark year in which the method was first established. "BP" was often used for corrected radiocarbon dates, "bp" for uncorrected dates (the "bp" first referred to "before present" or even "before physics"; see Taylor, *Radiocarbon Dating. An Archaeological Perspective*, p. 5). Today there are still no universal standards for reporting radiocarbon dates. Also see Gowlett, "The Archaeology of Radiocarbon Accelerator Dating."

27. Haas, Devine, Wenke, Lehner, Wolfli, and Bonani, "Radiocarbon Chronology and the Historical Calendar in Egypt."

28. Lehner, "Some Observations on the Layout of the Khufu and Khafre Pyramids"; idem., *The Pyramid Tomb of Hetep-heres and the Satellite Pyramid of Khufu.*

29. Michels, *Dating Methods in Archaeology;* and Tite, *Methods of Physical Examination in Archaeology.*

30. Sir W. M. F. Petrie formulated one of the earliest and mathematically most complete methods of relative seriation based on his analyses of ancient Egyptian tomb contents (see, e.g., his "Sequences in Prehistoric Remains"). Also see Dunnell, "Seriation Method and Its Evaluation."

31. Parsons, *Prehistoric Settlement Patterns in the Texcoco Region, Mexico.*

32. Drennan, "A Refinement of Chronological Seriation Using Nonmetric Multidimensional Scaling"; LeBlanc, "Microseriation. A Method for Fine Chronological Differentiation"; Wenke, "Western Iran in the Partho-Sasanian Period"; Marquardt, "Advances in Archaeological Seriation."

33. For example, Yellen, *Archaeological Approaches to the Present;* idem., "Cultural Patterning in Faunal Remains: Evidence from the !Kung Bushmen."

34. Kramer, *Village Ethnoarchaeology: Rural Iran in Archaeological Perspective;* Gould and Watson, "A Dialogue on the Meaning and Use of Analogy in Ethnoarchaeological Reasoning"; Wylie, "An Analogy By Any Other Name Is Just as Analogical."

35. Wendorf and Schild, eds., *Prehistory of the Eastern Sahara.*

36. Wenke, Long, and Buck, "Epipaleolithic and Neolithic Subsistence and Settlement in the Fayyum Oasis of Egypt"; Wenke and Lane, *Land of the Lake. 8000 Years of Human Settlement in Egypt's Fayyum Oasis.*

Bibliography

Adams, W. Y. 1988. "Archaeological Classification: Theory Versus Practice." *Antiquity* 61:40–56.

Aldenderfer, M. S., ed. 1987. *Quantitative Research in Archaeology.* Newbry Park, Calif.: Sage.

Aldenderfer, M. S. and R. K. Blashfield. 1984. *Cluster Analysis.* Beverly Hills: Sage.

Badekas, J., ed. 1975. *Photogrammetric Surveys of Monuments and Sites.* New York: Elsevier.

Baillie, M. G. L. 1982. *Tree-Ring Dating.* Chicago: University of Chicago Press.

Beck, C. and Jones, G. T. 1989. "Bias and Archaeological Classification." *American Antiquity* 54(2):244–62.

Binford, L. R. 1964. "A Consideration of Archaeological Research Design." *American Antiquity* 29:425–41.

———. 1985. "Human Ancestors: Changing Views of Their Behavior." *Journal of Anthropological Archaeology* 4:292–327.

Bodner, C. and R. M. Rowlett. 1980. "Separation of Bone, Charcoal, and Seeds by Chemical Flotation." *American Antiquity* 45:110–16.

Braun, D. P. 1983. "Pots as Tools," In *Archaeological Hammers and Theories*, eds., A. Keene and J. Moore. New York: Academic Press.

Brewer, D. J. 1987. Seasonality in the Prehistoric Faiyum Based on the Incremental Growth Structures of the Nile Catfish (Pisces: *Clarias*). *Journal of Archaeological Science* 14:459–72.

Browman, D. L. 1981. "Isotopic Discrimination and Correction Factors in Radiocarbon Dating." In *Advances in Archaeological Method and Theory, Vol. 4*, ed. M. B. Schiffer. New York: Academic Press.

Bunn, H. T. 1981. "Archaeological Evidence for Meat-eating by Plio-Pleistocene Hominids from Koobi Fora and Olduvai Gorge." *Nature* 291:574–77.

Bunn, H. T. and E. M. Kroll. 1986. "Systematic Butchery by Plio-Pleistocene Hominids at Olduvai Gorge, Tanzania." *Current Anthropology* 27(5): 431–52.

Carr, C., ed. 1989. *For Concordance in Archaeological Analysis*. Prospect Heights, Ill.: Waveland Press.

Crabtree, D. 1972. *An Introduction to Flintworking*. Occasional Papers of the Idaho State University Museum, no. 28. Pocatello.

David, N., J. Sterner, and K. Gavua. 1988. "Why Pots Are Decorated." *Current Anthropology* 29(3):365–89.

Deetz, J. 1967. *Invitation to Archaeology*. Garden City, N.Y.: Natural History Press.

Dimbleby, G. W. 1985. *The Palynology of Archaeological Sites*. London: Academic Press.

Doran, J. and F. Hodson. 1975. *Mathematics and Computers in Archaeology*. Cambridge: Harvard University Press.

Drennan, R. D. 1976. "A Refinement of Chronological Seriation Using Nonmetric Multidimensional Scaling." *American Antiquity* 41:290–320.

Dunnell, R. C. 1970. "Seriation Method and Its Evaluation." *American Antiquity* 35:305–19.

———. 1971. *Systematics in Prehistory*. New York: Free Press.

———. 1986. "Methodological Issues in Americanist Artifact Classification." *Advances in Archaeological Method and Theory* 9:149–207.

Gilbert, R. I., Jr., and J. H. Mielke, eds. 1985. *The Analysis of Prehistoric Diets*. New York: Academic Press.

Gould, R. A. and P. J. Watson. 1982. "A Dialogue on the Meaning and Use of Analogy in Ethnoarchaeological Reasoning." *Journal of Anthropological Archaeology* 1:355–81.

Gowlett, J. A. J. 1987. "The Archaeology of Radiocarbon Accelerator Dating." *Journal of World Prehistory* 1(2):127–70.

Haas, H., J. Devine, R. J. Wenke, M. E. Lehner, W. Wolfli, and G. Bonani. 1987. "Radiocarbon Chronology and the Historical Calendar in Egypt." In *Chronologies in the Near East*, eds. O. Avrenche, J. Evin, and P. Hours. *British Archaeological Reports* 379:585–606.

Hayden, B., ed. 1987. *Lithic Studies Among the Contemporary Highland Maya*. Tucson: University of Arizona Press.

Hole, F. and R. F. Heizer. 1973. *An Introduction to Prehistoric Archaeology*. 3rd ed. New York: Holt.

Jelinek, A. 1982. "The Tabun Cave and Paleolithic Man in the Levant." *Science* 216:1369–75.

Jensen, H. J. 1988. "Functional Analysis of Prehistoric Flint Tools by High-Power

Microscopy: A Review of West European Research." *Journal of World Prehistory* 2(1):53–88.

Joukowsky, M. 1980. *A Complete Manual of Field Archaeology*. Englewood Cliffs, N.J.: Prentice-Hall.

Keeley, L. H. 1980. *Experimental Determination of Stone Tool Use: A Microwear Analysis*. Chicago and London: University of Chicago Press.

Kimes, T., C. Haselgrove, and I. Hodder. 1982. "A Method for the Identification of the Location of Regional Cultural Boundaries." *Journal of Anthropological Archaeology* 1:113–31.

LeBlanc, S. A. 1975. "Microseriation. A Method for Fine Chronological Differentiation." *American Antiquity* 40:22–38.

Lehner, M. 1983. "Some Observations on the Layout of the Khufu and Khafre Pyramids." *Journal of the American Research Center in Egypt* 20:7–29.

———. 1985. *The Pyramid Tomb of Hetep-heres and the Satellite Pyramid of Khufu*. Mainz am Rhein: Philipp von Zabern.

Leonard, R. D. and G. T. Jones., eds. 1988. *Quantifying Diversity in Archaeology*. Cambridge: Cambridge University Press.

Leute, U. 1988. *Archaeometry*. New York: VCH Publishers.

Lewenstein, S. 1987. *Stone Tool Use at Cerros. The Ethnoarchaeological and Use-wear Evidence*. Austin: University of Texas Press.

Libby, W. F. 1955. *Radiocarbon Dating*. Chicago: University of Chicago Press.

Marquardt, W. H. 1979. "Advances in Archaeological Seriation." In *Advances in Archaeological Method and Theory*, vol. 1, ed. M. B. Schiffer. New York: Academic Press.

Michels, J. W. 1973. *Dating Methods in Archaeology*. New York: Seminar Press.

Mueller, J., ed. 1975. *Sampling in Archaeology*. Tucson: University of Arizona Press.

Parsons, J. R. 1971. *Prehistoric Settlement Patterns in the Texcoco Region, Mexico*. Ann Arbor: Memoir of the Museum of Anthropology, University of Michigan, N. 3.

Petrie, W. M. F. 1900. "Sequences in Prehistoric Remains." *Journal of the Anthropological Institute* 29:295–301.

Piperno, D. R. 1987. *Phytolith Analysis*. Orlando, Fl.: Academic Press.

Rathje, W. 1981. "A Manifesto for Modern Material Culture Studies." In *Modern Material Culture: The Archaeology of Us*, ed., R. Gould. New York: Academic Press.

Read, D. 1989. "The Substance of Archaeological Analysis and the Mold of Statistical Method: Enlightenment Out of Discordance?" In C. Carr, ed., *For Concordance in Archaeological Analysis*, pp. 45–86, Prospect Heights, Ill.: Waveland.

Reidhead, V. A. 1979. "Linear Programming Models in Archaeology." *Annual Review of Anthropology* 8:543–78.

Renfrew, J. M. 1973. *Palaeoethnobotany*. New York: Columbia University Press.

Rice, P. M. 1987. *Pottery Analysis*. Chicago: University of Chicago Press.

Sabloff, J. A., ed. 1981. *Simulations in Archaeology*. Albuquerque: University of New Mexico Press.

Sackett, J. R. 1982. "Approaches to Style in Lithic Archaeology." *Journal of Anthropological Archaeology* 1(1):59–112.

Schiffer, M. B. 1987. *Formation Processes of the Archaeological Record*. Albuquerque: University of New Mexico.

Sease, C. 1988. *A Conservation Manual for the Field Archaeologist*. Los Angeles: UCLA Institute of Archaeology, Archaeological Research Tools 4.

Semenov, S. 1964. *Prehistoric Technology*. Trans. M. W. Thompson. London: Cory, Adams & Mackay.

Shennan, S. 1988. *Quantifying Archaeology*. Orlando, Fl.: Academic Press.

Shipman, P. 1983. "Early Hominid Lifestyle: Hunting and Gathering or Foraging and Scavenging?" in *Animals and Archaeology*, Vol. 1, *Hunters and Their Prey*, eds. J. Glutton-Brock and C. Grigson. Oxford: British Archaeological Reports International Series, 163.

Stein, J. K. 1983. "Earthworm Activity: A Source of Potential Disturbance of Archae-ological Sediments." *American Antiquity* 48:227–89.

———. 1986. "Coring Archaeological Sites." *American Antiquity* 51:505–27.

Taylor, R. E. 1987. *Radiocarbon Dating. An Archaeological Perspective.* Orlando, Fl.: Aca-demic Press.

Thomas, D. H. 1974. *Predicting the Past: An Introduction to Anthropological Archaeology.* New York: Holt.

Tite, M. S. 1972. *Methods of Physical Examination in Archaeology.* London and New York: Seminar Press.

Vaughn, P. 1985. *Use-wear Analysis of Flaked Stone Tools.* Tucson: University of Arizona Press.

Weisner, P. 1983. "Style and Social Information in Kalahari San Projectile Points." *American Antiquity* 48(2):253–77.

Wendorf, F. and R. Schild, eds. 1980. *Prehistory of the Eastern Sahara.* New York: Aca-demic Press.

Wenke, R. J. 1987. "Western Iran in the Partho-Sasanian Period: The Imperial Trans-formation." In *The Archaeology of Western Iran,* ed. F. Hole. Washington, D.C.: Smithsonian Institution Press.

Wenke, R. J., J. E. Long, and P. E. Buck. 1988. "Epipaleolithic and Neolithic Subsis-tence and Settlement in the Fayyum Oasis of Egypt." *Journal of Field Archaeology* 15(1):29–51.

Wenke, R. J. and M. E. Lane, eds. In preparation. *Land of the Lake. 8000 Years of Human Settlement in Egypt's Fayyum Oasis.* Winona Lake, Ill.: Eisenbrauns.

Whallon, R. and J. A. Brown, eds. 1982. *Essays on Archaeological Typology.* Evanston: Center for American Archaeology Press.

Wheeler, M. 1954. *Archaeology from the Earth.* Baltimore: Penguin.

Wylie, A. 1982. "An Analogy By Any Other Name Is Just as Analogical." *Journal of Anthropological Archaeology* 1:382–401.

Yellen, J. 1977. *Archaeological Approaches to the Present.* New York: Academic Press.

———. 1977. "Cultural Patterning in Faunal Remains: Evidence from the !Kung Bush-men." In D. D. Ingersoll, J. Yellen, and W. MacDonald, eds., *Experimental Ar-chaeology.* New York: Columbia University Press.

Zegura, S. L. 1987. "Blood Test." *Natural History* 96(7):8–11.

The Origins of Culture

Man is an exception, whatever else he is. If it
is not true that a divine being fell, then we can
only say that one of the animals went entirely
off its head.

G. K. Chesterton (1874–1936)

Devoted filmgoers will recall in *2001: A Space Odyssey* the scene in which
our ape-like African ancestors of several million years ago awoke one
morning to find themselves inexplicably at the foot of a huge, black, perfectly smooth and rectangular object—which later reappeared in a twenty-
first century excavation on the moon. Stanley Kubrick's film of Arthur
Clarke's story used this monolith as a generic symbol of human evolution
and enlightenment: a descendant of the primates who first confronted the
monolith "discovers" tools by using an animal femur to smash in the skull
of another ape-man, and in bloody triumph, he throws the bone skyward,
where, as it spins, it is transformed into a rotating twenty-first century
space-station.

In *2001* the viewer is left to muse on the nature of the driving force of
human evolution, whether it is natural processes, God, extraterrestrial
beings, or something else. The anthropological approach to this issue has,
perhaps, much less poetry and mythic power to it than *2001*, but the
fundamental questions addressed are the same: How were we transformed
from apes to space-travelers, and what does it all mean?

These are, of course, ancient questions. Long ago a Greek mused on
the possibility that we are the progeny of spores blown here millions of
years ago through the illimitable reaches of space, and that the answers
to our questions of origins were, thus, in the stars. Even today some scientists believe that the evolution of life is so improbable that it could not

75

have happened in the few billion years the earth has existed, and that therefore life on earth derives from life elsewhere in the universe, perhaps from the accidental intersection of the earth's orbit with a cloud of complex chemicals brewed in some mighty stellar explosion. Alternatively, some people in every age have found great comfort in their sure and certain knowledge that we are the result of Divine Creation. One physicist has even raised the possibility that we are "the galaxy's way of evolving a brain."[1]

These many imaginative speculations may or may not be true, but the assumption of anthropology is that any search for our origins and the causes of our evolutionary history should begin with a consideration of real-world factors of climate, genetics, and culture, and with the careful sifting and analysis of the shattered fragments of our ancestors and their crude tools. Indeed, in their search for human origins, anthropologists have concentrated on such embarrassingly ordinary subjects as chimpanzee sexual behavior and the cuisine of African hunter-gatherers.

This chapter is a summary of what we have learned about the early stages of our history as a genus. It will become clear that anthropology has little to say about the *meaning* of our existence. The anthropological approach is unrelievedly scientific and objective, and it is probably worth noting that such an approach has never proved totally satisfactory to anyone. It seems part of human nature to speculate about our origins far beyond the limits of what we know about them scientifically.

The Nature of Culture

Intelligence . . . is the faculty of making artificial objects, especially tools, to make tools. (Henri Bergson, *L'Evolution Creatrice*)

What are we trying to explain when we speak of the origins of culture? Anthropologists restrict the term "cultural" to human beings (although chimpanzees are sometimes called protocultural), thus the origins of culture are essentially the origins of those human aspects that make us distinctive, unique forms of animal life. It is difficult to reduce humanity to this or that constellation of attributes and say that one has captured the essence of being human. For the Roman Catholic theologian, for example, a human fetus, whether 6 seconds or 6 months after conception, is human in the only way that really matters, being imbued with an immortal soul. But from an anthropological point of view, what is "culture"? What is it that distinguishes us from other life forms and thus must be the focus of our efforts to understand ourselves through the study of our past?

Most who address this question turn instinctively to attributes of the human mind. A traditional view is that humans are unique in their ability to manipulate symbols and that the evolution of this ability underlies all human achievements. Anyone who has seriously considered chimpanzees at play can feel the sense of kinship; the term protocultural seems hardly an exaggeration. But in analyzing the origins of culture, it is how we are *different* from other primates that is important, and in large part this difference is our total mentality, our creativity, intuition, logic, aesthetics—all the powers of the human mind. Leslie White argued, for example, that although chimpanzees are clever animals who can use tools and even can be taught to use plastic counters, computers, and sign languages to express simple emotions and desires, they are fundamentally different from us in two ways.[2] First, they can never use symbols at an abstract level that would give them any understanding whatever of concepts like "holy water" or the casting of a vote. Human virtuosity in manipulating symbols is most vividly illustrated, perhaps, by human aesthetic capacities. Chimpanzees, our most intelligent primate relatives, can smear canvasses with paint and in excellent imitation of "modern art," but they seem utterly untalented at more representational painting; and it seems highly unlikely that even chimpanzees can *respond* to aesthetics—can savor a Bach cantata or feel the vibrancy of a Van Gogh landscape.

In addition to their low-powered abilities to manipulate symbols, non-human animals also seem to pass little new learned knowledge from generation to generation. Japanese monkeys, for example, have been reported to have learned how to wash sand from handfuls of grain and then instruct younger members of each generation how to do this, but there is evidence that even such simple food-related behavior is partially genetically determined.[3] But even if it were not, such behavior is at such a low scale that it remains true that "the most characteristic part of being human is the ability to profit from the accumulated and transmitted experience of other human beings."[4]

But it is an unfortunate reality of anthropology that we will never know directly the symbolic capacities of our earliest ancestors: we must relate abstract concepts like "intelligence" to material things that we can find and measure. And for the first three million years of our genus, this means looking mainly at changes in the size and shape of hominid skulls and bodies, and at stone tools and other objects on which our ancestors imposed their intellects. These indirect reflections of evolving symbolic abilities constitute "culture" for the anthropologist.

The paradox that we have only one tool with which to try to understand the origins of the human mind, namely, the human mind, is a bit of cosmic whimsy appreciated by philosophers of all ages. There is an embedded circularity to this form of the analysis that has led some to kick stones, wondering if they are really there.

Readers interested in this topic and with much time to spare are referred to Volumes 2 through 9 of Father Frederick Copleston's rewarding *History of Philosophy*.

Early Studies of Human Origins

Until the eighteenth century, few people questioned the notion that humanity was directly created by a divine being, but by the end of the nineteenth century, both Darwin's ideas about the biological evolution of the human species and the discovery of stone tools in association with the bones of extinct animals in extremely ancient geological strata had convinced many scholars of the great antiquity of the human race. Nevertheless, so long as no bones were found that could be attributed to a human ancestor intermediate between ourselves and other primates in physical form, it was still possible to cling to the idea that humans were an exception: that we were an extremely old species, older than had previously been suspected, but that we had not evolved as other species had. Other scholars thought that even if we had an ancient evolutionary history, no human bones of this age could have survived.

Those who had studied Darwin and Lyell closely, however, and who were aware of the archaeological evidence of early humans in Europe knew it was just a matter of time before the first fossil "missing link" was found. The great French scholar (and full-time customs inspector) Boucher de Perthes, grown old and tired of waiting, offered a 200-franc reward to the discoverer of the first "antediluvian" man in France. His enterprising workers were soon "finding" human remains in many places—all put there, of course, by themselves in hopes of collecting the money.

It is ironic that the first premodern hominids had already been discovered some years before. A fragmentary Neandertal child's skull was found near Liege, Belgium, in 1829–30, and in 1848 work at a quarry on Gibraltar revealed a skull whose receding chin, heavy brow ridges, and thick bones we now recognize as a Neandertal. Neither the Belgium or Gibraltar finds excited much interest. In 1856, however, a skullcap (Figure 3.1) and some limb bones were found in a cave in the Neander Valley near Dusseldorf, Germany, and although these remains were dismissed by the great German anatomist Rudolf Virchow as those of a deformed human, Johann Karl Fuhlrott, the discoverer of these fossils, argued from the beginning that the remains were of an early form of human. But Virchow's opinion, and those of others who variously labeled it an ancient Celt, a victim of rickets, an idiot, or a Cossack, conspired to deny these bones their proper significance for many years. One French savant even

3.1 The remains of Neandertals were recovered from the Neander Valley, near Dusseldorf, Germany, in 1857. This evidence that ancient people were different from modern humankind was part of the assault by evolutionary ideas and evidence on traditional ideas about our place in the universe.

suggested that the huge brow ridges of the Neandertal came about because his deformed arm caused him such pain that he continually furrowed his brow, and the expression became ossified.

In 1886 two partial skeletons similar to the Neandertal specimen were recovered in a cave in Spy, Belgium, in direct association with stone tools and the bones of rhinoceroses, mammoths, and other animal species known to have been long extinct. Although Virchow also refused to accept these as ancient men, the tide of opinion had turned and scientists everywhere were soon looking eagerly for additional specimens of early hominids.

One of these was a young Dutch physician, Eugene Dubois, who spent years wandering the wilds of Sumatra without finding much of interest, and only when he arranged a transfer to Java did he make his great discovery. In 1890 he unearthed a fragment of a lower jaw, and over the next few years he recovered a skullcap and a femur (thighbone).

For the next thirty years, controversy raged over Dubois' find, with some authorities again claiming it to be a deformed freak or a giant chimpanzee, others recognizing it as an early form of our genus and a direct ancestor of modern humans. Dubois' fossil had an approximate brain volume of 1040 cubic centimeters, about a third less than modern people but far more than any living primate. Scholars of the time recognized that if this animal were in fact a human ancestor, it was an ancestor considerably different from ourselves and from "Neandertal man" (whose cranial capacity was slightly larger than ours).

In 1906 the German anatomist Gustav Schwalbe proposed three successive stages of hominid evolution: pithecanthropine (represented by Dubois' fossils), Neandertal, and modern. It was apparent, however, that if this were the correct succession of hominid forms, there would have been many intermediate types not yet found.

The recognition of a presumed fourth and earlier stage of hominid evolution did not come until the 1920s, when Raymond Dart discovered a nearly complete skull of a strange-looking child encased in stone quarried from a mine some 300 kilometers from Johannesburg, South Africa, at the Taung site. The skull indicated a brain volume much less than that of an individual of similar age of either pithecanthropine, Neandertal, or modern type, and the teeth and other physical characteristics convinced Dart he had found humankind's earliest, most primitive ancestor, which he labeled *Australopithecus africanus* ("Southern ape of Africa"). Dart's conclusions were discounted by some of the most influential scientists in Europe, but while Dart's claims were being challenged, another important find was made, this time in northeastern China, near Beijing, at a mining installation called Zhoukoudian (formerly written Choukoutien).

Fossilized bones from the area had been used for centuries as aphrodisiacs, and it is an interesting footnote to the study of fossil hominids in Asia that we must expect that many specimens of scientific interest may long since have been consumed in the form of aphrodisiac and medical potions.[5] A somewhat more scientific interest in these fossils was stimulated in 1921 when a single human-looking tooth was given to the English anatomist Henry Black. He recognized the tooth as belonging to an ancient form of hominid, and as a consequence excavations at Zhoukoudian were begun. Altogether, the remains of about forty individuals were found, including many skull fragments, and it was obvious that these individuals were similar in brain size, facial structure, and other characteristics to the hominid found by Dubois on Java. This supported Schwalbe's proposed pithecanthropine stage, and fossils of this type were accorded the name *Sinanthropus* or *Homo erectus* ("erect man").

3.2 Excavations at Zhoukoudian, China. The rope enclosures mark areas where hominid skulls were found.

Thus, by the 1930s four categories of early hominids had been recognized by at least some of the scientific community: australopithecines, *Homo erectus,* Neandertals, and modern humans. Since that time the arguments have focused not on whether humankind evolved from some sort of non-human ancestor, but on what these various ancestral forms were, how they are related genetically during their evolutionary histories, and what evolutionary mechanisms produced ourselves, *Homo sapiens sapiens,* from prehuman primates.

The first step in considering these questions is to look at the physical world in which our ancestors lived. For we must assume that these environments were powerful determinants of our evolutionary history.

The Ecological Context of Cultural Origins

In the bleak mid-winter
Frosty wind made moan
Earth stood hard as iron,
Water like a stone.
Snow has fallen, snow on snow,
Snow on snow,
In the bleak mid-winter
Long ago.
(Christina Rossetti)

One of the central facts of human physical and cultural evolution is that many of the crucial developments of our species occurred in Africa and southern Eurasia between about 2 million and 10,000 years ago, an interval known as the *Pleistocene* ("most recent") geological period. During the Pleistocene world climates fluctuated greatly but were, on average, cooler than today. The correspondence between this period of climatic change and our own origins has led scholars for many years to suspect that somehow Pleistocene climates directly shaped human evolution, perhaps by creating difficult and demanding environments that "selected for" larger and more intelligent hominids. But the relationship between climatic changes and human changes seems to have been a very complex one.

The many subdivisions of the past are often renamed, redated, and redefined. Average worldwide temperatures have fluctuated more rapidly in the last 14 million years than ever before, and a world as warm as our own has been a rarity during the last 2.5 million years. There have been, however, occasional interglacials—periods usually of about 10,000 years when temperatures rose almost to present-day levels. Even during the main glacial periods, there were interstadials, short warming phases when the temperature rose but did not reach today's level. Archaeologists use the term Holocene to refer to the last 10,000 years, but only time will tell if our age is a short interglacial or a long-term warming trend.

During the glacial periods themselves, ice sheets spread from the poles and from higher elevations to cover much of the higher latitudes. So much of the seas were locked in ice during the coldest periods that the sea level dropped by as much as 40 meters, making dry land of coastal areas that today are many meters below the sea. Land bridges formed between North America and Asia, Europe and Britain, and Southeast Asia and what are now offshore islands. This facilitated the spread of peoples throughout the world. Unfortunately, Pleistocene ice sheets have ground to powder many critical archaeological sites, and the rising seas of the postglacial periods have covered thousands of others.

During glacial periods many areas that are now quite temperate were either under ice or in permafrost conditions. Other areas, such as parts of China, were much drier than they are today. Elsewhere, in North America, for example, melting water from glaciers created a lush lake-forest environment.

The great climatic shifts of the Pleistocene seem to have been caused by a combination of factors, including fluctuations in solar radiation, mountain-building activity, and changes in the earth's atmosphere.[6] But in the next several decades the world may see the invalidation of the cliché, "everyone talks about the weather but nobody does anything about it." Various scientists have suggested that the hot summer and drought of 1988 were initial stages of the "Greenhouse Effect"—a dramatic warming of average world climates as a result of industrial pollution. Some meteorologists predict that by the early twenty-first century, and thus within the lifespan of many readers of this book, such coastal cities as New York will be partially flooded by rising sea levels.

We know that the first million years or more of the history of our genus were spent in the tropics of Africa and Asia, and we were probably all Africans until just a million years ago—late in our evolutionary story. Even if evidence (reviewed below) for our ancestors' invasion of Asia before a million years ago is confirmed, we must still look to the grasslands and forests of Africa for proof about our origins.

In many ways the history of our genus seems to tied to *savanna* environments.[7] Savannas are relatively flat, arid expanses with scattered trees and occasional water holes, and their mixed grasses, shrubs, and other plants usually support grazing animals such as zebras, buffalos, and gazelles. As we will see, the earliest fossil humans and their stone tools have been found near lakes and streams that cut through these savannas. The importance of savannas in our past may be overstated because of the difficulties of finding preserved remains in wetter, more forested areas, and some apparently early sites have finally been found in the forests of Africa; however, most scholars believe that the competitive struggles of life on the savannas of Africa shaped us in our early evolution, and there is also now evidence that the earliest *Homo sapiens sapiens*—people like us—evolved in Africa about 100,000 years ago. In sum, in what follows we will have to consider carefully the nature of African ecology.

Analyzing Cultural Origins

We have many sources of evidence and ideas about our origins including the study of our primate relatives, the fossil bones of our ancestors, and the early stone tools and other physical traces our forebears left. But before we look at the collected evidence, we must consider the assumptions and principles that tie them all together: evolutionary theory. The basic premise underlying the study of human origins is that over five million years ago our ancestors were animals something like chimpanzees and the forces of evolution changed us into what we are today. But what were these forces?

Evolutionary Theory

Biological evolutionary theory is an attempt to explain biological changes in life forms over time and space. We will see that one of the most spectacular changes in our own past is the evolution of average human cranial capacities from the approximately 600 cubic centimeters of three million years ago to the modern average at a little less than 1300 cubic centimeters. Biological evolutionary theory explains this pattern of change in terms of *natural selection:* the genetic potential for increased brain size was inherent in our primate ancestors; at a certain point increased brain size became more important in determining which animals had more offspring; and over the course of millions of years human brain size increased.

But how genetically and biologically does this kind of selection and change actually work? For evolution to operate there must be *variability* differences among individuals, so that over time such characteristics as increased brain size are selected for. The two main sources of genetic variety are *genetic recombination* and *mutation.* Within the nucleus of every animal cell are chromosomes, rope-like structures composed of deoxyribonucleic acid, or DNA. This DNA in turn is made of *nucleotides,* which include varying arrangements of chemical bases and a few other chemicals (Figure 3.3). The arrangement of these bases determines the physical structure of all living things, from AIDS viruses to elephants. DNA has the ability to reproduce itself, by splitting the helix down the middle and forming new ones. DNA contains all the information required to direct the operation of a cell, in combination with another nucleic acid, RNA (ribonucleic acid). Animal cells are constantly splitting *(mitosis)* and producing genetically identical copies of themselves, and thus growth and maintainance are accomplished. But in *meiosis,* egg and sperm cells are produced that carry only half the genetic code of the parent—in the case of humans, our twenty-three pairs of chromosomes total forty-six total chromosomes in each body cell, but each egg and sperm cell carries only half the total, so each new person receives twenty-three chromosomes each from his or her father and mother. The number of genetically unique offspring possible is thus determined by the equation $2^{23} \times 2^{23} = 2^{46} =$ about 70,000,000,000,000.[8] In addition, chromosomes in egg and sperm cells sometimes change through *crossing-over,* the exchange of chromosomal sections between chromosomes. Through these random assortments of genes through meiosis, crossing-over, and mating, a great deal of variety

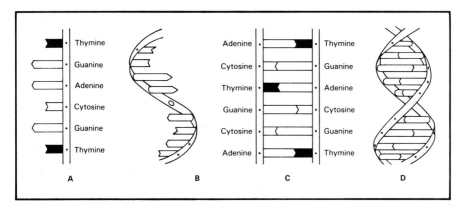

3.3A   The genetic code determining the biological characteristics of species lies in the sequence of bases that link two strands of nucleotides to form the double helix that is DNA. (A) Bases in part of one strand of nucleotides. (B) Part of one strand twisted. (C) Complementary bases joined. (D) Two strands joined and twisted.

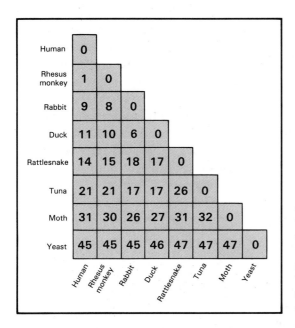

3.3B Amino-acid sequences. Organisms differ in the sequence of amino acids in protein cytochrome c, an enzyme used in energy production. The higher the number shown the greater the difference between any two.

is produced that natural selection can act on. Genetic variety is also produced by *mutations* in chromosomes—errors DNA makes in copying itself. Some mutations probably arise through the cosmic ray bombardment all living organisms are subjected to. Early work on DNA involved exposing hapless populations of fruitflies with X rays, producing monstrous combinations of eye colors, wing size, and other features. There is every reason to believe that industrial pollution, loss of ozone in the atmosphere, and other factors in the modern environment will lead to greater rates of mutation in all living things. But a great deal of genetic variety is produced simply by errors DNA makes in replicating itself.

Most genetic mutations seem to be "bad," in the sense that they reduce the chances of the individual carrying them to reproduce. Some mutations are "good," in that they increase reproductive potential, or "fitness." But what is good or bad in evolution is measured entirely by reproductive success. For example, sickle-cell anemia—a genetic condition in which red blood cells are misshapen—is a common disease in Africa and a few other places. People who are born with a certain combination of the genes for this condition often die in childhood; but those with another genetic combination involving this mutation are relatively unaffected. Research has shown that certain genetic arrangements of the gene associated with sickle-cell anemia offer protection against malaria—and thus the gene for sickle-cell anemia has persisted, because its presence in a population—although it results in the death of some people—helps others to live to reproductive age.

Genetic mutations can be perpetuated and spread through populations by various mechanisms. The variations in skin color in contemporary peoples, for example, seem to be the result of adaptations to different environments and of geographical barriers. The Sahara, the Himalayas, and other natural barriers have restricted the matings between people in different areas to the point that modern "races" evolved.

Changes in physical form can also result from *sexual selection:* once individuals chose mates nonrandomly with regard to some characteristics, evolutionary change would result. Gaudy colors and features, such as represented by the peacock, are probably the result of sexual selection. Some scientists believe that sexual selection was responsible for such features as the "exaggerated" size (by comparison to other primates) of human female breasts. Good point 66

Modern population genetics is a complex mathematical discipline, and it is largely beyond the scope of this book. For the study of cultural origins and the other physical changes we will consider here, including plant and animal domestication (Chapter 6), the important points to remember are that genetic variability is constantly arising through various mechanisms, and changing environments and conditions "select," or act on this variability through time to produce evolutionary change.

This kind of evolution can be demonstrated. A classic case is industrial melanism (coloration). A species of moth *(Biston betularia),* common in England, is marked in mottled gray and white in a pattern such that it is almost perfectly camouflaged when it rests on the lichen-covered bark of trees native to the English Midlands. But the Industrial Revolution produced so much soot that trees were blackened all over the range of these moths, reducing the effectiveness of the moths' camouflage against predation by birds. In 1840, however, a black *Biston betularia*—the result of a slight genetic mutation—was found near Manchester, and by 1895 it was discovered that 98% of the *Biston betularia* population in this region was black.[9] The protective advantages of a black wing color on soot-covered trees was subsequently demonstrated by releasing moths of various colors and recapturing them, to see what proportion had been eaten.

Similarly, most modern human physical characteristics are the result of complex genetic interactions that involve a subtle, long-term balancing of costs and benefits of genetic changes. For example, as professional basketball teams and the pygmies of central Africa demonstrate, humans can live perfectly normal lives as individuals anywhere when they are less than 4 feet and greater than 7 feet in height. So why, on the average, are people around the world about 5½ feet tall? And why do they vary sharply in average height by sex and geographical location? Why are we not all about 4 feet or 7 feet tall?

Like almost all human characteristics, the "environment"—in this case, nutrition—is an factor in human height. But height is also partially genetically controlled. And we must assume that we average between 5 and 6 feet and not between 7 and 8 feet or 3 and 4 feet because our average size made "evolutionary" sense in complex evolutionary equations. With selective breeding and adequate nutrition, a group of people could probably be bred to an average height of 7 feet or more, but the costs of maintaining that body mass obviously did not make evolutionary "sense"— given the advantages in reproductive success of larger size, or the advantages of being able to see over large bushes or win an infrequent match with a cave bear. And there is overwhelming evidence that contemporary death rates from cancer and other diseases are directly related to the fact that most of us savage our evolutionary heritage with diets that simply do not meet the needs of the omnivorous hunter-gatherers that natural selection fashioned us as.[10] Everyone of us is the direct descendant of tens of thousands of generations of people who were naturally selected to thrive on diets high in fiber and fruit, low on fat and salt—diets that involved eating hundreds of different species of animals and plants. Our brief 7000 years as farmers has had little effect on our basic physiology and metabolism, so to be in accord with our nature we should eat as wide-ranging hunter-foragers eat. Human physiology has been selected over millions of years to operate on almost anything—provided one balances occasional excesses with long-term diversity. Even Irish Coffee, that perfect end to a field archaeologist's day, containing as it does such basic food groups as fat, sugar, alcohol, and caffeine, can be tolerated so long as one scavenges widely among the other food groups the rest of the day.

Thus, we must assume that much of what we are today as a species is a result of a competitive struggle of the last several million years—an interval in which the 7000 years in which we have not lived as hunter-gatherers is insignificant. As walking, talking genetic packages we reflect what "worked" for hunters and gatherers, plus a smaller amount of genetic accidents. So although every characteristic of every person cannot be accounted for in terms of the value it represented in competition to pass on one's genes, a lot of what we are must be explained in exactly these terms. Thus, for example, our brain size averages about 1300 cubic centimeters because that size is a good balance between the advantages of greater intelligence and general cerebration and the high costs of large brains in terms of blood supply and the reduced mobility of the broadened female pelvis needed for successful live birth of the cerebral monstrosities that we are.

There are very real problems with the "adaptationist" views of human evolution—the idea that most or all of what we are is the product of selected accumulated successful mutations. A lot of what we are in fact may be random genetic variation, not the result of selection at all. And it is almost impossible to find some trait or behavior that *cannot* be explained as the solution to some kind of imagined evolutionary problem or opportunity.[11] It is also often difficult to identify exactly what the *focus* of selection was for complex physiological or anatomical evolution. Hamilton, for example, suggests that many of the sex differences between men and women and between humans and other apes are the result of long-term selection for high levels of sex hormones—not to exaggerate our sexuality but to give us the stamina we needed to be foragers who walked long and often to find food.[12]

Yet, as Tooby and DeVore note, much of what we are as humans, from our large brains and bipedal locomotion to our tool-use and symbolic capacities, involves the kinds of genetically based features that are complex and biologically "costly" in patterns suggesting that they are adaptations—the direct result of complex patterns of natural selection.[13]

With these simple ideas about evolution, it is easy to see why the "argument from design"—the conclusion that the universe had to have been created by a Divine Being because everything worked together in such harmony—was almost universally rejected by scientists from the late nineteenth century on. Biological evolution can shape a world of great complexity and, in a sense, beauty.

It is a great temptation to see "natural selection" as an active, quasi-intelligent force, shaping plants and animals for certain purposes and to meet certain requirements. But this is a fundamentally flawed perspective. Natural selection does not "know" what will be required at some later date. Natural selection is simply differential rates of reproduction.

It is easy to see from this why the "argument from design" was so difficult to give up: looking back from the perspective of the present, evolution seems to be directed toward a goal, such as larger human brain size. But evolution is simply a matter of what worked from individual to individual: it is the individual, and his or her genetic material, that either reproduces successfully or does not.

If this is so, what is the key to long-term evolutionary success in a lineage? One important element seems to be the maintenance of genetic variability. A classic example of this is the problem caused by the introduction of rabbits into Australia. Without a natural predator, the rabbits multiplied to the point of becoming a major crop pest. Efforts to control them by introducing a rabbit disease killed millions of them, but because of genetic variability a few individuals were naturally immune. They survived and quickly reestablished a population that was very resistant to this disease.

As the AIDS epidemic demonstrates, Life with a capital L is a constant struggle between life forms for survival. The AIDS virus's "strategy" involves rapidly changing forms and a late onset—afflicting many people after they have already reproduced. But almost certainly some people will be naturally resistant to the virus, and even if the epidemic sweeps the world, these resistant individuals could reestablish human populations.

The great potential of genetic variability is that, even if ethical considerations were not involved, it would make "eugenics" programs—selective breeding by human beings chosen on the basis of some particular traits— a bad idea. The whole history of biological evolution shows that for organisms such as ourselves, the best long-term strategy is to keep a great mix of genetic variability. It may be that natural resistance to skin cancer, exposure to asbestos and organic pollutants, or some newly mutated disease will be the key to our long-term survival as a species. Evolution is in no sense "over" for us just because we currently are doing rather well.

AIDS viruses and moths behave in ways determined by inherited genetic structures, and one might suspect that even if we could explain human physical characteristics in terms of similar evolutionary mechanisms, our behavior—the most important thing about us, in some ways—would not be within the compass of evolutionary theory. But evolutionists have recently made great strides in explaining the evolution of behavior in animals, and not just behavior under obvious and direct genetic control, like seasonal migrations of birds.[14] Archaeology, as noted in Chapter 1, has always searched for some way to use evolutionary theory to explain our cultural evolution and the appearance of cities, states, empires, etc. We will review some of these attempts in later chapters.

The Evidence of Cultural Origins

We have four basic categories of evidence for analyzing human origins: (1) *paleontology*, the study of ancient forms of animal life, including the ancestors of humankind; (2) *primatology*, the study of our contemporary nonhuman relatives, the other primates, whose behavior patterns may give us clues to the behavior of our own ancestors, and whose genetic composition can be compared through molecular biology with our own, to address questions of genetic relatedness and the rate of our common descent; (3) *archaeology*, the analysis of the archaeological record—the stones and bones and other tools used by our ancestors; (4) *ethnology*, particularly the study of contemporary or recent hunting and gathering peoples, whom we assume to be living in environments and patterns similar to those of our Paleolithic ancestors (although any contemporary human societies must be regarded as fundamentally different from those of our earliest ancestors).

The goal of anthropologists is to combine these four kinds of evidence to produce "models of cultural origins"—sets of related hypotheses about the factors that combined to change our ancestors from unremarkable primates to human beings.

Paleontological Evidence of Human Origins

Animal and plant life on this planet go back billions of years before the first humans appeared. Vertebrates—animals with internal skeletons—appeared only about 600 million years ago, marking a major evolutionary advance. Dinosaurs appeared perhaps as early as 200 million years ago and were widespread until the age of mammals, which began perhaps 100 million years ago. Since that time mammals, including ourselves, have radiated into most parts of the world.

One of the most difficult things for the people of the nineteenth century to accept was the idea that as a species we are the progeny of nonhuman and extinct primates, and they recoiled from the conclusion that we are the descendants of a small, pink-nosed, libidinous, insect-eating animal, whose modern form, the shrew, is on a pound-for-pound basis among the most ferociously effective predators known.

Taken as an overall sequence, from dinosaurs to tree shrews to ourselves, is there any trend in the evolution of animal life on this planet that would help us understand the appearance of culture and our own physical type?

One possible answer is suggested by the comparison of the ratio of brain size to body size in successive animal forms during the many millions of years before the first culture-bearing animals appeared. The anatomist

Table **3.1** Geologic Time Scale and Recent Life History of the P

Eras, Period	Epoch	Began millions of years ago	Duration in millions of years	Some impo in life of the
Cenozoic Era				
	Recent	.01	.01	Modern genera of animals with humans dominant.
Quaternary	Pleistocene	3(2.5–3.0)	3	Early humans and many giant mammals now extinct.
	Pliocene	10	7	Anthropoid radiation and culmination of mammalian specialization.
	Miocene	25	15	
	Oligocene	40	15	Expansion and modernization of mammals.
Tertiary	Eocene	60	20	
	Paleocene	70(±2)	10	
Mesozoic Era				
Cretaceous		135	65	Dinosaurs dominant to end; both marsupial and placental mammals appear; first flowering plants appear and radiate rapidly.
Jurassic		180	45	Dominance of dinosaurs; first mammals and birds; insects abundant, including social forms.
Triassic		225	45	First dinosaurs and mammal-like reptiles with culmination of laborinthodont amphibians.

3.4 This species of Southeast Asia tree shrew resembles closely the small insectivorous ratlike animals believed to be ancestral to all primates, including ourselves.

91

Henry Jerison has devised an encephalization index by dividing the total brain volume of each animal by the two-thirds power of its body size (the exponent was used because of the geometric relationship between surface area and volume in three-dimensional objects). This simple index thus represents a scaled ratio of brain volume to overall size.

Jerison's results (Figure 3.5) give us an answer of sorts to our questions. The increase in human brain size—from 500 to 1450 cubic centimeters in only a few million years—has been extraordinarily rapid, but, overall, we seem to be a continuation of a process that began at least 600 million years ago, a process involving long-term natural selection in some animal forms for increased brain-to-body ratios, and, presumably, mental capacity.

Why should there be this long-term evolution of larger brain–body ratios? We assume that it must be because of the reproductive advantages conferred by this development. But with regard to human brain size, Jerison stresses that

> the important point to remember is that very advanced behavior can be governed by very small amounts of brain tissue. The behavioral adaptations of the lower vertebrates are as remarkable as those of mammals in many ways. Encephalization in mammals, in primates, and in the human species is not easy to explain as a correlate of the refinement of behavior.
>
> Information-processing of a kind that could be done only by very large amounts of neural tissue must have been evolving in the mammals.[15]

3.5 Brain size plotted against body size for some 200 species of living vertebrates that represent an evolutionary sequence through time. The points connected by lines represent the extreme variations of measurements reported for *Homo sapiens sapiens*.

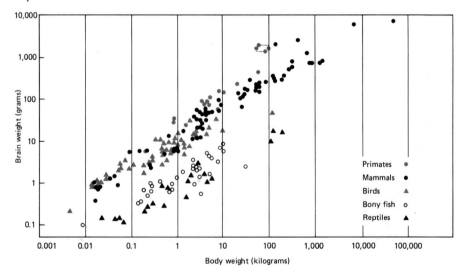

Most general explanations of why average human cranial capacities increased so drastically are linked to competition. A somewhat reductionist view of this is taken by Richard Alexander, who argues that "the human psyche evolved as a vehicle serving the genetic or reproductive interests of its possessors."[16] From this perspective, our huge brains evolved as a way to facilitate our ability to compete and pass on our genes:

> The only plausible way to account for the striking departure of humans from their predecessors and all other species with respect to mental and social attributes is to assume that humans uniquely became their own principal hostile forces of nature.[17]

According to Alexander, competition among humans is what produced increased brain size. Alexander argues that if humans become extinct it is possible that chimpanzees would become more human-like in their habits—that we have essentially driven them by competition into a specialization that they could evolve from if we were not here.

In many fictional treatments of the future, people are portrayed with enormous heads and correspondingly impoverished physiques. Is this likely, given Jerison's data? Human brain size in fact seems to have stabilized at about 1300 cubic centimeters over the last 100,000 years, but 100,000 years is insignificant in the span of animal life on our planet, and encephalization ratios may well continue to increase—or decrease—during the next millions of years. The size of the human female pelvis puts a limit on the size of the head at birth, but perhaps natural selection will produce people whose heads grow faster and larger during infancy. On the other hand, there is no evidence that in the contemporary world the size of an individual's head—and whatever differences in size might correspond to in contents—is in any direct way a determinant of how many children a person has. "Time will tell" is a cliché that still has some cogency in evolutionary studies. We too, no doubt, are in a sense "missing links" in the ancient evolutionary experiment that is the history of animal life on this planet.

One question we must address is, why are people so much smarter than they apparently need to be? As discussed below, we have conclusive evidence that our ancestors who had brains two-thirds our size successfully made stone tools and other implements, competed with a formidable array of other animals, and managed to colonize an area from Java to Spain. In the world of a half million years ago, or even 100,000 years ago, one needed a brain only a few hundred ccs bigger than a chimp to do very well, indeed. Why then did we evolve the capacities for quantum mechanics, Arabic verb forms, Puccini arias, and the poetry of Ezra Pound?

Once again it must be stressed that when we ask such questions we must look almost exclusively at our lives as hunter-gatherers, before 10,000 years ago. The last 10,000 years is far too short to have produced significant differences in our genetically based mentalities, compared to our Pleistocene forebears.

To return to the paleontological evidence, bits and pieces of various animals have been placed into our family tree as far back as twenty million years ago (Figure 3.6), and one of them may be directly related to us. This was *Ramapithecus* (named after "Rama," a figure in Hindu mythology, and "pithecus," the Greek word for ape). Fragments of ramapithicines have been found over a large area, from China to East Africa and into southern Europe, in geological strata dating to about twelve to twenty million years ago. We don't have enough fragments of ramapithicine bodies to determine much about their proportions or gait: the environments in which they are found suggest that some ramapithicines were mainly tree-climbers and arboreal feeders while others may have been more adapted to open country. The relatively small canines of the ramapithicines have always intrigued scientists, because ground-dwelling primates usually use canines for defense, aggressive displays, and ripping of tough foods. Ramapithicines may have used tools and other kinds of behavior in such a way that their need for canines was reduced.[18]

The status of *Ramapithecus* vis-à-vis ourselves remains the subject of a lively debate.[19] Molecular studies showing the divergence of humans and other apes only five or six million years ago has forced many to conclude that the ramapithicines—who are not known to be in Africa after twelve million years ago—were not hominids.[20]

Whichever of the primates of about twelve million years ago that we evolved from, we can consider how this evolution might have come about. Wolpoff suggests that these primates probably spread through the savannas and forests and

> presumably reduced competition through dietary (and eventually dental/gnathic) specialization and locomotor changes (true brachiation, knuckle walking) allowed an effective woodland/forest adaptation. . . . Precluded from these ecozones by competition, the hominid adaptation was to more open regions. Building on their ramapithicine inheritance, a combination of powerful masticatory apparatus, the probably rapid development of efficient bipedialism, the use of rudimentary tools and weapons (digging sticks, clubs) and a series of social changes possibly related to the recognition of extended kinship relations . . . allowed a wide range of difficult-to-gather and difficult-to-masticate foods to help form the basis of an effective adaptation to a unique open-country niche.[21]

To determine when the human line split from that of other primates, we can combine the fossil evidence with evidence from molecular biology. The most recent analyses of the sequences of DNA now strongly suggest that monkeys, gibbons, and orangutans split off from our line tens of

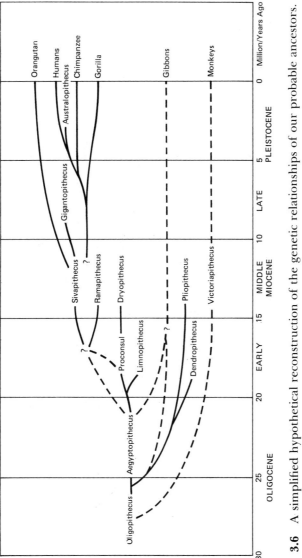

3.6 A simplified hypothetical reconstruction of the genetic relationships of our probable ancestors.

millions of years ago, followed by the gorilla about 11.0 to 7.7 million years ago, then the chimpanzees at about 7.7 to 5.5 million years ago, and finally, Australopithecines about 3.4 to 2.4 million years ago.[22] These age estimates and sequences are based on estimates of rates of change in nuclear genetic materials, and are known to vary widely in the animal kingdom (rates are much faster in rodents than in humans, for example). But several analytical techniques have produced similar estimates for the human/chimpanzee split. This poses a problem in that there is some evidence that not long after five million years ago our ancestors were bipedal, so the molecular clock may need to be re-calibrated in some way.[23]

In any case, a crucial period in the evolution of culture was between five million years ago, when the first australopithecines appeared, and a half-million years ago, by which time all the world's hominids belonged to a single genus, *Homo*, our direct ancestors.

Fragments of several hundred hominids have been found and dated to before a million years ago, and while much progress has been made in interpreting them, severe problems remain. It has been well and truly said that human paleontology shares with theology and extraterrestrial biology the peculiar trait that there are many more practitioners than objects of study.[24] This situation has given rise to many conflicting hypotheses.

Much of the story of cultural origins may have been "over" by 3.7 million years ago. Mary Leakey found human footprints of this age at Laetoli, Kenya, indicating that our ancestors already walked upright. And there is a sort of destiny in bipedalism: the hands are free to manipulate tools; the field of vision gives a slice of time and distance that rewards planning; and the arrangement of the limbs and cranium seems to require only additional brain tissue to "convert" a primate into something human.

But upright posture in the absence of tools—and we have no stone tools until about a million years after the Laetoli footprints—means something else. It means that an animal roughly of our body shape (but about two-thirds our size) and with our approximate powers of vision, smell, locomotion, etc., could compete in a world teaming with awesome predators. Without speed, claws, significant teeth, night vision, or protective coloration, these primates of 3.7 million years ago flourished, and there is some reason to suspect that they did so because they already possessed the critical elements of cultural behavior: intelligence, protective human family structures, and tool-use (perhaps like that of chimpanzees, who use sticks and unmodified stones).

If our ancestors of this era did indeed have these inchoate human capacities, then our answer to the question of why culture appeared may be

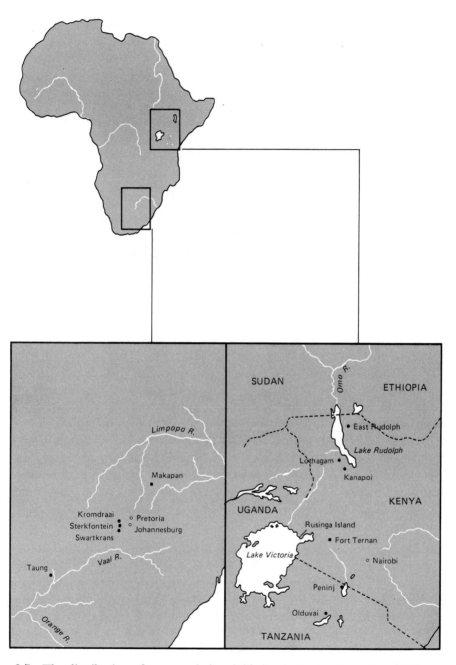

3.7 The distribution of some early hominid sites in East and southern Africa.

3.8 A reconstruction of an *Australopithecus afarensis*, the species to which "Lucy" belonged.

lost in undetectable climate changes that selected for bipedalism and life in savanna environments.

If we look for the ancestors of the hominids who left their tracks at Laetoli, we have only a few fragmentary candidates. In the Hadar region of Ethiopia we find the most substantial early remains, including "Lucy," a hominid named after a Beatles song by her discoverers, a group headed by Donald Johanson.[25] About three million years ago, Lucy strode upright through the river valleys of Ethiopia, as part of a group of primates who were around 1.2 to 1.5 meters tall, but who varied greatly in size and morphological characteristics. If Lucy and her kinfolk made stone tools, these implements have yet to be found.

Without tools, we cannot really know much about Lucy and her group—the "First Family" to use Johanson's label. By counting up the number of different bones, we can tell that there were at least thirteen people in Lucy's group, four of them children. A few bits of geological evidence and the absence of other animal bones near the hominids raise the possibility that the First Family was trapped in a ravine by a flash flood, but we'll never really know.

Despite their human-like posture, Lucy's and her friends' brain-to-body ratios were such that it is highly probable they could have led full mental and spiritual lives in any municipal zoo: there is no evidence they used tools or were cultural in significant ways. But they were *bipedal,* and this means that they were evolving in our direction, and the question is, what factors were responsible for the origins of bipedalism, and, once bipedalism was established, what factors produced *Homo,* our ancestor, from these bipedal apes?

Comparative Primatology and the Origins of Culture

> [M]any researchers view human and modern hunter-gatherers as a long corridor, where chimpanzees enter at one end and modern hunter-gatherers exit at the other.[26]

For us to use primate studies to investigate our origins, we must define clearly the ways in which we are different from them. Tooby and DeVore list six major areas in which we differ fundamentally from other apes:

> 1) bipedality; 2) situation-appropriate, intensive male parental investment and an increase in female parental investment; 3) an unparalleled degree of hunting and meat consumption; 4) a change in life history correlates: an extension of lifespan, an increase in the period of investment in offspring, a marked increase in the altriciality [i.e., the length of time the young remain in the direct care and physically with the parents] of human young; 5) an expansion of ecozones occupied, from tropical woodland and forest into savannah areas, but eventually including every other terrestial ecozone; 6) concealed ovulation with continuous sexual receptivity.[27]

Given that chimpanzees and ourselves share a common ancestor, perhaps as recently as five million years ago, how did these differences arise?

Some scholars think studying contemporary primates will tell us little about our origins; others see in them prototypical behavior for everything that is human. Teleki, for example, argues that nonhuman primates do virtually everything that people do, such as hunting game, sharing food, cooperating, carrying objects, having lifelong kinship ties, practicing labor division by sex, having incest prohibitions, making tools, exhibiting linguistic capacities, possessing long-term memory, and so on.[28]

A major problem in applying primatological studies to the question of human origins, aside from the fact that poachers are killing off the last remaining wild groups of these animals, is that observing them is not a purely scientific procedure. Observers see aggression, dominance, etc. partly as a projection of their own personalities, and thus models of cultural origins based on primate studies are inherently suspect.

To begin with the basic physical changes, to make a human out of a chimpanzee-like ancestor, the legs had to be lengthened greatly in relation to body size, the hip and knee joints had to be repositioned to allow bipedalism, and the spine had to develop an S-shaped curvature to cushion the brain against the shocks of bipedal locomotion. The face, jaw, and teeth had to be changed from the straight-row, prominent canines-and-incisors configuration of early primates to the curved human dental arch with reduced canines and incisors, and adapted to grinding rather than puncturing. In the long course of our evolution from apes, we lost most of our body hair, our sexual organs and behavior changed greatly, and our brains were greatly expanded and "rewired."

Consider first the simple matter of body size. A survey of numerous primate species indicates that a pronounced size difference between the sexes is relatively uncommon,[29] and thus "marked morphological and behavioral dimorphism is not a primitive characteristic of primates but has evolved in certain genera in relation to particular patterns of living."[30]

One clue as to what patterns of living these might have been can be seen in the size differences of the sexes in living nonhuman primates. Male and female gibbons, who spend most of their lives in trees, are almost identical in size; among chimpanzees, who spend some of their time on the ground, males are usually larger than females; and among gorillas, who are almost completely terrestrial, males average almost twice the size of females. Why should the physical size of the sexes show this association with terrain?

A probable factor in the evolution of both sexual dimorphisms and the overall increase in size of *Homo sapiens sapiens* of both sexes was the need for defense against predators. Our ancestors apparently evolved in open savanna-type environments where competition and predation by large cats and other carnivores would have been significant, and thus it is not at all surprising that part of our evolution in these environments would have been in the direction of greater physical size. A human with a rifle, or even a crude spear, would have some reasonable defenses, but until tool-use reached this point there would no doubt have been strong direct selection for increasing physical size. Our earliest tool-using ancestors were only about 3 or 4 feet tall, weighing perhaps 60 or 70 pounds—not really a formidable competitor, given the other animals in its range. Today, on the savannas of Africa, baboon populations are able to survive, despite occasional predation by leopards and lions, partly because they have retained impressive canine and incisor teeth (Figure 3.9). By ten million years ago, however, our ancestors had probably lost these dental defenses. Another factor may have been females' preference for mating with larger individuals—who may have had greater ability to provide food and protection.

But reasoning from the characteristics of a primate's physical environment directly to his or her physical characteristics and, especially, social behavior, is plagued with ambiguities and difficulties.[31] In some animal species, sexual dimorphisms seem to be linked to differences in feeding strategies of males and females. While this may have been of some impor-

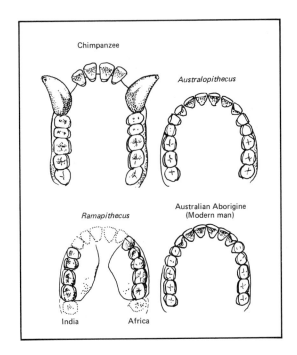

Chimpanzee

Australopithecus

Ramapithecus

Australian Aborigine
(Modern man)

India   Africa

3.9 Teeth preserve best of all body parts and are useful indicators of the diet and adaptation of primates.

tance to our early primate ancestors, the kinds of food eaten by males and females probably overlapped considerably soon after tool-use began and embryonic human social organizations formed.[32] Nonetheless, early hominid males and females would probably have been under increasingly different selective pressures as soon as males began specializing in hunting and females in gathering—specializations that may have appeared only comparatively late and sporadically.

Crook points out one sexual dimorphism in humans that may be tied to differential mortality rates: for every 100 female children conceived in modern human populations, 120 males are conceived, although of these only 105 survive to birth.[33] (Also, mortality is 25% higher for males in the first year of life.) And males have a stronger heart than females and more red blood cells per unit volume of blood. Again, one interpretation of this would be in terms of defense, female mating choices, and competition among males for mates.

Another aspect of sexual dimorphism among humans is particularly intriguing when seen in comparison with nonhuman primates: namely, the secondary sexual characteristics that differentiate human males and females. Among human females, breasts are significantly larger than among males—in fact, the size, shape, and other differences distinguishing male and female breasts are much more pronounced than in any other mammal species, and only in humans do breasts first enlarge at puberty rather than at the first pregnancy.[34] The erotic role of female breasts, with the role of breast-feeding in controlling fertility, suggests complex evolutionary relationships.[35] Rose Frisch has even suggested that the disproportionate size of the human female breast may have evolved because of its role "as a cooling fin to get rid of body heat."[36] Breast size cannot be explained simply in terms of function, since size seems to have almost no relationship to effective functioning in human beings. Similarly, the human penis is far larger, relatively and absolutely, than that of any other primate, including the gorilla. In addition, human females have softer skin and higher voices than males and have lost more of their body hair—although compared to the other species of primates, males also have remarkably little body hair.

Discussions of biological differences between hominid and human males and females are in some ways controversial. Male anthropologists have tended to stress "man the hunter" as a stereotype of our ancient ancestors, while women anthropologists have tended to emphasize "woman the gatherer." Some scholars have even proposed genetically and metabollically linked differences between men and women. Seymour Itzkoff, for example, notes that men and women in the United States consistently show similar mean scores on intelligence tests but that women's scores are less variable.[37] The distribution of IQ scores, he suggests, is one in which men and women both have normal bell-shape curves with the same approximate average, but fewer women are at the extreme high and low ends of these curves. He links this to the extra X chromosome women have—a source of relatively greater female biological stability in other aspects of physiology.

A primary difference between humans and other primates, and one that many people link directly to the human origins, is that human females do not exhibit estrus; that is, women's receptivity to sexual intercourse does not seem to increase to any large extent with the rise in estrogen levels that precedes ovulation. How can we account for this? One suggestion is that

> in all probability the evolutionary development of hair reduction, increased skin sensitivity and tactile changes involving skin tension were [sic] all associated with increasing the tactile sensations of coital body contact, especially in the frontal presentation. Likewise the breasts of young women taken together with other features (limb contour, complexion, and the like) seem to represent the main visual sexual releasers for the male. While the latter features may have been due to straight-forward intersexual selection by ancient males the former features have probably been selected in both sexes for their effect in improving sexual rewards, in inducing sexual love and in maintaining pair bonds.
>
> The same is also likely to be true for the presence of orgasm in women and the absence of the more typical mammalian estrus. The functional

significance of all these correlated changes is most plausibly seen within the context of the adaptations of seed-eating and of later partially carnivorous protohominids to open country life with associated shifts in social organization.[38]

Some scholars have emphasized female choice and selection in primate sexuality. As Small notes, "most female primates exhibit a pattern of sexual behavior that is excessive in terms of achieving a pregnancy and . . . are sexually assertive and often 'promiscuous'."[39] "Excessive" sexuality is difficult to define, however, given the short time the primate egg is fertile, the percentage of sperm that are ineffective, and other factors. And as Small notes, there is no reason to expect that evolution will have produced a fine-tuned sexuality in humans, when a sloppier system works reasonably well.

Our loss of the body hair that our primate relatives retained is a more radical development than it might seem. Naked skins may increase sexual sensitivity, and hair loss even seems to have helped, paradoxically, in water conservation, but naked skins increase risks of skin cancers and other diseases and they also open the body to vermin in a way that no chimpanzee must live with.[40]

Some of the basic differences between ourselves and other primates, and possible reasons for them, are evident if we compare baboons and chimpanzees. Baboons inhabit a diversity of environments, but our most reliable and complete information comes from savanna-adapted populations in central and east Africa. Within the baboon troop, the eldest males have almost unimpeded access to food and sex, and they protect their privileges with aggressive behavior. Younger males are allowed access to females only on the sufferance of more dominant males, although they are constantly testing this dominance by attempting to strike up meaningful relationships with concupiscent females. Often this will incite a dominant male to charge and make threatening gestures and rude remarks. Similarly, when desirable but limited food is available, such as fruit or a clutch of eggs, the dominant males and females take what they want first. A mother and child may share some foods, but there is no systematic food sharing between adult members of the group.

When baboon troops move across savanna or open environments, they typically position members according to age and sex. Adolescent males are on the periphery, dominant adult males are in the front-center, adult females are behind the adult males, and dependent young are stationed near the adult females. This is of course an excellent defensive formation, offering the best protection to the young and females, on whom the perpetuation of the group depends. There are many more adolescent males than necessary for reproduction, so although they would endure the most losses to predators, the stability of the group would not suffer much. Significantly, when the baboon troop moves into more forested areas, the positioning becomes much more fluid and dispersed, indicating, not too surprisingly, that terrain is an important determinant of behavior for baboons, and that savanna environments seem to put a premium on group cohesiveness and dominance hierarchies.

Some baboon groups seem to construct dominance hierarchies on the basis of length of residence in the troop, and in general baboons seem somewhat more variable and adaptable in dominance hierarchies and aggression than previously thought.[41]

Thus, in trying to explain how culture-bearing individuals developed from our primate ancestors, we must look for factors or conditions that would work against these dominance hierarchies and promote the distinctive human family structure, with its food sharing, division of labor, and pair bonding.

The relevance of baboons to our own ancestry may be somewhat limited because we are much more closely related to chimpanzees and our primate ancestors were probably much more adapted to forested environments than savannas.[42]

Chimpanzees also inhabit diverse environments, and generalizations about their behavior are as dangerous as in the case of baboons. The most stud-

3.10 Chimpanzees regularly hunt small game and share the meat in relationships based on age and sex. Some anthropologists believe that hunting and meat-sharing were important determinants in the origins of human family structures and sexual nature.

ied chimpanzees are those of the forests of Tanzania, especially those of the Gombe Stream area.[43] Many of these chimpanzee groups have been regularly fed by the people observing them, and most groups are cut off by modern development from their complete range and any interactions with more distant groups, so we cannot expect these primates to be perfect reflections of chimpanzee behavior in the true natural state. But it is striking that the chimpanzees of this area behave very differently from the savanna-dwelling baboons. Access to food and sex is much less rigidly controlled, and the entire dominance hierarchy seems to be less in evidence. In one instance where a female was sexually receptive for a record twenty-one days, the males of the chimpanzee group, young and old alike, stood in line for her favors with only minor apparent bickering.[44] This is not always typical behavior, and the adolescents were effectively barred from mating with some females in many cases, but in comparison with baboons, chimpanzees are absolute libertines in sexual matters. Strictly speaking, chimpanzees do not have true estrus cycles, but they do have periods of greater and lesser sexual receptivity.

Similarly, there is also a diminution of territoriality. Chimpanzees run for kilometers along trails through the forests, often alone or with one or two others, and individuals frequently leave one small group to go live with another. So relaxed are chimpanzees, in fact, about sex and territoriality that they have aroused envious feelings in some human observers. "Chimpanzees, it seems, successfully achieve what *Homo sapiens* radicals only dream of: peaceful, non-competitive, non-coercive, non-possessive, egalitarian, jealousy-free, promiscuous, non-tyrannical communes."[45]

This may be a somewhat idealized picture. Chimpanzees have elaborate threat rituals and do menace each other according to a well-developed hierarchy. In addition, chimpanzees are the only animals besides ourselves who regularly use objects as weapons, usually wooden clubs or thrown rocks and debris.[46] They fight over food and females, and sometimes over rank, and they are the only other primates known to practice cannibalism, having been observed on several occasions eating young chimps alive.[47]

Tanner notes that wild chimpanzees place nuts on flat roots or stones and then break them open by hitting them with sticks or rocks, suggesting that "a fairly intelligent ancestral ape could have taken the step from such opportunistic tool use during foraging, to tool-making and more regular use of tools in plant gathering."[48] She also notes that female chimpanzees use tools more often than males in nut-cracking, they generally use tools in more complex ways than males, and they teach their daughters to use tools and even supply them with tools.[49]

Books such as *African Genesis* by Robert Ardrey and *The Naked Ape* by Desmond Morris argue that much of our mentality, anatomy, and physiology is a result of millions of years of bloody, relentless hunting and killing. Indeed, if we look at chimpanzee predatory behavior, we see that in its emotional content and its implications, hunting and meat eating may not be just another way of getting food. Primatologist G. Teleki has observed the hunting behavior of chimpanzees, and he found that hunting, loss of estrus, economic reciprocity, and human family structure may be intimately interrelated.[50] He notes that adult males often cooperate in hunting small mammals and that there is a partial suspension of the dominance hierarchy during these times, as demonstrated by the fact that a low-dominance male sometimes leads the hunt. Significantly, although females and juveniles participate in hunting only rarely, they are often given meat by the males who make the kills, and females who are sexually receptive are more likely to get and eat meat.[51] Symons suggests that there may have been a pattern in early human evolution in which the amount of meat a female received from males was directly proportional to the length of time she was in estrus, and selection might have favored males who used surplus meat to increase their reproductive success at the expense of other males.[52]

But we must be cautious about reading our whole evolutionary history in these few observations; a great deal more research must be done before we can assume these important links between loss of estrus, hunting, and food sharing.[53]

In reviewing recent studies of hunting by chimpanzees, Teleki notes that cooperative behavior among chimpanzees while hunting has not been shown to produce more kills, and he suggests that difficulties in defining what is cooperative and in measuring effectiveness leave this point moot.[54] He concludes that sexual receptivity, kinship ties, and social status all determine how meat is distributed after a kill is made. Teleki suggests that there is sexual division of labor among chimpanzees that is not dependent solely on hunting behavior. He reviews various studies showing that female chimpanzees spend a much greater proportion of their time collecting termites and other invertebrates than males—a prelude, he argues, to the sharp division of labor in human hunting and gathering groups.

Some Models of Cultural Origins

Having compared nonhuman primates and ourselves, let us look at several attempts to formulate general explanations for the first stages of the evolution of culture. Many people have considered this problem, and it is a subject fraught with political implications, for it involves the basic nature of humanity and cultural change. No explanation for cultural origins is complete or conclusive, but many have points of interest.

Owen Lovejoy has recently argued that bipedalism as an evolutionary product is essentially a matter of sex and the economics of foraging.[55] As he notes, in some ways bipedalism seems an unlikely development, because it is a slow method of movement and an animal expends great amounts of energy just holding itself up. Also, because we evolved as a quadrupedal and semierect animal, bipedalism is in a sense an unnatural form of locomotion. Our spines evolved to serve as a cantilevered bridge-like structure and even though this structure was successfully modified into a bipedal erect form, we are heirs to many problems, including slipped disks and other back problems, bunions, hernias, poor leg and foot blood circulation, and a shortened birth canal that means difficult births for many mothers. If bipedalism is so bad, why should it appear? Lovejoy argues that all apes are potentially semierect animals by virtue of natural selection for "handedness"—the constant use of the hand to forage for food. And if one's evolutionary niche does not reward great bursts of speed over the short run, bipedalism is quite effective for long periods of methodical hunting and gathering.

Lovejoy notes that through the later history of the primates, there has been an evolution in the reproductive strategy. The trend has been toward decreasing the number of offspring a female has but increasing the parental investment in each one so that they have a better chance to survive and in turn reproduce. (Ecologists call this a "k-strategy," as opposed to the "r-strategy" of something like the housefly, which has thousands of offspring, only a few of which survive.)

The heart of Lovejoy's argument is that bipedalism was selected over quadrupedalism in our ancestors because it was a way to increase reproductive success. It allowed our ancestors to have more offspring while maintaining a high level of parental investment in the form of food provision, protection, and training. This was accomplished by evolving an economic adaptation and social organization that allowed the mother to spend less of her energy getting her own food and more of her time taking care of offspring. Lovejoy argues that one way females could successfully raise several children born within a short time interval would be to induce the male to provide the female with food on a regular basis, to take part in other tasks of raising many children, and to do it all without competing with other males to the point that the group would break up into murderous, insanely jealous individuals. This meant that the dominance hierarchy and competition common in other primate groups had to be suppressed. One way in which this might occur is if individuals began to differ in perceived "attractiveness," so that qualities of physical appearance and personality played a role in sexual desirability and feelings of "love."

If all this is true, bipedalism, which allows the male to carry back food, probably came before tool-use. Food-sharing, with or without tool-use and hunting, would tend to lengthen the period during which a female was

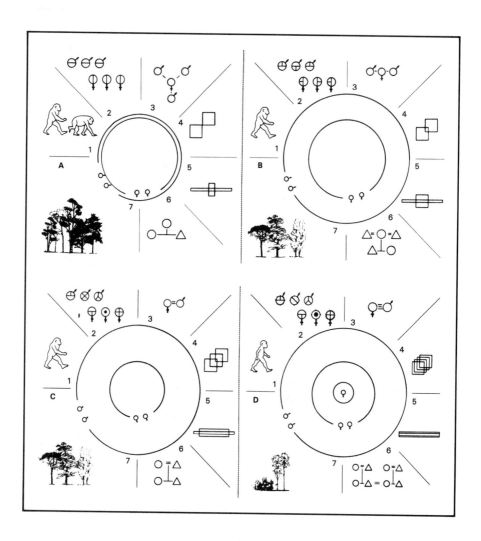

3.11 These four diagrams illustrate aspects of Owen Lovejoy's model of human origins. Reading from left to right, these diagrams reflect a sequence through time. Section 1 of each diagram traces the evolution of erectness itself. Section 2 reflects growing epigamic differentiation, the symbols indicating the growing distinctiveness of individuals, which led to growing selectivity in sexual pairing. This is reflected in the mating patterns shown in Section 3, where males at first mate indiscriminately with different females but later become mated partners of specific females. Section 4 depicts the changing patterns of birth spacing in infants. In Diagram A there is no overlap, reflecting the year or two spacing between infants. Diagrams B, C, and D show the progression to the human family of several young children. Section 5 indicates how these other changes were accompanied by changes in sexual activity. In each diagram the female menstrual cycle is represented by a long bar with the period of sexual receptivity indicated by the middle figure. In Diagram A virtually all sexual activity takes place in the females' period of estrus, but this changed until, in Diagram D, sexual receptivity is largely independent of sexually responsive, if sex served as an inducement to food sharing. And this was an important evolutionary change in more ways than one. Human females are fertile for only a relatively unpredictable three days a month, so intercourse at least a few times a week would be necessary for successful fertility rates.

As Figure 3.11 illustrates, Lovejoy ties all these cultural developments together in an evolutionary progression toward "culture."

Many people have criticized Lovejoy's model. As Kinzey notes, studies of chimpanzees show that it is the females, not the males, who most frequently share food with young chimps.[56] Another problem is that substantial differences in size between males and females in primates seem to associate with competition between males in polygamous mating patterns, whereas Lovejoy proposes monogamous patterns for early hominids.

Lovejoy's model is problematic, but the relationships he focuses on are probably central to any explanation of cultural origins. Bipedalism, intelligence, human family structure, and sexuality seem intimately interconnected in their evolution.[57] In some ways, human sexual differences seem the most radical contrasts between ourselves and other apes, and these differences seem to suggest slightly different evolutionary patterns for males and females. In an analysis of these arguments about the link between sexual behavior and other aspects of culture, Donald Symons has addressed the problem of human origins on the basis of a sustained comparison of sexuality in men and women.[58] Symons summarizes the major differences between men and women in this regard as follows: (1) Intra-

---

the estrus cycle, with essentially continuous receptivity. Section 6 shows how kinship relationships altered in concert with these other changes. In Diagram A the only kinship bonds are between mother and child. In Diagram B there is the beginning of a male female relationship but no true bonding. In Diagram D there are several such male-female pair bonds. All of these changes take place in the context of changing physical environments. The tropical forests of our prehuman ancestors gradually changed to the drier, more open savannas and woodlands. The circles in the centers of the four diagrams have to do with foraging and movement. The inner circle delimits the core area of the group, where females and dependent young spend their time. The outer circle represents the outer limits of the group, the farthest distance adult males forage away from the group. In Diagram A the two circles are nearly the same size because the female is not being given any food by the males and must do her own foraging. In Diagram B the male ranges more widely to remove himself from the core area and leave more food for the female, which she must have because she is beginning to have more dependent young to feed. In Diagram C this behavior intensifies. The male is now bringing back food for his mate, which requires him to forage more widely. She must stay closer to the core of the group because she has more dependent young. Diagram D shows the establishment of a true home base where the mother may leave her children in the care of relatives and forage more widely herself.

sexual competition generally is much more intense among males than among females, and in preliterate societies competition over women probably is the single most important cause of violence. (2) Men incline to polygyny, whereas women are more malleable in this respect and depending on the circumstances may be equally satisfied in polygynous, monogamous, or polyandrous marriages. (3) Almost universally, men experience sexual jealousy of their mates. Women are more malleable in this respect, but in certain circumstances women's experience of sexual jealousy may be characteristically as intense as men's. (4) Men are much more likely to be sexually aroused by the sight of women and the female genitals than women are by the sight of men and the male genitals. (5) Physical characteristics, especially those that come with youth, are by far the most important determinants of women's sexual attractiveness. Physical characteristics are somewhat less important determinants of men's sexual attractiveness; political and economic prowess are more important; and youth is relatively unimportant. (6) Much more than women, men are predisposed to desire a variety of sex partners for the sake of variety. (7) Among all peoples, copulation is considered to be essentially a service or favor that women render to men, and not vice versa, regardless of which sex derives or is thought to derive greater pleasure from sexual intercourse.[59]

Admitting the great plasticity and potential for variability of human sexual behavior, Symons thinks that there is a strong biological, genetic, profoundly evolutionary basis to these apparent behavioral differences between males and females and these cultural aspects of sexuality. And this is what one would expect, given our evolutionary origins. For millions of years the evolutionary advantage has been with men who have had intercourse with as many young women as they could—provided that the resulting progeny could be raised to reproductive age. Lovejoy's model was based on initial monogamous pairings, but ethnographic evidence suggests that human mating patterns have been primarily polygamous for millions of years. In evolutionary terms, the true loser is the male who provides for another male's offspring, giving rise to the observation that much of the social organization of human hunting and gathering societies is designed to allow a man to go on long hunting trips without returning and finding himself the provider for some other man's children.

Marshall Sahlins summarized this intertwining of sex, food, and human society as follows:

> Sexual attraction remains a determinant of human sociability. But it has become subordinated to the search for food, to economics. A most significant advance of early cultural society was the strict repression and canalization of sex, through the incest tabu, in favor of kinship, and thus mutual aid relations. Primate sexuality is utilized in human society to

reinforce bonds of economic, and to a lesser extent, defensive alliance. All marriage schemes are largely devices to check and regulate promiscuous behavior in the interest of the human economic scheme.[60]

John Tooby and Irven DeVore have recently suggested that the various models of cultural origins and subsequent human evolution seem most plausible if we assume that our ancestors of several million years ago began to increase the time they spent hunting or scavenging other animals: "if hunting were a major part of hominid foraging, it would elegantly and economically explain a large number of the unusual aspects of hominid evolution."[61]

They begin with *male parental investment*. As noted previously, it is highly unusual among primates for males to invest significant energy in providing their offspring with food, yet people do. Tooby and DeVore note that, whereas it is not economically efficient to carry low-calorie vegetable foods long distances back to one's mate and offspring, meat is a concentrated form of high-calorie nutrition that would make this an advantageous strategy.

Tooby and DeVore point out that "male coalitions" may have developed because of the advantages of group hunting. Reciprocity, sharing, and exchange, too, may be linked to hunting. Tooby and DeVore argue that vegetable foods offer few evident rewards to the kinds of exchange, sharing, and reciprocity typical of all human groups. Vegetable foods come in many different forms and sizes, and animals tend to gather them on the basis of energy costs and immediate needs. But meat, "unlike vegetable foods, comes in discrete quantities: an entire animal is either captured or lost. . . . variability in hunting success, and the fact that meat comes in chunked quantities often in excess of what the capturers can readily consume, provides a ready explanation for food sharing, food exchange and risk sharing through deferred reciprocation among the larger social group."[62]

Regarding the pronounced sexual division of labor among humans, Tooby and DeVore suggest that extreme differentiation in the kinds of foods males and females gather is not feasible without food exchange between males and females.

> If males changed from occasional to intensive hunting, one consequence would be the extreme sexual division of labor found among humans, with females exploiting the more [fixed, vegetable] food sources. . . . The evolution of this behavior requires no qualitative leaps from other primates: in chimpanzees . . . estrous females receive disproportionate shares of meat from hunts made by males.[63]

Tooby and DeVore also associate the early apparent emergence of the "home base" form of settlement in human groups with the consequences of a hunting adaptation. They suggest that a home base would be of little advantage to primates subsisting on vegetable foods. But for hunters, the sporadic, irregular, unpredictable nature of success at hunting would make a home base an advantage, in that it would be a place everyone knew to return to, and where meat could be shared. Similarly they see stone tool-use as probably an adjunct to the requirements of animal butchering and hunting.

Concerning the exaggerated size of the human brain, they note that the human brain is an extremely metabolically costly organ. "It may well be that this cost can only be justified if this expanding hominid brain, in turn, also makes rich new sources of food possible. We know that all carnivores have significantly higher encephalization quotients than do non-predators . . . that some nutrients found in meat are particularly important to brain tissues, and that these are hard to obtain and metabolically process from plant foods."[64] They also note that hunting may be linked to the "pongid-hominid" split, in that "the largest open country primates, savanna baboons, have been reported to have the highest rates of hunting for any non-human primate."[65] Finally they note that the wide geographic distribution of humans is only possible because of meat-eating and hunting—that in fact it is impossible for "modern humans to live exclusively on uncooked plant foods (e.g., cyanocobolamine deficiency)".[66]

These forms of small-game hunting and scavenging would have rewarded group cooperation, reduced dominance hierarchies, improved communication systems, and encouraged the development of stone tools for processing meat and vegetable foods.

In general summary of these models, we have seen that bipedalism, human sexuality and social networks, enlarged brains, terrain, hunting, and various other factors may all be related in complex patterns of interaction, resulting in the kinds of animals represented by "Lucy" and the other Hadar fossils. These hominids need not have been major tool-users to have been "cultural" in fundamental ways. Hunting and meat eating may have been important, but we have no direct evidence with which to test this.

The Australopithecine—Homo Transition (c. 3 to 1 Million Years Ago)

The earliest stone tools known are only between two and three million years old, and we might suspect that they are important markers of the transition to *Homo*. Primates of all kinds, including no doubt Lucy, use sticks and unmodified stones for a variety of purposes, but to make a stone tool requires that one "see" the finished artifact in the unworked lump of stone. Evidence suggests that the manufacture of stone tools was a rather late part of the whole transition to humanity, since bipedalism and even increasing brain size were well underway by the time the first stone tools apparently were made. But from that point on, stone tools may have been an important factor in the rate at which we changed and spread throughout the world. The importance of tools to our ancestors can perhaps best be appreciated if one imagines oneself standing about 4 feet tall on the Serengeti Plain of two million years ago, trying to resolve into its component parts a small antelope one has just killed or scavenged from a temporarily napping lion. Already harried by vultures and other scavengers, one tries to rip into the body with one's teeth and nails. Even with the relatively stout dentition of Lucy, it would have been difficult. The most nutritious parts, the liver, brain, and other internal organs, are protected by thick layers of skin, flesh, and bone that resist the puny tearing motions to which our ancestors would have been limited. But just a chip off one of the quartz pebbles abundantly scattered over this area would instantly have opened up life-saving rations for several individuals.

The first hominid to make and use stone tools was probably one of the australopithecine species or early *Homo*. Before looking at the stone tools and sites, however, it should be noted that the classification and determination of genetic relationships of early hominids involve several problems. Most physical anthropologists think that the australopithecines separated into two lineages about three million years ago, one of which—the more gracile (slight, slender) type—led to ourselves, while the other, more robust type became extinct. It was widely believed that the more robust forms of australopithecines—known as *Australopithecus boisei* in East Africa and *A. robustus* in southern Africa—became extinct because they did not use tools. But analyses of hominid remains in the South Africa cave of Swartkrans suggest the robust australopithecines may have used bone and stone tools.[67] The discovery of the "Black Skull," a robust australopithecine from Lake Turkana, Kenya, stained black by manganese deposits and dated to about 2.5 million years ago, set off a general reconsideration of australopithecine phylogeny. Among other issues, this skull raises questions about whether *Australopithecus afarensis,* represented by Lucy, was the primitive unspecialized australopithecine ancestral stock that many researchers thought it was.[68]

Physical anthropologists disagree heatedly about phylogenetic relationships and the economies of the various australopithecines, and particularly about at what point we can apply the genus name *Homo* to these animals, indicating our descent from them. One problem is that there is great variation in size and shape of some physical features of these early hominids. Marked diversity exists among living humans, too, of course, particularly in height, cranial volume, skin color, musculature, and other features. But if we consider the hominids of between about three and one million years ago, there seems to be more relative variation in tooth size and shape, cranial volume and shape, overall height, and other factors than is present among ourselves. And, significantly, these variations involve physical structures that seem intimately tied to the diet and activities of these animals, indicating perhaps very real differences and changes in their basic ecological adaptations.

There have been many attempts to arrange the earliest hominids in a phylogentic sequence. Skelton and his colleagues have summarized these in Figure 3.12.[69]

3.12   Six hypothetical reconstructions of the phylogenetic relationships among early hominids.

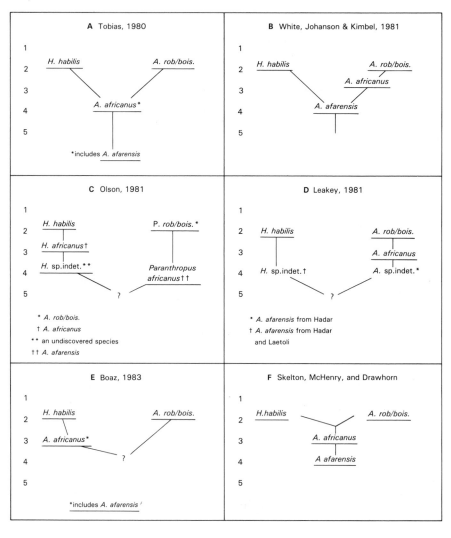

Some of these taxonomic disputes could be resolved if we had many full skeletons of the earliest *Homo,* but we may never have adequate samples of this hominid. Most anthropologists have used the category *Homo habilis* to refer to the species from which *Homo erectus* and thereby ourselves have descended, and the KNM-ER 1470 skull, dated to about 1.8 million years ago, is perhaps the best example of this category. *Homo habilis* is generally thought to have had a cranial capacity of about 660 cubic centimeters, halfway between some of the smaller australopithecines and *H. erectus,* and specimens have been identified from South Africa and Ethiopia, but its taxonomic position will likely change on the basis of future finds.[70] Hominid L894-I from Omo may be an early representative of *Homo,* but there are lingering questions about the date (2.33 million years ago), and less than 30% of this fossil is preserved.[71]

Even the taxonomic phylogenetic status of *Homo erectus,* whom most scholars have considered a homogeneous group, directly on our ancestral line, has been questioned. Hominids ranging from about two million to about 200,000 years in age have been assigned to *H. erectus.* There is considerable size and shape variability in *H. erectus,* and at this point there are Asian specimens that may be as old or older than anything yet found in Africa.[72] In 1984 a group headed by Richard Leakey found a *H. erectus* (known as KNM-WT 15000) near Lake Turkana[73], Kenya, that seems well-dated to about 1.6 million years ago and is virtually complete. Such a modern and "brainy" hominid at this time is somewhat surprising, given the other known fossils (this find is discussed in more detail in Chapter 4).

These many arguments about human lineage will doubtless continue for many more years, and all that really seems clear at this point is that from 3.0 to 1.5 million years ago, bipedal primates of about 1.25 to 1.75 meters in height roamed the grasslands and forests of Africa, the Middle East, and probably as far east as China and that at least some of these animals made tools.

Whether these different groups were interbreeding or entirely separate is unclear, as is the identity of which of them made stone tools. Any speculations about the ecological and social relationships among early hominids are hard to test because we have so little data and because evolutionary principles that apply to other primates may not apply to an early tool-using primate. Many archaeologists who saw *Quest for Fire* enjoyed it but faulted the movie on two points: about five different genera (or at least species) of humans were depicted as living at the same time, and most of them spent their lives trying to drive each other into extinction—except for one Neandertalish man and an almost ultramodern looking woman, whose miscegenation was fertile across species, perhaps generic, lines. The majority position in anthropology these days is that there were at most two or three different species of human beings at any one time and that, if studies of hunters and gatherers and of animal ecology are any guide, continuous battles to extinction between rival human species just never happened.[74]

For one thing, the organization of people into family units and their use of tools seem to have changed their selective environments so that the rate of speciation was much less than for other animals.[75]

But to understand fully how we became just one species, and which variants of human beings were tool users and which (if any) were not, we will need much larger fossil samples and much more powerful theoretical models of early hominid behavior. In the meantime, the reader can postpone memorizing Figure 3.12 in the sure and certain knowledge that these sequences will all be changed and expanded in the future.

Archaeological Evidence of the Origins of Culture

If we assume that one of the phylogenies in Figure 3.12 is correct, we can begin to look at the *archaeological record* to determine how our ancestors actually lived and changed over time. As always, the problem here is one of poor evidence. Perhaps the only way in which we will understand the details of our origins would be if we found several Pompeii-like sites where volcanic eruptions had trapped early humans some two to three million years ago, toward evening, say, around a water hole on the East African savanna. Such a catastrophe might have killed thousands of animals, including some hominids. Without relishing any disaster to even these our most distant hominid ancestors, such a volcanic event would be most scientifically useful if it happened just after these hominids had scavenged or hunted their food, butchered it with stone tools, and then dispersed into social groups to sleep off the meal.

Although no early hominid sites like this have been found, they may exist. In 1902 on Martinique about 28,000 people were killed within minutes when a volcanic eruption sent a landslide of ash, lava, and debris cascading over part of the island.

At present, however, our direct archaeological evidence about the initial period of the evolution of culture is extremely meager; we have only a few small sites, almost all of which are in southern and eastern Africa.

OLDUVAI GORGE

About five million years ago, a large lake covered about 130 square kilometers of what today is the Serengeti Plain, in central East Africa. Countless generations of animals, including, no doubt, our hominid ancestors, lived near this lake, and today their fossilized bones are thickly distributed through the black clay of the ancient shoreline and lake bed. About a half million years before the appearance of culture-bearing animals in this part of the world, the lake and the adjacent areas were the scene of dramatic geological activity. Volcanoes near the lake had been erupting for millions of years, and there was a particularly violent episode about a million and a half years ago, when a volcano to the south of the lake covered it with a layer of ash and molten rock some 5 meters thick. Similar eruptions occurred in succeeding millennia, and today we can measure the intensity of each volcanic episode by gauging the thickness of each superimposed ash level. Sometime during and after these volcanic eruptions, the climate of the area began a cycle of alternating dry and wet periods. These millions of years of volcanic activity and climatic change eventually covered the original lake and adjacent lands with a 100-meter-thick deposit of sand, ash, and lake sediments, the surface of which is the present Serengeti Plain. Ancient floods cut down through this plain at several points, creating Olduvai Gorge and revealing in its cliffs and floor many concentrations of stone tools and animal bones.

One such concentration, Site DK I, has been radiometrically dated to about 1.75 million years ago, making it one of the oldest in Olduvai Gorge. The site is composed of a layer several meters thick of bones, worked and unworked stone, and other debris. Because of the complex stratigraphy, it is difficult to separate this accumulation into different levels with any assurance that the divisions represent discrete hominid occupations.

The most prominent archaeological feature at DK I is a semicircle of stones (Figure 3.13) lying within a concentration of stone and animal bones. Measuring approximately 3 meters in diameter and made of chunks of vesicular basalt, this feature is interpreted by some as a foundation for a windbreak or some other temporary structure, but the evidence is inconclusive.

The tools found here are similar to the other materials from Beds I and II at Olduvai: hundreds of crudely flaked stones, some of them showing evidence of use, others simply by-products of the manufacturing process.

Site FLK I is particularly important because several hominid bone fragments were found in its various levels, including the remains of an early hominid that Louis Leakey named *Zinjanthropus*. This "Zinjanthropus floor" covers an area of more than 1036 square meters and contains several thousand pieces of worked stone and about 3500 identifiable bones of mammals. Many stone artifacts and pieces of shattered bone are concentrated in a "working area" some 5 meters in diameter, and a relatively clear arc-shaped area in the midst of all this debris suggests to some that there may have been a temporary shelter here.

A few other sites in Africa have been shown to have stone tools that

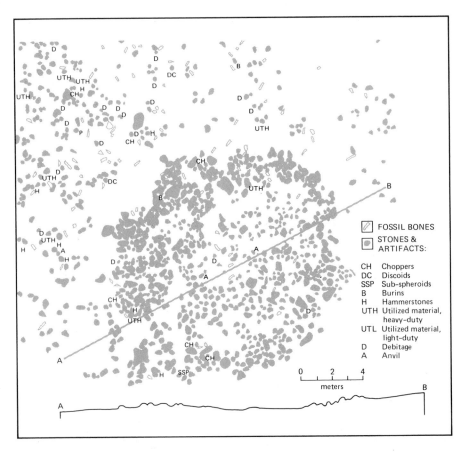

3.13 Site DK I. Plan of the stone circle and the remains on the occupation surface. Stones, including artifacts, are shown in gray. Fossil bones are shown in outline. This may have been a home base or a windbreak built by hominids about 1.7 million years ago, but the evidence for this is ambiguous.

date to about two million years ago, including Koobi Fora, where crude stone tools appear to be at least 2.3 million years old.

These early stone tools from Olduvai, Koobi Fora, and a few other sites provide some indications about the nature of our ancestors. Studies of the Olduwan tools by Nicholas Toth indicate that already by 1.7 million years ago our ancestors at Olduvai had developed a preference for right-handedness.[76] Based on the marks left that indicate how they held the stone cores when they hit them to produce stone flakes, the Olduvai hominids seem to have the same proportions of right-handers—90%—as do modern populations. Toth suggests that archaeologists may have been fundamentally wrong about the Olduvai tools in assuming that the big "core-tools" were the object of the tool making: he studied wear patterns on the putative "hand-axes" from Olduvai and concluded that they rarely show any wear indicative of use—rather it was the *flakes* that were the point of the tool making, and the "hand-axes" and or the core-tools were in many case the by-products of tool manufacture.

The question of what these tools were used for is a central element in debates about the nature of our early evolution.

Early Hominid Economies

With the evidence from Olduvai Gorge and the few other sites of this time period (c. 2.5–1.0 million years ago), we must account for, among other developments, the following: (1) the evolution of stone tool-use; (2) the approximate doubling of human brain size; (3) increased stature, and changes in facial architecture and other physical features; (4) the migrations of hominids throughout the warmer regions of the Old World.

Great controversies have erupted over the explanation of these changes, particularly about the *economic* basis of these transitions. What is at issue here is the very basis of almost all human evolution, since we must assume that we are mainly the physical products of the roughly two million years of natural selection on all the generations that connect us to the hominids of Olduvai Gorge. Arguments on these issues have centered on whether or not early *Homo* did no, some, or a lot of hunting, and how much like modern peoples they were in their social relationships.[77]

It is not surprising, perhaps, that in trying to understand what those selective forces and environments were, anthropologists have depended heavily on ethnographic studies of people who until recently lived as hunter-foragers in the savannas of southern and eastern Africa, particularly the Kalahari Bushmen. Some feel that the *behavior* of these modern-day hunter-gatherers may have almost nothing to tell us about our hominid ancestors.[78] But the Kalahari hunter-gatherers and others at least illustrate some of the possibilities open to our ancestors in this kind of environment.

Studies of African savanna ecology suggest that *scavenging* could have been a significant part of early hominid repertoires. African savannas would seem to reward animals who are opportunists and generalists when it comes to food: an early hominid might well have adopted the principle of "eat it if it moves, or has reasonably recently moved." Schaller and Lowther spent a few days during the dry season on the Serengeti Plain and found a lot of edible meat in the form of a dead buffalo (which did not appear to have been killed by a predator), some lion kills, and a few incapacitated animals that could have been killed.[79] Blumenschine found that scavenging would have been possible and even rewarding, presuming that our ancestors went for the brains and marrow of lion kills and other casualties.[80] He found that small animals, such as juvenile Thompson's gazelles, persisted as carcasses for only about a day before some scavenger totally consumed them, but larger animals, such as adult buffalo, remained as significant food sources for about four days, after which even if they were not eaten by lions, hyenas, vultures, etc., they were putrid to the point that their market value had plummeted.

Even without scavenging, our ancestors may have developed a way to subcontract some of the effort and risk of hunting to lions in a way that is right in line with our later history as one of the most selfish, exploitative, rapacious members of the animal family. O'Connell and his colleagues have shown that the Hadza, hunter-gatherers who lived in northern Tanzania, sometimes scavenge simply by scaring lions away from kills by making a lot of noise.[81]

The Kalahari hunter-gatherers keep their population densities low and thinly distributed. The low level of productivity of their hot, arid environment, as manipulated by their simple technology (bows, arrows, digging sticks, and so forth), requires that they spend most of the year in groups of twenty-five people or less, often on the move from one camp to another. Most of their diet is made up of vegetable products, tortoises, and other small game, but occasionally giraffes and other large animals are killed, usually through cooperative hunting by several males. Although females provide the bulk of the group's food by gathering and processing plants and eggs, nestlings, turtles, and other small animals, men do all the hunting of large animals among most known hunting-gathering societies, including the Bushmen and Aborigines.

Kalahari Bushmen also illustrate the value of division of labor. We all know that it is better, in complicated tasks, to work as an integrated group doing different things than for everyone to do everything, whether it be in getting a football downfield or building a supercomputer. The Bushmen split tasks into those that men do, those that women do, and those that everyone does, children included.

Hunter-gatherer economic life is also dominated by the principle of reciprocity: food and other resources are exchanged among kinsmen, balancing out the periodic shortages that may afflict any member or nuclear family in the band.

Like most hunting and gathering societies, the Bushmen are at least somewhat territorial. They move often but always within a relatively restricted region, usually 25 to 30 kilometers in all directions from a central water hole or home base. Clearly, some degree of territoriality is an advantage if resources are not uniformly distributed; it is efficient to know where reliable sources of flint, vegetables, game animals, and water are within one's territory, and the group forced out of its territory is faced with unpredictable supplies and, perhaps, the hostility of the group on whose territory it is trespassing. Thus, any models or reconstructions we make regarding our early hominid ancestors should incorporate the assumption that they were probably at least loosely territorial.

Hunter-gatherer population densities are directly determined by resource availability and are maintained by what seem to us harsh methods of population control. The aboriginal population density of most of Australia, for example, can be predicted with a high degree of accuracy simply from the amount of average rainfall in the various areas. This balance between population and resources is maintained mainly by marriage rules and female infanticide. Birdsell reports that 15 to 30% of all babies are killed, usually by the grandmother, who places her hand over the infant's mouth and nose as it is born so that it never draws its first breath.[82] She acts out a group's unstated decision that keeps a mother from carrying more than one infant while she works.

Studies of !Kung infant breast-feeding may reveal the powerful controls on fertility rates that are available to a group simply through natural mechanisms. !Kung women breast-feed their babies as often as sixty times every 24 hours, including during the night when the mother is asleep, and there is some evidence that this frequency is necessary to produce and maintain the hormonal levels that suppress ovulation; a study of women in Scotland shows that they usually nurse their babies only six times in 24 hours and this has less effect in suppressing ovulation.[83]

In her study of hunters and gatherers in New Guinea, Patricia Townsend observed that although women produce over 90% of one group's food, this society is strongly patriarchal.[84] Townsend found that the women in this group usually marry soon after puberty and have about six children. Malaria and other diseases kill off about 43% of the children early in life, but girls die from diseases at a much greater rate than boys, probably from malign neglect. If thought necessary, unwanted girls are killed by strangulation with a vine soon after birth.

Most hunter-gatherers keep their numbers at a level far below what could be supported in any average year, and it is the older people who remember the worst winters and the longest droughts that must be accommodated by the group's sizes. Someone said that when an old person dies it is like a library burning down, and this is true also among hunter-gatherers.

In summary, the Kalahari Bushmen may offer some general guidelines for interpreting the archaeological remains of our ancestors in these areas of Africa—even though we know that in mentality and many other ways they are profoundly different from early *Homo*.

How can we go about integrating all this ethnographic, archaeological, primatological, and paleontological evidence and use it to understand our origins? To begin with, in an evolutionary analysis we must concentrate on those factors that might have allowed some individuals and not others to have survived to pass on their genes, so that over time our brain size increased and all the other dimensions of the human animal were formed. And once we start examining differential reproduction, we must look directly at how food was procured, shared, and consumed.

Currently, a particularly heated debate is underway in archaeology between people who look at the same bones and stones and see, alternatively, conclusive evidence of hunting, conclusive evidence of scavenging, and no conclusive evidence at all.[85]

One might think that the simple conjunction of stone tools amidst piles of animal bones, some of which show obvious cut marks from being hacked at with stone tools, irresistibly would lead one to the conclusion that our ancestors butchered and ate these animals. But one archaeologist's stone-tool-cutting mark is another archaeologist's hyena-bite mark, and when it comes to the statistics of the extent and significance of different kinds of bone alteration, entirely different interpretations arise.

Why is it so important to know if these animals were hunted or scavenged? Here again we are dealing directly with the factors that shaped us physically and to some extent mentally. When anthropologists argue about the economy of early humans they are arguing about one of the driving forces of human evolution.

Early hominid hunting would have been very much different from that of our world. Modern rifles kill partially by shock: a deer hit almost anywhere with a high-powered bullet dies because the bullet's impact on the body compresses the blood and sends a shock wave of blood through the body, ripping out heart valves, sometimes pushing the eyes out of their sockets. It is rather a different undertaking to stick an elephant in the stomach with a spear and then wait hungrily until it bleeds to death. We know from studies of African pygmy groups, the Kalahari Bushmen, and others that people can hunt and kill large animals regularly with very crude spears and other tools, but the question is, when did our hominid ancestors begin systematic hunting as a major source of food?

Concerning the evidence from Olduvai Gorge—our only sample of the spatial distribution of stone tools and animals bones for the period of about 2.0 to 1.5 million years ago—opinions vary sharply. For example, Binford concluded that with regard to the *Zinjanthropus* floor, "both the skeletal-part frequency data and the cut-mark data are most consistent with and indicative of the scavenging of marginal foods from previously ravaged carcasses."[86] Bunn and Kroll, on the other hand, argue that the site may reflect substantial hunting, in combination with scavenging perhaps, and that there is some evidence of food sharing reflected in the distribution of animal parts.[87]

Even with high-powered magnification and careful laboratory experiments in cutting bones with stone tools, it would seem difficult to distinguish between human and animal activities. Hominids may have killed and butchered animals, and then, perhaps because their tools and butchering skills were not all that efficient, they may have left fairly meaty bones to be scavenged by hyenas and others.[88] Alternatively, lions or hyenas may have killed many animals and eaten them partially, after which people scavenged these carcasses.

Site DK I may have been a "home-base" for hominids, since it seems to have many different stone tools and animal bones. But the evidence is far from clear.[89]

Tooby notes that rewards for an early hominid scavenger may have been quite limited, consisting mainly of marrow, brain, and heads of medium-size animals left by lions near water courses during the dry season.[90] He concludes that

> The recent surge of interest in scavenging is tacitly predicated on the assumption that hunting is an advanced, difficult, and dangerous attainment, dependent on sophisticated abilities unlikely to appear until late in hominid evolution. Scavenging appears to be viewed as "easy" compared to hunting and putatively therefore a natural first step on the road to meat eating for inept early hominids. However, almost nothing about the ecology of scavenging supports this view.
> . . . although the role of scavenging in human evolution appears likely to have been selectively minor (i.e., a supplementary adjunct to hunting), its role in the creation of faunal assemblages may have been substantial because of the differential preservation bias of larger animal remains— those according to Blumenschine's research,[[91]] most likely to have been worth scavenging and also, because of their large size, least likely to have been acquired through hunting by Plio/Pleistocene hominids.[92]

Debates about the factors and circumstances that created the concentrations of stone tools and animal bones at Olduvai and other sites will doubtless continue for many years. There simply is no definitive way, at this point, to differentiate between marks made on bones by humans from those made by animals, or to determine in what condition (alive or dead) the animals were when they were butchered by humans.

When we look at the tools themselves, we see that they include quite a diversity of forms. These tools certainly are adequate for butchering animals, whether hunted or scavenged. The absence of "spear-points" or anything resembling them may suggest that these early humans had no means of killing large animals, but some of the tools may have been used as spear heads, or they could have used wooden spears—or, as noted above, they may not have hunted.

Early Hominids in Asia

Accepted ideas about the origins of hominids and the distribution of our ancestors in their earliest forms have been challenged by discoveries in Pakistan of stone tools in strata that are two million years old. The tools are exceptionally crude quartzite cobbles, but the discoverers of these objects consider the evidence "irrefutable" that they are tools, and they conclude that among the possible explanations of their find are that "(a) *Homo habilis* was distributed as far east as Pakistan, (b) *H. erectus* is an Asian lineage at least as old as *H. habilis,* or (c) another, as yet unidentified tool maker was responsible. One clear implication of this discovery, however, is that early tool making has an Asian as well as [sic] African dimension."[93]

Even if these Pakistani dates are not accurate, it is clear that early hominids were able to inhabit a great diversity of environments—they were a successful genus, expanding into new niches. The most conservative view and probably the most accurate one is that hominids evolved bipedalism, then tool-using and other cultural behavior in Africa, and then spread out along the warmer margins of the Old World in a long slow process of population growth and very gradual movements, as groups split and moved short distances away from one another, until hominids could be found from southern Africa to Java.

Finds from Pakistan, China, and elsewhere make it probable that our ancestors were distributed across the warmer regions of the world by at least 1.5 million years ago. The lowered seas of the Pleistocene would have facilitated this, opening rich coastal niches far out into the southeast Asian archipelagos.[94] But there are many problems with this simple reconstruction. If, for example, *Homo erectus* first appeared in Africa about 1.7 to 1.5 million years ago,[95] then any older dates for *H. erectus* from Asia—or anywhere else—would cause "considerable upheavals."[96]

Some scholars believe that this evidence already exists. Orchiston and Siesser date *H. erectus* in Java to 1.75 to 1.1 million years ago.[97] But Pope interprets the same evidence as indicating that all hominids in Java are younger than a million years.[98]

One of the most frustrating things about the record of *Homo erectus* in south Asia is that not a single fossil of this hominid has been found there in good association with stone tools.[99] A seemingly reliable date for artifacts in east Asia is provided by potassium-argon determinations from northern Thailand of about 800,000 to 600,000 years ago for a basalt layer overlying several stone cores.[100] There is also some rather questionable evidence that hominids had reached the frigid interior of north-central Asia by 700,000 years ago.[101]

In short, it remains at least possible that Asia was much more a scene of early hominid evolution than has been previously thought.

Summary and Conclusions

We are but "pulvis et umbra," Horace said, dust and shadow, and certainly this is true for the long line of fossils that connect us to our beginnings. Nor are the factors that have shaped us entirely clear. Climate changes, savanna adaptations, scavenging, hunting, tool-use—all these and many other factors seem to have been ingredients in our evolutionary

3.14   The increase in cranial capacity over time in our ancestors was enormous and rapid by evolutionary standards, but the rate of increase has slowed in the last 100,000 years, and we may have reached the limits allowed by the mechanics of human birth and locomotion.

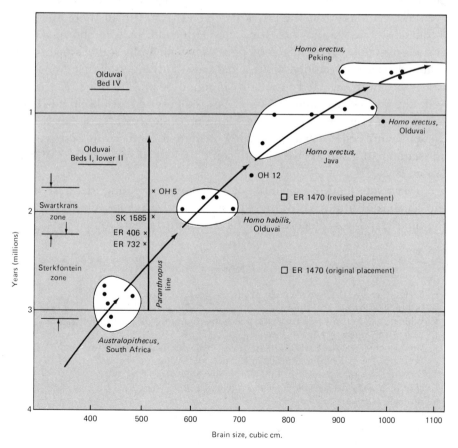

experiment, but the evidence is far too slight to identify the most important forces and circumstances in our past.

Even if we fight through all these issues of causation and taxonomy and eventually reach a consensus, we will be left with an equally important and profound question—the nature of the mode and tempo of human physical and cultural evolution. The question is this: Does evolution in such things as human cranial capacity, technological efficiency, administrative centralization, urbanization, and so forth proceed gradually, with approximately the same rate of change, or are there long periods of stability punctuated by periods of rapid change? As Robert McC. Adams phrased it in the context of early states, does the simile of a "ramp" or of a series of "steps" best describe our evolutionary history? [102]

This simple question involves some fairly murky political issues. Some fear that to consider human behavior as largely an adaptation and the result of natural selection means that human behavior is beyond social change, that aggression, racism, etc. may be genetically inherent. Some people think the Marxian concept of dialectical development means that even in such matters as the physical changes from the australopithecines to ourselves, we should look for a developmental pattern showing long periods of stability punctuated by periods in which physical characteristics changed relatively rapidly. Thus, Gould and Eldredge say that "no gradualism has been detected within any hominid taxon." [103] But J. E. Cronin and others who have analyzed the measurements on fossil hominids from four million years ago to ourselves found no evidence of "stasis" or "punctuation," and believe that the record of human evolution is best seen as an example of gradual change with some periods of varying rates of evolution. [104]

To have any kind of conclusive test of these competing ideas, we need the remains of hundreds of early hominids, at least some of which fall in the time "gaps" for which we now have no information.

Notes

1. Ferris, *Coming of Age in the Milky Way.*
2. White, *The Science of Culture.*
3. Wheatley, "Cultural Behavior and Extractive Foraging in *Macaca fascicularis.*"
4. Brace, *The Stages of Human Evolution,* p. 51.
5. Von Koenigswald, "Early Man in Java." See also Harrison, "Present Status and Problems for Paleolithic Studies in Borneo and Adjacent Islands," pp. 54–55.
6. Bowen, *Quaternary Geology;* Flint, *Glacial and Quaternary Geology.*
7. Harris, *Human Ecology and Savanna Environments.*
8. Birdsell, *Human Evolution.*
9. Ibid.
10. Cohen, "Diet and Cancer."
11. Lewontin, "Sociobiology as an Adaptationist Program."

12. Hamilton, "Revising Evolutionary Narratives: A Consideration of Alternative Assumptions about Sexual Selection and Competition for Mates."
13. Tooby and DeVore, "The Reconstruction of Hominid Behavioral Evolution," p. 194.
14. Krebs and Davies, *Introduction to Behavioral Ecology.*
15. Jerison, "The Evolution of the Mammalian Brain," pp. 142–43.
16. Richard Alexander, quoted in Lewin, "Debate over Emergence of Human Tooth Pattern," p. 668.
17. Alexander, quoted in Lewin, "Debate over Emergence of Human Tooth Pattern."
18. See Jolley, "The Seed-Eaters: A New Model of Hominid Differentiation Based on a Baboon Analogy."
19. See Wolpoff, "*Ramapithecus* and Hominid Origins"; Sarich "Comment on M. Wolpoff, '*Ramapithecus* and Hominid Origins' "; Zihlman and Lowenstein, "Comment on M. Wolpoff, '*Ramapithecus* and Hominid Origins.' "
20. Zihlman and Lowenstein, "Comment on M. Wolpoff, '*Ramapithecus* and Hominid Origins.' "
21. Wolpoff, "*Ramapithecus* and Hominid Origins," pp. 508–9; see also Kurland and Beckerman, "Optimal Foraging and Hominid Evolution: Labor and Reciprocity."
22. Diamond, "DNA-based Phylogenies of the Three Chimpanzees."
23. Tooby and DeVore, "The Reconstruction of Hominid Behavioral Evolution," p. 205.
24. Pilbeam and Gould, "Size and Scaling in Human Evolution."
25. Johanson and Edey, *Lucy: The Beginnings of Humankind.*
26. Tooby and DeVore, "The Reconstruction of Hominid Behavioral Evolution," p. 203.
27. Ibid., pp. 208–9.
28. Teleki, "The Omnivorous Diet and Eclectic Feeding Habits of Chimpanzees in Gombe National Park, Tanzania," p. 339.
29. Napier and Napier, *A Handbook of Living Primates.*
30. Crook, "Sexual Selection, Dimorphism, and Social Organization in Primates," p. 235.
31. Dunbar, *Primate Social Systems;* Kinzey, *The Evolution of Human Behavior: Primate Models.*
32. Pianka, *Evolutionary Ecology.*
33. Crook, "Sexual Selection, Dimorphism, and Social Organization in Primates."
34. Anderson, "The Reproductive Role of the Human Breast."
35. Ibid.
36. Frisch, "Comment on Anderson, P. 'The Reproductive Role of the Human Breast,' " p. 32.
37. Itzkoff, *Why Humans Vary in Intelligence.*
38. Crook, "Sexual Selection, Dimorphism, and Social Organization in Primates," p. 254.
39. Small, "Female Primate Sexual Behavior and Conception," p. 81; idem., *Female Primates. Studies by Women Primatologists.*
40. McArthur and Clark, "Body Temperature and Heat and Water Balance."
41. Strum and Mitchell, "Baboon Models and Muddles."
42. Kinzey, *The Evolution of Human Behavior: Primate Models,* p. xiii.
43. Goodall, *The Chimpanzees of Gombe.*
44. van Lawick-Goodall, "Some Aspects of Aggressive Behavior in a Group of Free-living Chimpanzees."
45. Van den Berghe, "Sex Differentiation and Infant Care: A Rejoinder to Sharlotte Neely Williams," p. 772.
46. van Lawick-Goodall, "Some Aspects of Aggressive Behavior in a Group of Free-living Chimpanzees."
47. van Lawick-Goodall, "The Behavior of Free-living·Chimpanzees in the Gombe Stream

Area"; idem., "Some Aspects of Aggressive Behavior in a Group of Free-living Chimpanzees"; idem., "The Behavior of Chimpanzees in Their Natural Habitat."

48. Tanner, "The Chimpanzee Model Revisited and the Gathering Hypothesis," p. 20.

49. Tanner, "The Chimpanzee Model Revisited and the Gathering Hypothesis," citing Boesch and Boesch, "Sex Differences in the Use of Natural Hammers," and Boesch and Boesch, "Possible Causes of Sex Differences in the Use of Natural Hammers by Wild Chimpanzees."

50. Teleki, "The Omnivorous Chimpanzee."

51. Ibid.

52. Symons, *The Evolution of Human Sexuality,* pp. 96–141.

53. For a somewhat different perspective, see Martin and Voorhies, *The Female of the Species.*

54. Teleki, "The Omnivorous Diet and Eclectic Feeding Habits of Chimpanzees in Gombe National Park, Tanzania."

55. Lovejoy, "Hominid Origins: The Role of Bipedalism."

56. Kinzey, *The Evolution of Human Behavior: Primate Models,* p. x.

57. Tooby and DeVore, "The Reconstruction of Hominid Behavioral Evolution," pp. 215–17.

58. Symons, *The Evolution of Human Sexuality.*

59. Ibid., pp. 27–28.

60. Sahlins, *Stone Age Economics,* p. xxx.

61. Tooby and DeVore, "The Reconstruction of Hominid Behavioral Evolution," p. 223.

62. Ibid., pp. 223–24.

63. Ibid., p. 224.

64. Ibid., p. 225.

65. Ibid.

66. Ibid., p. 226.

67. Bower, "Retooled Ancestors."

68. Reviewed in Clark, "Some Thoughts on the Black Skull."

69. Skelton et al., "Phylogenetic Analysis of Early Hominids."

70. Stringer, "Middle Pleistocene Hominid Variability"; see also Clark, "Some Thoughts on the Black Skull."

71. Eckhardt and Piermarini, "More on Phylogenetic Analysis of Early Hominids."

72. Stringer, "Middle Pleistocene Hominid Variability."

73. Brown, Harris, and Walker, "Early *Homo erectus* Skeleton."

74. Swedlund, "The Use of Ecological Hypotheses in Australopithecine Taxonomy."

75. Brace, "Biological Parameters and Pleistocene Hominid Lifeways."

76. Toth, "The Oldowan Reconsidered."

77. Binford, "Human Ancestors."

78. Lovejoy, "Models of Human Evolution."

79. Schaller and Lowther, "The Relevance of Carnivore Behavior to the Study of Early Hominids."

80. Blumenschine, "Characteristics of the Early Hominid Scavenging Niche."

81. O'Connell et al., "Hadza Scavenging: Implications for Plio/Pleistocene Hominid Subsistence."

82. Birdsell, "Ecological Influences on Australian Aboriginal Social Organization."

83. Anderson, "The Reproductive Role of the Human Breast," p. 31.

84. Townsend, "New Guinea Sago Gatherers: A Study of Demography in Relation to Subsistence."

85. Binford, "Fact and Fiction about the Zinjanthropus Floor: Data, Arguments, and Interpretations"; Bunn and Kroll, "Reply to L. Binford 'Fact and Fiction about the Zinjanthropus Floor: Data, Arguments, and Interpretations' "; Zeleznik, Grele, Pollack, and Aloni, "On Systematic Butchery by Plio/Pleistocene Hominids."

86. Binford, "Fact and Fiction about the Zinjanthropus Floor: Data, Arguments, and Interpretations," p. 135.
87. Bunn and Kroll, "Reply to L. Binford 'Fact and Fiction about the Zinjanthropus Floor: Data, Arguments, and Interpretations.'"
88. Potts, "On Butchery by Olduvai Hominids."
89. Reviewed in Isaac, "The Archaeology of Human Origins: Studies of the Lower Pleistocene in East Africa 1971–1981."
90. Tooby, "Comment on *Characteristics of the Early Hominid Scavenging Niche*, by R. J. Blumenschine."
91. Blumenschine, "Characteristics of the Early Hominid Scavenging Niche."
92. Tooby, "Comment on *Characteristics of the Early Hominid Scavenging Niche*, by R. J. Blumenschine," pp. 399–400.
93. Dennell et al., "Late Pliocene Artefacts from Northern Pakistan," p. 498; idem., "Early Tool-Making in Asia: Two Million Year Old Artefacts in Pakistan."
94. Bellwood, "The Prehistory of Island Southeast Asia: A Multidisciplinary Review of Recent Research," Fig. 1.
95. Rightmire, "*Homo erectus* and Human Evolution in the African Middle Pleistocene."
96. Bellwood, "The Prehistory of Island Southeast Asia: A Multidisciplinary Review of Recent Research," p. 183.
97. Orchiston and Siesser, "Chronostratigraphy of the Pleistocene Fossil Hominids of Java."
98. Pope, "Evidence on the Age of the Asian Hominidae"; see also Bellwood, "The Prehistory of Island Southeast Asia: A Multidisciplinary Review of Recent Research," p. 183.
99. Bellwood, *The Prehistory of the Indo-Malaysian Archipelago.*
100. Pope et al., "Earliest Radiometrically Dated Artifacts from Southeast Asia."
101. Larichev et al., "Lower and Middle Paleolithic of Northern Asia: Achievements, Problems, and Perspectives."
102. Adams, *The Evolution of Urban Society.*
103. Gould and Eldredge, "Punctuated Equilibria: The Tempo and Mode of Evolution Reconsidered."
104. Cronin et al., "Tempo and Mode in Hominid Evolution."

Bibliography

Adams, R. McC. 1966. *The Evolution of Urban Society.* Chicago: Aldine.
Anati, E. and N. Haas. 1967. "The Hazorea Pleistocene Site: A Preliminary Report." *Man* 2:454–56.
Anderson, P. 1983. "The Reproductive Role of the Human Breast." *Current Anthropology* 24(1):25–45.
Andrews, P. J. 1984. "The Descent of Man." *New Scientist* 1408:24–25.
Arambourg, C. 1967. "Le deuxime mission scientifique de l'Omo." *L'Anthropologie* 71:562–66.
Ardrey, R. 1961. *African Genesis.* New York: Atheneum.
Arens, W. 1979. *The Man-Eating Myth.* New York: Oxford University Press.
Barnard, A. 1983. "Contemporary Hunter-Gatherers: Current Theoretical Issues in Ecology and Social Organization." *Annual Review of Anthropology* 12:193–214.
Bar-Yosef, O. 1975. "Archeological Occurrences in the Middle Pleistocene of Israel." In *After the Australopithecines*, eds. K. W. Butzer and G. Isaac. The Hague: Mouton.
Bellwood, P. 1986. *The Prehistory of the Indo-Malaysian Archipelago.* Orlando, Fla.: Academic Press.

———. 1987. "The Prehistory of Island Southeast Asia: A Multidisciplinary Review of Recent Research." *Journal of World Prehistory* 1(2):171–224.

Binford, L. R. 1977. "A Review of *Olorgesailie: Archaeological Studies of a Middle Pleisto-cene Lake Basin in Kenya*, by G. L. Isaac." *Journal of Anthropological Research* 33(4):493–502.

———. 1981. *Bones: Ancient Men and Modern Myth.* New York: Academic Press.

———. 1985. "Human Ancestors: Changing Views of Their Behavior." *Journal of Anthropological Archaeology* 4(4):292–327.

———. 1988. "Fact and Fiction about the Zinjanthropus Floor: Data, Arguments, and Interpretations." *Current Anthropology* 29(1):123–35.

Birdsell, J. B. 1972. *Human Evolution.* Chicago: Rand McNally.

———. 1979. "Ecological Influences on Australian Aboriginal Social Organization." In *Primate Ecology and Human Origins: Ecological Influences and Social Organization,* eds. I. S. Bernstein and E. O. Smith. New York: Garland.

Blumenschine, R. J. 1987. "Characteristics of the Early Hominid Scavenging Niche," *Current Anthropology* 28(4):383–407.

Boaz, N. T. 1979. "Hominid Evolution in Eastern Africa during the Pliocene and Early Pleistocene." *Annual Review of Anthropology* 8:71–85.

Boesch, C. and H. Boesch. 1981. "Sex Differences in the Use of Natural Hammers by Wild Chimpanzees: A Preliminary Report." *Journal of Human Evolution* 10:565–83.

———. 1984. "Possible Causes of Sex Differences in the Use of Natural Hammers by Wild Chimpanzees." *Journal of Human Evolution* 13:415–40.

Bowen, D. Q. 1978. *Quaternary Geology.* Oxford: Oxford University Press.

Bower, B. 1988. "Retooled Ancestors." *Science News* 133:345.

Brace, C. L. 1967. *The Stages of Human Evolution: Human and Cultural Origins.* Engle-wood Cliffs, N.J.: Prentice-Hall.

———. 1979. "Biological Parameters and Pleistocene Hominid Life-ways." In *Primate Ecology and Human Origins: Ecological Influences and Social Organization,* eds. I. S. Bernstein and E. O. Smith. New York: Garland.

———. 1981. "Tales of the Phylogenetic Woods: The Evolution and Significance of Evolutionary Trees." *American Journal of Physical Anthropology* 56:411–29.

Bromage, T. G. and M. C. Dean. 1985. "Re-evaluation of the Age at Death of Imma-ture Fossil Hominids." *Nature* 317:525.

Brown, F., J. Harris, and A. Walker. 1985. "Early *Homo erectus* Skeleton from West Lake Turkana, Kenya." *Nature* 316:788–92.

Bunn, H. T. 1981. "Archaeological Evidence for Meat-Eating by Plio-Pleistocene Hom-inids from Koobi Fora and Olduvai Gorge." *Nature* 291:574–77.

Bunn, H. T. and E. M. Kroll. 1988. "Reply to L. Binford 'Fact and Fiction about the Zinjanthropus Floor: Data, Arguments, and Interpretations.'" *Current Anthropology* 29(1):135–49.

Butzer, K. W. 1975. "Geological and Ecological Perspectives on the Middle Pleisto-cene." In *After the Australopithecines,* eds. K. W. Butzer and G. Isaac. The Hague: Mouton.

———. 1982. "The Paleo-ecology of the African Continent: the Physical Environments of Africa from the Earliest Geological to Later Stone Age Times." In *The Cam-bridge History of Africa,* Vol. 1, ed. J. D. Clark. Cambridge, England: Cambridge University Press.

Cachel, S. 1975. "A New View of Speciation." In *Australopithecus Paleoanthropology: Mor-phology and Paleoecology,* ed. R. H. Tuttle. The Hague: Mouton.

Cashdan, E. 1983. "Territoriality among Hunter Foragers: Ecological Models and an Application to Four Bushman Groups." *Current Anthropology* 24(1):47–66.

Clark, G. A. 1988. "Some Thoughts on the Black Skull: An Archaeologist's Assessment of WT-17000 *(A. boisei)* and Systematics in Human Paleontology." *American An-thropologist* 90(2):357–71.

Cohen, L. A. 1987. "Diet and Cancer." *Scientific American* 257(5):42–48.

Cronin, J. E., N. T. Boaz, C. B. Stringer, and Y. Rak. 1981. "Tempo and Mode in Hominid Evolution." *Nature* 292:113–22.

Crook, J. H. 1972. "Sexual Selection, Dimorphism, and Social Organization in Primates." In *Sexual Selection and the Descent of Man 1871–1971*, ed. B. Campbell. Chicago: Aldine.

Deacon, H. J. 1975. "Demography, Subsistence, and Culture During the Acheulian in Southern Africa." In *After the Australopithecines*, eds. K. W. Butzer and G. Isaac. The Hague: Mouton.

Delson, E., ed. 1985. *Ancestors. The Hard Evidence*. New York: A. R. Liss.

Dennell, H. M., H. M. Rendell, and A. E. Hailwood. 1988a. "Late Pliocene Artefacts from Northern Pakistan." *Current Anthropology* 29(3):495–98.

———. 1988b. "Early Tool-Making in Asia: Two Million Year Old Artefacts in Pakistan." *Antiquity* 62:98–106.

Diamond, J. M. 1988. "DNA-based Phylogenies of the Three Chimpanzees." *Nature* 332:685–86.

Dunbar, R. I. 1988. *Primate Social Systems*. Croom Helm: Cornell University Press.

Eckhardt, R. B. 1988. "On Early Hominid Adaptation and Heat Stress." *Current Anthropology* 29(3):493.

Eckhardt, R. B. and A. L. Piermarini. 1988. "More on Phylogenetic Analysis of Early Hominids." *Current Anthropology* 29(3):493–94.

Ferris, T. 1988. *Coming of Age in the Milky Way*. New York: William Morrow.

Fialkowski, K. 1986. "A Mechanism for the Origin of the Human Brain: A Hypothesis." *Current Anthropology* 27(3):288–90.

Finkel, D. J. 1981. "An Analysis of Australopithecine Dentition." *American Journal of Physical Anthropology* 55:69–80.

Fisher, E. 1979. *Woman's Creation*. New York: McGraw-Hill.

Flint, R. F. 1971. *Glacial and Quaternary Geology*. New York: Wiley.

Foley, R. 1987. "Hominid Species and Stone-tool Assemblages: How Are They Related?" *Antiquity* 61:380–92.

Freeman, L. G. 1975. "Acheulian Sites and Stratigraphy in Iberia and the Meghreb." In *After the Australopithecines*, eds. K. W. Butzer and G. Isaac. The Hague: Mouton.

Frisch, R. E. 1983. "Comment on Anderson, P. 'The Reproductive Role of the Human Breast.' " *Current Anthropology* 24(1):25–45.

Garn, S. and W. Block. 1970. "The Limited Nutritional Value of Cannibalism." *American Anthropologist* 72:106.

Goodall, J. 1986. (Also see van Lawick-Goodall.) *The Chimpanzees of Gombe*. Cambridge, Mass.: Belknap (Harvard University Press).

Gould, S. J. 1977. *Ever Since Darwin*. New York: Norton.

Gould, S. J. and N. Eldredge. 1977. "Punctuated Equilibria: The Tempo and Mode of Evolution Reconsidered." *Paleobiology* 3:115–51.

Grimaud, D. 1980. "Les Paritaux de l'Homo erectus comparison avec ceux des Pithecanthropes de Java." *Bulletins et Memoires de la Societe XIIIe Serie no 3.*

Hamilton, M. E. 1984. "Revising Evolutionary Narratives: A Consideration of Alternative Assumptions about Sexual Selection and Competition for Mates." *American Anthropologist* 86(3):651–62.

Harding, R. S. and G. Teleki, eds. 1981. *Omnivorous Primates*. New York: Columbia University Press.

Harris, D. R., ed. 1980. *Human Ecology and Savanna Environments*. New York: Academic Press.

Harrison, T. 1978. "Present Status and Problems for Paleolithic Studies in Borneo and Adjacent Islands." In *Early Paleolithic in South and East Asia*, ed. F. Ikawa-Smith. The Hague: Mouton.

Hill, J. H. 1978. "Apes and Language." *Annual Review of Anthropology* 7:89–112.

Hinterberger, J. 1988. "Crude Food." *Seattle Times/Seattle Post-Intelligencer*, February 28, 1988, "Pacific" section, p. 4.

Howell, F. C. 1982. "Origins and Evolution of the African Hominidae." In *The Cambridge History of Africa*, Vol. 1, ed. J. D. Clark. Cambridge, England: Cambridge University Press.

Howells, W. 1973. *Evolution of the Genus Homo*. Reading, Mass.: Addison-Wesley.

Hughes, A. R. and P. V. Tobias. 1977. "A Fossil Skull Probably of the Genus Homo from Sterkfontein, Transvaal." *Nature* 265:310–12.

Ikawa-Smith, F., ed. 1978. *Early Paleolithic in South and East Asia*. The Hague: Mouton.

Ingold, T. 1988. *The Appropriation of Nature*. Iowa City: University of Iowa Press.

Isaac, G. Ll. 1975. "Sorting Out the Muddle in the Middle: An Anthropologist's Post-Conference Appraisal." In *After the Australopithecines*, eds. K. W. Butzer and G. Isaac. The Hague: Mouton.

———. 1977. *Olorgesailie: Archaeological Studies of a Middle Pleistocene Lake Basin in Kenya*. Chicago: University of Chicago Press.

———. 1978. "Food-Sharing and Human Evolution: Archaeological Evidence from the Pre-Pleistocene of East Africa." *Journal of Anthropological Research* 34:311–25.

———. 1982. "The Earliest Archaeological Traces." In *The Cambridge History of Africa*, Vol. 1, ed. J. D. Clark. Cambridge, England: Cambridge University Press.

———. 1984. "The Archaeology of Human Origins: Studies of the Lower Pleistocene in East Africa 1971–1981." *Advances in World Archaeology* 3:1–87.

Itzkoff, S. 1987. *Why Humans Vary in Intelligence*. Ashfield, Mass.: Paideia.

Janzen, D. 1976. "Why Bamboos Wait So Long to Flower." *Annual Review of Ecology and Systematics* 7:347–91.

Jelinek, A. J. 1977. "The Lower Paleolithic: Current Evidence and Interpretations." *Annual Review of Anthropology* 6:11–32.

Jerison, H. J. 1973. *Evolution of the Brain and Intelligence*. New York: Academic Press.

———. 1983. "The Evolution of the Mammalian Brain as an Information-Processing System." *Advances in the Study of Mammalian Behavior*, eds., J. F. Eisenberg and D. G. Kleinman, Special Publication No. 7, The American Society of Mammologists.

Johanson, D. C. and M. A. Edey. 1981. *Lucy: The Beginnings of Humankind*. New York: Simon and Schuster.

Johanson, D. C. and T. D. White. 1979. "A Systematic Assessment of Early African Hominids." *Science* 203:321–29.

Jolley, C. J. 1970. "The Seed-Eaters: A New Model of Hominid Differentiation Based on a Baboon Analogy." *Man* 5:6–26.

Kinzey, W. G., ed. 1987. *The Evolution of Human Behavior: Primate Models*. Albany: State University of New York Press.

Klein, R. G. 1977. "The Ecology of Early Man in Southern Africa." *Science* 197:115–26.

———. 1989. *The Human Career*. Chicago: University of Chicago Press.

Kranz, G. 1975. "The Double Descent of Man." In *Australopithecus Paleoanthropology: Morphology and Paleoecology*, ed. R. H. Tuttle. The Hague: Mouton.

Krebs, J. and N. B. Davies. 1981. *Introduction to Behavioral Ecology*. Oxford: Blackwell.

Kurland, J. A. and S. J. Beckerman. 1985. "Optimal Foraging and Hominid Evolution: Labor and Reciprocity." *American Anthropologist* 87(1):73–93.

Larichev, V., U. Khol'ushkin, and I. Laricheva. 1987. "Lower and Middle Paleolithic of Northern Asia: Achievements, Problems, and Perspectives." *Journal of World Prehistory* 1(4):415–64.

Lawick-Goodall, J. van. 1968. "The Behavior of Free-living Chimpanzees in the Gombe Stream Area." *Animal Behavior Monographs* 1:161–311.

———. 1971. "Some Aspects of Aggressive Behavior in a Group of Free-living Chimpanzees." *International Social Science Journal* 23:89–97.

———. 1973. "The Behavior of Chimpanzees in Their Natural Habitat." *American Journal of Psychiatry* 130:1–12.

Leakey, M. G. and R. E. Leakey, eds. 1978. *Koobi Fora Research Project: Vol. 1, The Fossil Hominids and an Introduction to Their Context, 1968–1974.* Oxford: Clarendon Press.

Leakey, R. E. and R. Lewin. 1977. *Origins.* New York: Dutton.

Lewin, R. 1987. "Debate over Emergence of Human Tooth Pattern." *Nature* 235:748–50.

Lewontin, R. C. 1979. "Sociobiology as an Adaptationist Program." *Behavioral Science* 24:5–14.

Lovejoy, O. 1980. "Hominid Origins: The Role of Bipedalism." *American Journal of Physical Anthropology* 52:250.

———. 1982. "Models of Human Evolution." *Science* 217:304–5.

Lumley, H. de. 1969. "A Paleolithic Camp at Nice." *Scientific American* 225:42–59.

———. 1975. "Cultural Evolution in France in its Paleoecological Setting During the Middle Pleistocene." In *After the Australopithecines,* eds. K. W. Butzer and G. Isaac. The Hague: Mouton.

McArthur, A. J. and J. A. Clark. 1987. "Body Temperature and Heat and Water Balance." *Nature* 326:647–48.

Martin, K. and B. Voorhies. 1975. *The Female of the Species.* New York: Columbia University Press.

Morris, D. 1967. *The Naked Ape.* London: Jonathan Cape.

Napier, J. R. and P. H. Napier. 1967. *A Handbook of Living Primates.* London: Academic Press.

Newsweek, "Tracking the Sasquatch," September 21, 1987, pp. 71–73.

O'Connell, J. F., K. Hawkes, and N. B. Jones. 1988. "Hadza Scavenging: Implications for Plio/Pleistocene Hominid Subsistence." *Current Anthropology* 29(2):356–63.

Okladnikov, A. P. and G. A. Pospelova. 1982. "Ulalinka, the Oldest Palaeolithic Site in Siberia." *Current Anthropology* 23(6):710–12.

Orchiston, D. W. and W. G. Siesser. 1982. "Chronostratigraphy of the Pleistocene Fossil Hominids of Java." *Modern Quaternary Research in Southeast Asia* 1:131–50.

Penny, D., L. R. Foulds, and M. D. Hendy. 1982. "Testing the Theory of Evolution by Comparing Phylogenetic Trees Constructed from Five Different Protein Sequences." *Nature* 297:197–200.

Pfeiffer, J. E. 1978. *The Emergence of Man.* 3rd ed. New York: Harper & Row.

Pianka, E. 1974. *Evolutionary Ecology.* New York: Harper & Row.

Pilbeam, D. 1975. "Middle Pleistocene Hominids." In *After the Australopithecines,* eds. K. W. Butzer and G. Isaac. The Hague: Mouton.

———. 1982. "New Hominid Skull Material from the Miocene of Pakistan." *Nature* 295:232–34.

Pilbeam, D. and S. J. Gould. 1974. "Size and Scaling in Human Evolution." *Science* 186:892–901.

Poirier, F. 1973. *Fossil Man: An Evolutionary Journey.* St. Louis: Mosby.

Pope, G. G. 1983. "Evidence on the Age of the Asian Hominidae." *Proceedings of the National Academy of Sciences USA* 80:4988–92.

Pope, G. G., S. Barr, A. Macdonald, and S. Nakabanlang. 1986. "Earliest Radiometrically Dated Artifacts from Southeast Asia." *Current Anthropology* 27(3):275–79.

Potts, R. 1987. "On Butchery by Olduvai Hominids." *Current Anthropology* 28(1):95–96.

Rightmire, G. P. 1980. "*Homo erectus* and Human Evolution in the African Middle Pleistocene." In *Current Argument on Early Man,* ed. L. K. Konigsson. Oxford: Pergamon.

Sagan, C. 1977. *The Dragons of Eden. Speculations on the Evolution of Human Intelligence.* New York: Ballantine Books.

Salt, G. W., ed. 1984. *Ecology and Evolutionary Biology.* Chicago: University of Chicago Press.

Sahlins, M. 1972. *Stone Age Economics*. Chicago: Aldine.

Sarich, V. 1982. "Comment on M. Wolpoff, '*Ramapithecus* and Hominid Origins.' " *Current Anthropology* 23(5):501–22.

Schaller, G. B. and G. R. Lowther. 1969. "The Relevance of Carnivore Behavior to the Study of Early Hominids." *Southwestern Journal of Anthropology* 25:307–41.

Schwalbe, G. 1906. *Studien zur Vorgeschichte des Menschen*. Stuttgart: Scheizerbart.

Schrire, C., ed. 1984. *Past and Present in Hunter-Gatherer Studies*. New York: Academic Press.

Sevink, J., E. H. Hebeda, H. N. Priem, and R. H. Verschure. 1981. "A Note on the Approximately 730,000-Year-Old Mammal Fauna and Associated Human Activity Sites Near Isernia, Central Italy." *Journal of Archaeological Science* 8:105–6.

Shapiro, H. L. 1974. *Peking Man*. New York: Simon and Schuster.

Simons, E. B. and D. R. Pilbeam. 1965. "Preliminary Revision of the Dryopithecinae (Pongidae, Anthropoidea)." *Folia Primatologica* 3:81–152.

Skelton, R. R., H. M. McHenry, and G. M. Drawhorn. 1986. "Phylogenetic Analysis of Early Hominids." *Current Anthropology* 27(1):21–43.

Small, M. F. 1988. "Female Primate Sexual Behavior and Conception." *Current Anthropology* 29(1):81–100.

Small, M. F. ed. 1984. *Female Primates. Studies by Women Primatologists*. New York: A. R. Liss.

Smith, F. H. and G. C. Ranyard. 1981. "Evolution of the Supraorbital Region in Upper Pleistocene Fossil Hominids from South-central Europe." *American Journal of Physical Anthropology* 53:589–610.

Smith, F. H. and F. Spencer, eds. 1984. *The Origins of Modern Humans*. New York: A. R. Liss.

Smith, P. E. L. 1982. "The Late Paleolithic and Epi-Paleolithic of Northern Africa." In *The Cambridge History of Africa*, Vol. 1, ed. J. D. Clark. Cambridge, England: Cambridge University Press.

Stringer, C. B. 1985. "Middle Pleistocene Hominid Variability and the Origin of Late Pleistocene Humans." In *Ancestors: The Hard Evidence*, ed. E. Delson. New York: Liss.

Strum, S. C. and W. Mitchell. 1987. "Baboon Models and Muddles." In *The Evolution of Human Behavior: Primate Models*, ed. W. G. Kinzey. Albany: State University of New York Press.

Speth, J. and D. Davis. 1976. "Seasonal Variability in Early Hominid Predation." *Science* 192:441–45.

Swedlund, A. 1974. "The Use of Ecological Hypotheses in Australopithecine Taxonomy." *American Anthropologist* 76:515–29.

Symons, D. 1979. *The Evolution of Human Sexuality*. New York: Oxford University Press.

Tanner, N. M. 1981. *On Becoming Human: A Model of the Transition from Ape into Human and the Reconstruction of Early Human Social Life*. Cambridge, England: Cambridge University Press.

———. 1987. "The Chimpanzee Model Revisited and the Gathering Hypothesis." In *The Evolution of Human Behavior*, ed. W. G. Kinzey. Albany: State University of New York Press.

Teleki, G. 1973. "The Omnivorous Chimpanzee." *Scientific American* 228:32–47.

———. 1981. "The Omnivorous Diet and Eclectic Feeding Habits of Chimpanzees in Gombe National Park, Tanzania." In *Omnivorous Primates*, eds. R. S. Harding and G. Teleki. New York: Columbia University Press.

Testart, A. 1982. "The Significance of Food Storage among Hunter-gatherers: Residence Patterns, Population Densities, and Social Inequalities." *Current Anthropology* 23(5):523–37.

Thorne, A. G. and M. Wolpoff. 1981. "Regional Continuity and Australasian Pleistocene Hominid Evolution." *American Journal of Physical Anthropology* 55:337–49.

Tobias, P. 1980. "A Study and Synthesis of the African Hominids of the Late Tertiary and Early Quaternary Periods." In *Current Arguments on Early Man,* ed. L. -K. Konigsson. Oxford: Pergamon Press.

Tooby, J. 1987. "Comment on *Characteristics of the Early Hominid Scavenging Niche,* by R. J. Blumenschine." *Current Anthropology* 28(4):383–407.

Tooby, J. and I. DeVore. 1987. "The Reconstruction of Hominid Behavioral Evolution through Strategic Modeling." In *The Evolution of Human Behavior: Primate Models,* ed. W. G. Kinzey. Albany: State University of New York Press.

Toth, N. 1985. "The Oldowan Reconsidered: A Close Look at Early Stone Artifacts." *Journal of Archaeological Science* 12:101–20.

Townsend, P. K. 1971. "New Guinea Sago Gatherers: A Study of Demography in Relation to Subsistence." *Ecology of Food and Nutrition* 1:19–24.

Van den Berghe, P. L. 1972. "Sex Differentiation and Infant Care: A Rejoinder to Sharlotte Neely Williams." *American Anthropologist* 74:770–72.

Vertes, L. 1965. "Typology of the Buda Industry. A Pebble-tool Industry from the Hungarian Lower Paleolithic." *Quaternaria* 7:185–95.

Von Koenigswald, G. H. R. 1975. "Early Man in Java: Catalogue and Problems." In *Australopithecus Paleoanthropology: Morphology and Paleoecology,* ed. R. H. Tuttle. The Hague: Mouton.

Washburn, S. L. 1968. "Discussion." In *Man the Hunter,* eds. R. Lee and I. DeVore. Chicago: Aldine.

Washburn, S. L. and R. L. Ciochon. 1974. "Canine Teeth: Notes on Controversies in the Study of Human Evolution." *American Anthropologist* 76:765–84.

Wheatley, B. P. 1988. "Cultural Behavior and Extractive Foraging in *Macaca fascicularis.*" *Current Anthropology* 29(3):516–19.

White, L. 1949. *The Science of Culture.* New York: Grove Press.

White, T., D. Johanson, and W. Kimbell. 1981. "*Australopithecus africanus:* Its Phyletic Position Reconsidered." *South African Journal of Science* 77:445–70.

Wolpoff, M. 1971. "Vertesszollos and the Presapiens Theory." *American Journal of Physical Anthropology* 35:209–16.

———. 1982. "*Ramapithecus* and Hominid Origins." *Current Anthropology* 23(5):501–22.

Wood, B. 1987. "Who Is the 'Real' *Homo habilis?*" *Nature* 327:187–88.

Woodburn, J. 1982. "Egalitarian Societies." *Man* (NS) 17:431–51.

Yellen, J. E. 1976. "Settlement Patterns of the !Kung: An Archaeological Perspective." In *Kalahari Hunter-Gatherers: Studies of the !Kung San and Their Neighbors,* eds. R. Lee and I. Devore. Cambridge, Mass.: Harvard University Press.

Zeleznik, S., A. W. Grele, J. Pollack, and I. Aloni. 1988. "On Systematic Butchery by Plio/Pleistocene Hominids." *Current Anthropology* 29(1):151–53.

Zihlman, A. and J. Lowenstein. 1982. "Comment on M. Wolpoff, '*Ramapithecus* and Hominid Origins.'" *Current Anthropology* 23(5):501–22.

The Origins of
Homo sapiens sapiens

Man is the missing link between anthropoid apes
and human beings.

Anonymous

American and Russian astronauts alike seem to have been forcefully impressed by our common humanity and fate as they looked down at the earth's green and blue beauty in the isolation of a colorless and seemingly lifeless universe. A somewhat similar perspective might be taken when we look back at our common human ancestry. Recent genetic research suggests that everyone in the world today may have had a common ancestor who lived just 140,000–200,000 years ago. And we can even look back at Lucy of three million years ago (see p. 98) and other early hominids of the first millions of years of our line with a recognition of their embryonic humanity. But it is in the crucial interval of two to one million years ago that our genus, *Homo*, becomes the dominant hominid, and only after about 400,000 years ago did humans appear whom we can relate to ourselves by calling them, too, *Homo sapiens*. We reserve the ultimate accolade of "people like us," in the sense of *Homo sapiens sapiens*, for only some of the humans who lived after 100,000 years ago, and it is not until about 30,000 years ago that we alone, *Homo sapiens sapiens*, constitute humanity.

If we truly want to follow the Socratic dictum to know ourselves, at least in the sense of ourselves as a biological species, we must examine these past two million years and try to understand what forces and factors produced us. As farmers and city-dwellers we are only a few thousand years old, so everything of importance that we are biologically is the result of the long thousands of millennia, the millions of years that we spent under the tutelage of natural selection, as we evolved into our modern form. We

136

are, simply put, largely what "worked" throughout our great antiquity. Natural selection is no master craftsman; there is much random variation and seemingly haphazard solutions to evolutionary problems. But we must suppose that the general trends of our evolution, the radically increased brain size and the reshaping of the human face, and the technological innovations of the past two million years are responses to the demands of the worlds in which our ancestors lived. We can then ask: What were these forces and factors that produced modern humanity?

Two general scenarios about our origins are currently being debated that form a framework in which research on questions about our origins is conducted. One notion is that about 1.5 million years ago a generic *Homo* ancestor of ours spread out across the warmer latitudes of Africa, the Middle East, Asia, and possibly the southernmost fringe of Europe. With the passage of the millennia these groups began to diverge somewhat as they adapted to local and different environments, but across the whole range of *Homo* they were evolving toward *Homo sapiens* because of gene flow that connected all human groups to some extent, and because they were under similar evolutionary selective forces, as generalized hunter-foragers, so that they all converged at about 30,000 years ago as one species, *Homo sapiens sapiens*—but with the differences that distinguish modern Europeans, Asians, and Africans from each other as souvenirs of their somewhat different routes to *Homo sapiens sapiens.* A contrasting scenario is that our *Homo* ancestors spread across the warmer ranges of the Old World after 1.5 million years ago and began to diverge considerably as they adapted to different environments and as distance and natural barriers (e.g., the Sahara Desert, the Himalayas, the Alps) restricted gene flow among various groups. Then, perhaps 140,000 years ago, *Homo sapiens* evolved in one place (probably Africa), and spread across the world, displacing most groups, driving some into extinction, and absorbing a small fraction of the others through intermarriage.

From *Homo* to *Homo erectus:* c. 2.0–.5 million years ago

For some decades, scholars have believed that groups of early *Homo* had colonized much of east, central, and south Africa by about 1.5 million years ago, and had begun to colonize the warmer regions of the Middle East and Asia at least by one million years ago, and probably long before. The discovery of what appear to be stone tools in Pakistan that seem to date to two million years ago would greatly change our views of early hominid evolution, but controversies about these Pakistani finds will probably continue for some years. What seems clear, however, is that by shortly

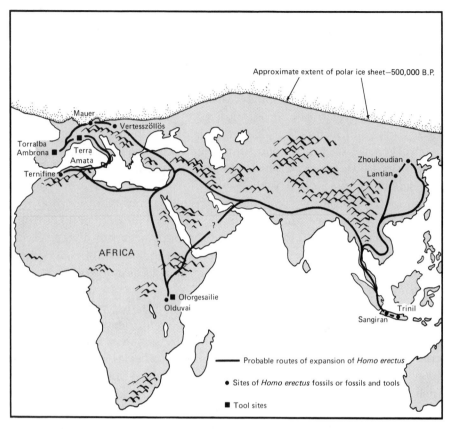

4.1   Distribution of *Homo erectus* sites. *Homo erectus* was the first hominid to invade temperate climates and may have been more widely distributed than is represented here. Land bridges, now submerged, facilitated *H. erectus* movements.

after a million years ago, our ancestors had colonized most of the Old World habitats that we, as the tropical animals all hominids are, would have preferred. For as tropical animals we could not possibly have survived northern winters without clothing, fire, and tools.

After about a million years ago, however, it is apparent that our ancestors were beginning to invade more northern environments. If there is any answer as to why they would begin to move out of the tropical environments they had evolved in, it is probably in the nature of the hunting-gathering band. Ethnographic studies suggest that human hunting-gathering bands with primitive technologies need vast areas to support themselves, and also that one response to growing populations is for bands to split and for "daughter" groups to establish themselves in open territories on the original group's boundary. A few hundred thousand years of this, coupled with slow evolution in the efficencies of tools would have populated the warmer ranges of the Old World.

At many times in the past million years the climate of northern Europe and Asia was much warmer than it is at present. So warm were areas of Germany and northern France about 800,000 or 900,000 years ago that monkeys swarmed through dense forests there. But at other times the icy fingers of glaciers drew down out of the Alps and other mountains, and great expanses of northern Europe were tundra. To invade the more northern climates, and to survive there during periods of expanding glaciers, would have required control of fire. Fire would have been vital, not just for keeping warm, but for cooking food and, perhaps more important, for evicting bears and other carnivores from caves and rock shelters—the safest and warmest places to live in northern latitudes.

The problems of living in northern climates go beyond simply keeping warm. In winter most of the plants suitable for human consumption die, and an animal like ourselves has only two ways to get food: store it, or hunt, fish, or scavenge other animals. On the African savannas and in other warm environments, women supply most of the food in hunting and gathering cultures, and even pregnant women, children, and the aged can gather much of their own food all year long. But such self-reliance would not have been possible in northern latitudes, where snow covered the ground for five to six months of the year.

Between about one million and 35,000 years ago, from the first occupation of northern areas until most of Europe and northern Asia were inhabited by humans, the versatility and efficiency of tools increased, and people became larger, bigger-brained, and "modern" in physical form.

It now appears unlikely that simply the migrations into temperate environments directly *caused* these physical and cultural changes; and by now it will not surprise the reader to learn that everything else about our ancestors' spread into temperate latitudes is also a matter of dispute, ranging from when they did it, to how they did it, to who did it.

Let us begin with the problem of which hominid actually did make the first forays into the Middle East and Asia, and then into the north. The most likely candidate would seem to be *H. erectus*. Alas, physical anthropologists have had a great crisis of faith with regard to *H. erectus:*

> Does *Homo erectus* exist as a true taxon or should it be sunk into *Homo sapiens*? Is it a palaeospecies that exists . . . as a segment of the line that emerged from *Homo habilis* and gave rise to *Homo sapiens*? Is *Homo erectus* an extinct form that had no part to play in the evolution of *Homo sapiens*? Is *Homo erectus* a good example of a 'statis event' . . . with little or no change in its form during its existence? Is there a clear cut example of *Homo erectus* in the European fossil record of man? Finally, are the Asian forms so far removed from the evolution of *Homo sapiens* in Africa to call into question the existence of *Homo erectus sensu stricto* in Africa at all?[1]

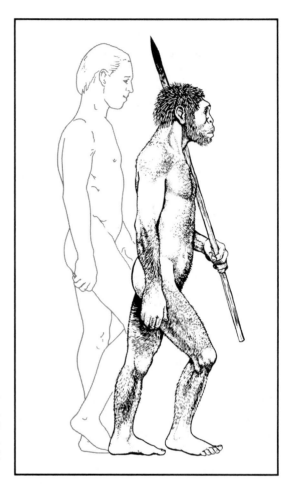

4.2 *Homo erectus,* shown beside modern man for scale, averaged between 1.5 and 1.8 m (5–6 ft) in height and weighed between 40 to 72.2 kg (88–160 lb).

Most of the fossils classified as *Homo erectus* come from Africa and Asia, although some anthropologists also group some European fossils in this class. *Homo erectus* individuals averaged just less than five feet tall (about 1.3 meters, although an African example may have been 1.6 meters),[2] walked fully erect (though its body and legs differed in various proportions and shapes from ours), and had a cranial capacity that ranged between about 700 and 1225 cubic centimeters.[3] Compared to us, these hominids had flat skulls, large brow ridges, thick cranial bones, and almost no chins (Figure 4.3).

It used to be assumed that *Homo erectus* was generally the same animal over most of its range, and that modern humans evolved out of the worldwide mix of these hominids. But as noted above, we must consider at least the possibility—which is supported by much recent research[4]—

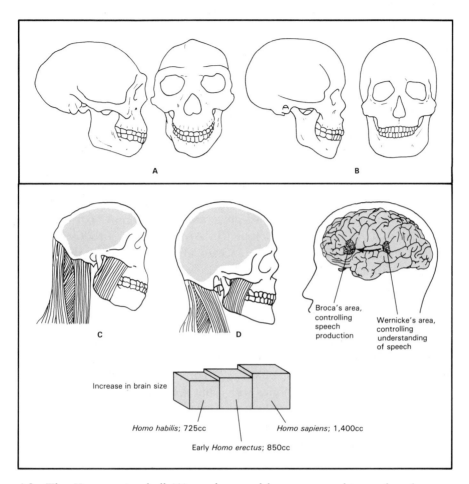

4.3   The *Homo erectus* skull (A) was long and low, compared to modern humans (B), with heavy brow ridges, no chin, and protruding jaws. In the transition from *Homo erectus* (C) to modern humans (D) the brain enlarged and the muscles controlling the head and neck were changed. One must presume that there were also changes in the internal structure of the brain, such as in the development of speech centers.

that at one time, perhaps a half a million years ago, the hominids living in, say, China, were quite different from those living in Europe and Africa, and that all these groups evolved in somewhat different ways and patterns. "According to this view the archaic sapiens forms of Europe and Asia, including the neanderthals, are local continuations of their regional populations of '*H. erectus*', with only the African forms giving rise to modern *H. sapiens*."[5]

There may be some evidence for these distinctive regional patterns of evolution in that the "hominid artifacts of the Pleistocene are patterned geographically and temporally in ways that mirror to some extent the variability in fossil hominids"[6] (Figure 4.4).

To sort out these patterns and construct explanations of what happened to our ancestors in the critical period from about a million years ago to the end of the Pleistocene, about 10,000 years ago, we must look at some of the many hundreds of sites and fossils that date to this period.

Some of the earliest *H. erectus* and their stone tools seem to have first appeared in Africa. At Olduvai Gorge, for example, Olduvai Hominid 9, from Upper Bed II, has been radiometrically dated to between one million and 900,000 years ago and seems to be a *Homo erectus*. Recent finds from West Turkana, in northern Kenya, may be even more convincing as evidence that *Homo erectus* developed in Africa. The WT15000 skeleton has been dated with some confidence to about 1.6 million years ago.[7] WT15000 appears to be the remains of a twelve-year-old boy who already may have been 1.64 meters (5′ 3 ⅕″) tall.

No stone tools were found with WT15000, but Acheulian-style (named after the French site at Acheul) stone tools associated with *Homo erectus* in Europe at a much later date (c. 600,000 years ago at the earliest) were apparently already present in many other East African sites more than a million years ago.[8] For example, at Koobi Fora, on Lake Turkana in Kenya, *Homo* remains and animal bones with what seem to be butchering marks have been found in substantial numbers, and there are small clusters of stone tools, some in concentrations about 5 to 10 meters in diameter with from about ten to one hundred artifacts.[9]

The hand-axes and other stone tools associated with the early African *Homo erectus* are a major improvement on the tools associated with the australopithecines of the previous million years, and this developing technology may have enabled hominids to radiate from warmer latitudes into colder, more demanding climates. Some think *Homo erectus* was primarily a hunter and that the demands of the hunting way of life partially "drove" human evolution toward ourselves; others doubt *H. erectus* did much more than scavenging and opportunistic hunting.

A critical site in this dispute is Olorgesailie, located near Nairobi, Kenya. Here are many small concentrations of stone tools and bones spread out along a peninsula in an extinct lake. Most of the tools are cleavers and hand-axes, and some show considerable chipping and blunting wear. Mixed in with them are bones from several species of large mammals, including a hippopotamus and, curiously, sixty-three individuals of an extinct species of baboon (but no hominids). There are no hearths or burned bones, but microscopic pieces of charcoal have been found all over the site. Whether these were the result of human or natural activity remains un-

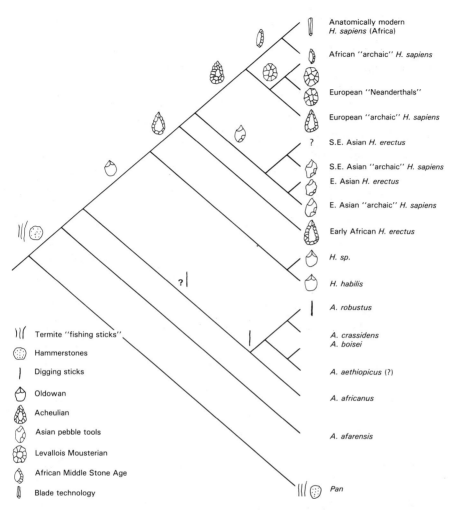

Anatomically modern
H. sapiens (Africa)

African "archaic" *H. sapiens*

European "Neanderthals"

European "archaic" *H. sapiens*

? S.E. Asian *H. erectus*

S.E. Asian "archaic" *H. sapiens*
E. Asian *H. erectus*

E. Asian "archaic" *H. sapiens*

Early African *H. erectus*

H. sp.

H. habilis

A. robustus

A. crassidens
A. boisei

A. aethiopicus (?)

A. africanus

A. afarensis

Pan

Termite "fishing sticks"

Hammerstones

Digging sticks

Oldowan

Acheulian

Asian pebble tools

Levallois Mousterian

African Middle Stone Age

Blade technology

4.4 A cladogram of human evolution. The lines reflect degrees of genetic and cultural relationships and the styles of stone tools found with different hominids.

clear. Potassium-argon dating of the Olorgesailie formation yields an age of about 480,000 years.[10]

Glyn Isaac suggested that ancient hominids—presumably *Homo erectus*—encircled a troop of baboons, perhaps at night, spooked them by making a lot of noise, and then systematically clubbed them to death as they tried to escape. Pat Shipman, too, has interpreted this site as the result of a hunting episode.[11] If Olorgesailie is a case of hominids hunting baboons, as some suggest, it is possible we have been underestimating the linguistic and physical prowess of *Homo erectus*. It is instructive to try to imagine oneself about two-thirds our size, going out at night with stones and clubs to kill sixty-three baboons.

But Lewis Binford has questioned whether Olorgesailie—and most other early hominid sites—were places where hominids killed and butchered animals. In general he argues that with present techniques we cannot reliably distinguish between human hunting/butchering and other kinds of natural and cultural forces and circumstances. And he specifically suggests that the evidence from Olduvai and Olorgesailie may well be remains left by hominids scavenging the kills of other animals, mainly for bone marrow, which they obtained by smashing bones with stone tools. He sees no evidence that early hominids shared food in complex patterns like humans, or that they foraged from home bases.[12]

A site outside of Africa in the same time range as Olorgesailie is 'Ubeidiya, 3 kilometers south of the Sea of Galilee. About fourteen distinct archaeological assemblages have been uncovered at this site, all dating to no later than 640,000 years ago, and probably much earlier, perhaps around 900,000 years ago.[13] The tools found here seem very similar to those from Middle and Upper Bed II at Olduvai Gorge, being mainly choppers, spheroids (rounded stones), hand-axes, and used flakes.

Various other sites and fossils dating to the period of from about a million to 300,000 years ago have been found in a broad swath across the warmer regions between Africa and Java (Figure 4.5), but these are relatively rich environments in which the amount of food available in winter and summer would not have been much different. But as these early *Homo* spread out along the bottom of the temperate zone, they would have encountered a very different world. As Binford notes,

> I cannot imagine that the earliest hominids to radiate into temperate settings practiced storage strategies, but one can imagine them under pressure to become at least seasonal predators. . . . In addition, the increased shelter requirements in the temperate zone would render it even more reproductively advantageous to reduce the mobility of females and their offspring, providing still further advantages to increases in provisioning behavior on the part of males.[14]

As any winter resident of the North knows, movement into northern climates means getting whiter and fatter. Up to 80% of the energy value of food goes simply to maintain body temperatures, and it is an inescapable fact of physics that spheres lose heat more slowly than any another shape of equal volume. Even with the pervasive migrations of people during the last century, the relationship between body shape and mean temperature—as epitomized by the Watusi and the Eskimo—is strong.[15]

The white skins of northern peoples have to do most directly with their dependency on Vitamin D, which is present in some foods but can also be synthesized in humans by the action of sunlight on skin. Whiter shades of skin are more susceptible to cancer, acne, and psoriasis than darker tones, so the selection for whiter skins—necessary to maintain Vitamin D production—as people moved farther north was not an unmixed blessing. It is also probable that these later invasions of cold climates gave rise to many subtle genetic changes. Even such apparently minor matters as blood types may have been developing at this time in response to factors we cannot isolate. Blood type O, for example, seems to be more resistant to various diseases and even certain kinds of bone fractures,[16] but we don't know how or when the different blood types evolved.

An important site for the problem of human penetration of temperate environments is Zhoukoudian (literally "Dragon Bone Hill"), a cave located 43 kilometers southwest of Peking in a range of limestone hills.

4.5   Some important *Homo erectus* sites in Africa.

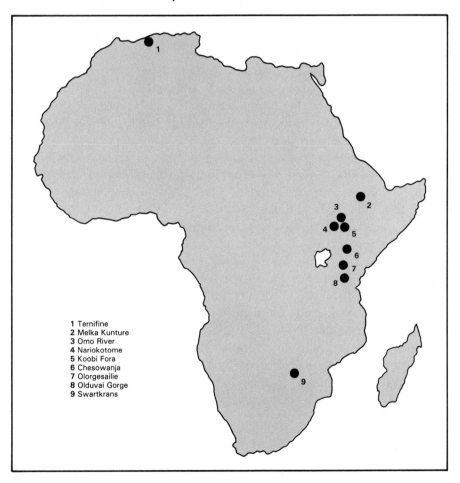

1 Ternifine
2 Melka Kunture
3 Omo River
4 Nariokotome
5 Koobi Fora
6 Chesowanja
7 Olorgesailie
8 Olduvai Gorge
9 Swartkrans

Using explosives, excavators managed to blast out and examine thousands of cubic meters of collapsed cave debris at the site between 1927 and 1937, and these excavations and later research revealed the remains of more than forty hominids, as well as over 100,000 stone tools, countless animal bones, and many hearths and ash layers, all well stratified in a deposit that is an astonishing 50 meters deep. Not all this was cultural debris—cave bears and other animals alternated with hominids in occupying the cave and they probably brought in many of the animals. But Zhoukoudian has more superimposed occupational layers than any other known *Homo erectus* site. Analysis of the fauna and hominids and various forms of dating have produced conflicting age estimates, ranging from 730,000 to 128,000 years,[17] but an age of 300,000 to 400,000 years ago seems most likely.[18]

The fourteen skullcaps, six skull bones, ten jaw fragments, one hundred forty-seven teeth, and assorted arm, leg, and hand bones found at Zhoukoudian all appear to have come from *Homo erectus*. Brain volumes average about 1040 cubic centimeters—somewhat larger than the *H. erectus* from Java—and teeth sizes fall between ourselves and australopithecines (actually, they are only slightly larger than contemporary native Australian populations). Based on the few leg-bone fragments recovered, it is estimated that the Zhoukoudian hominids averaged about 5' 1" in height—which may seem short, but is significantly larger than australopithecines and only an inch or two less than the average height of most people of just a few hundred years ago.

Many authorities regard Zhoukoudian as a base camp from which hominids hunted and to which they brought back their kills to be cooked and eaten. The site includes thousands of stone tools and animal bones, mainly from deer, but also elephants, rhinoceroses, beavers, bison, boars, and horses. Some hackberry seeds from Zhoukoudian are the oldest known vegetable remains from an archaeological site, and they probably survived in the rubbish of Zhoukoudian only because they were burnt and thereby preserved. Such finds remind us that we see only a distorted picture of early hominid diets: they probably ate vastly greater quantities of plant foods than meat, but only when they used fire to cook plants do we find the traces of vegetable foods.

But the issue of fire at Zhoukoudian is somewhat problematic. Binford and Stone argue that most of the bones at the site were deposited by animals who lived in dens there (e.g., bears) and that much of the presumed evidence for fires may in fact be bat dung:

> [T]here are no recognizable characteristics of hominid hunting in the remaining Zhoukoudian faunal collections. All positive characteristics point instead to hominid scavenging. . . . We conclude that previous interpretations presenting *H. erectus* as a big-game hunter are no longer defensi-

ble. The use of fire may well have characterized the later occupations at Zhoukoudian, but none of these glimpses of the past justify the *assumption* that the Zhoukoudian occupants were organized like modern hunters.[19]

The thousands of stone tools at Zhoukoudian, then, were made and used primarily to butcher scavenged animals, according to Binford and Stone.

One other aspect of Zhoukoudian they challenge is the possibility of cannibalism, which many scholars thought was evident in the condition of the human bones at the site. Not a single skull from this site had an attached face, and the base of each skull was broken, perhaps to get at the brains. But Binford and Stone found little convincing evidence of intentional butchering of human bones with stone tools.

Arens, too, rejects the notion that cannibalism was a common practice for very long in any culture, ancient or recent.[20] He may well be right, because ecologically, systematic cannibalism is a poor food-procurement strategy.[21] Compared to almost any other animals, people are hard to catch in relation to the food they supply, and one risks depleting the stock extremely quickly if cannibalism is at all frequent. Still, we must consider at least the possibility that *Homo erectus* practiced ritual cannibalism, as did many other human groups, even though there is little about the rest of the artifacts and remains of *H. erectus* to suggest a rich ritual life.

Some of these questions about the Zhoukoudian humans could be answered if we still had their bones, but all the hominid remains disappeared while being transferred from Peking to an American ship during the Japanese invasion of China prior to World War II. Although there are some mysterious indications that they have survived, to date no progress has been made in locating any of the fossils.[22] Fortunately, at the time of their discovery the great German anatomist Franz Widenreich made excellent plaster casts of them all and described them in superlative detail.

In addition to the Zhoukoudian materials, East Asia, South Asia, and the Middle East have produced a variety of sites and hominid fossils that date to between 1.5 million and 0.5 million years. Olorgesailie, Zhoukoudian, and Koobi Fora remain our largest, most intensively studied sites of this crucial interval.

Early Sites in Europe

Zhoukoudian may offer evidence that our ancestors had mastered the use of fire and clothing to adapt to northern climates, and data from Europe of this same period shows hominids were penetrating temperate latitudes all across the Old World.

The earliest known archaeological site in Europe may be Vallonet Cave, on the coast of southeastern France, where a small concentration of stone tools and broken animal bones has been tentatively dated to about 950,000

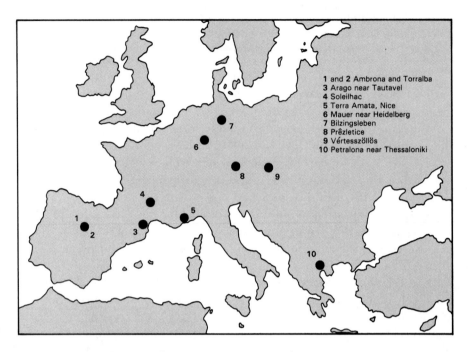

1 and 2 Ambrona and Torralba
3 Arago near Tautavel
4 Soleilhac
5 Terra Amata, Nice
6 Mauer near Heidelberg
7 Bilzingsleben
8 Přêzletice
9 Vértesszöllös
10 Petralona near Thessaloniki

4.6   Some important *Homo erectus* sites in Europe.

years ago (but the date is still somewhat questionable).[23] The tools from Vallonet look rather like the Olduwan implements; they consist of five crude choppers made from fist-sized pebbles and four flake tools. A number of apparently worked animal bones were also found, and the occupants of this small cave probably brought in antlers shed by deer.[24] The most abundant animal bones are from extinct forms of wild cow and bear, and there are also remains from antelopes, deer, boars, rhinoceroses, elephants, horses, hippopotami, seals, and a monkey. Several bones appear to have been deliberately broken and a few flakes were struck off the end of a rhinoceros leg bone, but the large number of bear bones in the cave suggests that most of these animals were probably not brought back to the cave by hominids; cave bears no doubt carried their prey back to dens in these caves, and the presence of many of these bones is probably a result of the bears' activities. Frost-cracked rocks and annual layers of ceiling-fall at Vallonet reflect at least some periods of cold, but there is no evidence of any hearths or fires. These would not have been necessary during the warmer parts of the year, of course, particularly since Vallonet is located at the coast, where the Mediterranean greatly moderates the climate. The presence in the cave of the bones of a monkey also suggests a fairly moderate climate.

A more securely dated early European site is Isernia, in central Italy, where chipped limestone tools and animal bones have been found beneath a geological formation dated to about 730,000 years ago.[25]

Perhaps our best evidence of European adaptations hundreds of thousands of years ago comes from Torralba and Ambrona, located about 1.5 kilometers apart in a deep valley 150 kilometers northeast of Madrid. Given the many difficulties of discriminating between human scavenging and hunting, no single site can be taken as a convincing evidence of the general nature of early human life in Europe. But many regard Torralba and Ambrona as evidence that at least some groups in Middle Pleistocene Europe engaged in big-game hunting. Excavations at Torralba in the early 1960s by F. Clark Howell exposed about 300 square meters of stratified archaeological deposits, from which were collected hundreds of pollen samples, several thousand stone tools, and countless animal bones (but no human remains). The kinds of pollen found in these remains have convinced some scholars that Torralba dates to about 400,000 years ago, but some think it is only about 200,000 years old. Pollen analysis indicates the area was a cool, swampy valley when the site was inhabited.[26]

The remains of at least thirty elephants were found at Torralba, as well as about twenty-five deer, twenty-five horses, ten wild oxen, and several rhinoceroses. Almost all the skeletons were disarticulated, and many bones were smashed and split, perhaps in an effort to get at the marrow. Nearly all the elephant skulls are missing, as are many of the other bones bearing the most edible cuts of meat—suggesting they were moved some distance to be butchered, and perhaps indicating the efforts of a large, cooperating work force.

Bits of charcoal were found mixed with the bones and stones at Torralba, raising the possibility that these people had driven the animals into the swampy areas of the site by lighting fires. Scattered among the charcoal and other debris were preserved bits of wood, and some have speculated that wood may have been brought to the site to cook the butchered meat. Not a single hearth, ash concentration, or depression has been found, however.

The process of driving animals into the swamps and killing and butchering them would have been quite a spectacle, with great clouds of smoke, shrieking, demented animals, and running, shouting hominids. But how could these tiny, stupid, ill-equipped people have killed these huge animals? Not a single stone spear-point was found at the site, but Freeman suggests that the animals were either stoned to death with the many fragments of rock found amid the bone or dispatched with wooden spears. It is a bit difficult to envisage any human, let alone *H. erectus*, about to stone three or four large elephants to death, but if they did, it would have been a fantastic Hitchcockian scene played out in this Spanish valley hundreds of thousands of years ago.

The stratigraphy at Torralba is complex, and the site may represent not one, but many different hunting episodes. In fact, various archaeologists have suggested that many of the animals at Torralba and Ambrona show no signs of butchering, that the evidence for the intentional use of fire is slight, and that small groups of humans scavenging some of the animals could have created the site.[27]

Another interesting early European site is Terra Amata.[28] In 1965 a bulldozer operation in an alley named Terra Amata in Nice, on the Mediterranean coast of France, uncovered twenty-one discrete sites dating to about 300,000 B.P., when the level of the Mediterranean was 25 meters higher than it is today, and the climate somewhat cooler and more humid. The inhabitants of Terra Amata constructed large huts, ranging from 8 to 15 meters in length and from 4 to nearly 6 meters in width, most of them oval shaped and estimated to have sheltered ten to twenty people. Nothing remains of the huts themselves, but their pattern is clearly evident in the long lines of postholes where supporting logs were driven into the sand, as well as in the rows of stones evidently used to brace the walls. Inside each hut was a hearth, and the floors of the huts were thickly covered with ash and the residue of organic debris, except for some areas close to the fire that were relatively free of debris, perhaps indicating places where people slept. Flat limestone blocks within the huts may have been seats or convenient places to prepare food.

The outlines of these various huts are so neatly superimposed and separated by such thin layers of sand that they probably reflect the same or closely related groups coming back each year to the same spot to build their huts and exploit local resources. Analysis of fossilized human feces revealed pollen from plants that bloom in the spring and early summer, suggesting a temporary occupation each year at this time. Most of the animal bones found at the site were from stags, elephants, boars, ibex, and rhinoceroses, but there were also bones of birds, turtles, rabbits, rodents, and a few fish and shellfish.

The stone tools from the site are mainly the large bifacially worked Acheulian tools found elsewhere in Africa and Europe at this time, and there are also a few bone tools. Unfortunately, not a single human remain has been found at Terra Amata, but it is likely that we see here the work of *Homo erectus*. A single footprint was uncovered during excavations, and its modern shape and length indicate a person about 5' 1" in height.

Terra Amata's occupants were probably hunters and gatherers who ranged widely with the changing seasons, collecting berries, nuts, and other foods, foraging for eggs, small game, and shellfish, and hunting and scavenging mid-size and even large game.

Another important northern European site is Vértesszöllös, a rock quarry west of Budapest, Hungary. A recent series of uranium dates put the site at about 185,000 years,[29] but it could be as old as 350,000. When the site was first occupied, it was on the banks of some hot springs, and several layers of human occupational debris have been found near these springs. Excavations in the 1960s uncovered about 3,000 stone tools, many smashed and burned animal bones, and the occipital bone from one hominid and a few teeth from another.[30]

The estimated cranial capacity of between 1115 and 1437 cubic centimeters is large for a *Homo erectus,* and some consider this fossil to be an archaic form of *Homo sapiens.*[31]

No hearths have been found at the site, but there are burned bones, indicating the use of fire.

The Middle/Upper Paleolithic Transition
(c. 300,000 – 35,000 B.P.)

In his tool-use, social system, and economy, *Homo* of several hundred thousand years ago, as we know them from Zhoukoudian, Vértesszöllös, and the other sites described above, seems to have been similar to modern hunters and gatherers in day-to-day economy, group size, and other basics, yet there is something alien about the creature. We look for artifacts expressing ritual or complex symbolism, but not a single figurine, wall painting, or rock carving can be securely attributed to *Homo erectus.* Later, beginning at least 30,000 years ago, people made exquisitely crafted stone tools, some so delicately worked that even moderate use would ruin them— tools that must have been made in part simply for the pleasure of creating something beautiful. But the tools of *Homo erectus* are undeviatingly simple, efficient, utilitarian objects.

Perhaps even more revealing, there are no known *Homo erectus* burials or ritual dispositions of corpses. For at least the last 30,000 years, death has almost everywhere been an occasion for the outpouring of human emotion, and even the simplest hunters and gatherers during this span usually disposed of their dead by digging a hole and placing a few stone tools or bits of shell in with the body; but not a single *Homo erectus* anywhere in the world appears to have been even intentionally buried, let alone sent off to the next world with a few provisions and expressions of goodwill.

These various absences of stylistic behavior among *Homo erectus* can be interpreted in several different ways. *Homo erectus,* with his brain about two-thirds the size of our own, may simply have lacked the mental equipment to generalize and symbolize his experiences as we do. *Homo erectus*'s language skills, in particular, may have been quite limited.[32] On the other hand, *Homo erectus* may have had the potential for stylistic, religious, and social impulses but lived in circumstances that did not elicit such expressions.

It is difficult to overstate the importance of the evolution of the capacity for aesthetic, ritual, and social feelings, for as we will see, it was precisely these mental characteristics that made possible the rise of great civilizations. Thus, we are particularly concerned in this chapter with the conditions under which these feelings first appeared (as reflected in the archaeological record) and with their concurrent important cultural developments.

The differences between *Homo erectus* and *Homo sapiens*, and the time period from 300,000 to 30,000 years ago in which they occur, are generally referred to as the "Middle/Upper Paleolithic transition."[33] This "transition" is visible in radical changes in human physical characteristics and material culture, including: (1) an increase of average human brain size from about 1100 to about 1400 cubic centimeters (although local variability was high, and the modern range of normal brain functioning is at least 1000 to 2000 cubic centimeters); (2) changes in human facial architecture and other physical characteristics, culminating in the appearance of modern *H. sapiens sapiens;* (3) increased human population numbers and densities—again with considerable local variation; (4) many technological innovations, including the bow and arrow, atlatl (throwing stick), bone and wood tools of diverse types, and techniques for extracting a relatively great amount of cutting edge from a given amount of stone; (5) increased aesthetic expression, in figurines, usually of bone or stone, beautiful wall paintings and rock carvings, burial techniques, and in objects used for personal adornment; (6) a shift from generalized hunting patterns to concentrations in some areas on gregarious herd mammals like deer, reindeer, and horses; and (7) the appearance of artifact styles and trade in exotic items that bespeak the first manifestation of some sort of regional "ethnic" identity[34] that exceeds by a wide margin the local band society—in short, changes that may reflect the "total restructuring" of social relationships during the Middle/Upper Paleolithic transition.[35]

The obvious question is: How were our ancestors changed by time and circumstance in these highly significant ways?

To begin with the physical types of hominids that made these transitions, Chris Stringer and others and Erik Trinkhaus have arranged most of the fossils between 500,000 and 30,000 into an order based on morphology—though the chronology is confused and great differences in date can be found in these categories.[36]

Grade 1 hominids may be as old as 500,000 years, but dating these fossils has proved difficult in most cases. Our best Grade 1 fossil specimens come from Europe. Trinkhaus notes that some of these have been classified as *H. erectus* and others seem to have Neandertal traits, but as a group they seem to be consistent with a gradual change toward modern human forms (for example, increased size of the cranial vault and decreasing mean dimensions of molars).

In Europe these Grade 1 hominids are followed by a group that seems to fall between the grades in that they are quite archaic in some characteristics but also resemble slightly the Neandertals, who generally seem later in time than the Grade 1 group. A good example of these intermediate Grade 1–2 fossils is from Swanscombe, England, along the Thames River, not far from London. In 1935 workers in a cement plant uncovered a cranial bone from a gravel bank, and a year later another cranial bone fragment was found nearby that articulated perfectly with the first bone. Later, during excavations connected with preparations for the Allied invasion of France in 1944, another bone from the same skull was found just 25 meters from the site of the first find. It is very possible, incidentally, that more hominid bones were included in the gravel used to make concrete for floating docks during the D-Day operation.[37]

In the same gravel layers that produced these bones, excavators recovered the bones of extinct forms of elephants, deer, rhinoceroses, and pigs which, together with subsequent chemical analysis and geological evidence, dated the Swanscombe fossils to an interglacial period about 225,000 years ago, when the abundance of horses, elephants, rhinoceroses, and other big-game species would have made England an ideal place for generalized hunting and gathering groups.[38] Nor is there any problem explaining how these hominids would have gotten there, since Britain and Ireland were physically joined to Europe by a land bridge at various times during the Pleistocene.

The Swanscombe cranial remains are probably those of a woman of twenty to twenty-five years of age, with a cranial capacity of about 1325 cubic centimeters—well within the range of modern humans, but on the small side.[39] Hand-axes roughly similar to those of the Acheulian assemblages of France and Africa are among the most frequent tools in the level where the skull was found, but lower levels contain only flakes and choppers. Similar flakes and choppers have been found elsewhere in England and are commonly referred to as the Clactonian assemblage.[40] A wooden object that looks like the shaped end of a spear that was found at Clacton is the earliest wooden artifact recovered anywhere, dating to 400,000 to 200,000 years ago. Stone projectile points are not found at Swanscombe, Zhoukoudian, or any other site prior to about 150,000 years ago, and thus the wooden spear fragment—if that is what it was used for—may be a clue to how these Middle Pleistocene peoples managed to kill animals. If animals were trapped in bogs, they could have been killed by multiple stab wounds with wooden spears—although it would not have been pleasant work.

But Lewis Binford's analyses of cut-marks and other features of the animal bones from Swanscombe led him to suggest that "the Swanscombe fauna represents hominid scavenging at the source of the carcass."[41]

The Swanscombe hominid is usually classed with another fossil from Steinheim, Germany. This cranium, dated to about 250,000 years ago, probably belonged to a young woman whose brain size and facial features place her between *H. erectus* and ourselves. Unfortunately, no artifacts were found with the Steinheim skull, so we cannot compare the site with the material from southern England. Nonetheless, the physical differences between this individual and *Homo erectus* indicate that the transition from *H. erectus* to *H. sapiens* was well underway by 300,000 to 250,000 years ago and was taking place in more than one part of western Europe.

Excavations in a cave site in the French Pyrenees unearthed a skull (the Arago skull) and two mandibles (Figure 4.7) dated to about 200,000 years ago that seemed to fill the gap between *Homo erectus* and the European Neandertals.[42] The skull possesses some morphological characteristics of *H. erectus* in the Far East but lacks the incipient sagittal crest usually found

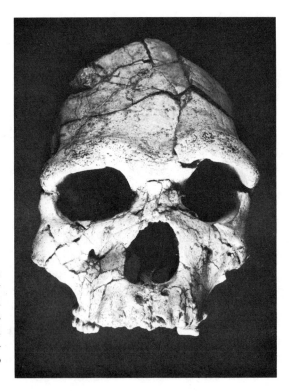

4.7 The Arago skull, from France. This approximately twenty-year-old man had more robust facial features than many Neandertals and is probably intermediate between *Homo erectus* and *Homo sapiens*.

in these populations. The large size of the teeth and mandible and the structure of the chin seem to foreshadow the features of the "classic" (western European) Neandertal.

Late *Homo* remains have also been found at Bilzingsleben in eastern Germany, along with a somewhat atypical assemblage of very small tools with much larger implements, all quite different from the typical Acheulian assemblage. In fact, Svoboda has argued that the many small tools found at Arago, Vértesszöllös, and Bilzingsleben developed out of adapting to cold, open landscapes.[43]

Lewis Binford's analysis of the French site of Abri Vaufrey led him to conclude that the people who occupied this cave 200,000 years ago used fire and regularly processed large animals for meat, but here again he sees a scavenging rather than a hunting pattern.[44]

In general, tool technology, diet, site locations, and average group size in Europe about 200,000 years ago do not seem much different from those of several hundred thousand years earlier, as represented at Zhoukoudian and Torralba-Ambrona. Clearly, however, population densities were increasing, and as people moved into more diverse niches there was an increasing variety in the stone tools associated with them.

In searching for causes of the increasing brain size and other changes between 400,000 and 100,000 years ago, we might note that rates of evolutionary change frequently seem to be higher along the margins of a species' range. This may have been the case with *Homo,* as bands of these hominids probed far into England, northern Europe, and perhaps northern Eurasia, and began to specialize in various forms of hunting, scavenging, and gathering. The Swanscombe and Steinheim individuals, with their nearly modern brain size, may be reflections of these developments along the northern periphery. Gene flow in most hunting and gathering societies is sufficiently high that these changes in brain size and facial architecture would probably have been quickly disseminated over a wide area. Michael Day suggests that the Swanscombe fossil, for example, can best be regarded as a female example of the transitional form between *Homo erectus* and *Homo sapiens* that was at the root of the evolution of European Neandertals.[45]

But the "margins" of a cultural animal like early humans can be culturally and technologically—not just geographically—defined. As noted below, brain sizes may have increased most rapidly in Africa, not on the world's cold periphery. Brain tissue has a relatively high "cost": it consumes great amounts of energy and oxygen. Also, the birth of large-brained offspring requires pelvic bone structure that reduces maternal mobility. Since *Homo erectus* was obviously an efficient scavenger, hunter (at least of small game), forager, and toolmaker, perhaps the increased brain size was related to increasing emotional capacities rather than improvements in the problem-solving abilities that were important in hunting or tool-making. Great advantages would accrue to a Pleistocene hunting and gathering group that could organize itself as part of a social network involving many different bands and hundreds of individuals, and perhaps the increasing brain size had to do with the selective advantage of being able to generalize emotions to scores of "kinsmen."

In any case, the great variability of cranial capacity among "normal" people today and the fact that human brain size seems generally to have increased quite uniformly up to about 100,000 years ago should warn against simplistic explanations of this phenomenon.

To explain the relatively slow rate of technological change between 2.0 and 0.1 million years ago, we must reflect on the fact that not only were our ancestors of this era less intelligent, there were also many fewer of them. Technological innovation is not a simple product of the number of minds available to create new ideas, but a strong relationship exists between population numbers and innovation in the simple hunting-gathering economies of the early and middle Pleistocene. Even as late as 500,000 years ago, there were probably only a million people in the entire world. Also, people of this era tended to live much shorter lives. Few survived past thirty years of age, and—although few adolescents of any age have believed it—people learn a great deal and retain considerable creativity past thirty.

In any case, we have to be careful about facile interpretations of Pleistocene life based mainly on stone tools. As Ralph Holloway noted, "if we were to focus only on [the Australian Aborigines'] stone tools, we might have to conclude that these people, whose kinship systems we still have trouble understanding, had no language or culture."[46]

From *Homo sapiens* to *Homo sapiens sapiens:*
Our last 100,000 Years

The last 100,000 years of our long evolutionary history pose as many puzzles as the earlier eras, even though we have somewhat more data. The main problem is basically this: during the last 100,000 years at least two different kinds of humans seem to have lived, the Neandertals and modern *Homo sapiens sapiens*. The problem is that we still do not know how they were related; and until we understand how they were related biologically and evolutionarily, we will not be able to comprehend the driving forces of our evolution in the past 100,000 years.

As we saw in the evidence from Swanscombe, Steinheim, and other sites, by about 200,000 to 300,000 years ago, hominids with nearly modern cranial capacities seem to have lived across much of Europe and probably in much of the rest of the Old World, as well. But from about 200,000 years ago, the diversity of humans and their relationships in time and space become very difficult to sort out.

The Neandertals

Nothing about the Neandertals seems simple. Evidence suggests they lived between about 130,000 and 35,000 years ago, but their precise period in history is difficult to define. Most of them seemed to live in Europe and western Asia, some as far east as Soviet Central Asia and in much of the Middle East, but some of their characteristic physical features seem to appear in fossil remains as far away as Africa and east Asia.[47] The "Classic," or western, Neandertals were different from us and from their contemporaries in Africa, east Asia, and Australia in various physical characteristics, but scholars disagree on the extent and significance of these differences. Some think the Neandertals could speak with about as much fluency as we do, others think the Neandertals did not have the mentality or vocal apparatus for normal human speech. Scholars disagree perhaps most heatedly about what "happened" to the Neandertals: that is, are we descendants of the Neandertals, or did they lose out in competition with our ancestors?

The fossil that gave this stage of human evolution its name was found in the Neander Valley in southwestern Germany in 1856, and because it belonged to the first premodern human identified, the Neandertals received much of the initial hostility to the concept of human evolution (earlier Neandertal finds were not recognized as such). From the beginning anthropologists, clergy, and others held that the Neandertals were an aberrant stage in human development, not directly related to our own, presumably superior, ancestors.

C. Loring Brace has argued that the initial classification of Neandertals as off the main line of our own evolution was rooted in the errors of nineteenth-century French paleontology.[48] The chief villain in Brace's history of Neandertal studies is French paleontologist Marcellin Boule, who between 1911 and 1913 published studies that depicted the Neandertals as bow-legged, slouching, simian-looking individuals who were neither very intelligent nor agile. He did not actually state that Neandertals couldn't walk and chew gum at the same time, or that they drooled incessantly, but Boule used the words ape-like, primitive, and inferior so frequently that neither he nor later scholars were anxious to claim Neandertals as ancestors.

Before and after Boule's publications, however, some scholars suspected that the Neandertals were the connecting link between *Homo erec-*

tus and at least some populations of *Homo sapiens sapiens*. In 1957 a conference on the Neandertal produced evidence that Neandertal brain size on the average was larger than that of some modern human groups and that there were no grounds for concluding that their brains were structurally inferior or that they did not walk fully erect. In fact, it was suggested that "if he could be reincarnated and placed in a New York subway—provided that he were bathed, shaved, and dressed in modern clothing—it is doubtful he would attract any more attention than some of its other denizens."[49]

Yet there are differences between ourselves and the Neandertals. The characteristics most frequently used to define them are: (1) a receding or virtually absent chin; (2) large cheekbones and prominent brow ridges curving over the eye orbits and connecting across the bridge of the nose; (3) prognathism (protruding lower face); (4) a strong masticatory apparatus, including larger front teeth than are found in most modern human populations; (5) short (average of perhaps 5 feet) but powerful stature, with thick and slightly curved long bones; and (6) a cranial capacity within

4.8 This artist's reconstruction of a Neandertal band on the move depicts these individuals as somewhat more brutish in appearance than they probably were.

the range of modern humans, though slightly larger on average for the classic "Western" type.

Erik Trinkhaus has shown how the shapes and sizes of Neandertal teeth and heads vary in many obvious and also subtle ways from our own.[50] He interprets the projecting mid-face of the Neandertals as a feature selected by natural selection in part "to facilitate the use of the anterior teeth as a vise."[51] Neandertals would have had no problems with candied apples or corn on the cob, had they been available. They probably ate meat by the stuff-and-cut method common among contemporary Eskimos, stuffing their mouths and then cutting of the excess meat with a stone knife before chewing.

Based on their legs and lower bodies, Trinkhaus concluded that the

> overall impression of Neandertal locomotor anatomy is . . . one of great strength, as with the upper limb, but also one adapted for endurance for prolonged locomotion over irregular terrain. In fact, the level of diaphyseal hypertrophy of their femora and tibiae suggests that they spent a significant portion of their waking hours moving across the landscape . . . far more than did early modern humans.[52]

The Neandertals would have been a bit short to be good offensive football linemen, but they would have made superstar wrestlers or baseball catchers, given their tremendous arm strength (though their ability to grip the ball might have been less than ours).

The Neandertals were adept stone toolmakers. Most of their tools belong to the Mousterian stone-tool industry (named after the site of Le Moustier in southern France), which includes several distinctive stylistic and functional elements. There are scores of Mousterian sites in the Dordogne region of southwestern France, including cave sites, rock shelters, and "open-air" locations. One of the largest and most complex Mousterian sites in this area is a cave in the Combe Grenal Valley, near the Dordogne River. François Bordes uncovered sixty-four superimposed occupational levels in this cave, spanning the period from about 85,000 to 45,000 years ago, with few long periods of abandonment. The lowest levels contained tools resembling the Acheulian tools found at Swanscombe, but all later levels had the classic Mousterian tools usually associated with Neandertals. More than 19,000 Mousterian implements were collected and analyzed from this cave, and the tools from different levels contrast sharply. Some levels contained many small flake-like pieces of stone, while others had concentrations of scores of "toothed" or "denticulated" tools. Moreover, analysis of the different levels revealed that certain types of tools tended to be spatially associated with a number of other types. That is, levels containing a relatively high number of projectile points would usually contain relatively large numbers of scrapers and flakes—but few denticulates.

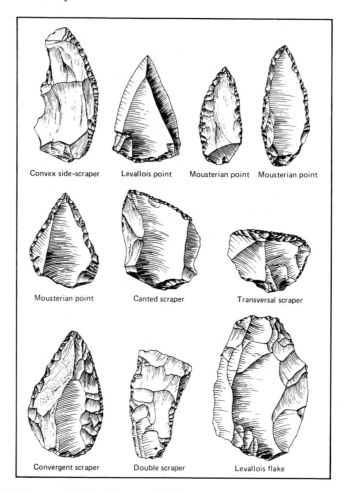

Convex side-scraper   Levallois point   Mousterian point   Mousterian point

Mousterian point   Canted scraper   Transversal scraper

Convergent scraper   Double scraper   Levallois flake

4.9 Some typical tool types of the Mousterian period. The Neandertals made effective but not particularly elaborate stone tools for a variety of purposes.

This diversity of tools may seem relatively unimportant, but it was the focus of a long debate that involved issues fundamental to the New Archaeology of the past two decades (see Chapter 1). The specific case of the Mousterian tools became part of a larger question: How are we to measure and interpret variability in ancient artifacts? For the Mousterian tools, François Bordes spent years in excavation and analysis to establish a typology that has been the framework for much of the work done on this period. Bordes classified all Mousterian tools into four categories, based on the relative frequencies of certain types.[53] Bordes considered various explanations of the diversity of Neandertal tools, such as that this variability reflected different time periods, climates, or seasons of the year, but on the basis of the archaeological evidence he rejected these possibilities and eventually concluded that the four different clusters of tools are the remains of four distinct cultural traditions, or "tribes," which developed certain kinds of tool manufacture and retained these distinctive expressions over the 30,000 years of the Mousterian period.

Bordes's vision of the Mousterian was one in which different tribes of Neandertals wandered much of the Old World for generation after generation, through tens of thousands of years, each group maintaining its unique styles of tool manufacture and meeting the others infrequently and usually with hostility.

This vision was questioned by Lewis and Sally Binford, who assumed that Mousterian tool variability was largely a reflection of the different tasks Neandertals had to perform to meet successfully the demands of their environment.[54]

They tried to test their interpretation by a statistical analysis of Mousterian tools from three widely separated sites: the Jabrud Rock Shelter (near Damascus, Syria); Mugharet es-Shubbabiq Cave, in Israel; and an "open-air" station near Houpeville, France. Each site contained several different levels, representing different occupations; the total number for all three sites was sixteen. Lithics from each site were classified in terms of Bordes's system and statistically analyzed for evidence that these groups of tools were used for different economic activities, rather than simply representing stylistic traditions. Factor analysis was the statistical method used to determine which of the tool types were usually found in close proximity to one another in the various levels of the different sites. On this basis they defined several different "tool kits," whose presumed functions included tool preparation, wood-working, butchering, and various other tasks.

The Binfords' study stimulated interest in statistical analysis of archaeological data, and since that time many archaeologists have worked at developing statistical applications to archaeological data that would help create a truly scientific form of archaeology. But the exact nature of Neandertal tool-making variability and subsistence has remained a matter of dispute. It seems highly likely that much of the variability in the tools is related to the specific functions they were used for, of course. But reconstructing these functions has involved many problems of interpretation. Even whether or not the Neandertals were systematic hunters has been disputed. Binford concludes that regular hunting of mid- to large size animals like deer and reindeer became an important part of human economies just prior to the appearance of *Homo sapiens sapiens*. In an analysis of the bones from Combe Grenal, for example, Binford found in some levels evidence that he thinks is "good evidence that the majority of the moderate-body-sized

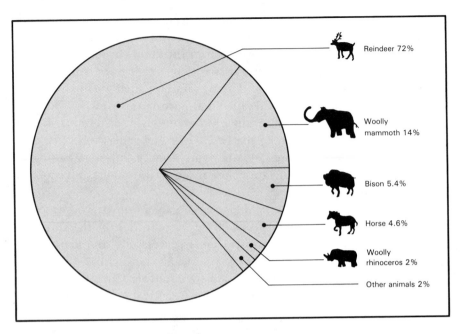

Reindeer 72%

Woolly mammoth 14%

Bison 5.4%

Horse 4.6%

Woolly rhinoceros 2%

Other animals 2%

4.10 Neandertal economy included at least some hunting in some areas. This figure shows the percentages of different prey killed 55,000 years ago at the summer open campsite near Salzgitter-Lebenstedt in West Germany.

animals [e.g., deer] at Combe Grenal were *hunted for meat* [emphasis Binford's]."[55] But we still have only small samples of Neandertal economy. And although archaeologists used to believe that only Neandertals made Mousterian-style tools, it is now clear that they made tools in other styles toward the end of their time in history, and so we can expect that the Mousterian sites will not tell us everything about Neandertal economy.

NEANDERTAL CULTURE AND SOCIETY

As if the slurs cast on Neandertal intelligence and posture by early archaeologists were not enough, some anthropologists have questioned whether or not Neandertals were able to produce the range of sounds necessary for normal human speech. P. Lieberman and E. Crelin reconstructed the vocal apparatus of Neandertals using a computer simulation based on the measurement of a classic Neandertal, and using the vocal tracts of chimpanzees and human infants for comparison, they concluded that western European Neandertals would not have been able to make some vowels, such as /e/, and, perhaps, some labial and dental consonants, such as /b/ and /d/.[56] Others doubt that the Neandertals could speak at all.

However, since parrots have been taught to make most of the sounds of many languages, it is difficult to see much significance in differences between Neandertal and modern vocal anatomy; and there is no conclusive evidence that osteological dimensions and topography are certain indicators of speech abilities.[57]

Archaeologists have long believed that, whatever their verbal fluency, the Neandertals were at least human enough to bury their dead. Excavations at La Chapelle-aux-Saints, in France, revealed a Neandertal corpse apparently laid out in a shallow trench, with a bison leg placed on his chest, and the trench filled in with bones, tools, and other debris—perhaps representing offerings of meat and implements. At La Ferrassie, France, a Neandertal "cemetery" was found, where a man, a woman, two children, and two infants seem to have been buried. A flat stone slab was on the man's chest, the woman was in a flexed position, and, toward the back of the cave, the skull and skeleton of one of the children were buried in separate holes, about one meter apart. At Monte Cicero, in Italy, a Neandertal skull—with a hole cut into the base—was placed in the middle of a circle of stones. At Teshik-Tash, in Siberia, a Neandertal child was buried in a grave with goat skulls whose horns seemingly had been jabbed into the ground. At Shanidar Cave in Iraq, the soil near a Neandertal's body contained massive quantities of flower pollen. Ralph Solecki, the excavator, and the palynologist, Arlette Leroi-Gourhan, concluded that the skeleton had been buried with garlands of flowers.

But in these and other cases the evidence is at least somewhat equivocal. Even in the best of circumstances, archaeological excavation is a messy, ambiguous business of judging the significance of faint changes in the texture and color of sediments, and the Neandertal cave burials are notoriously difficult to excavate and interpret. Such caves have been homes not just to people but to hundreds of thousands of generations of cave bears, rodents, and other animals, and the natural processes of roof fall and sediment accumulation make for a confusing stratigraphic sequence. Thus, where the excavator may see a Neandertal carefully buried in a pit and a rock slab placed on his chest, another may observe evidence of a Neandertal who simply expired and was covered with rocks falling from the cave roof and by other debris. Thus some scholars see no evidence that any Neandertals were intentionally buried, while others believe that the Neandertals invested death with human-like emotion and ritual.[58]

Overall, the evidence seems to suggest at least some Neandertal emotional investment in death and burial, but in any case there is evidence that Neandertals were not insensitive to the plight of the handicapped. Some Neandertals evidently suffered terribly from arthritis or had lost limbs and so could not have contributed much to the group's food supply.

4.11 John Shea (1988) has shown that many pointed stone tools indicate the typical damage patterns of high-speed impact, such as would be expected on spear tips. This fracture pattern is illustrated on this Levallois point from Hayonim Cave, Israel. Such evidence suggests that people in this area 50,000 to 100,000 years ago did considerable systematic big-game hunting.

Yet, they must have been supported by the rest of their society. Despite these touching displays of societal concern, there is some evidence that Neandertals killed, butchered, and perhaps ate one another. At Krapina in Yugoslavia, excavations revealed twenty Neandertals, men, women, and children, whose skulls and long bones had been smashed and split in suspicious ways. But Trinkhaus questions whether cannibalism was actually practiced.[59]

All Neandertals were apparently hunters and gathers, but they must have varied considerably throughout their range in the kinds of resources they exploited. The archaeological record is no doubt biased, because most Neandertal sites found and excavated are those made evident by masses of animal bones associated with stone tools; the remains of plant foods and wooden tools, of course, do not preserve nearly so well and are not as easily found.

Neandertals were probably like recent human hunters and gatherers in habits and abilities. Population densities appear to have been low, and it is likely that most Neandertals lived with the same group of twenty-five or fifty people their whole lives, from time to time meeting other bands for mate exchanges. They were skilled hunters, locked into seasonal migrations with the animals they hunted, but in most habitats they probably foraged widely for eggs, birds, plants, and other small resources. They competed quite successfully with other predators for game but must have occasionally lost out to the zoological carnival of horrors whose ranges they shared. Giant cave bears, saber-toothed cats, and wolves occasionally "selected out" an unfortunate Neandertal: "Some days you eat the bear, some days the bear eats you" was probably no empty cliché to them.

Neandertals and Homo sapiens sapiens

Having reviewed the Neandertals' life and times, we now must come back to the complex problem of how they relate to us. The Neandertals may well have been primarily a western European, cold-adapted physical type, reproductively isolated to some degree from other populations, so we must consider who was living in the rest of the world, outside the Neandertals' range.[60] We must also considered why the Neandertals disappeared.

Few aspects of Neandertal existence have aroused such interest as their demise, primarily because it was originally thought that the Neandertals "disappeared" as a physical type after about 30,000 years ago. No Neandertal bones have been found postdating this time, and in some sites tool types widely believed to be associated with Neandertals are overlain with levels containing tools different in style.

The Neandertals successfully lived over much of Europe and the Middle East for tens of thousands of years—why should they have vanished?

Three possible explanations for their disappearance are currently being considered seriously by anthropologists, and Michael Day summarizes these as follows.[61] The *Neandertal Phase of Man* hypothesis places the Neandertals between *Homo erectus* and modern humans as one of the direct ancestral forms of our line, and implies that people of European ancestry today are directly related genetically to the Neandertals. Few hominid fossils from the period between 45,000 and 25,000 years ago in western Europe have been found, so we do not know much about what was happening to the physical form of these populations during this interval. C. L. Brace has suggested that the improved efficiency of Mousterian tools greatly relaxed the selective pressures for a heavy masticatory apparatus, which is a primary difference between Neandertal and modern skeletal material.[62] Most Neandertal dentition shows evidence of extreme wear on the front teeth, so the process may have been only beginning in the early Mousterian period.

The major differences between Neandertals and ourselves are in head and face shapes and sizes, but the skeleton of the Neandertals, especially the limbs and hip bones, is also different in size and morphology from that of modern humans. The torsional strength of Neandertal leg bones, for example, is about twice that of moderns.[63] Tools may have had a role in changing the selective pressures on the teeth, but how could tools or any other factor have altered the skeletal parts? At this point we simply don't know what selective forces might have caused such changes, or if they did. Some scholars, such as Fred Smith, maintain that the fossils from south and central European Neandertals all exhibit morphological changes in the direction of modern hominids.[64]

The *Preneandertal Hypothesis* suggests that "the Neandertals arose from a Preneandertal stock that became progressively specialized for resisting cold, [and] underwent severe natural selection and restricted gene flow that led to 'classic' Neandertal isolates."[65] According to this idea, the Neandertals evolved their distinctive features as part of adapting to cold climates and had so little genetic connection to people living elsewhere in the world that they became distinctive. Eventually, however, they were absorbed or displaced by more modern-looking humans.

A third hypothesis—which many anthropologists now reject—is the *Presapiens Hypothesis*. In this view, the people represented by the Swanscombe skull and similar fossils of more than 200,000 years ago were not directly related to the Neandertals and gave rise to modern humans, while the Neandertals were a distinct group that eventually became extinct.[66] As Day notes, it has become apparent that the Swanscombe, Steinheim, and other fossils on which the Presapiens Hypothesis is based are not anatomically as modern as previously believed and seem to have some Neandertal characteristics.[67]

These various ideas about the relationship of Neandertals and other early *Homo sapiens*, and the relationship of both to ourselves, have been greatly affected by several recent discoveries and analyses.

The first of these involves genetic analyses that were done to answer an ancient and intriguing question: How many generations ago did we have a common ancestor? It is an inescapable fact of genetics that all the people alive in the world are related, and that at some point an individual existed whom we can all claim as an ancestor. It's easy to appreciate the manner in which such questions can become permeated with racism or egalitarianism or other political or sociological agendas. In the early twentieth century, in fact, various scholars argued that the separate "races" of humankind evolved separately from nonhuman primates—thereby forcing any common ancestor back millions of years. Similarly, in our more liberal age, many would like to see the genetic "brotherhood of man" converge in the more recent past, making racial physical differences late and inconsequential.

Some scholars now believe on the basis of genetic evidence that all of us alive today have a common female ancestor who lived only about 200,000 years ago. This "Eve Hypothesis," associated with the research of Allan Wilson, Rebecca Cann, and others, is based on the study of DNA taken from the mitochondria, which are features in human cells where energy to keep the cell functioning is produced.[68] Unlike nuclear DNA, mitochondrial DNA is inherited only through the mother. Mitochondrial DNA is subject to random mutations, which are expressed as minor mistakes in copying the genetic code that are then passed on to the next generation.

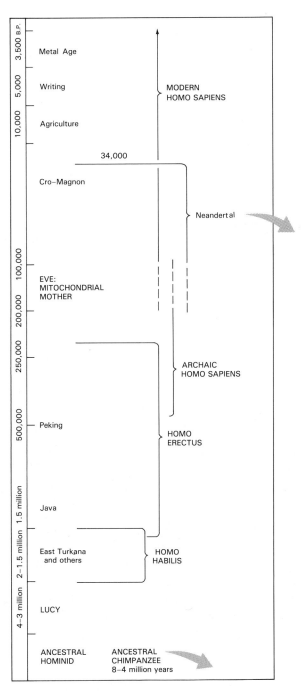

4.12 One reconstruction of human origins. The split of the Neandertals from the main human line remains controversial.

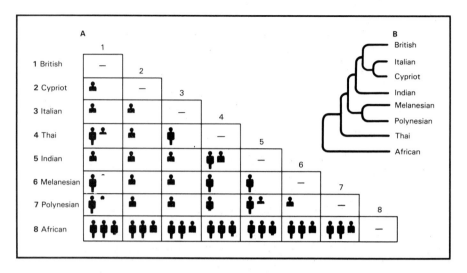

4.13 (A) Analyses of DNA have been used to suggest genetic distances among contemporary groups of people. In this figure the genetic distances between groups are based on a sample of 600 individuals from eight groups. Between any two groups, the fewer the symbols the shorter the genetic distance and the closer the presumed relationship. In (B) these data are arranged in a family tree. These data suggest all modern humans are genetic descendants of one small inbred group of prehistoric Africans.

Thus, we would not even expect two people from, for example, the same small isolated group of hunter-foragers in highland New Guinea to have identical mitochondrial DNA. And we would expect that differences in mitochondrial DNA would increase the longer any two people are separated in terms of a common ancestor, since the minor genetic mutations in mitochondrial DNA would accumulate over time.

To test the relatedness of contemporary human groups, one group of researchers led by Rebecca Cann collected the placentas of 147 women in the United States, Asia, Europe, the Middle East, New Guinea, and Australia, and then separated the mitochondrial DNA from these samples and compared them.[69] Allan Wilson has done similar research[70], as have others, and most of these studies conclude that everyone living today is descended from a southern African woman who lived between 140,000 and 200,000 years ago.

These genetic studies are based on arguable assumptions about rates of mutation in mitochondrial DNA, and other difficulties make their conclusions questionable. But even if these age estimates of a common ancestor are considerably off, we still have the problem of determining how our common ancestor evolved into ourselves and managed to spread across the world.

One possibility is that an archaic form of *Homo* spread from Africa across the world more than a million years ago, and then became somewhat locally specialized, like the cold-adapted Neandertals and the Asian forms of *Homo erectus;* eventually, however, if we accept the genetic studies described above, modern *Homo sapiens sapiens* arose in Africa or perhaps China and replaced all other forms, including the Neandertals, without much gene exchange among populations.

The alternative, already mentioned, is that people all over the world were evolving in the direction of *Homo sapiens sapiens,* and that the local characteristics of these populations account for some of the differences we see in modern peoples. The relatively large noses and angular features of modern Europeans, for example, contrast somewhat with the features of Asians, and it is certainly possible this represents the genetic influence of Neandertals in Europeans and the archaic Asian forms of *Homo sapiens* in contemporary Asian peoples. If, like modern hunter-gatherers, these ancient peoples practiced *exogamy*—the social rule that requires each person to marry someone outside their own group of twenty-five or so—then gene flow across the world might have been quite regular, especially given the relatively small total world population.

But some recent archaeological evidence makes this and other interpretations difficult to accept. Helene Valladas and her colleagues used thermoluminescence to date twenty burnt flint tools from Qafzeh, a cave near Nazareth, in Israel, in strata that contained the bones of modern-looking *Homo sapiens.*[71] They have concluded on the basis of this analysis that a primitive form of modern humans lived here 92,000 years ago, about 30,000 years before Neandertals inhabited this region. Valladas et al. speculate that Neandertals came into the Middle East about 60,000 years ago, perhaps migrating into the Mediterranean areas as glaciers expanded during the Pleistocene. Thus, if modern humans evolved in Africa long before they reached Qafzeh, as the DNA evidence might suggest, then the Qafzeh region might have been a contact area between Neandertals and early modern humans—since the Neandertals lived all over Europe 125,000 years ago.

The Qafzeh dates raise many questions. As Erik Trinkhaus noted, if early modern humans reached the Middle East 92,000 years ago, they must have lived there in some relationship to the Neandertals for tens of thousands of years, since Neandertal remains in the Middle East dating to the period between 60,000 and 36,000 are well documented.[72] Milford Wolpoff said that the Qafzeh material "is the first direct evidence that Neanderthals could breed with people from other parts of the world."[73]

He suspects that Neandertals and early *H. sapiens* interbred to produce *Homo sapiens sapiens*. Others doubt that the Neandertals and moderns interbred and suspect that modern humans eventually displaced the Neandertals and drove them into extinction. But, if so, why did it take 50,000 years for the moderns of Qafzeh to spread into Europe, which had already been occupied for 40,000 years or more by Neandertals—their presumed biological and social inferiors?[74]

Human skulls from other caves near Qafzeh, especially Skhul and Tabun, raise other problems. Tabun Cave's complex strata have yielded various hominid fragments, dated variously to between about 60,000 and 40,000 years ago (some earlier dates have been obtained).[75] Other hominid fragments were found in Skhul Cave, just a few meters away. Although some levels of Skhul seem contemporaneous with some levels at Tabun, Skhul is generally regarded as at least several thousand years younger than Tabun. The Tabun hominids are primarily Mediterranean

4.14   Early *Homo sapiens* in Asia. Most of these sites are estimated to be between 100,000 and 50,000 years old.

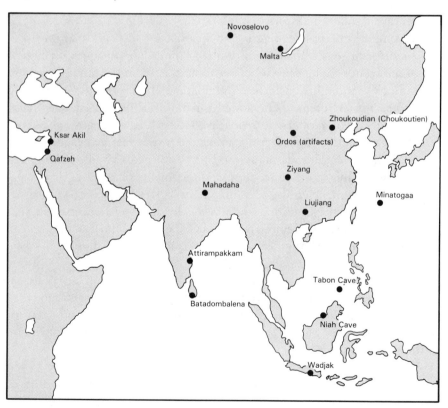

types of Neandertals, whereas the Skhul hominids are virtually modern in appearance. Some authorities see these two sites as evidence for local evolution of moderns from Neandertals, others consider it a case of hybridization, still others see it as a case of two distinct groups living side by side.[76]

A study of the Tabun Cave stone tools by Arthur Jelinek found that the stone tools here seemed to have changed gradually over a very long period, suggesting an indigenous development of "modern" stone tools from those associated with Neandertals.[77]

John Shea has identified what clearly seem to be stone spear points in 50- to 60,000-year-old levels of Kebara Cave, in 90- to 100,000-year-old layers of Qafzeh Cave, and at other sites as well. These stone points show the distinctive fracture patterns typical of impact (Figure 4.11). Shea's evidence casts doubt on the argument by Binford and others that Middle Pleistocene peoples rarely hunted larger animals with spears or other tools, and that in that lack of technologically assisted hunting they were different from modern *Homo sapiens sapiens*.[78]

Tabun, Skhul, Qafzeh, and other Middle Eastern sites probably mark the eastern limits of Neandertals, and it will probably be years before the relationships among these fossils and industries are clarified. But in recent years the context in which these Middle Eastern sites have been interpreted was changed by discoveries in Africa.

A critical African site in the transition to *Homo sapiens sapiens* may be Klasies River Mouth Cave, in South Africa; its interpretation has generated more argumentative articles than almost any other recent site.[79]

The excavators of this site, R. Singer and J. Wymer, concluded that the *Homo sapiens sapiens* mandible found here is "the earliest date for modern man so far from any site in the world," and that the many stone points found at the site indicate substantial hunting with hafted spears;[80] Binford interprets the same data as indicative of a "regular, directional trend in which hunting of small species and the young of large species increased simultaneously with a decrease in scavenging of marginal parts from carcasses of moderate to large animals previously ravaged by carnivores."[81]

Early hominid fossils of *Homo sapiens* have been found at Omo (Ethiopia), Laetoli (Tanzania), and Border Cave (South Africa), and these may represent the modern *Homo sapiens sapiens* whose genes spread through the Middle East and to the rest of the world, and from which we are all descended. This would then represent "a clinal pattern of the spread of modern *H. sapiens* from southern Africa through the Near East to Central and finally Western Europe."[82]

But as Fred Smith goes on to point out, there are many problems with this scenario. The African dates are almost all suspect, there seems to be an indigenous transition to modern *Homo sapiens sapiens* characteristics in some aspects of human cranial morphology in central and southern Europe, and it is difficult to demonstrate the kind of mechanisms of change that would explain the many post-cranial changes in these hominids.[83]

Also, in some ways it is difficult to imagine that the Neandertals were killed off directly by expanding populations of *Homo sapiens sapiens*. This theme is skillfully used by William Golding in his novel *The Inheritors*, where peaceful, egalitarian, vegetarian Neandertals face oblivion at the hands of villainous, meat-eating, beer-drinking *Homo sapiens sapiens*. Yet throughout the ages humans have expressed a fine democratic spirit in sexual affairs, and wherever different "races" have coexisted, they have interbred.

A very slight advantage in intelligence, fertility, or any other important aspect of life could easily lead to the extinction of subspecies of early humans, if they maintained genetic isolation, but such isolation seems to be unlikely, based on what we know of human behavior.

The confusion of the various models about the demise of the Neandertals and the emergence of modern *Homo sapiens sapiens* is made even more complex by archaeological discoveries that challenge traditional ideas about which sites are those left by Neandertals and those left by *H. sapiens sapiens*. This issue goes back to what Harvey Bricker has called the "classical" model of the Upper Paleolithic.[84] This model had its origins when, in 1868, five very ancient-appearing human skeletons were found in a rock shelter during a road-widening project near Les Eyzies in southern France. The first Neandertals had been discovered a few years before, and uneasy feelings about our descent from such barbaric-looking creatures had already begun to surface. But the bones from near Les Eyzies—named Cro-Magnon man after the rock shelter where they had been found—proved to be from individuals very much like modern Europeans in physical form. Here was an extremely ancient man, but of a race with which nineteenth-century Europeans could feel a strong sense of kinship and even pride. The discovery of these respectable ancestors stimulated great interest in prehistory, and amateur archaeologists soon began to pillage sites all over Europe.

French archaeologists initially defined the Upper paleolithic into periods based on specific types of tools, such as "end-scrapers," "burins," and long blade-like tools, and various kinds of bone, antler, and ivory artifacts; and because these tools were not known in many areas, including Southeast Asia and sub-Saharan Africa, they concluded that these areas did not "have" an Upper Paleolithic, even suggesting that some of these areas were culturally "retarded."

Recent research shows that old ideas about the direct equation between types of stone tools and types of hominids are probably wrong. The earliest stone-tool industry usually associated with the Upper Paleolithic, the

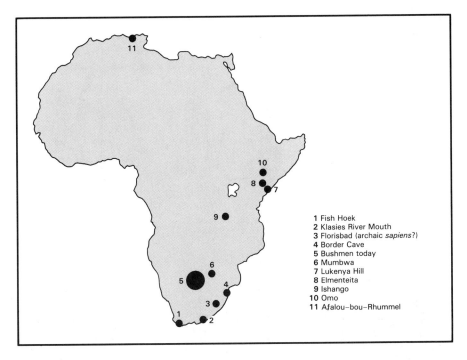

1 Fish Hoek
2 Klasies River Mouth
3 Florisbad (archaic *sapiens?*)
4 Border Cave
5 Bushmen today
6 Mumbwa
7 Lukenya Hill
8 Elmenteita
9 Ishango
10 Omo
11 Afalou–bou–Rhummel

4.15 Early *Homo sapiens* sites in Africa. The dates of some sites are controversial. Most are thought to be between 120,000 and 60,000 B.P.

Chatelperronian (c. 35,000 to 32,000 B.C.), has been found in association with a Neandertal skeleton at the site of Saint-Cesaire.[85] As Trinkhaus notes,

> If the Chatelperronian industry was produced solely by Neandertals and all the Aurignacian assemblages were the products of early modern humans . . . then there must have been temporal overlap of these two human groups in western Europe, given the contemporaneity and interstratification of Chatelperronian and early Aurignacian assemblages. . . . Even if a direct biological-industrial association cannot be assumed, it is still evident that Neandertals were present in western Europe less than [35,000 years ago] and that early modern humans were present there by at least [30,000 years ago].[86]

The significance of this is that 5000 years is a very short time for the physical changes observed between Neandertals and modern humans to have occurred.

In a systematic reanalysis of the Neandertal problem, Richard Klein concluded that

> . . . both the question of what happened to the Neanderthals and the closely related question of where and when modern people originated

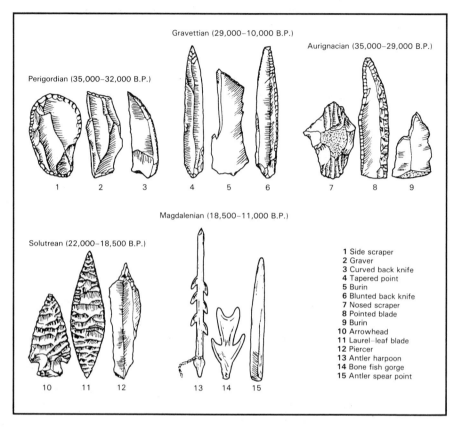

Gravettian (29,000–10,000 B.P.)

Aurignacian (35,000–29,000 B.P.)

Perigordian (35,000–32,000 B.P.)

1   2   3   4   5   6   7   8   9

Magdalenian (18,500–11,000 B.P.)

Solutrean (22,000–18,500 B.P.)

1 Side scraper
2 Graver
3 Curved back knife
4 Tapered point
5 Burin
6 Blunted back knife
7 Nosed scraper
8 Pointed blade
9 Burin
10 Arrowhead
11 Laurel-leaf blade
12 Piercer
13 Antler harpoon
14 Bone fish gorge
15 Antler spear point

10   11   12   13   14   15

4.16   European toolkits, 35,000 to 11,000 years ago. The increasingly diversified economies of the late Pleistocene are reflected in increasingly diverse and sophisticated tools kits, compared to earlier periods. (Not drawn to scale.)

remain inconclusively resolved. However, on present evidence, it is certainly reasonable to hypothesize that modern people first appeared in Africa. They may have spread early on to the Near East, if in fact the Near East was not essentially part of the basic African area in which they first evolved. Initially, their behavioral capabilities differed little from those of the Neanderthals, but eventually, perhaps because of a neurological change that is not detectable in the fossil record, they developed a capacity for culture that gave them a clear adaptive advantage over the Neanderthals and all other nonmodern people. The result was that they spread throughout the world, physically replacing all nonmoderns, of whom the last to succumb were perhaps the Neanderthals of western Europe.[87]

In summary of these many questions about our evolutionary history during the past 120,000 years, it is evident that the traditional ideas have been severely strained by recent finds and analyses. We may have originated as a species in one place, Africa or Asia probably, and then spread out to displace all other humans with little or no breeding between these groups. Alternatively, there may have been rapid and complex gene flows that kept most or all humans evolving in the same direction for the past several hundred thousand years. As is always the case in anthropology, more research—and some spectacular and lucky finds—will be required to resolve these questions.

Late Pleistocene Adaptations

If we are ever fortunate enough to find people of 30,000 years ago well-preserved in some glacier (the odds against this are considerable), we would probably see few if any differences between him or her and us. Over much of the world human dentition has become smaller during the past 30,000 years and some subtle changes associated with this have occurred, but not much else was different. We might wonder why human brain size stabilized around 1400 cubic centimeters and the answer is probably in the physical difficulties of giving live birth to an infant with cranial potential much larger than this. Also, after language abilities developed, people were somewhat "sheltered" from direct selection of increased brain size, in that any inventions, ideas, etc. could be conveyed to every member of the group, so that the whole population could take advantage of the intelligence of its most gifted members, and thus greater intelligence was probably a less direct factor in determining reproductive success.

With the appearance of modern humans after about 35,000 years ago, hunting and gathering societies began to exist in almost every part of the Old World, and in Australia and the Americas as well. Oscillating sea levels had some role, since in some periods land bridges were formed that allowed people to walk from southeast Asia to many areas that are now islands, such as Java, Sumatra, and Borneo. Rising sea levels may also have forced people into migrations: Thiel argues that people on islands in danger of being submerged by rising waters made the colonizing trips to Australia and elsewhere as early as 50,000 years ago.[88]

The chronology of the Upper Paleolithic has been clarified recently by the publication of a series of accelerator radiocarbon dates.[89] France seems to have been densely occupied during this period, particularly near the confluence of the Dordogne and Vezere rivers. This lovely part of the world is a well-watered, heavily forested limestone formation that is honeycombed with caves and rock shelters offering excellent places to live. Mammoths, horses, and many other animals were hunted by these Upper Paleolithic peoples, but the reindeer was the staff of life: at many sites 99% of all the animal bones found belonged to reindeer; reindeer hides provided clothing and coverings for shelters; reindeer antlers were the hammers, or the "batons," used to produce the long elegant blades for which these people are justly famous; and reindeer bone was the raw material for fish gorges, needles, awls, and other important tools.

Reindeer travel long distances each year as they follow the grazing lands from one climatic zone to another. Thus, Upper Paleolithic peoples of southern France could exploit through the reindeer herds land they had never seen; the reindeer would browse their way to the far north each year and then return to southern France for the winter, at which time they could be harvested.

Average group size may have been relatively large during the Upper Paleolithic in Europe, the Middle East, and other areas because of the requirements of hunting large gregarious mammals such as reindeer, horses, and wild cattle. The Neandertals apparently specialized in mammoths, rhinoceroses, and other animals whose movements are not easily manipulated by people and whose habits are, in some cases, more solitary than those of reindeer, horses, and wild cattle. With these latter species, an efficient hunting technique is the drive, where many people work together to stampede a herd over a cliff or into a bog. Such mass slaughter also requires many people to process the carcasses, and a large group would also have been advantageous in these circumstances as a means of defending particularly favorable places along migration routes.

The overall population also increased in some parts of Europe—and probably much of Eurasia—during the late Upper Paleolithic. Several factors were probably important in this population growth. The stone-tool technology of this period, with its indirect percussion and punch-blade techniques, was vastly more efficient than previous industries. Spear throwers, or atlatls, were also in common use—a very significant innovation, considering the heavy reliance on big game. It is estimated that an atlatl increases the range of a short spear from about 60 meters, if thrown by hand, to about 150 meters. The bow and arrow eventually added significantly to hunting effectiveness. Some of the earliest evidence of bows and arrows comes from the Stellmoor site, near Hamburg, Germany, where about a hundred wooden arrows dating to approximately 10,000 years ago were recovered. But the bow-arrow combination was probably invented independently many times and perhaps long before 10,000 years ago.

Life in the Upper Paleolithic was somewhat more severe than is often imagined. From a sample of seventy-six Upper Paleolithic skeletons drawn from sites in Europe and Asia, Vallois found that less than half these individuals had reached the age of twenty-one, that only 12% were over forty, and that not a single female had reached the age of thirty.[90] In fact, the distribution of ages and sexes represented by these skeletons was not significantly different from what one might expect from a comparable sample of Neandertals.[91] But even worse, many skeletons evidenced rickets, malnutrition, and other diseases and deformities. Not content with nature's provisions for population control, Upper Paleolithic peoples seem also to have occasionally slaughtered each other. At the site of Sandalja II (12,000 years ago), near Pula, Yugoslavia, for example, the skeletal remains of twenty-nine people were found in a smashed and splintered condition. Elsewhere, there is unmistakable evidence of wounds from arrows and spears.

By about 14,000 years ago, the people of western Europe had developed fish traps to harvest the countless salmon that migrate up the rivers there each year. This relatively late exploitation of fish in Europe has a parallel in prehistoric southeastern North America, where Native Americans lived for thousands of years subsisting primarily on deer, mussels, and a variety of plant foods, almost totally ignoring the myriad fish in nearby lakes and streams. If salmon were present in great numbers in European rivers during the Mousterian and early Upper Paleolithic, their exploitation may have been blocked by the terms of human adaptation to reindeer and other animals. Reindeer and other game would have been a more dependable resource in the sense that at least some of these animals would have been available year round, while the salmon would have been sharply seasonal. Salmon runs, in fact, might have conflicted with the scheduling of reindeer hunting, and as a consequence these peoples may have been far from the river, exploiting different resources, at the time the salmon were most available. Perhaps even more important, the successful exploitation of salmon would have required technological readaptation on a major scale. Catching salmon one by one would not have been especially productive; their real utility probably came only after nets, fish weirs, drying racks, smoking racks, and other largely nonportable technology came into common use.

The slow growth of worldwide human population density through most of the Pleistocene would seem to suggest that Upper Paleolithic populations were not in any sense "driven" by growing population size to exploit new resources, such as reindeer and salmon. Rather, it seems the reverse: as people began to devise ways to exploit rich salmon streams, reindeer herds, and other resources, larger groups could be supported. Just a slight increase in fertility or the number of offspring who lived on the average to reproductive age would in the long run produce vastly greater densities.

And it was not just in France and the richer parts of Europe that populations were growing in the late Pleistocene. Olga Soffer has documented, for example, the ingenuity of generations of peoples as they adapted to the harsh winters of the central Russian Plain.[92] One of the most amply documented Upper Paleolithic cultures in eastern Europe is the Kostenki-Bershevo culture centered in the Don River Valley, about 470 kilometers southeast of Moscow. About 25,000 to 11,000 years ago, the Kostenki-Bershevo area was an open grassland environment, with no rock shelters, caves, or other natural habitations, and with very little wood available for fires. People here left a variety of archaeological sites, including base camps, where pit houses were constructed by digging a pit a meter or so deep, ringing the excavation with mammoth bones or tusks, and then draping hides over these supports. The savage winters of Pleistocene Russia must have required constantly burning fires, and the great quantities of bone ash found at these sites indicate that these fires were often fed with mammoth bones in lieu of very scarce wood. Some excavated pit houses were relatively large, with many hearths, suggesting that several families may have passed the winter together. The people of Kostenki subsisted primarily through big-game hunting, mainly of mammoths or horses, with an occasional wild cow or reindeer. Numerous wolf and fox bones at these sites probably reflect the hunting of these animals for their fur for clothing. Like their Upper Paleolithic counterparts else-

4.17 This reconstruction of a 27,000-year-old settlement at Dolni Vestonice (Czechoslovakia) illustrates how people used shelters to colonize the frigid plains of eastern Europe.

4.18 Cast of Venus figurine from Willendorf, Austria. Late Pleistocene peoples appear to have celebrated human fertility in much of their art.

where, the Kostenki culture manufactured a variety of decorative items, including "Venus" figurines (representations of women, usually with exaggerated sexual characteristics).

Throughout much of central and eastern Europe, big-game hunting was the main subsistence basis for thousands of hunting and gathering bands, much as it had been for tens of thousands of years.[93] But the mammoth hunters of central and eastern Europe probably did not aggregate into relatively large groups, as the people in western Europe and the Near East at this time are thought to have done. Mammoths probably were not as gregarious as reindeer, horses, or wild cattle, and their hunting may have required only a few men working together.

Until recently, few Upper Paleolithic sites were known in East Asia. Excavations at Zhoukoudian revealed levels dating to about 10,000 years ago containing approximately seven individuals—all of whom had been killed, but apparently not eaten. One individual had clearly died from an arrow or small spear wound to the skull, and another had been beaten about the head with a large stone. Elsewhere, two skulls have been retrieved from Wadjah, in central Java, but dating these has proved difficult.

Hundreds of late Pleistocene sites have been found in Japan. Dating these sites is difficult, but the classic European Upper Paleolithic blade and burin industries are well represented in Japan, particularly in the northern areas across from Siberia.

The earliest known stone tools from Southeast Asia may be those of the Sonviian assemblage from northern Vietnam, dating to about 23,000 B.P.,[94] but the Hoabinihian lithics of about 12,000 years ago are the earliest widespread lithic Upper Paleolithic industry.

One of the most curious aspects of late Pleistocene cultures in Asia is the movement of people from Asia into Australia and New Guinea (which were connected by a land bridge). Between 50,000 and 30,000 years ago, the ancestors of the Australian Aborigines somehow managed to cross at least a few kilometers of open ocean to reach Australia. It is perhaps not so unlikely that an occasional boat of fisherfolk was shipwrecked on the New Guinea-Australia coast, but computer simulations that take into account normal fertility rates and genetic diversity of modern populations suggest that more than just a boatload or two of colonists founded that country's present aboriginal population.[95] Two distinct groups of ancient people have been found there.[96] Of the various possible reasons for this morphological divergence, the most likely now seems to be colonization by two different groups, one a more robust type, by 50,000 years ago, the other, a more gracile people, before 20,000 years ago.

Like their contemporaries in America, the late Pleistocene peoples in Australia lived at a time when many large and small animal species were becoming extinct, and their possible role in this extinction pattern remains a matter of controversy.

Late Pleistocene peoples of the Americas are considered in the next chapter.

Life, Art, and Ritual in the Upper Paleolithic

In 1868, near the Spanish port of Santander, a hunter's dog fell into a crevice in some boulders, and in rescuing the animal the hunter moved some rocks, revealing the opening of a cave. The owner of the land on which the cave (known as Altamira) was located, a Spanish nobleman and amateur archaeologist, eventually began to excavate the cave floor. He found some stone artifacts, but, according to the story, was unaware of the paintings in the cave until his twelve-year-old daughter visited the site and glanced at the ceiling. In the glow of her lantern she saw beautiful visions of animals. The central painting is of a group of about twenty-five animals, mainly bison, with a few horses, deer, wolves, and boars. Roughly life size, these paintings were done in rich browns, yellows, reds, and blacks, and the natural configuration of the cave ceiling had been used to emphasize the shape of the animals. The rounded haunch of a bison, for example, was painted over a natural bulge in the stone ceiling, creating a three-dimensional effect.[97]

Scholarly reception to the Altamira discoveries was almost uniformly negative. Some respected prehistorians even hinted that Don Marcelino, their discoverer, had hired an art student to fake these paintings, while another scholar dismissed them as simply the expression "of a mediocre student of the modern school." So abused by critics was the Don that eventually he padlocked the cave, and he died in 1888 without having seen his discoveries accepted as true Paleolithic expressions. Years later, when many more paintings and other art works had been discovered, the antiquity of Altamira was finally acknowledged, and most of these paintings are now given dates between about 34,000 and 12,000 years ago. Analysis shows that the colors were produced by mixing natural mineral pigments, such as ocher and manganese dioxide, with a binder (blood, urine, vegetable juice, or something similar), and that they were either brushed on with an implement made of animal hair or applied by making a kind of crayon from the pigments and lubricant. Some painting may also have been done by using a pipe to blow the powdered pigments on a surface prepared with animal fat.

Many of these paintings were executed in the dark recesses of caves, by light provided by lamps made of stone bowls filled with animal fat, with a wick made of lichens, grass, or juniper.[98]

During World War II paintings on a scale comparable to those of Altamira were discovered at Lascaux Cave, in France. Researchers estimate that the Lascaux paintings date from about 34,000 to 12,000 years ago, and were done on many different occasions within this span. Many varieties of animals are depicted here, including some one hopes were imaginary. The animals are often painted as if they are in motion, and the general effect is very impressive (Figure 4.19). One of the many curious things about these and other Upper Paleolithic cave paintings is that while the animals are depicted in very real, very representational terms, the figures of humans are either simple stick drawings or else weird half-humans, half-animals.

Since the Altamira and Lascaux discoveries, rock carvings, paintings, and other Paleolithic representations have been found all over the world, and many scholars have tried to assess what these pictures and paintings mean. We must accept at the onset that we can never really know the thoughts of these long-dead Paleolithic artists. The same might be said of Vincent Van Gogh or any other artist, of course, given that aesthetic

4.19   Lascaux Cave paintings may have been rituals designed to increase the chances of successful hunting, or simple expressions of aesthetic sensitivity, but they are superb art whatever their purpose.

expressions can never be fully, rationally, comprehended, even by the artist. But neither Van Gogh's work nor Paleolithic art can be expected to be random with regard to theme, technique, or style. They can be expected to tell us something about their authors. And in any case, there is something profoundly unsatisfying about analyses of ancient peoples based only on stone tools, hut foundations, and other techno-environmental residues. Most people, including archaeologists, wish to know the "minds" of ancient peoples, and in few ways do the minds of ancient peoples seem so accessible as in their art.

People apply the term Paleolithic art to a great range of materials, including cave paintings, rock carvings, sculpted and carved animal bones, ivory statuettes, and baked clay objects. As Margaret Conkey noted,[99] one should probably not imagine that all these expressions were fundamentally aesthetic in nature, in the sense that we think of aesthetics as removed from economic function. An early interpretation of the great cave paintings was that they were sympathetic magic, done to ensure success in hunting and other activities. By picturing animals with spears stuck in them or as caught in traps, Upper Paleolithic people may have thought they increased their chances of killing and trapping these animals. Many of the paintings are in small, hidden passages where working conditions were very cramped, suggesting that these pictures were not created for the pleasure of the general viewing public. Then again, many paintings are superimposed on one, two, or even more older ones, indicating perhaps that these efforts were ritual in nature, not simply artistic. And it may be significant that animals most likely to be feared in hunting, such as mammoths, are portrayed more frequently than less dangerous prey.[100] In fact, the most common themes of these Upper Paleolithic artists were food and sex, with food receiving most of the attention.

The disemboweled bulls, prancing deer, and other hunting scenes, plus the popularity of the penis and vulva motifs, suggest to some that these earliest of Spanish and French impressionists were men. But recently Elizabeth Fisher has argued that students of cave art have concealed the high frequency of female sex organs represented and thus the implication that many of the artists were probably women.[101] Line markings that some archaeologists have considered calendrical devices Fisher thinks may be records of menstrual periods.

It would not be surprising, however, if the many incisings and markings that one finds on Upper Paleolithic artifacts and in caves are in fact calendrical or astronomical in nature.[102] A Pleistocene hunting and gathering group surprised by the onset of winter or the seasonal unavailability of plants and animals would not only have been invincibly stupid, it would have been in deep trouble.

These economic interpretations of Paleolithic art were challenged by Andre Leroi-Gourhan, who plotted the relative frequencies of lions, mammoths, bisons, reindeer, and other animals in caves with many such representations, and concluded that these paintings were invested with cosmological significance—reflecting in various ways the patterns in which Paleolithic peoples ordered their world.[103]

In recent years various alternative explanations of Paleolithic art have been suggested. Lewis-Williams and Dowson have argued that much of Paleolithic art was the product of people in "entopic," or altered, states of consciousness—either through drugs or meditation.[104] They note that ethnographic studies of contemporary hunter-gatherers show wide use of hallucinogens, trances, and such altered states to produce paintings and carvings. They also argue that it seems a feature of human neuroanatomy that images perceived in altered states of consciousness include both "real" representational forms and fantastic nonrepresentational forms, and that people tend to project these images on walls and ceilings in their minds.

Thus, "Tracing projected mental images with a finger in the sand or on the soft wall of a cave to experience them more fully would have 'fixed' them and would have been an initial step in the history of art. They were merely touching them and marking *what was already there*."[105]

Others have argued that the floresence of Paleolithic art was a result of the maturing mentalities of our ancestors, as they replaced the less-intelligent Neandertals.[106]

The study of Paleolithic visual imagery and art is a demonstration of a point made well by Hodder, that archaeology is not necessarily a neutral discipline, in which analyses are scientific and culture-free.[107] Paleolithic art has often been a "Rorschach test" in the sense that modern-day observers have tried to read into it the mind and spirit of primitive humans, but they perhaps have learned more about their own psyches than about the primitive. In any case, as Conkey notes, " 'paleolithic art' [is] an extremely diverse and abundant repertoire of material culture that cannot be accounted for by any inclusive umbrella except perhaps as 'cultural'."[108]

Yet John Halverson argues that "If, as I suppose, Paleolithic art had no practical function, it may be the first cultural work of mankind to be freed from praxis and therefore belongs in the general category of play and in the specific category of play of the mind."[109]

Finally, it is interesting to note that Colin Renfrew has suggested that

4.20   Many cave paintings include realistic depictions of sexual organs, and some scholars interpret some abstract cave paintings as stylized sexual representations. Leroi-Gourhan here illustrates three groups of symbols for each sex in normal, simplified, and derived forms.

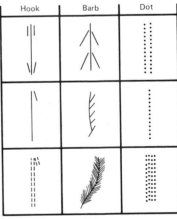

Female   Male

4.21 After the end of the Pleistocene about 12,000 years ago, hunter-gatherers spread into almost all of the world's environments and adapted to local environments. These remains of East African (Somalia) burials date to between about 8100–5400 B.P. and are some of the earliest known African examples of burials with grave goods—in this case, skulls of the lesser kudu.

the authors of much of the Upper Paleolithic art of Europe may have spoken a language ancestral to Indo-European.[110]

Summary and Conclusions

We began this chapter with our early ancestors spread out along the warm margins of Africa, the Middle East, and Asia, making a reasonable living with little more than crude stone tools, intellectual abilities far poorer than our own, and probably only the rudiments of human social organization. We end this chapter, in effect, with ourselves, in the form of our late Pleistocene relatives of just 12,000 years ago.

If economy is destiny, to some extent, then the motive factors of human evolution in this long period must remain a matter of doubt and ambiguity. Hunting in temperate climates has long been suspected as an important part of this evolutionary mix, but early *Homo sapiens* sites in Africa and Asia make this unlikely, as does the problematical nature of the temperate zone archaeological sites. Lewis Binford concludes that "At present the inevitable conclusion seems to be that regular, moderate- to large-mammal hunting appears simultaneously with the foreshadowing changes occurring just prior to the appearance of fully modern man. . . . Systematic hunting of moderate to large animals appears to be a part of our modern condition, not its cause."[111] Even if this turns out not to be entirely accurate, it remains unclear how hunting and the other economic activities of the Pleistocene might have changed our ancestors into ourselves.

And we should also take note of the fact that even though we are physically very much like our ancestors of 12,000 years ago, we are enormously different culturally. Human minds and lives are so directly determined by cultural, not just physical aspects of life, that in a cultural sense we are not at all the same species as the human hunter-gatherers of the late Pleistocene. Yet we are in many ways a Pleistocene animal living in a cultural universe that is tremendously different from that of the Pleistocene. Millions of years of evolution have shaped us physically as animals that do well as omnivorous hunter-foragers living active lives in small groups. Perhaps this explains why modern life inflicts ulcers, early heart attacks, obesity, and neuroses on so many city-dwelling Westerners. To be in accord with nature—that is, to live a life for which evolution has shaped us—we should probably eat a varied diet, live closely in a small group, and walk a lot. Other than that, however, we are the supreme generalists, adaptable to everything from a life as a North Pole seal hunter to that of an advertising account executive.

Notes

1. Day, *Guide to Fossil Man*, p. 409.
2. Ibid., p. 238.
3. Ibid., p. 410.
4. See, e.g., Smith and Spencer, eds., *The Origins of Modern Humans*.
5. Foley, "Hominid Species and Stone-tool Assemblages: How Are They Related?," p. 382.
6. Ibid., p. 387.
7. Day, *Guide to Fossil Man*, p. 235.
8. Reviewed in Day, *Guide to Fossil Man*, pp. 177–178, for Olduvai Gorge.

9. Isaac, "The Archaeology of Human Origins," p. 60.
10. Isaac, "Sorting Out the Muddle in the Middle—An Anthropologist's Post-Conference Appraisal," p. 504.
11. Shipman, "Early Hominid Lifestyle."
12. Binford, *Bones, Ancient Men and Modern Myth*, p. 294; idem., "Human Ancestors."
13. Bar-Yosef, "Archeological Occurrences in the Middle Pleistocene of Israel."
14. Binford, "Human Ancestors," p. 314.
15. Roberts, "Body Weight, Race and Climate."
16. Jorgensen, "A Contribution of the Hypothesis of a Little More Fitness of Blood Group O."
17. Reviewed in Day, *Guide to Fossil Man*, pp. 368–69.
18. Day, *Guide to Fossil Man*, p. 369.
19. Binford and Stone, "Zhoukoudien: A Closer Look," p. 469.
20. Arens, *The Man-Eating Myth*.
21. Garn and Block, "The Limited Nutritional Value of Cannibalism."
22. Shapiro, *Peking Man*.
23. de Lumley, "Cultural Evolution in France in Its Paleoecological Setting During the Middle Pleistocene," p. 752; Howell, "Observations on the Earlier Phases of the European Lower Paleolithic."
24. de Lumley, "Cultural Evolution in France in Its Paleoecological Setting During the Middle Pleistocene."
25. Sevink et al., "A Note on the Approximately 730,000-year-old Mammal Fauna and Associated Human Activity Sites near Isernia, Central Italy."
26. Freeman, "Acheulian Sites and Stratigraphy in Iberia and the Maghreb," p. 664.
27. Freeman, "Acheulian Sites and Stratigraphy in Iberia and the Maghreb," p. 682, suggests large groups of hunters; but see Binford, "Human Ancestors," and James, "Hominid Use of Fire," for contrasting views.
28. de Lumley, H., "A Paleolithic Camp at Nice"; Villa, *Terra Amata and the Middle Pleistocene Archaeological Record of Southern France*.
29. Day, *Guide to Fossil Man*, p. 100.
30. Reviewed in Howell, "Observations of the Earlier Phases of the European Lower Paleolithic."
31. Reviewed in Day, *Guide to Fossil Man*, p. 102.
32. Bender, "Comment on G. Krantz (*CA* 21:773–79)."
33. White, "Rethinking the Middle/Upper Paleolithic Transition."
34. Wobst, "Stylistic Behavior and Information Exchange."
35. White, "Rethinking the Middle/Upper Paleolithic Transition."
36. Stringer et al., "The Significance of the Fossil Hominid from Petralona, Greece"; Trinkhaus, "Evolutionary Continuity among Archaic *Homo sapiens*."
37. Pfeiffer, *The Emergence of Man*, p. 173.
38. Day, *Guide to Fossil Man*, p. 20.
39. Ibid., p. 22.
40. Roe, *The Lower and Middle Paleolithic Periods in Britain*.
41. Binford, "Human Ancestors," pp. 316–17.
42. de Lumley, "Ante-Neanderthals of Western Europe."
43. Svoboda, "Lithic Industries of the Arago, Vértesszöllös, and Bilzingsleben Hominids: Comparison and Evolutionary Interpretation."
44. Binford, "Human Ancestors," p. 318. See also Rigaud, "Le Paleolithique en Perigord."
45. Day, *Guide to Fossil Man*, p. 23.
46. Holloway, "Comment on 'On Depiction and Language.'"
47. Trinkhaus, "The Neandertals and Modern Human Origins."
48. Brace, "The Fate of the 'Classic' Neanderthals: A Consideration of Human Catastrophism."
49. Straus and Cave, "Pathology and the Posture of Neanderthal Man."

50. Trinkhaus, "The Neandertals and Modern Human Origins."
51. Ibid., p. 203.
52. Ibid., p. 205.
53. Bordes, *Typologie du Paleolithique Ancien et Moyen.*
54. Binford and Binford, "A Preliminary Analysis of Functional Variability in the Mousterian of Levallois Facies."
55. Binford, "Human Ancestors," p. 319.
56. Lieberman and Crelin, "On the Speech of Neanderthals."
57. Carlisle and Siegel, "Some Problems in the Interpretation of Neanderthal Speech Capabilities."
58. For a review of the evidence and dissenting comments on this review, see Gargett, "Grave Shortcomings: The Evidence for Neandertal Burial"; Chase and Dibble, "Middle Paleolithic Symbolism"; Brace, "Review of Shanidar: *The First Flower People,* by R. Solecki," p. 86.
59. Trinkhaus, "The Neanderthals and Modern Human Origins."
60. Howells, "Neanderthal Man: Facts and Figures," p. 405.
61. Day, *Guide to Fossil Man,* pp. 414–16.
62. Brace, *The Stages of Human Evolution: Human and Cultural Origins.* See also Smith and Ranyard, "Evolution of the Supraorbital Region in Upper Pleistocene Fossil Hominids from South-central Europe."
63. Lovejoy and Trinkhaus, "Strength of Robusticity of the Neanderthal Tibia."
64. Smith, "Upper Pleistocene Hominid Evolution in South-Central Europe: A Review of the Evidence and Analysis of Trends."
65. Day, *Guide to Fossil Man,* p. 415. See also Trinkhaus and Howells, "The Neanderthals."
66. Boule, *Les Hommes Fossiles.*
67. Day, *Guide to Fossil Man,* p. 415.
68. See Tierney et al., "The Search for Adam and Eve."
69. Cann et al., "Mitochondrial DNA and Human Evolution."
70. Wilson was quoted by T.A. Maugh II in the *Los Angeles Times* (reprinted in *The Seattle Times,* Thursday, October 5, 1989) as concluding on the basis of his most recent research that "We are all derivative of one of the !Kung lineages," at a date he now estimates as 142,000 years ago.
71. Valladas et al., "Thermoluminescence Dating of Mousterian 'Proto-Cro-Magnon' Remains."
72. Trinkhaus, quoted in B. Bower, "An Earlier Dawn for Modern Humans?" p. 138.
73. Wolpoff, quoted in B. Bower, ibid.
74. Bower, "An Earlier Dawn for Modern Humans?" p. 138.
75. Day, *Guide to Fossil Man,* p. 109.
76. Ibid., p. 118.
77. Jelinek, "The Tabun Cave and Paleolithic Man in the Levant"; Jelinek, "The Middle Palaeolithic in the Southern Levant, with Comments on the Appearance of Modern *Homo sapiens.*"
78. Shea, "Spear Points from the Middle Paleolithic of the Levant."
79. See, for example, Binford, *Faunal Remains from Klasies River Mouth;* Binford, "Reply to J. F. Thackeray 'Further Comment on Fauna from Klasies River Mouth' "; Singer and Wymer, *The Middle Stone Age at Klasies River Mouth in South Africa;* Thackeray, "Further Comment on Fauna from Klasies River Mouth."
80. Singer and Wymer, *The Middle Stone Age at Klasies River Mouth in South Africa.*
81. Binford, "Reply to J. F. Thackeray 'Further Comment on Fauna From Klasies River Mouth,' " p. 60.
82. Smith, "Upper Pleistocene Hominid Evolution in South-Central Europe: A Review of the Evidence and Analysis of Trends," p. 698.
83. Ibid.
84. Bricker, "Upper Paleolithic Archaeology."

85. Leveque and Vandermeersch, "Les Decouvertes de Restes Humains."
86. Trinkhaus, "The Neandertals and Modern Human Origins," p. 198.
87. Klein, *The Human Career*. I thank Richard Klein for kindly supplying me with an excerpt from his book, which was still in press when this manuscript was being completed.
88. Thiel, "Early Settlement of the Philippines, Eastern Indonesia, and Australia-New Guinea: A New Hypothesis."
89. Mellars et al., "Radiocarbon Acclerator Dating of French Upper Paleolithic Sites."
90. Vallois, "The Social Life of Early Man: The Evidence of the Skeletons."
91. Ibid.
92. Soffer, *The Upper Paleolithic of the Central Russian Plain.*
93. See, for example, Klein, *Ice-Age Hunters of the Ukraine;* Soffer, *The Upper Paleolithic of the Central Russian Plain.*
94. Bellwood, *The Prehistory of the Indo-Malaysian Archipelago.*
95. White and O'Connell, "Australian Prehistory. New Aspects of Antiquity."
96. Thorne, "The Arrival of Man in Australia."
97. Prideaux et al., *Cro-Magnon Man*, pp. 93–94.
98. De Beaune, "Paleolithic Lamps and Their Specialization: A Hypothesis."
99. Conkey, "New Approaches in the Search for Meaning? A Review of Research in 'Paleolithic Art.'"
100. Rice and Paterson, "Cave Art and Bones: Exploring the Interrelationships," p. 98.
101. Fisher, *Woman's Creation.*
102. Marshack, "Some Implications of the Paleolithic Symbolic Evidence for the Origins of Language."
103. Leroi-Gourhan, "The Evolution of Paleolithic Art."
104. Lewis-Williams and Dowson, "The Signs of All Times: Entopic Phenomena in Upper Paleolithic Art."
105. Ibid., p. 215 (emphasis theirs).
106. See, for example, Pfeiffer, "Cro-magnons Were Really Us, Working Out Strategies For Survival."
107. Hodder, *Reading the Past.*
108. Conkey, "New Approaches in the Search for Meaning? A Review of Research in 'Paleolithic Art,'" p. 422.
109. Halverson, "Art for Art's Sake in the Paleolithic," p. 71.
110. Renfrew, "Archaeology and Language," p. 439.
111. Binford, "Human Ancestors," p. 321.

Bibliography

Arens, W. 1979. The Man-Eating Myth. New York: Oxford University Press.

Bar-Yosef, O. 1975. "Archaeological Occurrences in the Middle Pleistocene of Israel." In *After the Australopithecines*, eds., K. W. Butzer and G. L. Isaac. The Hague: Mouton.

Bar-Yosef, O., B. Vandermeersch, B. Arensburg, P. Goldberg, H. Laville, L. Meignen, Y. Rak, E. Tchernov, and A. Tillier. 1986. "New Data on the Origin of Modern Man in the Levant." *Current Anthropology* 27(1):63–65.

Barnard, A. 1983. "Contemporary Hunter-Gatherers: Current Theoretical Issues in Ecology and Social Organization." *Annual Review of Anthropology* 12:193–214.

Bellwood, P. 1986. *The Prehistory of the Indo-Malaysian Archipelago*. Orlando, Fla.: Academic Press.

———. 1987. "The Prehistory of Island Southeast Asia: A Multidisciplinary Review of Recent Research." *Journal of World Prehistory* 1(2):171–224.

Bender, M. 1983. "Comment on G. Krantz (*CA* 21:773–79)." *Current Anthropology* 24(1):113.

Binford, L. R. 1968. "Post-Pleistocene Adaptations." In *New Perspectives in Archeology*, eds. L. Binford and S. Binford. Chicago: Aldine.

———. 1981. *Bones. Ancient Men and Modern Myth*. New York: Academic Press.

———. 1982. "Comment on Rethinking the Middle/Upper Paleolithic Transition." *Current Anthropology* 23:177–81.

———. 1984. *Faunal Remains from Klasies River Mouth*. New York: Academic Press.

———. 1985. "Human Ancestors: Changing Views of Their Behavior." *Journal of Anthropological Archaeology* 4(4):292–327.

———. 1986. "Reply to J. F. Thackeray ('Further Comment on Fauna From Klasies River Mouth' *Current Anthropology* 27(5):511–12)." *Current Anthropology* 27(1):57–62.

Binford, L. R. and S. Binford. 1966. "A Preliminary Analysis of Functional Variability in the Mousterian of Levallois Facies." In *Recent Studies in Paleoanthropology*. *American Anthropologist* special publication, pp. 238–95.

Binford, L. and N. Stone. 1986. "Zhoukoudien: A Closer Look." *Current Anthropology* 27(5):453–75.

Boaz, N. T. 1982. "Comment on F. H. Smith 'Upper Pleistocene Hominid Evolution in South-Central Europe: A Review of the Evidence and Analysis of Trends.'" *Current Anthropology* 23(6):667–703.

Bordes, F. 1961a. *Typologie du paleolithique ancien et moyen*. Bordeaux: Publication de l'Institut de Prehistoire de l'Universite de Bordeaux.

———. 1961b. "Mousterian Cultures in France." *Science* 134:803–10.

———. 1968. *The Old Stone Age*. London: Weidenfeld & Nicholson.

———. 1972. *A Tale of Two Caves*. New York: Harper & Row.

———. 1978. "Typological Variability in the Mousterian Layers at Pech de l'Aze II and IV." *Journal of Anthropological Research* 34:181–93.

Bordes, F. and D. de Sonneville-Bordes. 1970. "The Significance of Variability in Paleolithic Assemblages." *World Archaeology* 2:61–73.

Boule, M. 1923. *Les Hommes Fossiles: Elements de Paleontologie Humaine*. 2nd Ed. Paris: Masson et Cie.

Boule, M. and H. Vallois. 1932. *Fossil Men*. London: Thames and Hudson.

Bower, B. 1988. "An Earlier Dawn for Modern Humans?" *Science News* 133:138.

Brace, C. L. 1964. "The Fate of the 'Classic' Neanderthals: A Consideration of Human Catastrophism." *Current Anthropology* 5:3.

———. 1967. *The Stages of Human Evolution: Human and Cultural Origins*. Englewood Cliffs, N.J.: Prentice-Hall.

———. 1975. "Review of *Shanidar: the First Flower People*, by R. Solecki." *Natural History* 80:82–86.

Bricker, H. 1976. "Upper Paleolithic Archaeology." *Annual Review of Anthropology* 5:133–48.

Brose, D. and M. Wolpoff. 1971. "Early Upper Paleolithic Man and Late Paleolithic Tools." *American Anthropologist* 73:1156.

Brothwell, D. 1961. "Upper Pleistocene Human Skull from Niah Caves, Sarawak." *Sarawak Museum Journal* 9:323.

Butzer, K. W. and G. L. Isaac, eds. 1975. *After the Australopithecines*. The Hague: Mouton.

Campbell, B. 1985. *Humankind Emerging*. 3rd Ed. Boston: Little, Brown.

Cann, R. L., M. Stoneking, and A. C. Wilson. 1987. "Mitochondrial DNA and Human Evolution." *Nature* 325:31–36.

Carlisle, R. C. and M. I. Siegel. 1974. "Some Problems in the Interpretation of Neanderthal Speech Capabilities. A Reply to Lieberman." *American Anthropologist* 76:319–22.

Chase, P. G. and H. L. Dibble, 1987. "Middle Paleolithic Symbolism: A Review of Current Evidence and Interpretations." *Journal of Anthropological Archaeology* 6:263–96.

Conkey, M. 1980. "The Identification of Prehistoric Hunter-Gatherer Aggregation Sites: The Case of Altamira." *Current Anthropology* 21:609–30.

———. 1987. "New Approaches in the Search for Meaning? A Review of Research in 'Paleolithic Art.' " *Journal of Field Archaeology* 14(4):413–30.

Day, M. H. 1986. *Guide to Fossil Man*. 4th Ed. Chicago: University of Chicago Press.

De Beaune, S. A. 1987. "Paleolithic Lamps and Their Specialization: A Hypothesis." *Current Anthropology* 28(4):569–77.

Dubois, E. 1921. "The Proto-Australian Fossil Man of Wadjak, Java." *Proceedings: Koninklijke Nederlandse Akademie van Wetenschappen* 23:1013.

Edey, M. A. and Editors of Time-Life. 1972. *The Missing Link*. New York: Time-Life.

Falk, D. 1975. "Comparative Anatomy of the Larynx in Man and the Chimpanzee: Implications for Language in Neanderthal." *American Journal of Physical Anthropology* 43:123–32.

Farb, P. and G. Armelagos. 1980. *Consuming Passions*. Boston: Houghton Mifflin.

Fisher, E. 1979. *Woman's Creation*. New York: McGraw-Hill.

Foley, R. 1987. "Hominid Species and Stone-tool Assemblages: How Are They Related?" *Antiquity* 61:380–92.

Frayer, D. W. 1981. "Body Size, Weapons Use, and Natural Selection in the European Upper Paleolithic and Mesolithic." *American Anthropologist* 83:57–73.

Freeman, L. G. 1975. "Acheulian Sites and Stratigraphy in Iberia and the Maghreb." In *After the Australopithecines*, eds., K. W. Butzer and G. L. Isaac. The Hague: Mouton.

Freeman, M. 1971. "A Social and Economic Analysis of Systematic Female Infanticide." *American Anthropologist* 73:1011–18.

Frolov, B. A. 1978–1979. "Numbers in Paleolithic Graphic Art and the Initial Stages of Development of Mathematics." *Soviet Anthropology and Archeology* 17:41–74.

Gabow, S. L. 1977. "Population Structure and the Rate of Hominid Brain Evolution." *Journal of Human Evolution* 6:643–65.

Gamble, C. 1986. *The Palaeolithic Settlement of Europe*. Cambridge, England: Cambridge University Press.

Gargett, R. H. 1989. "Grave Shortcomings: The Evidence for Neandertal Burial." *Current Anthropology* 30(2):157–90.

Garn, S. and W. Block. 1970. "The Limited Nutritional Value of Cannibalism." *American Anthropologist* 72:106.

Gould, S. J. 1980. *The Panda's Thumb*. New York: Norton.

Gron, O. 1987. "Seasonal Variation in Maglemosian Group Size and Structure." *Current Anthropology* 28(3):303–27.

Halverson, J. 1987. "Art for Art's Sake in the Paleolithic." *Current Anthropology* 28(1):63–89.

Hodder, I. 1986. *Reading the Past: Current Approaches to Interpretation in Archaeology*. Cambridge, England: Cambridge University Press.

Holloway, R. L., Jr. 1989. Comment on "On Depiction and Language." *Current Anthropology* 30(3):331–32.

Howell, F. C. 1961. "Isimila: A Paleolithic site in Africa." *Scientific American* 205:118–31.

———. 1965. *Early Man*. New York: Time-Life.

———. 1966. "Observations on the Earlier Phases of the European Lower Paleolithic." *American Anthropologist* 68(2)(pt.2):88–201.

Howells, W. W. 1975. "Neanderthal Man: Facts and Figures." In *Australopithecus Paleonanthropology. Morphology and Paleoecology*, ed. R. H. Tuttle. The Hague: Mouton.

Isaac, G. 1975. "Sorting Out the Muddle in the Middle—An Anthropologist's Post-Conference Appraisal." In *After the Australopithecines*, eds., K. W. Butzer and G. Isaac. The Hague: Mouton.

———. 1984. "The Archaeology of Human Origins: Studies of the Lower Palaeolithic in East Africa, 1971–1981." *Advances in World Archaeology* 3:1–89.

James, S. R. 1989. "Hominid Use of Fire in the Lower and Middle Pleistocene: A Review of the Evidence." *Current Anthropology* 30(1):1–26.

Jelinek, A. 1982a. "The Tabun Cave and Paleolithic Man in the Levant." *Science* 216:1369–75.

———. 1982b. "The Middle Palaeolithic in the Southern Levant, with Comments on the Appearance of Modern *Homo sapiens*." In *The Transition from Lower to Middle Palaeolithic and Origin of Modern Man*, ed. A. Ronen. B.A.R. International Series 151.

Jorgensen, G. 1977. "A Contribution of the Hypothesis of a Little More Fitness of Blood Group O." *Journal of Human Evolution* 6:741–44.

Klein, R. G. 1973. *Ice-Age Hunters of the Ukraine*. Chicago: University of Chicago Press.

———. 1989. *The Human Career*. Chicago: University of Chicago Press.

Larichev, V., U. Khol'ushkin, and I. Laricheva. 1987. "Lower and Middle Paleolithic of Northern Asia: Achievements, Problems, and Perspectives." *Journal of World Prehistory* 1(4):415–64.

Leakey, M. D. 1975. "Cultural Patterns in the Olduvai Sequence." In *After the Australo-pithecines*, eds. K. W. Butzer and G. L. Isaac. The Hague: Mouton.

Leroi-Gourhan, A. 1968. "The Evolution of Paleolithic Art." *Scientific American* 209(2):58–74.

———.1982. *The Dawn of European Art*. Cambridge, England: Cambridge University Press.

Leveque, F. and B. Vandermeersch. 1980. "Les Decouvertes de Restes Humains dans un Horizon Castelperronien de Saint-Cesaire (Charente-Maritime)." *Bulletin de la Societe Prehistorique Francaise* 77:35.

Lewin, R. 1987. "Debate over Emergence of Human Tooth Pattern." *Nature* 235:748–50.

Lewis-Williams, J. D. and T. A. Dowson. 1988. "The Signs of All Times: Entopic Phenomena in Upper Paleolithic Art." *Current Anthropology* 29(2):201–46.

Lieberman, P. E. and E. S. Crelin. 1971. "On the Speech of Neanderthals." *Linguistic Inquiry* 2:203–22.

Lieberman, P. E., E. S. Crelin, and D. H. Klatt. 1972. "Phonetic Ability and Related Anatomy of the Newborn and Adult Human, Neanderthal Man, and the Chimpanzee." *American Anthropologist* 74:287.

Littlefield, A., L. Lieberman, and L. T. Reynolds. 1982. "Redefining Race: The Potential Demise of a Concept in Physical Anthropology." *Current Anthropology* 23(6):641–55.

Lovejoy, O. and E. Trinkhaus. 1980. "Strength of Robusticity of the Neanderthal Tibia." *American Journal of Physical Anthropology* 53:465–70.

Luguet, G. 1930. *The Art and Religion of Fossil Man*. New Haven: Yale University Press.

Lumley, H. de. 1969. "A Paleolithic Camp at Nice." *Scientific American* 225:42–59.

———. 1969. *Le Paleolithique Inferieur et Moyen du Midi Mediterraneen dans son Cadre Geologique, Vol. 1: Ligurie Provence*. Paris: Editions du Centre National de La Recherche Scientifique.

———. 1971. *Le Paleolithique Inferieur et Moyen du Midi Mediterranean dans son Cadre Geologique, Vol. 2: Bas—Languedoc, Roussillon, Catalogue*. Paris: Editions du Centre National de la Recherche Scientifique.

———. 1975. "Cultural Evolution in France in Its Paleoecological Setting During the Middle Pleistocene." In *After the Australopithecines*, eds. K. W. Butzer and G. L. Isaac. The Hague: Mouton.

Lumley, H. de and M-A. de Lumley. 1971. "Decouverte de restes humains anteneandertaliens dates du debut du Riss la Caune de l'Arago (Tautavel, Pyrenees-Orientales)." *Comptes Rendus de l'Academie des Sciences de Paris* 272:1739–42.

Lumley, M-A. de. 1973. *Anteneandertaliens et neandertaliens du Bassin Mediterraneen Occidental Europeen.* Etudes Quaternaires, Memoire n. 2. Universite de Provence.

———. 1975. "Ante-Neanderthals of Western Europe." In *Australopithecus Paleoanthropology: Morphology and Paleoecology,* ed. R. H. Tuttle. The Hague: Mouton.

Marshack, A. 1976. "Some Implications of the Paleolithic Symbolic Evidence for the Origins of Language." *Current Anthropology* 17:274–82.

———. 1984. "The Ecology and Brain of Two-Handed Bipedalism: an Analytic, Cognitive, and Evolutionary Assessment." In *Animal Cognition,* eds. H. Terrace, H. Roitblat, and T. Bever. Pp. 491–511. Hillsdale, N.J.: Lawrence Erlbaum.

Mellars, P. A., H. M. Bricker, J. A. Gowlett, and E. E. M. Hedges. 1987. "Radiocarbon Acclerator Dating of French Upper Paleolithic Sites." *Current Anthropology* 28(1):128–33.

O'Kelley, M. J. 1988. *Early Ireland.* Cambridge, England: Cambridge University Press.

Pfeiffer, J. E. 1978. *The Emergence of Man.* New York: Harper & Row.

———. 1986. "Cro-Magnons Were Really Us, Working Out Strategies For Survival." *Smithsonian* (October):74–85.

Pilbeam, D. 1975. "Middle Pleistocene Hominids." In *After the Australopithecines,* eds. K. W. Butzer and G. L. Isaac. The Hague: Mouton.

Poirier, F. E. 1973. *Fossil Man.* St. Louis: Mosby.

Prideaux, T. and the Editors of Time-Life. 1973. *Cro-Magnon Man.* New York: Time-Life.

Protsch, R. 1982. Public Lecture. Seattle.

Renfrew, C. 1988. "Archaeology and Language." *Current Anthropology* 29(3):437–68.

Rice, P. C. and A. L. Paterson. 1985. "Cave Art and Bones: Exploring the Interrelationships." *American Anthropologist* 87(1):94–100.

Rigaud, J.-P. 1982. "Le Paleolithique en Perigord: Les Données du Sud-Ouest Sarladais et Leur Implications." Ph.D. Dissertation, University of Bordeaux, France.

Roberts, D. F. 1953. "Body Weight, Race and Climate." *American Journal of Physical Anthropology* NS 11:533–58.

Roe, D. A. 1981. *The Lower and Middle Paleolithic Periods in Britain.* London: Routledge & Kegan Paul.

Ronen, A., ed. 1982. "The Transition from Lower to Middle Palaeolithic and Origin of Modern Man." B.A.R. International Series 151.

Shapiro, H. 1974. *Peking Man.* New York: Simon and Schuster.

Shea, J. J. 1988. "Spear Points from the Middle Paleolithic of the Levant." *Journal of Field Archaeology* 15(4):441–56.

Shipman, P. 1983. "Early Hominid Lifestyle: Hunting and Gathering or Foraging and Scavenging?" In *Animals and Archaeology: Hunters and Their Prey,* eds. J. Clutton-Brock and C. Grigson. Oxford: British Archaeological Reports.

Sieveking, A. 1979. *The Cave Artists.* London: Thames & Hudson.

Sigmon, B. A. and J. S. Cybulski, eds. 1981. *Homo Erectus Papers in Honor of Davidson Black.* Toronto: University of Toronto Press.

Singer, R. and J. Wymer. 1982. *The Middle Stone Age at Klasies River Mouth in South Africa.* Chicago: University of Chicago Press.

Smith, F. H. 1982. "Upper Pleistocene Hominid Evolution in South-Central Europe: A Review of the Evidence and Analysis of Trends." *Current Anthropology* 23(6):667–703.

Smith, F. H. and G. C. Ranyard. 1981. "Evolution of the Supraorbital Region in Upper Pleistocene Fossil Hominids from South-central Europe." *American Journal of Physical Anthropology* 53:589–610.

Smith, F. H. and F. Spencer, eds. 1984. *The Origins of Modern Humans.* New York: Liss.

Soffer, O. 1985. *The Upper Paleolithic of the Central Russian Plain.* New York: Academic Press.

Soffer, O., ed. 1987. *The Pleistocene Old World: Regional Perspectives.* New York: Plenum Press.

Straus, L. G. 1982. "Carnivores and Cave Sites in Cantabrian Spain." *Journal of Anthropological Research* 38:75–96.

Straus, W. and A. Cave. 1957. "Pathology and the Posture of Neanderthal Man." *Quarterly Review of Biology* 32:348.

Stringer, C. 1984. "Human Evolution and Biological Adaption in the Pleistocene." In *Human Evolution and Community Ecology,* ed. R. Foley. London: Academic Press.

Stringer, C. et al. 1979. "The Significance of the Fossil Hominid from Petralona, Greece." *Journal of Archaeological Science* 6:235–53.

Stringer, C. et al. 1984. "The Origin of Anatomically Modern Humans in Western Europe." In *The Origins of Modern Humans,* eds., F. Smith and F. Spencer. New York: Liss.

Svoboda, J. 1987. "Lithic Industries of the Arago, Vértesszöllös, and Bilzingsleben Hominids: Comparison and Evolutionary Interpretation." *Current Anthropology* 28(2):219–27.

Swedlund, A. 1974. "The Use of Ecological Hypotheses in Australopithecine Taxonomy." *American Anthropologist* 76:515–29.

Thackeray, J. F. 1986. "Further Comment on Fauna from Klasies River Mouth." *Current Anthropology* 27(5):511–12.

Thiel, B. 1987. "Early Settlement of the Philippines, Eastern Indonesia, and Australia-New Guinea: A New Hypothesis." *Current Anthropology* 28(2):236–41.

Thorne, A. 1980. "The Arrival of Man in Australia." In *The Cambridge Encyclopedia of Archaeology,* pp. 96–100. New York: Crown Publishers/Cambridge University Press.

Tierney, J., with L. Wright, and K. Springen. 1988. "The Search for Adam and Eve." *Newsweek,* January 11, pp. 46–52.

Tobias, P. and G. Von Koenigswald. 1964. "Comparison Between the Olduvai Hominines and Those of Java and Some Implications for Phylogeny." *Nature* 204:515.

Trinkhaus, E. 1982. "Evolutionary Continuity among Archaic *Homo sapiens.*" In *The Transition from Lower to Middle Palaeolithic and Origin of Modern Man,* ed. A. Ronen. Oxford: B.A.R. International Series 151.

———. 1983. *The Shanidar Neandertals.* New York: Academic Press.

———. 1986. "The Neandertals and Modern Human Origins." *Annual Review of Anthropology* 15:193–218.

———. 1988. Quoted in B. Bower, "An Earlier Dawn for Modern Humans?" *Science News* 133:138.

Trinkhaus, E. and W. W. Howells. 1979. "The Neanderthals." *Scientific American* 241:118–33.

Valladas, H., J. L. Reyss, J. L. Joron, G. Valladas, O. Bar-Yosef, and B. Vandermeersch. 1988. "Thermoluminescence Dating of Mousterian 'Proto-Cro-Magnon' Remains from Israel and the Origin of Modern Man." *Nature* 331:614–16.

Vallois, H. 1961. "The Social Life of Early Man: The Evidence of the Skeletons." In *Social Life of Early Man,* ed. S. Washburn. Chicago: Aldine.

Verraeersch, P. M. 1979. *Elkab Il: l'Elkabian epipaleolithique de la Vallee du Nil Egyptien.* Brussels: Publications du Coruite des Fouilles Belges en Egypte.

Villa, P. 1983. *Terra Amata and the Middle Pleistocene Archaeological Record of Southern France.* Berkeley: U.C. Publications in Anthropology, Vol. 13, University of California Press.

White, J. P. and J. F. O'Connell. 1982. *A Prehistory of Australia, New Guinea and Sahul.* Sydney: Academic.

White, R. 1982. "Rethinking the Middle/Upper Paleolithic Transition." *Current Anthropology* 23:169–91.

Windels, F. 1965. *The Lascaux Cave Paintings.* London: Faber and Faber.

Wobst, H. M. 1974a. "Boundary Conditions for Paleolithic Social Systems: A Simulation Approach." *American Antiquity* 39:147–78.

———. 1974b. "The Archaeology of Band Society: Some Unanswered Questions." *American Antiquity* 39:v–xiii.

———. 1977. "Stylistic Behavior and Information Exchange." In *Papers for the Director: Essays in Honor of James B. Griffin.* Ann Arbor: University of Michigan Museum of Anthropology Anthropological Papers 61.

Wolpoff, M. 1968. "Climatic Influence on the Skeletal Nasal Aperture." *American Journal of Physical Anthropology* 29:405–23.

———. 1970. "The Evidence for Multiple Hominid Taxa at Swartkrans." *American Anthropologist* 72:56–57.

———. 1971. "Vértesszöllös and the Presapiens Theory." *American Journal of Physical Anthropology* 35:309.

———. 1988. Quoted in B. Bower, "An Earlier Dawn for Modern Humans?" *Science News* 133:138.

5

The First Americans

At break of day the shore was thronged with
people all young . . . all of good stature, fine
looking. . . . I was anxious to learn whether
they had any gold, as I noticed that some of
the natives had rings hanging from holes in
their noses. . . . I tried to get them to go for
some, but they could not understand they were
to go.
Christopher Columbus (13 October 1492)

Columbus and most of the European "discoverers" of the New World
were not surprised to find "Indians" there, because they thought they had
landed in India, or perhaps Japan. But as soon as Europeans realized
they were not in the Orient and became aware of the rich diversity of
New World cultures, they began to struggle with the problem of the origin of the Native Americans. The Bible, final authority for most Europeans of this era, was strangely silent on the very existence of this "second-
earth," so Europeans began speculating on how the "Indians" could have
reached the New World from the Garden of Eden, where they were assumed to have originated. Early explorers were greatly impressed by
"similarities" between Egyptian and aboriginal American cultures, such as
the great pyramids to be found in Egypt, Mexico, and the Mississippi
Valley, and so some concluded that the Native Americans were descendants of Ham, one of Noah's sons, who was also thought to have been the
father of the Egyptians.[1]

Another popular idea was that the Native Americans were descendants
of the "lost tribes of Israel," Jews who had been evicted from Palestine by
the Romans (film fans will remember this idea in *Cat Ballou*). The "lost
tribes" idea eventually was incorporated into the doctrines of the Church
of the Latter-Day Saints, whose Book of Mormon explains how the Native
Americans were remnants of tribes of Israel that had come to the New
World by ship hundreds of years before Columbus. Others believe that

196

5.1 An early photograph of a Native American. Physical resemblances between Americans and Asian peoples reflect their common origins.

Native Americans are descendants of people who fled Atlantis or Mu—islands some think destroyed thousands of years ago by volcanic eruption.

But already by 1590 a Spanish Jesuit, José de Acosta, had proposed that Native Americans had come from Asia, and by 1781 Thomas Jefferson could describe the peopling of the New World in terms we know today to be largely accurate:

> Late discoveries of Captain Cook, coasting from Kamschatka to California, have proved that if the two continents of Asia and America be separated at all, it is only by a narrow straight. So that from this side also, inhabitants may have passed into America; and the resemblance between the Indians of America and the eastern inhabitants of Asia, would induce us to conjecture, that the former are the descendants of the latter, or the latter of the former: excepting indeed the Eskimaux, who, from the same circumstances of resemblance, and from identity of language, must be derived from the Groenlanders, and these probably from some of the northern parts of the old continent.[2]

In their search for clues to the origins of Native Americans, Jefferson and others of his age were struck by the close physical resemblance the Native Americans had to Asian races with their dark brown eyes, black, coarse, straight hair, and, relative to Europeans, widely spaced cheekbones. But differences were also apparent: except for the Eskimos and Aleuts, most aboriginal Americans had less pronounced epicanthic folds (part of the eyelid) that distinguish East Asian populations, and many Native Americans had relatively more prominent noses than east Asian people.

Europeans also noted that the New World included an impressive diversity of languages and cultures, that none of the languages of Native Americans bore much resemblance to Old World languages, and that many American languages were unrelated and mutually unintelligible to their speakers. Thus, it seemed certain that although the aboriginal Americans were probably descendants of Asian peoples, they must have lived in America for a long time for such physical and linguistic divergence to have occurred.

Today the view that the first Americans came from East Asia via a land route is generally accepted, but intense debates continue about when the first migrations occurred and how these intrepid migrants managed to traverse some of the world's most demanding environments.

Initial Colonization of the Americas

In a sense, the questions of when the first people reached North America and their precise cultural origins are less important than their subsequent history in the New World. Whenever they first arrived, New World peoples give us a comparative case—a long and rich culture history that can be compared with that of the Old World, revealing perhaps the general nature of cultural dynamics.

But archaeologists have found irresistible the question of when people first came to the New World. To an extent, the archaeological search for the "First Americans" is a game anyone in the New World can play, for the present candidates for the earliest known American sites include occupations in Alaska, Pennsylvania, Venezuela, and southern Chile (Figure 5.3). Thus, almost any resident of the New World can hope to find an older site than those now known.

That people have lived in the New World for at least 12,000 years has been evident for a long time. In 1926 a cowboy was riding along the edge of a gully in New Mexico, near the town of Folsom, when he discovered some "arrowheads" and animal bones protruding from a layer of soil about 6 meters beneath the surface of the plain. Eventually, his find came to the

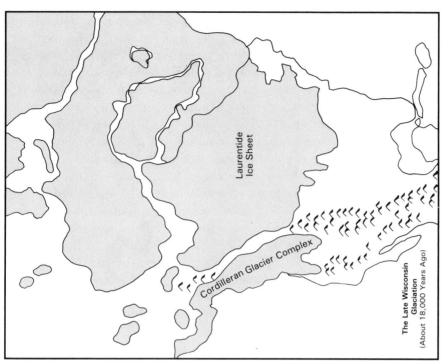

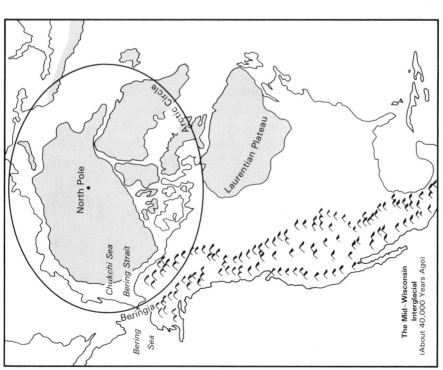

5.2 Between 60,000 and 12,000 years ago, world climate warmed several times, opening routes between the Arctic and more southern latitudes of the New World. Two possible periods of colonization are during the Mid-Wisconsin Interglacial about 40,000 years ago, and after the Late Wisconsin Glaciation, about 18,000 years ago.

199

attention of J. D. Figgins, director of the Colorado Museum of Natural History, who began a long series of excavations at this site.[3] A primary find was a long, "fluted" (having two long strips removed near the base and parallel to the length of the point) "projectile point" that was embedded in the ribs of a species of bison that had been extinct for about 10,000 years. So skeptical were most archaeologists and others of the idea that humans had been in the New World that long that Figgins insisted on excavating the site with a committee of archaeologists there to watch his every move, so that no one could claim the evidence had been faked.

In 1932 another important find was made, tnis time near the town of Clovis, New Mexico, and here too large blade tools were discovered in association with extinct animals. But at this site artifacts somewhat different from those at Folsom (Figure 5.9) were found in a layer beneath some "Folsom points." Analysis suggested a date of about 12,000 years ago for the earliest Clovis-style artifacts, and within a few years artifacts similar in size, shape, and style to the Clovis points were discovered at many different places in North America. Since then, stone points resembling Clovis artifacts have been found in every state of the Union, as well as far north into the Arctic and deep into South America.

So the disputes about early Americans really focus on the questions of whether or not anyone was here before the Clovis cultures, that is before about 12,000 years ago.

The main lines of evidence in this context are: (1) environmental evidence about the land bridge over the Bering Sea that the colonists are presumed to have crossed and the kinds of environments they encountered in their migrations; (2) biological similarities and differences between Asian and American peoples; (3) linguistic similarities and differences of Asian and American peoples; and (4) the archaeological record.

All students of the problem of New World colonization use these same categories of evidence, but they disagree profoundly in their conclusions: some think it most likely that people came to the New World only after about 10,000 to 12,000 B.P.; others that it was between 20,000 and 70,000 years ago; and there are still some who think an initial date of 80,000 to 150,000 years ago is possible.[4]

Why people came to the Americas has never been an issue. People, like other animal species, seem "naturally" to penetrate their entire available range. The "budding-off" process described in the previous chapter is probably applicable to the colonization of the Americas. Ethnographic studies tell us that hunter-forager groups often increase to a certain size level, then a fight develops among some members of the group, and the social unit splits, with some members going off into new territory. The long winter nights of north Asia doubtless offered ample opportunity to brood on what a jerk one's cousin was, and by tens of thousands of years

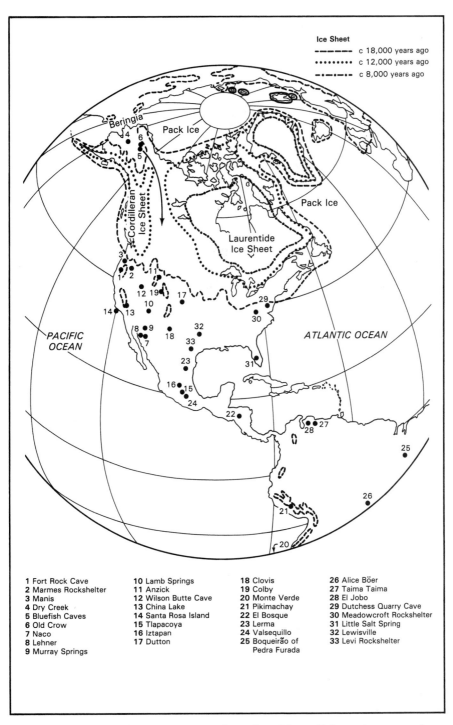

Ice Sheet
- ▬ ▬ ▬ ▬ c 18,000 years ago
- • • • • • • c 12,000 years ago
- ▬ • ▬ • ▬ • c 8,000 years ago

Beringia

Pack Ice

Cordilleran Ice Sheet

Laurentide Ice Sheet

Pack Ice

PACIFIC OCEAN

ATLANTIC OCEAN

1 Fort Rock Cave	10 Lamb Springs	18 Clovis	26 Alice Böer
2 Marmes Rockshelter	11 Anzick	19 Colby	27 Taima Taima
3 Manis	12 Wilson Butte Cave	20 Monte Verde	28 El Jobo
4 Dry Creek	13 China Lake	21 Pikimachay	29 Dutchess Quarry Cave
5 Bluefish Caves	14 Santa Rosa Island	22 El Bosque	30 Meadowcroft Rockshelter
6 Old Crow	15 Tlapacoya	23 Lerma	31 Little Salt Spring
7 Naco	16 Iztapan	24 Valsequillo	32 Lewisville
8 Lehner	17 Dutton	25 Boqueirão of	33 Levi Rockshelter
9 Murray Springs		Pedra Furada	

5.3 Some of the earliest known American sites. Most of these sites range from 20,000 to 10,000 years ago.

of group-fissioning, population growth, and small group social interactions, the world was populated.

Environmental Evidence

Any explanation of the colonization of the New World from Asia must consider the kinds of physical environments these colonists would have faced, and recent research has challenged many traditional ideas on this point.

Today, Eskimos using skin boats easily cross the 90 kilometers of open sea separating Siberia and America, and recently an American woman slathered herself with grease and actually swam from Alaska to Siberia. But such a sea crossing would not have been necessary during much of the Pleistocene. During periods of glacial advance within the last million years, enormous quantities of water were converted to ice, lowering the sea level sufficiently to expose a 1500- to 3000-kilometer-wide expanse of the floor of the Bering Sea. This land bridge—usually referred to as Beringia (Figure 5.3)— was probably available at least four times in the last 60,000 years.

5.4   Several archaeological sites in Siberia may be of cultures ancestral to the first Americans.

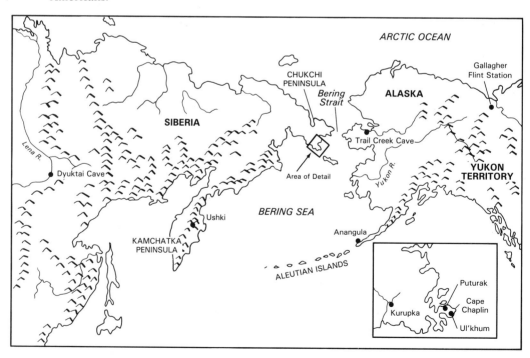

The kinds of resources Beringia would have offered hunter-foragers would have depended on the period in which the migrants crossed. At various times, ocean currents created an ice-free, tundra-covered connection from eastern Siberia across the land bridge and into central Alaska. These conditions are reflected in the many non-Arctic-adapted animal species that crossed from Asia to America during the Pleistocene. Prior to 10,000 years ago, species of deer, bison, camels, bears, foxes, mammoths, moose, caribou, and even rodents crossed from Siberia into the New World. Going the other way—from America to Asia—were foxes, woodchucks, and, during the early Pleistocene, the ancestors of modern forms of horses, wolves, and other animals.[5]

In some periods Beringia would not have been particularly rich—there is even the possibility of dust storms, so dry and barren was most of it.[6]

People could possibly have reached the New World during the mid-Wisconsin interglacial, a warmer period about 60,000 to 25,000 years ago. We know that people of this age made a sea crossing of at least a few miles to get to Australia (Chapter 12). In winter people could have walked from Siberia most of or all of the way to Alaska on the ice. Interior Alaska and Canada were relatively rich environments in the mid-Wisconsin interglacial, and they would have had a clear ice-free run all the way to South America at this time. Pollen cores from easternmost Beringia suggest that from 30,000 to 14,000 years ago, the time when most archaeologists think the first Americans arrived, the "landscape [of Beringia] consisted of relatively bare polar desert or fell-field tundra, a rocky terrain sparsely vegetated by herbs and dwarf shrubs. This suggests that the late Wisconsian environment in this part of Beringia was as harsh as the modern high Arctic."[7] Fladmark also notes that there would have been relatively few big game animals in Beringia during the colder periods. But he points out that the coasts of Beringia could have supported rich resources in the form of vegetation, seabirds, salmon, whales, seals, and many other species.

It used to be thought that southward migrations could have occurred only a few times, because it was assumed that for most of the Pleistocene the way would have been blocked by coalescing ice sheets. Even if these ice sheets did not completely bar the way south, most authorities thought that a narrow open corridor between them would have had too little food to support groups of hunters and gatherers—who require large territories to support even low population densities.

But some authorities now believe that the two or three major glaciers of North America reached their peaks at different times, formed more rapidly than previously thought, and were perhaps thinner than earlier estimates.[8] If so, even during the coldest periods of the late Pleistocene, about 22,000 to 18,000 years ago, North America would not have been so formidable a place for human colonization and migration as previously thought.

The various expansions and contractions of the glaciers would have displaced people and animals, and as Fladmark notes in some cases people in southern Canada and the northern United States may even have been forced back toward Beringia as the glaciers expanded.[9]

The exact path the first Americans took is not known, but the most likely would be along the coast of the land bridge, then into Alaska north of the Brooks Range, up the Yukon River Valley, then into the Mackenzie River Valley, and from there southward, along the eastern slopes of the Rockies and on into the Dakotas and then further southward. Along the coast of the land bridge were probably abundant resources in the form of fish, birds, eggs, invertebrates, and many plant foods. If groups did come this way, it is unlikely that we shall ever be able to document their journey archaeologically, for the rising seas of the post-Pleistocene era have submerged the ancient shoreline. A Pacific coast route is technically possible, but here again we have no evidence because such sites would lie deep beneath the coastal seas.[10]

We might expect that if the immigrants crossed the land bridge primarily along the coast, then the groups migrating south would have retained some elements of the generalized hunting, fishing, and foraging economies required in the intertidal zone. If, on the other hand, they came across the middle of the land bridge, a greater reliance on big-game hunting would have been essential. Whatever their adaption, the migrants were almost certainly not constantly on the move southward. The budding-off process described above is the usual one in cases when new species move into empty niches, and many thousands of years may have passed between the time the first human set foot on the land bridge and the year the first groups arrived in the mid-latitudes of North America.

North winds blowing off the northern glacial ice sheets made the climate of Pleistocene North America much different from today's. Much of Nevada and Utah was covered by Lake Bonneville, of which only a shrinking remnant (Great Salt Lake) remains. Wyoming, Iowa, and other parts of the Great Plains were vast pine, spruce, and tamarack forests and lush open grasslands. To the south, the area between the Mississippi River and the Rockies was a verdant mosaic of grasslands, lakes, and birch and alder forests. In eastern North America huge expanses of coniferous forests stretched from the edge of the glacial ice sheets to the lower Ohio River Valley.

The animals inhabiting this wilderness of 14,000 to 12,000 years ago closely approximate a modern hunter's vision of paradise. Giant moose, 3 meters and more in height, could be found in many of the wetter areas,

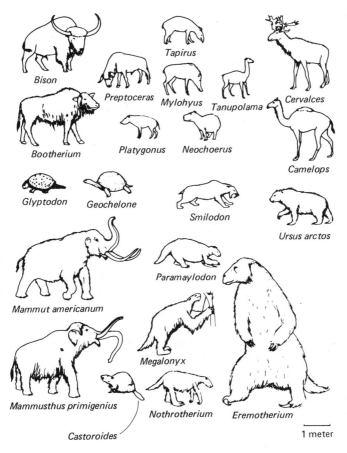

5.5 Some large mammals of Pleistocene America. The scale preserves relative size.

along with *Castoroides,* a species of beaver as large as a modern bear. Along the woodland edges of the southeastern United States were large populations of giant ground sloths, ungainly creatures fully as tall as modern giraffes. In more open country were vast herds of straight-horned bison, caribou, musk-oxen, and mammoths—some species stood 4 meters high at the shoulders. In the more forested areas of the East and South were the mammoth's cousin, the mastodon, a more solitary animal than the mammoth and apparently a browser rather than a grassland grazer. Amid all these large creatures were rabbits, armadillos, birds, camels, peccaries, and other animals. And the carnivores that such a movable feast attracted were equally impressive. Packs of dire wolves roamed most of the New World, as did panthers as large as modern lions and two species of saber-toothed cat, one about the size of a lion.[11] Thus, the first Americans were hardly entering an "empty niche," as these ferocious predators no doubt provided stiff competition for people trying to specialize in hunting.

Central and South America were also rich game preserves during and just after the Pleistocene, but in the rainforests of the Amazon Basin, the great coniferous forests of the North American South and East, and a few other locations, so much energy was in the form of inedible and unnutritious vegetation (cellulose) that there would have been few resources for primitive hunters and gatherers.

Human Anatomical Evidence

By 100,000 years ago Old World peoples seem to have had all the skills at making clothes, fire, and tools that would have been necessary for the trek to the New World. But intensive research has failed to turn up any New World human skeletons that cannot be comfortably fitted within our own species, *Homo sapiens sapiens*. Because the first *H. sapiens sapiens* appear in the Near East and Europe about 35,000 years ago, we might assume that the aboriginal Americans must have come over since then, but we do not know when the first *Homo sapiens sapiens* appeared in the East Asian areas—they may have been there long before they were in the Near East or Europe.[12]

It is uncertain when some East Asian populations developed epicanthic folds, small noses, and relatively flat facial profiles, but in view of the prominent noses of most Native Americans and their non-Mongolian eye shapes, it is apparent that either the Americans are only distantly related to these populations, or that they emigrated before the Asian populations developed their distinctive features. We have very little evidence on which to estimate rates of evolutionary change in these superficial physical characteristics, but the overall similarity of New World populations suggests that the natural selective pressures that produce variation in these characteristics have not had very long to operate.

C. Turner, in an analysis of teeth from various populations, produced the cluster-diagram presented in Figure 5.6. Teeth preserve well, have morphological characteristics under relatively simple genetic control, and seem to change relatively quickly under changing selective conditions. Figure 5.6 indicates the relative frequency in various groups of incisor "shoveling," the degree to which the lingual surface of incisors is curled or ridged. On this trait alone, one would clearly see North Chinese ancestry for Americans.[13] By comparing large samples of teeth on many different measurements, Turner concluded that: (1) New World groups are more like Asians than like Europeans; (2) all New World groups resemble each other more than they do most Old World populations; (3) dental variation

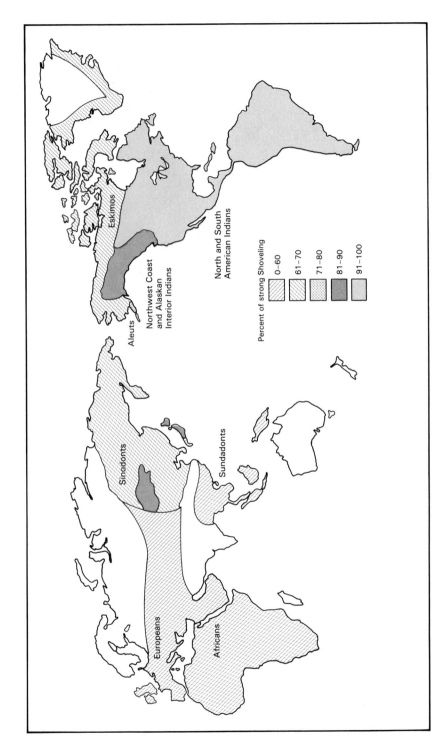

5.6 Variations in human teeth shapes reflect genetic ancestry. Incisor "shoveling"—the curved cross-sectional shape of incisors— occurs most frequently in northeast Asian and New World peoples and distinguishes them from the rest of the world. This evidence supports the view that the New World was colonized by three successive migrations from Asia.

is greater in North America than in South America; (4) there are three "clusters" of New World peoples.[14]

As Turner notes, these characteristics seem to make sense if there were three distinct migrations into the Americas, all from North China. It is very difficult, however, to estimate rates of change in these kinds of physical features, and thereby to estimate how long ago the migrations to the Americas began.

The dental evidence for these separate groups is generally supported by studies of blood constituents, but systematic work on this problem has only just begun.[15]

Linguistic Evidence

Various scholars have tried to estimate the date and patterns of New World colonization by examining Native American languages. Just as resemblances of English, Greek, Persian, and other languages point toward a common Indo-European original language, the Native American languages have been screened for signs of common origin. A recent and controversial analysis of this problem is that by Joseph Greenberg, who argues that New World languages reflect three periods of colonization, the earliest, which is associated with Amerind, then a group of what are called Na-Dene speakers, and finally, at a later time, Eskimo-Aleut speakers.[16] Each of these three language groups probably originated from separate Old World language families, and there is some (controversial) evidence that all the world's languages can be traced to a single language family.[17] But like studies of teeth, estimates of the time of New World colonization on the basis of language changes must be considered quite speculative.

Archaeological Evidence

It used to be widely believed that the first wave of immigrants to the New World stayed mainly on the western side of North America, colonizing the area east of the Mississippi River only about 10,000 years ago, but some archaeological evidence suggests that the East may have been occupied much earlier. Of particular interest are the so-called "pebble-tool" complexes that have been found in Tennessee, Alabama, the Ohio Valley, and elsewhere. These crude tools have been found in both open-air sites and rock shelters, and occasionally in levels containing the Clovis-type points dated to about 12,000 years ago in the Great Plains. Some archaeologists think these pebble tools were used by people entering the North American East 20,000 or more years ago.[18] The land surfaces of much of the East are millions of years old and preservation of artifacts is generally poor, so it is not surprising that we do not have good carbon-14 dates for these early cultures. The ancient date given to these tools is based on their simplicity and extreme patination in comparison to other artifacts, and also on their resemblance to crude pebble tools and other implements found in Peru and radiocarbon dated (controversially) to between 19,600 and 14,150 B.C. Some have characterized the earliest New World cultures as the Pre-projectile Point Stage.[19] The absence of projectile points and the simplicity of these early artifacts may mean that the earliest immigrants were not highly specialized big-game hunters, which suggests that they came over during the warmer periods of the Pleistocene as generalized hunters and gatherers. We cannot be certain, of course, because successful big-game hunters (like *Homo erectus*) used crude stone tools; but these pebble tools do not look as if they were specialized killing or butchering implements. Nor have they been found with butchered animal bones or in environmental circumstances suggesting specialized hunting.

Except for this pebble-tool complex, however, to a large degree, arguments about the date of colonization of the New World come down to one's belief in specific radiocarbon dates. The radiocarbon method has been shown to be broadly accurate around the world. The problem is that any single radiocarbon date can be misleading, because of any number of factors, ranging from natural contamination through oil seeps or coal deposits, to incompetent laboratory technicians. When the first radiocarbon dates were produced on Egyptian materials, for example, Egyptologists rejected these "scientific" dates because they conflicted with a decidedly unscientific farrago of evidence from king-lists, artifact styles, etc. But the Egyptologists were right. Until archaeologists discovered that radiocarbon dates had to be corrected for past fluctuations in the amounts of radiocarbon in the atmosphere, these radiometric dates were wrong by centuries.

So we must regard the New World dates with proper skepticism, and put our faith only in broad patterns. Thus the dates in Table 5.1 seem to rule out many sites that were once thought evidence of early American occupations.[20] But some sites, such as Monte Verde and Meadowcroft (discussed below), look promising.

If radiocarbon dates are always somewhat suspect, the only other obvious basis on which to date New World colonization would seem to be a comparison of Old World artifact styles and sites and the earliest New World sites and implements. But when we try to do this, we find that similarities between North Asian and early American stone tools are very much in the eye of the beholder.

Table **5.1**   Accelerator Dates for Some Human Remains in the New World

Lab No.	Site	Radiocarbon Date
OxA-186	La Jolla	5600 ± 400
OxA-188	Del Mar	5400 ± 120
OxA-774	Del Mar	5270 ± 100
OxA-189	Laguna	5100 ± 500
OxA-154	San Diego Site W-12	8470 ± 140
OxA-153	UCLA 1425	4950 ± 150
OxA-152	UCLA 1425	4850 ± 150
UCR-1437A (conventional)	Sunnyvale	4390 ± 150
AA-50 (UCR-1437A)	Sunnyvale	3600 ± 600
AA-51 (UCR-1437D)	Sunnyvale	4650 ± 400
AA-52 (UCR-1437B)	Sunnyvale	4850 ± 400
OxA-187	Sunnyvale	6350 ± 400
AA-610A	La Jolla shores II	4820 ± 270
AA-610B	La Jolla shores II	5370 ± 250
AA-611	La Jolla shores II	6330 ± 250
AA-283	Yuha	1650 ± 250
AA-284	Yuha	3850 ± 250
AA-295	Yuha	2820 ± 200
(Chalk River Lab.)	Taber	3550 ± 500
OxA-773	Taber	3390 ± 90

NORTH ASIA

We might expect to find a long trail of archaeological sites stretching from northeastern Siberia along the sea floor of the now submerged land bridge into central Alaska, and then southward, between the ice caps. If these early immigrants were big-game hunters, we would expect many sites to be concentrations of animal bones, hearths, and flint tools appropriate for killing and preparing these animals. Alas, the gap between the real and ideal—so persistent in all archaeological research—is particularly great in this case. Some sites do fit this pattern, but the evidence is far from conclusive. First of all, adequate surveys of northeastern Siberia have never been completed. Aside from the rigors of the Arctic climate, the politics of doing surveys amid ICBM missile silos and radar stations have deterred most Western scholars.

Present evidence suggests that before about 30,000 years ago, Siberia had extremely low population densities or was unsettled altogether. The great frigid interior swamps and forests of Siberia may have barred human colonization, at least until huts had been developed. In any case, the earliest known sites in northeastern Siberia date to about 23,000 years ago and reflect a life focused on the hunting of mammoths and other large animals. Projectile points, scrapers, and burins—tool types associated with big-game hunters all over the Old World—are the most frequent implements at Siberian sites, such as Mal'ta, Afontova gora II, and other sites of this period. Stone tools from Mal'ta in deposits radiocarbon dated to 14,750 years ago do not include Clovis-style points but in other ways are quite similar.[21] Haynes has even raised the possibility that the Mal'ta site represents an intermediate point in migrations of eastern European people across Siberia and into the New World—which he thinks is plausible because of the similarities between Clovis assemblages and late Pleistocene eastern European industries.[22]

N. Dikov's work on the shores of Lake Ushki offers at least some evidence about possible American ancestors. Dikov found the remains of what appears to be a large camp, in which people lived in hide or wood shelters. The center of one camp area was a deep burial pit in which the poorly preserved fragments of a body were found along with about a thousand beads of soft stone, some pendants of the same material, and some amber beads. The whole area of the burial was covered with red ocher, used probably to simulate blood or a living complexion on bodies.

Radiocarbon dating places the site about about 14,300 to 13,600 years ago, but other dating methods and considerations of groundwater contamination have led Dikov to estimate the date as at least 1500 to 2000 years older. For purposes of linking this site to the Americas, it is significant that Dikov found stone points in the bifacial, leaf-shape form that resemble those discovered in the Great Basin of North America at about 13,000 to 11,000 years ago. Bifacial leaf-shape points have also been found at the Dyuktai Cave site, on the Alden River in Siberia. These points probably date to about 15,000 years ago and also appear in Alaskan and Western Canadian sites.

Dikov reviewed the evidence for other sites on the way to what would have been Beringia and concluded that nothing "challenges the conservative view that the significant population movements from Asia occurred after the peak of the last glaciation, which was some 18,000 years ago."[23]

NORTH AND SOUTH AMERICA

The evidence is also somewhat sketchy on the American side of the land bridge. The Alaskan and northern Canadian climates make surveys very difficult, and farther south the constant waxing and waning of the glacial ice sheets and the vast riverine systems that drained them have thoroughly chewed up much of the land along the Canadian corridor. No doubt many sites in these areas have long since been scoured away by ice or water.

A spatulate-shaped "flesher" fashioned of a caribou bone from the Old Crow area of northeastern Alaska was once thought to be quite old, but it has recently been redated to only about 1400 years ago. Some apparent

5.7A Wooden hut foundations at Monte Verde, Chile. Radiocarbon dates for these and other organic materials at this site average about 13,000 B.P.

bone tools found elsewhere in the Old Crow area, however, have recently been dated by the accelerator method to 22,000 to 43,000 years ago, and other apparent bone tools were found in strata an astounding 40 feet below these dated tools.[24] These tools were in association with extinct forms of animals and in strata such that archaeologists must seriously consider the possibility that people were in this part of the world a full 100,000 years ago.[25]

Ironically, the site with perhaps the best claim to a pre-12,000 B.P. date in the Americas is among the farthest south, Monte Verde, in south-central Chile (Figure 5.7). Here Tom Dillehay and his crew have excavated a camp site that has been radiocarbon-dated to about 13,000 years ago, and below the levels of that age are layers of tools and debris that some think are 33,000 years old.[26]

Monte Verde is particularly interesting because, unlike most other sites of this age, it is not a cave or rock shelter. It is an open-air site located in a cold, wet forested area that was covered by a peat bog and thereby extremely well preserved. A piece of mastodon flesh was found intact, plus a human footprint, log foundations for huts, animal skins, numerous plant remains, and many other remains. Almost all the bones from the site were from seven mastodons, which appear to have been killed elsewhere and carried home in large segments. Scores of plants were eaten, and remains of some plants with medicinal properties were found in the camp debris. Trade with other groups was indicated by the presence of salt, bitumen, non-native plants, and other commodities. Dillehay estimates that thirty to fifty people lived at Monte Verde at one time, with an

5.7B Reconstruction of the community at Monte Verde, Chile. Evidence from Monte Verde may be the strongest yet found for human occupation of the New World before 12,000 years ago.

impressive degree of occupational specialization in wood-working, stone-tool manufacture, and other skills.

Monte Verde, if it truly is as old as the radiocarbon dates suggest, is one of the most important sites in the New World. If people were here at 13,000, they must have been in the New World far earlier; if they lived in the almost sedentary communities that Monte Verde seems to represent, many models of hunting-foraging economies may not apply to much of the New World's earliest populations. And if the layers of the site dated by radiocarbon methods to 34,000 B.P. are really that old, then most of the debates about New World colonization have been beside the point.

The Monte Verde dates have received some support in the form of radiocarbon dates of hearths from sites near Pedra Furada, in eastern Brazil, where numerous stone tools and animal bones were found with charcoal in stratified layers that yielded a consistent series of twelve dates, from about 32,000 to 17,000 years ago.[27]

Some of the earliest dates for a New World site come from Pikimachay Cave in Peru, where archaeologist Richard MacNeish excavated crude stone tools and animal bones that have been radiocarbon-dated as early as 22,000 years ago.[28] The association between these stone tools and some forms of extinct animals is unmistakable in 12,000-year-old levels of the site, but

5.8   Human foot impression in mud at Monte Verde,
Chile. Associated sediments
have been radiocarbon dated
to about 13,000 years ago.

some archaeologists doubt that its earliest levels represent human occupation, and all early Peruvian dates seem at least somewhat equivocal.[29]
The evidence for extensive penetration of most of South America by 11,000
B.P. is overwhelming, however, even in difficult environments like the interior of Brazil.[30]

One of the most controversial and potentially important early New World
sites is Meadowcroft Rock Shelter, near Avella, Pennsylvania, where many
layers of alluvial sediments were deposited by a tributary of the Ohio River.
Interspersed in these alluvial sediments are many clear indications of human occupation, including prismatic flint blades, basketry, and hearths.
Millennia of reoccupation left a 16-foot thick "layercake" of tools, bones,
baskets, and other debris.

The question is, how far back does Meadowcroft's earliest occupation
date? The site was excellently excavated, and there is no question about
the provenience of the artifacts. A bit of charred basketry from near the
bottom was radiocarbon-dated to 19,600 +/− 2,400 and 19,150 +/− 800
years B.P. Charcoal from a level beneath the basketry dates to as old as

37,000 years ago, but there were no associated artifacts. Unfortunately, there is some suggestion that the samples may have been contaminated by groundwater and that the site's stratigraphy has been misunderstood.[31] The kinds of animals whose bones were found at the site also pose some problems, in that they do not seem to be animals one would expect this close to the glaciers that would have pushed far down into the northeast United States in the late Pleistocene.

The stone tools of Meadowcroft's earliest levels are similar to the typical Clovis assemblage of 12,000 years ago, but some blade sizes and types are different. Adovasio has compared the Meadowcroft lithics to those from other sites and suspects that they may represent a link between Old World tool industries and the early cultures of North America, such as the Clovis.[32]

But if the earlier dates for Meadowcroft are accepted, it means that people were in the continental United States before the last glacial advance, which implies that the ice sheets were in fact no barrier and that the scarcity of sites found is a matter of inadequate searching, low population densities, and the postdepositional destruction of sites.

If people entered the New World 30,000 or more years ago, it might seem surprising that more archaeological sites have not been found dating to this period; after all, hundreds of Old World sites date to this age. But population densities in Europe and other parts of the Old World had been slowly building for hundreds of thousands of years before people entered the New World. In addition, most European sites of this age are found in caves and rock shelters, and there are relatively few of these along the probable route into the New World. Alternatively, the first Americans may have worked out adaptations to cold climates that did not require caves. Their ancestors must have done this, because there are no caves along the Siberian-Beringia-Alaska route.

Another problem in locating early sites in eastern North America is that in much of the lower Mississippi area, alluviation has been so great that many such sites would be buried under tens of meters of soil. And in the rest of the South and East, because the glacial ice sheets never reached these areas, the present land surface is millions of years old. Thus, soft-drink bottles and ancient artifacts can be found in virtually the same soil, because accumulation of soil over these areas has been so slow for millions of years. Additionally, bones and other datable materials do not preserve well under these humid, exposed conditions.

Early American Economies

For many thousands of years after people reached mid-continental and southern latitudes in the New World, population densities were probably extremely light, and most bands no doubt stayed in the same general mixed forest and grassland environments their ancestors had adapted to in more northern areas. By about 12,000 years ago, however, they began to display greater diversity in subsistence strategies as they evolved adaptions to a greater range of environments.

Archaeologists think their reconstructions of these early American cultures may be somewhat inaccurate because of the nature of the archaeological record they left. The easiest archaeological sites to find are those with large stone tools and the bones of large animals, and many early New World sites are of this type: dozens of stone tools intermixed with mammoth, bison, and other bones, often near streams or bogs where hunters ambushed these animals. But there may be many less obvious sites—those not yet found because they are not marked by masses of animal bones or impressive stone tools. In 1987, for example, a Clovis site in an apple orchard (the Ritchie Site) was found when holes were dug for new trees and beautiful, exceptionally large stone-points were unearthed. But if an earlier site had been there made up of crude pebble tools, it is highly improbable that it would ever have been recognized. In short, extreme biases are likely in our recognition of the early American archaeological record. Many archaeologists question, for example, the presumption that even the Clovis and Folsom peoples were mainly big-game hunters, although there is some evidence that groups along the tundra zone, in New York and Massachusetts specialized in caribou.[33] Meltzer argues that in much of the southeast United States, the people who left the many fluted points in this area were generalized foragers who probably used many of the points as hafted knives, not as spear-points.[34]

Similarly, several sites dating to about 11,000 B.C. have been found in Mexico, most of them composed of the debris from killing and butchering mammoths. The consistency with which mammoths are found in these sites may seem to reflect an economy specialized in hunting large animals, but most of the people of Mexico at this time were probably generalized hunters and gatherers and only because of the great size and preservation of the mammoth bones do we find so many of these sites. As MacNeish noted, these early hunters probably killed one mammoth in a lifetime—and never stopped talking about it.

But there is no question that some people did kill and eat big animals. Alan Bryan described excavations at Taima-taima, in Venezuela, in which the skeleton of a young mastodon was found with a large stone point inside the pubic cavity, and cut-marks inside the rib cage that suggest that some ancient hunter crawled into the animal's carcass to remove internal organs.[35] The mastodon's head was missing, and cut-marks showed where steaks had been removed from a foreleg. Leg bones of a larger mastodon had been used as a butcher's block to hack up the younger one. Near the skeleton and preserved by the wetness of the site was a mass of vegetable

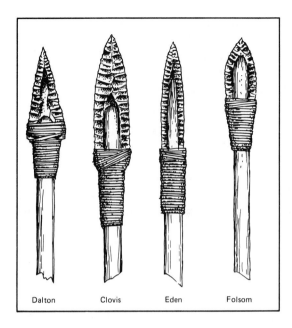

Dalton   Clovis   Eden   Folsom

5.9. Some types of early American projectile points.

matter whose sheared ends and compaction suggests that it was the mastodon's stomach contents and was dated by the radiocarbon method to no later than 13,000 years ago.

Perhaps the most direct way to demonstrate hunting of large animals by early Americans is to find residues of animal tissue on the tools themselves, and in a few cases this is possible. At least one of the large fluted points at the Ritchie Site in eastern Washington State was found to have traces of blood on it—bovine blood, in fact, probably from a mammoth.

Other evidence of hunting comes from the Arctic, such as the Dry Creek, Trail Creek, and similar occupations, most of which date to between 11,000 and 8000 years ago. These sites seem to be hunting camps in unglaciated areas and are marked by bifacial knives, tiny blades, and the bones of horses, bison, hares, ducks, and other animals. By this time, these Arctic hunters may have been pursuing a dying way of life, as the glaciers and the rich animal life they supported were slowly withdrawing.[36]

Shortly after 8000 B.C. at least some groups in the American West were undoubtedly specializing in hunting big game. They often practiced "jump hunting," in which many people cooperate to stampede bison herds or other animals over a precipice, killing them by the score. Where there were no convenient precipices, animals were driven into natural cul-de-sacs, where they could be easily killed. In sites reflecting these practices, archaeologists find the bones of hundreds of bison, many showing clear butchering marks. These hunters took cuts of meat from these animals in a way that showed a shrewd appreciation of their nutritional value.[37] By drying the meat, these hunters could accumulate large food reserves, and the skin, hide, and bones of the bison had many uses.

The earliest distinctive and widely distributed tool complex found in the North American East is distinguished by the presence of large stone projectile points. Most such points are between 12 and 30 centimeters long and occasionally have been "fluted." Formerly, many archaeologists thought these eastern points were imitations of the Folsom and Clovis points, but it is now evident that these larger points occur as early—and probably earlier—in the eastern woodlands as on the Plains and are more numerous in the East: Alabama alone has yielded more points than the entire western half of North America.[38]

Pleistocene Extinctions

The spread of human hunting and gathering societies over the New World after 12,000 years ago, at the end of the last glacial period, coincides with the extinction of many animal species, and by about 10,000 years ago, all or most of the mammoths, mastodons, long-horned bison, tapirs, horses, giant ground sloths, dire wolves, camels, and many other creatures had disappeared. Extinction is, of course, a natural evolutionary development and can be accounted for by known biological processes. But the number of animal species that became extinct in the New World and their apparently rapid rate of extinction has led some to conclude that human hunters forced many New World animals into extinction shortly after the Pleistocene.[39]

But people may have had little or nothing to do with these extinctions. In a sample of thirty-five early sites in North America, the animal bones found were mainly those of mammoths and bison—species that did in fact become extinct; but many of the other large animals that became extinct during this period are either not found at all in these sites or in very limited numbers. Perhaps even more significant, game species were not the only ones to die out. Donald Grayson has shown that numerous species of birds also became extinct at this time, and it is difficult to believe that people could have played any role in this.[40] Also, some "big-game" species such as the mastodon lived in the North American South and East while humans were there and are known to have perished shortly after the end of the Pleistocene, but their bones have rarely been found in association with evidence of human activities. Thus, it would seem that at least some larger animals became extinct without much human assistance. Finally, we might also note that there is no archaeological evidence that the hunting practices most likely to lead to animal extinctions, such as drives and jumps, were ever used during the period, some 10,000 to 8000 years ago, when most of the larger species became extinct.

Kelly and Todd present an alternative to the overkill model, arguing that the Clovis peoples were unlike most known hunter-foragers in that they may have had very little processing technology for food storage, relying instead on proficient hunting using high-quality flint bifacial tools, and shifting their range as they exhausted the resources of an area.[41] They suggest that the pattern of Pleistocene extinctions would have been similar even if people had never entered the New World.

If we rule out human hunting as the most important factor in these extinctions, what alternative explanation can we give? Many have been suggested, but none is really very satisfactory. Clearly, the immediate post-Pleistocene period was one of radical climate change, and no doubt this was of some importance. Of thirty-one genera of mammals that became extinct, however, only seven had entered the New World during the last 70,000 years, and thus all the others had managed to adapt to the climatic changes of previous interglacial periods, which were fully as dramatic as those after the last glacial retreat. Why did they become extinct after the last glacial period?

Some have suggested that New World animals might have been decimated by diseases introduced from the Old World during one of the later intervals when the Bering land bridge route was open—a situation reminiscent of the frightful casualties inflicted on aboriginal Americans by smallpox, measles, and other diseases introduced by the Europeans. Such epidemics, however, usually do not cause species to become extinct: they decimate local populations over a large area but generally leave small pockets of resistant individuals who eventually reestablish the species.

As Grayson has pointed out, it is entirely possible that these Pleistocene extinctions were the results of climatic changes and human interactions of fairly subtle and presently unknown dimensions.[42]

By about 5000 B.C., the glaciers had retreated to the point that the flora and fauna of the eastern United States were very similar to what they are today—except where changed by human activity—and there was a broad cultural readaption to the changing environments.

Generally, the early Holocene in the Americas was a time of cultural specialization, as technologies, social systems, and subsistence systems were evolved to meet diverse and changing environments. The full length of the New World was inhabited shortly after 12,000 B.P., with sites radiocarbon dated to this early period now known all the way south into southern Argentina.[43] Human population densities rose in many parts of the New World, and, as we will see in Chapter 15, by 7000 years ago some hunting and gathering societies were already in the process of domesticating plants and beginning the transition to the agricultural, complex cultures of late prehistoric America.

Conclusion

Despite the evidence for early sites, such as Monte Verde, most archaeologists still consider the actual date of the New World colonization to be undemonstrated. We know that people were here in substantial number by 12,000 years ago, and within a few thousand years they had occupied most of the more productive New World areas and were beginning to diversify culturally.

Notes

1. Stewart, *The People of America*, p. 60.
2. Quoted in Stewart, *The People of America*, p. 70.
3. Figgins, "The Antiquity of Man in America."
4. Reviewed in Irving, "Context and Chronology of Early Man in the Americas."
5. Haag, "The Bering Strait Land Bridge," p. 269.
6. Cwyner and Ritchie, "Arctic Step-Tundra: A Yukon Perspective."
7. Fladmark, "Getting One's Berings," p. 14.
8. Ibid.
9. Ibid., p. 16.
10. Fladmark, *British Columbian Prehistory;* idem., "Getting One's Berings."
11. Martin and Wright, *Pleistocene Extinctions: The Search for a Cause*, pp. 32–33.
12. Bellwood, *The Prehistory of the Indo-Malaysian Archipelago*.
13. Turner, "Tell-tale Teeth."
14. Turner, "Tell-tale Teeth"; Greenberg, Turner, and Zegura, "The Settlement of the Americas: A Comparison of the Linguistic, Dental, and Genetic Evidence," p. 484.
15. Zegura, "Blood Test."
16. Greenberg, Turner, and Zegura, "The Settlement of the Americas: A Comparison of the Linguistic, Dental, and Genetic Evidence."
17. Ruhlen, "Voices from the Past."
18. Dragoo, "Some Aspects of Eastern North American Prehistory: A Review 1975."
19. Krieger, "Early Man in the New World."
20. Gowlett, "The Archaeology of Radiocarbon Acclerator Dating."
21. Haynes, "Geofacts and Fancy," p. 12.
22. Ibid.
23. Dikov, "On the Road to America," p. 15.
24. Irving, "New Dates from Old Bones," p. 12.
25. Ibid.
26. Dillehay, "By the Banks of the Chinchihuapi."
27. Guidon, "Cliff Notes."
28. MacNeish, *The Science of Archaeology?*, p. 203.
29. See, for example, Vescelius, "Early and/or Not-so-Early Man in Peru: Guitarrero Cave Revisited."
30. Schmitz, "Prehistoric Hunters and Gatherers of Brazil."
31. Dincauze, "The Meadowcroft Papers"; cf. Adovasio et al., "The Meadowcroft Papers: A Response to Dincauze."

32. Adovasio and Carlisle, "Pennsylvania Pioneers."
33. Meltzer, "Late Pleistocene Human Adaptations in Eastern North America."
34. Ibid., p. 43.
35. Bryan, "The First Americans."
36. Hopkins et al., *Paleoecology of Beringia.*
37. Speth, "Les strategies alimentaires des chasseurs-cueilleurs."
38. Dragoo, "Some Aspects of Eastern North American Prehistory: A Review 1975."
39. Martin and Wright, *Pleistocene Extinctions;* Haynes, "Elephant Hunting in North America"; Moismann and Martin, "Simulating Overkill by Paleoindians"; Martin and Klein, *Quaternary Extinctions. A Prehistoric Revolution.*
40. Grayson, "Pleistocene Avifaunas and the Overkill Hypothesis."
41. Kelly and Todd, "Coming into the Country: Early Paleoindian Hunting and Mobility."
42. Grayson, "Death by Natural Causes."
43. Orquera, "Advances in the Archaeology of the Pampa and Patagonia."

Bibliography

Adovasio, J. M. and R. C. Carlisle. 1987. "The First Americans." *Natural History* 95(12):20–27.

Adovasio, J. M., J. Donahue, J. D. Gunn, and R. Stuckenrath. 1981. "The Meadowcroft Papers: A Response to Dincauze." *Quarterly Review of Archaeology* 2:14–15.

Aikens, C. M. 1978. "Archaeology of the Great Basin." *Annual Review of Anthropology* 7:71–87.

Bada, J. L. and P. M. Helfman. 1975. "Amino Acid Racemization Dating of Fossil Bones." *World Archaeology* 7:160–73.

Bellwood, P. 1986. *The Prehistory of the Indo-Malaysian Archipelago.* Orlando, Fla.: Academic Press.

Berger, R. 1975. "Advances and Results in Radiocarbon Dating. Early Man in North America." *World Archaeology* 7:174–84.

Bettinger, R. L. 1977. "Aboriginal Human Ecology in Owens Valley: Prehistoric Change in the Great Basin." *American Antiquity* 42:3–17.

Bryan, A. 1987. "The First Americans." *Natural History* 96(6):6–11.

Chard, C. S. 1956. "The Oldest Sites of Northeast Siberia." *American Antiquity* 21:405–9.

———. 1960. "Routes to Bering Strait." *American Antiquity* 26:283–85.

Claiborne, R. and the Editors of Time-Life. 1973. *The First Americans.* New York: Time-Life.

Cwyner, L. C. and J. C. Ritchie. 1980. "Arctic Step-Tundra: A Yukon Perspective." *Science* 208:1375–77.

DeJarnette, D. L. 1967. "Alabama Pebbletools: The Lively Complex." *Eastern States Archaeological Federation Bulletin* 26.

deTerra, H. 1949. "Early Man in Mexico." In *Tepexpan Man,* by H. deTerra, J. Romero, and T. D. Stewart. New York: Viking Fund Publications in Anthropology 11:11–86.

Dikov, N. N. 1977. *Monuments in Kamchatka, Chukotka, and the Upper Reaches of the Kolyma: Asia Joining America in Ancient Times.* Moscow: Nauka.

———. 1987. "On the Road to America." *Natural History* 97(1):12–15.

Dillehay, T. D. 1987. "By the Banks of the Chinchihuapi." *Natural History* 96(4):8–12.

Dincauze, D. 1981. "The Meadowcroft Papers." *The Quarterly Review of Archaeology* 2:3–4.

Dragoo, D. W. 1976. "Some Aspects of Eastern North American Prehistory: A Review 1975." *American Antiquity* 41:3–27.

Figgins, J. D. 1927. "The Antiquity of Man in America." *Natural History* 27:229–39.

Fladmark, K. 1986. *British Columbian Prehistory.* Chicago: University of Chicago Press.

———. 1987. "Getting One's Berings." *Natural History* 95(11):8–19.

Frison, G. C., M. Wilson, and D. J. Wilson. 1976. "Fossil Bison and Artifacts from an Early Altithermal Period Arroyo Trap in Wyoming." *American Antiquity* 41:28–57.

Frison, G. C., R. L. Andrews, J. M. Adovasio, R. C. Carlisle, and Robert Edgar. 1986. "A Late Paleoindian Animal Trapping Net from Northern Wyoming." *American Antiquity* 51(2):352–61.

Frison, G. C. and L. C. Todd, eds. 1987. *The Horner Site.* Orlando, Fla.: Academic Press.

Gowlett, J. A. J. 1987. "The Archaeology of Radiocarbon Acclerator Dating." *Journal of World Prehistory* 1(2):127–70.

Grayson, D. K. 1977. "Pleistocene Avifaunas and the Overkill Hypothesis." *Science* 195:691–93.

———. 1987. "Death by Natural Causes." *Natural History* 96(5):8–13.

Greenberg, J. H., C. G. Turner II, and S. L. Zegura. 1986. "The Settlement of the Americas: A Comparison of the Linguistic, Dental, and Genetic Evidence." *Current Anthropology* 27(5):477–97.

Greenman, E. F. 1963. "The Upper Paleolithic in the New World," *Current Anthropology* 4:41–91.

Griffin, J. B. 1960. "Some Prehistoric Connections Between Siberia and America." *Science* 131:801–12.

———. 1967. "Eastern North American Archaeology: A Summary." *Science* 156:175–91.

Guidon, N. 1987. "Cliff Notes." *Natural History* 96(8):6–12.

Guthrie, R. D. 1980. "The First Americans? The Elusive Arctic Bone Culture." *The Quarterly Review of Archaeology* 1:2.

Haag, W. G. 1962. "The Bering Strait Land Bridge." *Scientific American* 206:112–23.

Harris M. 1977. *Cannibals and Kings: The Origins of Culture.* New York: Random House.

Hassan, F. A. 1980. "Prehistoric Settlements in Egypt." In *The Sahara and the Nile,* eds. M. Williams and H. Fasure. Rotterdam: Balkema Press.

Hayden, B. 1981. "Research and Development in the Stone Age. Technological Transitions among Hunter-Gatherers." *Current Anthropology* 22:519–48.

Haynes, C. V., Jr. 1969. "The Earliest Americans." *Science* 166:709–15.

———. 1971. "Time, Environment, and Early Man." In *Papers from a Symposium on Early Man in North America, New Developments: 1960–1970,* ed. R. Shutler, Jr. *Arctic Anthropology* 8:3–14.

———. 1974. "Elephant Hunting in North America." *New World Prehistory: Readings from Scientific American,* eds. Ed. Zubrow et al. San Francisco: Freeman.

———. 1987. "Geofacts and Fancy." *Natural History* 97(2):4–12.

Hopkins D. M., ed. 1967. *The Bering Land Bridge.* Stanford: Stanford University Press.

Hopkins, D. M., J. V. Matthews, Jr., C. E. Schweger, and S. B. Young. 1982. *Paleoecology of Beringia.* New York: Academic Press.

Irving, W. N. 1985. "Context and Chronology of Early Man in the Americas." *Annual Review of Anthropology* 14:529–55.

———. 1987. "New Dates from Old Bones." *Natural History* 96(2):8–13.

Jairazbhoy, R. A. 1974. *Old World Origins of American Civilization.* 2 vols. London: Rauman and Littlefield.

Jennings, J. D. 1964. "The Desert West." In *Prehistoric Man in the New World,* eds. J. D. Jennings and E. Norbeck. Chicago; University of Chicago Press.

———. 1974. *The Prehistory of North America.* 2nd Ed. New York: McGraw-Hill.

Jennings, J. D. and E. Norbeck, eds. 1964. *Prehistoric Man in the New World.* Chicago: University of Chicago Press.

Kelly, R. L. and L. C. Todd. 1988. "Coming into the Country: Early Paleoindian Hunting and Mobility." *American Antiquity* 53(2):231–44.

Krieger, A. D. 1964. "Early Man in the New World." In *Prehistoric Man in the New World,* eds. J. D. Jennings and E. Norbeck. Chicago: University of Chicago Press.

Larichev, V., U. Khol'ushkin, and I. Laricheva. 1987. "Lower and Middle Paleolithic of Northern Asia: Achievements, Problems, and Perspectives." *Journal of World Prehistory* 1(4):415–64.

Lively, M. 1965. "The Lively Complex: Announcing a Pebble Tool Industry in Alabama." *Journal of Alabama Archaeology* 11:103–22.

Lyell, C. 1863. *Principles of Geology.* London: Murray.

MacNeish, R. S. 1971. "Early Man in the Andes." *Scientific American* 4:36–46.

———. 1978. *The Science of Archaeology?* North Scituate, Mass.: Duxbury Press.

MacNeish, R. S., ed. 1973. *Early Man in America.* San Francisco: Freeman.

Madsen, D. B. and M. S. Berry. 1975. "A Reassessment of Northeastern Great Basin Prehistory." *American Antiquity* 40:391–405.

Martin, P. S. 1973. "The Discovery of America." *Science* 179:969–74.

Martin, P. S. and J. E. Guilday. 1967. "A Bestiary for Pleistocene Biologists." In *Pleistocene Extinctions: The Search for a Cause,* eds. P. S. Martin and H. E. Wright, Jr. New Haven: Yale University Press.

Martin, P. S. and R. Klein, eds. 1984. *Quarternary Extinctions. A Prehistoric Revolution.* Tucson: University of Arizona Press.

Martin, P. S. and H. E. Wright, Jr., eds. 1967. *Pleistocene Extinctions: The Search for a Cause.* New Haven; Yale University Press.

Mehringer, P., Jr. 1977. "Great Basin Late Quaternary Environments and Chronology." In *Models in Great Basin Prehistory: A Symposium,* ed. D. D. Fowler. Desert Research Institute Publications in the Social Sciences 12:113–68.

Meltzer D. J. 1988. "Late Pleistocene Human Adaptations in Eastern North America." *Journal of World Prehistory* 2(1):1–52.

Meltzer, D. J. and J. I. Mead 1983. "The Timing of Late Pleistocene Mammalian Extinctions in North America." *Quaternary Research* 19:103–35.

Moismann, J. E. and P. S. Martin. 1975. "Simulating Overkill by Paleoindians." *American Scientist* 63:304–13.

Moratto, M. J. Foreword by F. A. Riddell. Contributions by D. A. Frederickson, C. Raven, and C. N. Warren. 1984. *California Archaeology.* New York: Academic Press.

Mueller-Beck, H. J. 1966. "Paleo-Hunters in America Origins and Diffusion." *Science* 152:1191–1210.

Orquera, L. A. 1987. "Advances in the Archaeology of the Pampa and Patagonia." *Journal of World Prehistory* 1(4):333–413.

Reeves, B. O. R. 1971. "On the Coalescence of the Laurentide and Cordilleran Ice Sheets in the Western Interior of North America." In *Aboriginal Man and Environments on the Plateau of Northwest America,* eds. A. Stryd and R. A. Smith. Calgary: University of Calgary Archaeological Association.

Ruhlen, M. 1987. "Voices from the Past." *Natural History* 96(3):6–10.

Schmitz, P. E. 1987. "Prehistoric Hunters and Gatherers of Brazil." *Journal of World Prehistory* 1(1):53–126.

Speth, J. 1987. "Les strategies alimentaires des chasseurs-cueilleurs." *La Recherche* 18(190):894–903.

Stewart, T. D. 1973. *The People of America.* New York: Scribner's.

Turner, C. G., II. 1987. "Tell-tale Teeth." *Natural History* 96(1):6–10.

Tuohy, D. R. 1968. "Some Early Lithic Sites in Western Nevada." In *Early Man in Western North America,* ed. C. Irwin-Williams. Portales: Eastern New Mexico University Press.

Vescelius, G. S. 1981. "Early and/or Not-so-Early Man in Peru: Guitarrero Cave Revisited." *Quarterly Review of Archaeology* 2:8–13, 19–20.

Warren, C. and A. Ranere. 1968. "Outside Danger Cave: A View of Early Men in the Great Basin." In *Early Man in Western North America,* ed. C. Irwin-Williams. Portales: Eastern New Mexico University Press.

Wheat, J. B. 1972. "The Olsen-Chubbuck Site. A Paleo-Indian Bison Kill." *Memoirs of the Society for American Archaeology,* no. 26, pt 2.

Williams, B. J. 1974. "A Model of Band Society." *American Antiquity* 39, pt. 2, Memoirs 29, pt 2.

Zegura, S. L. 1987. "Blood Test." *Natural History* 96(7):8–11.

6

The Origins of Agriculture

> The greatest events come to pass without any
> design; chance makes blunders good. . . . The
> important events of the world are not deliber-
> ately brought about: they occur.
>
> George C. Lichtenberg

The several million years separating the first tool-using hominids of Africa and the French cave artists of just 20,000 years ago encompassed a
period of momentous change. Human brain size trebled, crude stone implements were replaced by an impressive array of specialized tools, and
our ancestors eventually were able to colonize most of the world.

But in one important respect all Pleistocene societies, from the barely
human hominids of two million years ago to the creative forager-collectors
of just 12,000 years ago, were alike: they made their living through hunting and gathering, that is, through *nonagricultural* economies. Every Pleistocene society hunted and gathered most of its own food, made all or
most of their own tools, and in basic economy and society was probably
much like every other society.

We can look back, perhaps even with a bit of pride, at the adaptability
and durability of our ancient hunting-gathering forebears. Our ancestors
lived as hunter-gatherers for more than 20,000 centuries, and managed
to make a living in every kind of environment, from the icy Arctic tundra
to the jungles of the Amazon. Even today, a few hunter-gatherer societies
persist on the margins of modern states. But the hunting-gathering way
of life was doomed to extinction when the first farmers appeared. Just
after 10,000 years ago, people in a few widely separated areas of the world
began to domesticate a few plant and animal species and at about the
same time they built villages, lived in them year-round, and began to subsist mainly on plants they cultivated and animals they kept. In most envi-

225

ronments these agricultural economies were able to produce much greater amounts of food than hunting and gathering could, and much more reliably. Today hunter-gatherers exist only in the Kalahari Desert, the Arctic, and the few other places where hunting and gathering is still more productive than agriculture.

All the major civilizations throughout history have been based on the cultivation of one or more of just six plant species: wheat, barley, millet, rice, maize, and potatoes. Domesticated animals are also important, but it is these six plant species that provide most of the energy to "run" humanity. And if, for some reason, we all had to go back and live on undomesticated plants and animals, most of us would surely starve, despite our microwave ovens, nuclear energy, and computers.

Agriculture has had such tremendous importance in human affairs that one is tempted to the logical fallacy of assuming that the factors that first produced agricultural economies must be equally momentous. They may have been; at this point we do not know. Perhaps only slight changes in climate, population density, or technology set off the "agricultural revolution."

We know approximately where and when the most important domesticates first appeared, but much of modern archaeology has been devoted to analyzing *why* they appeared. For one of the particularly striking characteristics of the "agricultural revolution" is that it was, by archaeological standards, so rapid and widespread: for millions of years our ancestors subsisted solely on the proceeds of hunting and gathering, yet within just a few thousand years, between about 10,000 and 3500 years ago, people all over the world without any apparent connection began growing crops and establishing agricultural economies based on potatoes in the Andes, maize in Mexico, wheat in the Middle East, rice in China, and many other crops in other places.

Why did people become farmers after so many long years of hunting-foraging? And why did agriculture appear all over the world at about the same time?

Domestication, Agriculture, and Sedentary Communities

The terms domestication, agriculture, and sedentary communities are used rather loosely by most people, but here they have specific connotations. *Domestication* can be defined as a "coevolutionary process in which any given taxon diverges from an original gene pool and establishes a symbiotic protection and dispersal relationship with the animal feeding upon it."[1] When people domesticate plants and animals they do so by active interference in the life cycles of these species in such a way that subse-

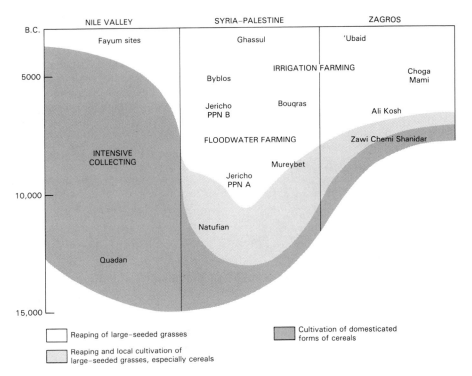

| | NILE VALLEY | SYRIA–PALESTINE | ZAGROS |

6.1   The world was forever changed by mutualistic relationships that evolved between people and large seeded grasses sometime before about 15,000 years ago. Agriculture was only the last stage of the evolution of these relationships, but the large and reliable harvests made possible by the domestication of wheat and barley were the basis on which ancient and modern civilizations evolved.

quent generations of these organisms are in more intimate association with, and often of more use to, people. By interfering in these life cycles, people have produced plants and animals that are less "fit" than their undomesticated relatives: domesticated maize, for example, no longer has an effective natural mechanism for seed dispersal, because the seeds are all clustered on a cob that, without human intervention, usually remains tightly attached to the plant. Similarly, a variety of sheep in Southwest Asia has been selectively bred over the last several thousand years so that its tail is a 5- to 8-pound mass of fat, making it necessary, it is said, for people to help these animals mate.

Such mutualistic relationships are not unique to humans and their plant and animal food, nor do they depend on human intellect. David Rindos points out that African cultivator ants

> prepare special beds, generally of plant debris, cut-up leaves, flowers, and excrement, in special chambers in the ant nest. The ants are meticulous

about growth conditions within the chamber; numerous ventilation passages are dug, and these are opened or closed to regulate both temperature and humidity. To construct the beds, the ants chew the substrate material to make a pulpy mass and deposit it in layers in the chamber. The bed is then planted with propagules from previously maintained beds. Constant care is given the beds. The ants remove alien fungi and add anal and salivary secretions which apparently have a positive effect on the growth of the fungi. These cultivation activities encourage the production, by the fungus, of small whitish round bodies, the so-called kohlrabi structures. These structures are the principal food of the ant colony.[2]

One does not see tiny scarecrows or silos among these ants, but Rindos' point is well taken: in terms of their ecological relationship with the sustaining species, ants do not differ significantly from us. And if this is true, the import of his observation is obvious: what we consider the uniquely human activities of domestication and agriculture are in fact somewhat common ecological relationships that can be "fixed" by natural selection without regard to the intelligence or volition of the organism involved.

If one looks at domestication from the point of view of cereals, for example, people have proved to be excellent devices for cereals to conquer the world. Wheat, which in its undomesticated forms was limited to small areas of the Middle East, has managed to extend its range from South America to central Russia, all by "domesticating" humans in such a way that they became dependent on it. As much as we are the creators of modern plant and animal hybrid domesticates, then, we are their creation; we are locked together in coevolutionary embrace.

Archaeological evidence indicates that, at least initially, people began to domesticate many plant and animal genera not out of the desire or intention of making these plants more useful to people, but rather out of relatively simple changes in exploitation patterns, brought on by changing ecological and cultural conditions. Any plant or animal species and population regularly eaten or used by people will reflect this relationship in its genetic characteristics, and thus all through the Pleistocene hunters and gatherers had some effect on the genetic makeup of various plant and animal species. But the low population densities and mobile way of life of the period before about 10,000 years ago kept people from exploiting plant and animal populations with sufficient intensity to perpetuate the kinds of mutations represented by maize and fat-tailed sheep.

Domestication has usually been a long-term process, not something that an occasional genius thought of and then applied; it has been a process in which the physical characteristics of plant and animal species change as these species' relationships to their human consumers change. And human intentions to domesticate plants and animals are neither archaeologically retrievable nor logically important to this notion of domestication.

As David Rindos points out, "People could not create the variation [in plants and animals] that would permit domestication, they could only select; and they could not have known how important the products of their selection would become."[3]

Nor do people's "taste" preferences seem to be important. One must recognize that maggotty meat, fresh cow blood, rotted fish-heads, raw sheep eyes, charred goat-intestines, tuna casserole, pot roasts simmered in packaged onion soup—all these are today savored with as much gusto as anything from the restaurants of Florence or Paris. Economic conditions usually are the major determinants of food preferences. The reason our local supermarkets are currently offering special prices on New Zealand lamb and not Labrador retrievers has more to do with the economics of raising these animals (and the economic conditions of the ancient societies in which they were first domesticated) than with their intrinsic tastes. The determination of food preferences other than inherent subtleties and complexities of tastes and textures is illustrated, for example, by the crowds of Americans who queue for hamburgers at Rome's McDonald's, less than 100 meters from the equally priced 37-item antipasto buffet at the Ristorante a la Rampa.

The key concept in the notion of *agriculture* involves human efforts to modify the environments of plants and animals to increase their productivity and usefulness. By weeding gardens, fashioning scarecrows, penning pigs, and a thousand other farming tasks, people over the millennia have sought to get more food out of their labors. Hunter-gatherers also modify the environments of plants and animals, if only by collecting and eating them; but agriculture is the systematic modification of environments to increase productivity. This definition of agriculture, like that of domestication, is relative. At the one extreme are the simple efforts a group might make to suppress weeds near a stand of wild wheat; at the other are contemporary agribusinesses where crops are grown on precisely leveled fields and treated with pesticides and fertilizers, and where chickens live out unremarkable lives in cages until a computer determines that their ratio of food-consumption and growth reaches a critical—and fatal—value.

In terms of what happened in history, one of the most important things about agriculture is that it produces not just great amounts of food, it supplies reliable and predictable amounts of food, and thereby allows population densities to rise and people in many areas of the world to live year round in the same place—in *sedentary communities*. Such communities probably first appeared in Southwest Asia as much as 1000 years before agriculture or domesticated wheat or barley, while in Mesoamerica at least five plant species were in the process of domestication several thousand years before the first sedentary or agricultural communities appeared.

Sedentary life usually involves the construction of permanent structures. For a million years and more before agriculture, people built crude shelters and lived in caves, but in almost every case they moved to other camps for part of the year. When our ancestors began living year-round in the same villages, a complex set of relationships between people, plants, and animals was established that had never existed before. Village life imposes a certain kind of psychological reality on people that hunters and gatherers do not have: it alters fertility rates, increases the chances of epidemic diseases, and encourages family-unit economic production.

There is a tremendous "inertia" in village agriculture, once it is established. In only a few environments can people move easily from agricultural to other kinds of economies, for the essence of agriculture is that one concentrates one's energies on a very limited number of species that are very reliable. In most agricultural economies, activities that conflict with agricultural production are soon abandoned. In Denmark, for example, analysis of human bone composition shows that in one area along the ocean shore, people before 4000 B.C. lived mainly on marine resources, but after that time they subsisted almost entirely on agricultural products, even though their villages were within a few hundred meters of the rich fish and other resources of the shore.[4]

People cannot always be considered entirely rational consumers who put together economies that are optimally efficient, and we cannot explain agricultural origins simply as the natural outcome of people who search through all the possibilities and put together the best economy based on their technology. But *everything* cultural is subjected to natural selection to some extent, and so we must consider agriculture to be the outcome of something that "worked." In this regard, archaeologists have been considering "optimal foraging strategies." This concept is an outgrowth of evolutionary ecology studies in which mathematical analyses are made of the feeding strategies of various species. Implicit in most of these analyses is the assumption that natural selection and competition are inevitable conditions resulting from the nature of biological reproduction in finite environments.[5]

The "optimal" part of the concept of optimal foraging strategies refers to the assumption that natural selection over time tends to promote "efficiency" in the strategies that species evolve in their choice of foods, the amount of time they spend pursuing their prey, and the size of the group in which they pursue it. Optimal foraging theory is used to predict on the basis of environmental variables the kinds of feeding strategies that will evolve. As applied to people, optimal foraging strategies are complicated by the existence of technology; the efficiencies of their grinding stones, flint sickle blades, clay storage silos, etc., are factors in the choices any group has in getting food from its environment.

Arthur Keene has used optimal foraging theory to construct a computerized mathematical model of the hunting and gathering practices of people in southeastern Michigan between 4000 and 1000 B.C.[6] He estimated the costs and benefits of almost every resource in this area—from bears to berries—and then combined this information with demographic and technological estimates of these ancient peoples to "predict" where archaeological sites would be found and the kinds of technological and plant and animal remains that would be found in these sites. Keene notes that models like his have various uses in archaeology, such as in designing surveys and testing hypotheses about ancient economies.

Optimal foraging theory underscores an important point about human adaptations: although people choose this or that blend of resources for reasons that are forever beyond the reach of archaeology, natural selection does not "care" what the people thought about their world. Natural selection fixes foraging strategies on the basis of the costs and benefits of these strategies, regardless of the hopes, fears, ideologies, and preferences of the people involved.

Early Domestication and Agriculture:
The Late Pleistocene Background

We must assume that, to some extent, the causes of agriculture are to be found in the thousands of years of hunting-gathering that immediately preceded the first agricultural societies. A critical interval may be the period between 15,000 and 8000 years ago—a time of major climatic changes for much of the world. In western Europe population densities shifted as the herds of reindeer and horses that once supported many hunting bands moved northward with the retreating glaciers. Some people moved with them, but others worked out subsistence strategies stressing plants, smaller game, and fish. Salmon became especially important in Europe as traps, drying racks, and other tools were developed to make salmon exploitation a reliable way to make a living. In Southwest Asia, parts of Africa, and parts of the Americas, some late Pleistocene and early post-Pleistocene peoples began to eat more small game, fish, waterfowl, clams, wild cereals, and similar foods; elsewhere big-game hunting specializations persisted. In North America, for example, some groups centered their lives around vast bison herds.[7]

Where a shift to smaller, more varied resources was made, technologies also changed. The bow and arrow and throwing stick replaced the stabbing spears, and new tools were developed to dig plants, trap wild-fowl, and prepare and cook this broader diet. Small, simple geometric stone tools predominated in many areas (Figure 6.2). The world about 12,000 years ago was relatively diverse culturally, as some groups remained big-

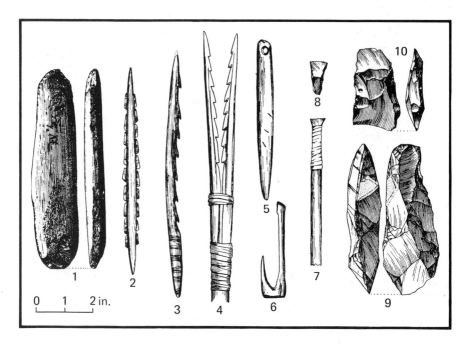

6.2   After about 10,000 years ago, people all over Europe began to exploit a much greater diversity of plants and animals than did their Pleistocene predecessors. This shift is reflected in Mesolithic fishing and hunting equipment: (1) limpet hammer, (2) bone fish-spear with microlith barbs, southern Sweden, (3) barbed point in red deer antler, c. 7500 B.C., Star Carr, Yorkshire, (4) leister prongs of Eskimo fishermen (shows how the barbed point may have been used), (5) net-making needle (?), and (6) bone fish hook, Denmark; (7), (8) microliths or transverse arrowheads, one found in peat hafted in wood with sinew binding, Denmark; (9) core-axe with transversely sharpened edge, Sussex; (1) flake-axe, Denmark.

game hunters while others took up fishing, intensive foraging, and other pursuits. Thus, a great diversity of plants and animals was being exploited with varying intensities and technologies in a wide range of climates. Out of this vast mixture of peoples, plants, animals, and places the first domesticates and farmers appeared (see Figure 6.3).

But which of these groups became agriculturalists and why? Although all the people before about 10,000 years ago were hunter-gatherers, this term covers a wide range of economies.[8] Much of the variability in hunter-gatherer adaptations seems to be linked to *food storage*.[9] Alain Testart notes that some hunter-gatherers store large amounts of food, others do not, and that storing hunter-gatherer societies exhibit three characteristics—sedentarism, a high population density, and the development of socioeconomic inequalities—which have been considered typical of agricultural societies and possible only with an agricultural way of life. Furthermore, their economic cycle—massive harvest and intensive storage of a seasonal resource—is the same as that of societies based on the cultivation of cereals. The difference between storing hunter-gatherers and agriculturalists lies in whether the staple food species are wild or domesticated: this proves to be only a minor difference, since it does not affect the main aspects of

6.3   Early archaeological occurrences of some important Old World and New World domesticates. Domestication is a process, not an event, and these specific sites represent only some early occurrences of species domesticated over wide areas. Many other species have been domesticated.

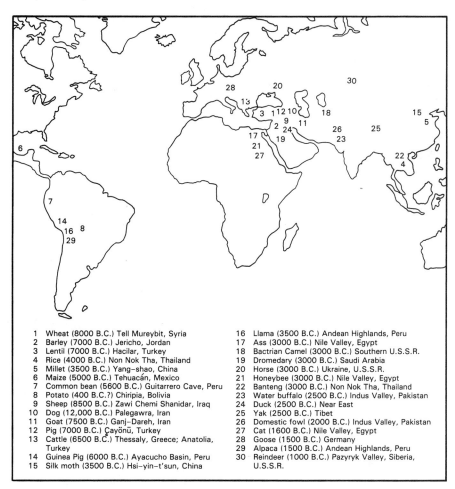

1	Wheat (8000 B.C.) Tell Mureybit, Syria	16	Llama (3500 B.C.) Andean Highlands, Peru
2	Barley (7000 B.C.) Jericho, Jordan	17	Ass (3000 B.C.) Nile Valley, Egypt
3	Lentil (7000 B.C.) Hacilar, Turkey	18	Bactrian Camel (3000 B.C.) Southern U.S.S.R.
4	Rice (4000 B.C.) Non Nok Tha, Thailand	19	Dromedary (3000 B.C.) Saudi Arabia
5	Millet (3500 B.C.) Yang–shao, China	20	Horse (3000 B.C.) Ukraine, U.S.S.R.
6	Maize (5000 B.C.) Tehuacán, Mexico	21	Honeybee (3000 B.C.) Nile Valley, Egypt
7	Common bean (5600 B.C.) Guitarrero Cave, Peru	22	Banteng (3000 B.C.) Non Nok Tha, Thailand
8	Potato (400 B.C.?) Chiripia, Bolivia	23	Water buffalo (2500 B.C.) Indus Valley, Pakistan
9	Sheep (8500 B.C.) Zawi Chemi Shanidar, Iraq	24	Duck (2500 B.C.) Near East
10	Dog (12,000 B.C.) Palegawra, Iran	25	Yak (2500 B.C.) Tibet
11	Goat (7500 B.C.) Ganj–Dareh, Iran	26	Domestic fowl (2000 B.C.) Indus Valley, Pakistan
12	Pig (7000 B.C.) Çayönü, Turkey	27	Cat (1600 B.C.) Nile Valley, Egypt
13	Cattle (6500 B.C.) Thessaly, Greece; Anatolia, Turkey	28	Goose (1500 B.C.) Germany
14	Guinea Pig (6000 B.C.) Ayacucho Basin, Peru	29	Alpaca (1500 B.C.) Andean Highlands, Peru
15	Silk moth (3500 B.C.) Hsi–yin–t'sun, China	30	Reindeer (1000 B.C.) Pazyryk Valley, Siberia, U.S.S.R.

society. Agriculturalists and storing hunter-gatherers together are neatly
in opposition to non-storing hunter-gatherers. The conclusion to be drawn
is that it is certainly not the presence of agriculture or its absence which
is the relevant factor when dealing with such societies, but rather the
presence or absence of an economy with intensive storage as its corner-
stone.[10]

Thus, in looking for causes of agricultural economies, we must consider
kinds of adaptations where storage is a potential factor. This relates directly to certain kinds of foods. Cereals store well, but many tubers do
not. Sheep and other animals can also be considered a form of food storage, since one simply feeds them excess foods until the need arises to eat
them.

In looking for the origins of agriculture, we can look for environments
where food storage was an early and important option. It is probably also
of significance, as Kent Flannery notes, that the major seed crops that
supported the first farmers and remain the basis of modern economies,
including wheat, barley, millet, and rice, appear to have derived from
wild ancestors that were "third-choice" foods: plants that were usually more
difficult to gather and process than other wild plants and thus were probably first eaten in quantity because people had to, not because they wanted
to.[11] On the other hand, most of these third-choice foods are easily storable, plentiful, easy to grow, and, as annuals, are genetically maleable—
altering the selective forces on them each year can quickly change them
genetically.

The Origins of Domestication, Agriculture, and Sedentary Communities in Southwest Asia

The best known and perhaps the world's first case of the origins of domestication and agriculture occurred in Southwest Asia, and involved
peoples and environments ranging from Afghanistan to Greece, over a
time period of 10,000 years.

Millions of years ago movements of the earth's crust forced the Arabian
Peninsula toward the stable Iranian Plateau, compressing the land in between so that it is pleated like the folds of an accordion. At the end of the
Pleistocene, the uplands of the "Fertile Crescent" (Figure 6.4) supported
large herds of wild sheep, goats, cattle, and pigs and, in many areas, dense
stands of wild wheat and barley. In lower elevations and wetter regions
were lakes and streams with abundant supplies of waterfowl and fish.

The cold climate and the lowered ocean levels of the late Pleistocene
meant that many parts of Southwest Asia that had been forests became
dry steppe country. In these climates the "Mediterranean" vegetation that

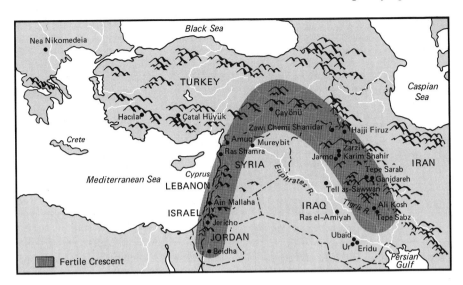

6.4 The "Fertile Crescent" and some important early preagricultural and agricultural sites.

includes grasses ancestral to modern cereals would have been greatly restricted, mainly to upland areas where there was winter rainfall sufficient to support the open-woodlands habitats where these grasses are found. But with the end of the Pleistocene 10,000 years ago, the warmer, wetter weather patterns would have extended these cereals' ranges into the areas of Southwest Asia where the first domesticated plants and agricultural communities later appeared.[12] Also, the rising sea levels would have submerged from 10 to 20 kilometers of coastal plains around much of Southwest Asia, perhaps forcing people into new areas. There was a maximum rise of about 120 meters in sea level between the height of Pleistocene cold about 18,000 B.P. and 10,000 B.P., but the average annual rise would have been only 7 to 8 millimeters, so one should not imagine late Pleistocene hunter-foragers scurrying in front of unexpected floods.[13]

Since there were many previous glacial fluctuations that did not produce agriculture, these climatic variations are obviously not a complete explanation of agricultural origins; but it is not without significance for the timing and location of agricultural origins that they coincided with these environmental changes.

To understand the origins of agriculture in Southwest Asia, we have to understand the domestication of wheat, barley, sheep, and goats, for these have been the primary food sources in this area from about 9000 years ago to the present day. Lentils and other legumes may have been the earliest important food plant domesticated in Southwest Asia,[14] but it was wheat and barley, sheep and goats that underlie Southwest Asia's cultural evolution.

If you were a peasant agriculturalist forced to farm the wild ancestors of wheat and barley, you would have had a hard time. Wild wheat, for example, has many limitations as a food crop. Its distribution is sharply limited by temperature, soil, and moisture. As a result, stands of these grains can be widely scattered and therefore difficult to harvest. Much greater efficiency could be attained if these plants could be adapted to a

6.5 Domestication of wheat, one of the world's most important crops, involved both human manipulation and natural hybridization between related genera. Human intervention appears to have been aimed at producing free-threshing, non-shattering varieties. The simplest wheats are "diploid," meaning that they have two sets of seven chromosomes. Hybridization with related species produced tetraploid wheats, with six sets of chromosomes, which occur only in cultivated species of wheat. By mixing genetic material from various species, early farmers produced forms of wheat that could adapt to diverse habitats.

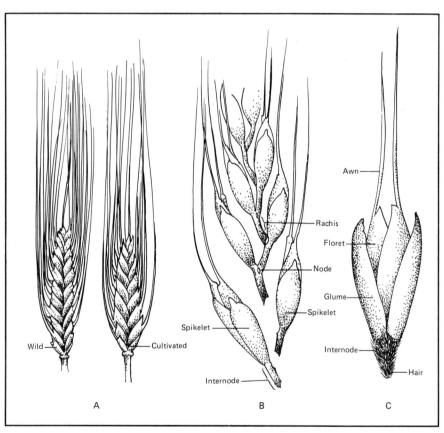

greater variety of temperature, soil, and moisture ranges—especially to the torrid summers of the lowlands, where the rivers offered the potential of irrigation agriculture.

Also, the rachis, the segment of the stalk to which the kernels of wheat or barley are attached, becomes extremely brittle as wild wheat ripens. This brittleness is essential to the successful propagation of these plants because it allows the seeds to be separated from the plant and dispersed by the merest touch of an animal or simply the force of the wind. The head of the plant becomes brittle gradually, from top to bottom, and seed dispersal is spread over one to two weeks. Although this is advantageous for the plant because it prevents the seeds from sprouting in a dense mass of competing seedlings, it poses problems for the human collector. If the grain is gathered when quite ripe, the slightest contact will cause the rachis to fall off, so harvesting with a sickle is difficult—although holding the stalk over a basket and tapping it with a stick works. If, on the other hand, the grain is harvested before it is fully mature, excess moisture in the unripe kernels will cause them to rot in storage. A plant with a tougher rachis and on which the kernels ripen at about the same time would clearly be more useful to people.

Another problem with wild wheat is that the kernels are enclosed in very tough protective husks, called glumes. These protect the seeds from frost and dehydration, but primitive threshing often will not separate the seeds from the glumes, and the human digestive tract cannot break down their tough fibers. Thus a cereal with less tough and less developed glumes would be more digestible.

Also, each stalk of wild grain has only two rows of kernels. Domesticated varieties have six rows, rendering them much more productive as a food resource, and wild species had to change in this direction before it was profitable to invest energy in sowing, cultivating, and harvesting wheat and barley in many areas—particularly in those areas where natural conditions were not optimal for wild cereals.

To base your economy on wheat and barley would also require some specialized tools, for cutting, transporting, threshing, storing, milling, and baking. A full-time cereal farmer is best off living year round in a village or town because of the costs of moving large quantities of grain and because successful harvesting requires that collectors be near stands at very precise times. In contrast to many plants, wild grain can be collected only during the few days when the plants are ripening, and even then there is considerable competition from birds and other predators. Another consideration is that women and children in cereal-gathering societies could contribute a great deal to the food supply, whereas in societies specializing in hunting, women and children have less direct economic return.

Given all these liabilities and required cultural adjustments, one might wonder why and how these cereals were domesticated. In fact, this seems to have been a relatively rapid process in Southwest Asia, occurring within a few thousand years, in several native grasses, including wild barley (*Hordeum spontaneum*), wild einkorn (*Triticum boeoticum*), and wild emmer wheat (*Triticum dicoccoides*), each with different habitats and characteristics.

At the same time these plants were being domesticated, about 10,000 years ago, sheep, goats, pigs, and cattle were also being domesticated. In some areas, such as the southern Mesopotamian Alluvium, permanent settlement was not possible until domesticated animals were available to supply the fats and proteins that were not readily obtainable from any other source. Domesticated animals also provided a way of converting highland grasses, weeds, shrubs, surplus grain stubble, and other plants into storable, portable, high-quality foods and other usable products. Later, some animals, such as cattle, horses, and donkeys, provided draught and transport power.

The Archaeological Record of Plant and Animal Domestication in Southwest Asia

Archaeological sites throughout Southwest Asia during the late Upper Pleistocene, from about 20,000 to 16,000 B.C., are monotonously alike in their concentrations of stone tools, ash, and the bones of large, hoofed mammals. Almost all the meat eaten by people came from just a few species of ungulates (hoofed mammals), mainly gazelles and wild cows. Based on the tools and other artifacts from Southwest Asian sites of this period, it appears that the basic social unit was a band of about fifteen or twenty people comprising several families who season after season moved through this area hunting animals and gathering plants.

The Levantine cultures of between 18,000 and 15,000 years ago are called *Kebaran,* and they were followed by the *Geometric Kebaran* cultures (named, as so many ancient peoples are, after stone tools from a particular site) from about 15,000 to 12,500 B.P., and the *Natufian* cultures of 12,500 to 10,000 years ago.[15]

In general, the people of the Levant and elsewhere in Southwest Asia up to about the time of the Natufian cultures (that is, until about 12,500 years ago) are thought to have been relatively simple hunter-forager groups. The Neve David site, on the western slopes of Mt. Carmel, in Israel, suggests, however, that people in the Geometric Kebaran Period may have had more complex social lives than one would gather from the other sites of this era. Neve David seems to have been home to a rather large group, a "macroband." It has the full range of ground stone tools, it contains two small stone structures, one in a circular form about 2 meters in diameter, and a grave has been found at the site in which a body was interred with apparent ritual use of grinding stones.[16]

Early studies of Kebaran sites and those of later periods led several scholars to the conclusion that there was a "broad spectrum revolution" between 20,000 and 10,000 years ago, in which people in Southwest Asia and many other parts of the world began eating a broader range of foods.[17] In addition to the wide range of plant and animal species found in some sites toward the end of this period, more efficient and varied tools seem to have been used as these millennia passed. Barbed spears and arrows, bows, knives made by setting obsidian and flint flakes into bone or wood, and other tools indicate a greater range of subsistence activities. Some animal species were exploited systematically for the first time, and the grinding stone, "sickles," and other new tools indicate that vegetable foods, including wild cereal grasses, may have been important parts of the diet. Minor local trade in obsidian and sea shells was carried on, and substantial huts appeared in some areas. Apparently, population densities were slowly increasing in some areas—although the archaeological evidence for this is not at all clear.[18]

But in some ways, the broad-spectrum revolution and the agricultural revolution may together form an example of that "great tragedy of Science—the slaying of a beautiful hypothesis by an ugly fact," as Thomas Huxley put it.

In a long and thorough review of the evidence from Southwest Asia, archaeologist Donald Henry came to two conclusions:

> First, the proposal of a gradual broadening of the economy leading to a more secure subsistence base, the emergence of sedentary communities, and a growth in population can be rejected. Such a trend is not indicated by the pre-Natufian subsistence evidence. Secondly, the emergence of large sedentary communities, as represented by the Natufian, can best be explained by the intensive exploitation of cereals and nuts. Specialization, as opposed to diversification, characterized the Natufian economy. Although small animal species did form a larger portion of the diet in Natufian times than earlier, these food sources furnished an exceedingly small, almost negligible, part of the overall diet. Instead the specialized hunting of gazelle and the intensive collection of cereals and nuts contributed the greatest part of the subsistence base.[19]

Some evidence of the specialized gazelle hunting that Henry mentions has been provided by Legge and Rowley-Conwy, who have shown that 11,000 years ago, in the Euphrates Valley in Syria, at the site of Tell Abu Hureya, people intensively hunted gazelles.[20] Of the 60,000 identifiable animal bones found at this site, about 80% are from gazelles—and from gazelles of all sizes and ages and both sexes. Near the site are the traces of stone enclosures, and Legge and Rowley-Conwy have concluded that for thousands of years the people of Tell Abu Hureya drove herds of gazelle into these enclosures and killed them all. Many of these enclosures have been found and it is possible that eventually the gazelle were driven so close to extinction that people turned to sheep and goats, and in the process domesticated them.

Detecting the process of animal domestication on the basis of archaeological data is a complex matter on which archaeologists disagree, but the concept of domestication in an archaeological sense usually involves three classes of evidence.[21] Just the presence of an animal species outside its natural range, such as the presence of highland species of sheep in lowland environments, may indicate herding. Second, in most animals, morphological changes occur as domestication progresses. In sheep and goats, the size and shape of horns are major factors in reproductive success, since the larger males fight with their horns to establish breeding hierarchies. As humans domesticated these animals, however, they relaxed selective pressures for large, strong horns (Figure 6.6), and the size and shape of horns changed (morphological changes can also occur at the microscopic level). A third indication of domestication may be evidence of abrupt increases in the number of some species relative to others that cannot be accounted for by natural causes (for example, at many sites in Southwest Asia at about 8000 B.C. the proportion of sheep and goat bones increased dramatically, relative to other animals).

The dog is no doubt the oldest animal domesticate in Southwest Asia, probably having been fully domesticated between 22,000 and 10,000 B.C. Hunters and gatherers the world over are known to have remarkably unsentimental ideas about pets, and we should probably see the early domestication of dogs as a result of a symbiotic, utilitarian relationship. Dogs probably served as watchdogs, assisted in the hunt, and were eaten as a starvation food.

In economic terms the most important domesticated animals in Southwest Asia are sheep and goats. From about 9000 B.C. to the present, most meat, milk, and hide products used in Southwest Asia have come from these animals.

The first evidence of sheep domestication may be the presence of sheep bones in Neolithic (c. 9000 to 6000 B.C.) settlements in the Jordan Valley. These bones reflect no morphological changes in the direction of domestication, but sheep and goats are not native to this area, and their presence here probably reflects intentional introduction.

Ethnographic studies of modern pastoralists suggest that they typically sell or eat 50% or more of the animals born each year, principally the males, since the females can be kept for reproduction and only a few males are required to service the breeding population. Once a male has reached two years of age, any further investment of food or time in him yields little additional return. Members of archaeological projects spend-

Phase	Site	Zone	Cross section quadrilateral	Cross section lozenge-shaped	Medially flat, but untwisted	Medially concave, helical twist	Too young or too broken to diagnose
Bayat	TS	A_1				1	
Bayat	TS	A_2				1	
Mehmeh	TS	A_3			1	3	1
Mehmeh	TS	B_1			1		
Mehmeh	TS	B_2				1	1
Mehmeh	TS	B_3					1
Khazineh	TS	C_1					1
Khazineh	TS	C_2					
Khazineh	TS	C_3					3
Sabz	TS	D			2	1	6
Mohammad Jaffar	AK	A_1	4	1		1	7
Mohammad Jaffar	AK	A_2					8
Ali Kosh	AK	B_1		1	2		3
Ali Kosh	AK	B_2	11	7	8		27
Bus Mordeh	AK	C_1	2?	2			
Bus Mordeh	AK	C_2		2			7

Ca. 3700 B.C. — Ca. 7500 B.C.

6.6 The essence of archaeological interpretation is to discern patterns of change in the archaeological record over time. Here, the goat horn cores found at some sites in western Iran indicate that the shape of these horns was changing rapidly between 7500 B.C. and 3700 B.C. One explanation is that in the process of domestication the selective forces that kept these horns quadrilateral—and therefore relatively strong—were being relaxed, probably as a result of the domestication process. Farmers kept goats in herds and selectively bred them, and the strength of the horns apparently was no longer an important determinant of the reproductive success of these goats.

241

Table **6.1**   Plant Remains from the Middle East

Lab No.	Site	Plant	Radiocarbon Date
OxO-170	Abu Hureyra	*Triticum boeoticum*, E261	10,600 ± 200
OxO-171	Abu Hureyra	*Triticum boeoticum*, E313	10,600 ± 200
OxO-172	Abu Hureyra	*Triticum boeoticum*, E326	10,900 ± 200
OxO-386	Abu Hureyra	*Triticum boeoticum*, E276	10,800 ± 160
OxO-397	Abu Hureyra	*Triticum boeoticum*, E286	10,420 ± 150
OxO-541	Rakafet	vetch, *Vicia ervilia*	2,760 ± 200
OxO-543	Ain Mallaha	*Triticum aestivum*	220 ± 100
OxO-388	Can Hasan	grain of *Triticum dicoccum*	7,910 ± 160
OxO-392	Can Hasan	rachis of *Triticum durum*	250 ± 90
OxO-393	Wadi Hammeh 27	charred seeds, incl. legumes	11,920 ± 150
OxO-394	Wadi Hammeh 27	charred seeds, *Chenopodium?*	12,200 ± 160
OxO-507	Wadi Hammeh 27	charred seeds, incl. legumes	11,950 ± 160
OxO-389	Nahal Oren	*Triticum*, 307.0/33	2,940 ± 120
OxO-390	Nahal Oren	*Triticum*, 307.0/32–33	33,000
OxO-395	Nahal Oren	*Triticum*, 308.0/28	3,100 ± 130
OxO-396	Nahal Oren	humics of OxA-395	6,650 ± 190

ing their first season in the Middle East are often surprised to find that they eat chicken almost every day but almost never get lamb, even though the villages swarm with sheep. It is mainly because lamb and mutton are relatively expensive. In 1988 an Egyptian archaeologist with our project paid the quivalent of $55 for a sheep he bought to be slaughtered for an important religious holiday. Even before the distortions introduced by modern urban markets, the economic importance of sheep and goats was mainly in the milk, cheese, yoghurt, and wool obtained from these animals. A well-kept animal can supply many times its own weight in nutritious, storable foods each year for several years, so only young males are regularly slaughtered for meat.

Evidence from two sites near Zawi Chemi Shanidar in Iraq indicates that similar selective slaughtering may have been underway as early as 9000 B.C. Prior to 12,000 B.C., only about 20% of the animals killed and eaten at one of the sites were immature, but by 8650 B.C., 44 to 58% of the sheep and 25 to 43% of the goats eaten were immature when butchered.[22]

Shortly after 10,000 B.C. there were still no full-time farmers in Southwest Asia, but sedentary communities appeared for the first time, such as those of the Natufian culture, represented by scores of sites located in a wide strip of land running from southern Turkey to the edge of the Nile delta. At about 10,000 B.C. intensive collectors and hunters in this area subsisted largely on gazelle hunting, fishing, and the collection of wild cereals. The importance of cereals in these communities is reflected in their archaeological debris, particularly by the large numbers of sickle blades, many of which have a glossy sheen from continued contact with the rough stems of cereal plants. Some of the most recent radiocarbon dates for these communities are presented in Table 6.1,[23] showing a pattern of wheat use beginning about 12,000 years ago.

While some Natufian peoples retained a mobile way of life, others established sedentary communities, such as at Ain Mallaha, near Lake Huleh, Israel, which between 9000 and 8000 B.C. comprised about fifty huts, most of them circular, semisubterranean, rock-lined, and from 2.5 to 9 meters in diameter. Mortars and pestles litter the site and occur in most huts, and storage pits were found both in individual huts and in the compound's interior.

Kent Flannery has noted that many contemporary African peoples also live in compounds of circular huts and that most such societies share several characteristics: only one or two people are usually housed in each hut; many of the huts are not residential, but are used for storage, kitch-

6.7 An early Natufian house, situated on virgin soil at Tell as-Sultan, Jericho. The round shallow pits in the floor were probably storage pits, and the stone querns to the left were probably used for grinding cereals.

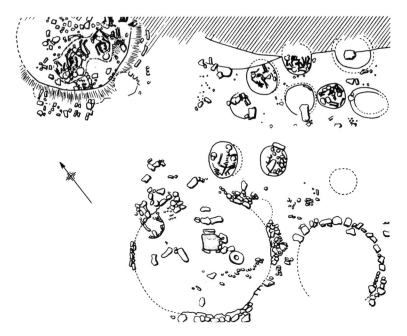

6.8 Simplified plan of an early settlement at Ain Mallaha (Israel). Compounds of circular huts such as those at Ain Mallaha were widespread in Southwest Asia after about 8000 B.C., but by 6000 B.C. had been superseded largely by villages of rectangular huts.

ens, stables, and the like; huts are often placed in a circle around a cleared space; food storage is usually open and shared by all occupants; and, perhaps most important, the social organization of the typical compound, like that of hunting and gathering groups, usually consists of six to eight males, each associated with from one to three women and their respective children, and there is a strong sexual division of labor.[24]

Flannery argues that settlements of adjacent rectangular buildings—which he calls villages—have advantages over settlements of circular buildings—which he calls compounds. The former are more easily enlarged because rooms can simply be added on, whereas increasing the number of circular residences rapidly increases the diameter of the settlement to an unwieldy size. Villages are also more defensible than compounds for a number of reasons. But the primary difference is in their respective capacities for intensification of production. In compounds, storage facilities are open and shared, and the basic economic unit is the group; but in villages the basic unit is the family, which maintains its own storage of supplies and thus has greater incentives for intensification of production.[25]

If Flannery is correct, the transition that occurred between 9000 B.C. and 7000 B.C. from compounds of circular structures to villages of rectangular rooms is a reflection of changes in the social organization of the Greater Mesopotamian peoples, with the nuclear family gradually replacing the hunting and gathering group as the unit of economic production. And although the circular-building tradition continued for several thousand years in parts of Southwest Asia, it was eventually entirely supplanted by rectangular-unit villages.

To return to the archaeological evidence for early domestication, it is evident that Natufian culture contrasted in many ways with that of its predecessors. Sickles, querns (hand mills), mortars, pestles, pounders, and other ground stone tools occur in abundance at Natufian sites, and many such tools show signs of long, intensive use. Fish hooks and gorges and net sinkers attest to the growing importance of fish in the diet in some areas. Stone vessels indicate an increased need for containers, but there is no evidence of Natufian clay working or pottery.

6.9   Common implements for grinding and preparing grain in Iraq, between 7000 and 4000 B.C. The ceramic husking tray above was used to strip grain from chaff; the heavy stone quern and round pestle below were used to grind grain to flour.

Compared to the Pleistocene cultures of the Levant, there appears to have been considerable social change in the Natufian. Cleverly carved figurines of animals, women, and other subjects occur in many sites, and Natufian period cave paintings have been found in Anatolia, Syria, and Iran. Trade in shell, obsidian, and other commodities seems to have been on the rise, and we suspect that exchange of perishables, such as skins, foodstuffs, and salt, was also increasing. With the increased importance of wild cereals in the diet, salt probably became for the first time a near necessity: people who eat a lot of meat get many essential salts from this diet, but diets based on cereals can be deficient in salts. Salt was probably also important as a food preservative in early villages.

More than 200 Natufian burials have been found, most of them simple graves set in house floors. Grave goods are infrequent, but some burials indicate concern with the philosophical implications of death; at Ain Mallaha skeletons were buried with their heads wedged between two stones and their joints covered with large stones, "to ensure perhaps that the deceased would not rise from his grave."[26] Some of the earliest evidence for domesticated grain cultivation in the Levant comes from the lower levels of Jericho (c. 8350 to 7350 B.C.), next to the springs in the center of this oasis. At some time during this period domestic forms of wheat and barley were cultivated in quantity here. Neither wild wheat nor barley appears to have been native to the arid wastelands that surround the site, so these grains were probably brought down from the uplands of the Jordan Valley and grown at Jericho, perhaps as wild species initially. No domestic animals were used in this period, but wild gazelles, goats, cattle, and boars were intensively hunted. Two thousand or more people probably lived at Jericho at any time between 8350 and 7350 B.C., and although the earlier communities were apparently unwalled, around 7350 B.C. the inhabitants built a massive stone wall, 3 meters thick, 4 meters high, and perhaps 700 meters in circumference. Asphalt, sulphur, salt, and a little obsidian seem to have been traded, but in moderate quantities. The Biblical references to the collapse of Jericho's walls may have to do with the earthquakes that frequently in the past flattened the town.

At the same time that agricultural economies were evolving in Palestine, specialized nomadic economies were also probably developing.[27] And people living on the flanks of the Zagros and Taurus mountains were making the transition to sedentary communities based on intensive plant collection. One of the earliest such communities was at Tell Mureybit, on the Euphrates River east of Aleppo, Syria. There, at about 8200 to 8000 B.C., people built circular stone huts, similar in almost every respect to the circular huts at Ain Mallaha. Charred wild einkorn seeds have been recovered from Tell Mureybit, as well as the remains of wild barley, lentils, bitter vetch, pistachios, toad rush, and possibly peas. Most of these plants

6.10 Figurines from the early agricultural village of Ain Ghazal, Jordan.

can be found locally, but wild einkorn and barley are not native to this area and in fact can be found in natural stands no nearer than the Anatolian hills some 100 to 150 kilometers to the northwest.[28] The impracticality of moving large amounts of grain this distance suggests that Tell Mureybit may be one of the earliest agricultural settlements in Southwest Asia, that here and in adjacent areas intensive collectors first tried to plant, cultivate, and harvest their own fields of grain. Tell Mureybit is a deep site, and its many levels of construction, first of circular compounds of crude huts, then larger rectangular villages, suggest the success of this experiment.

Soon after 8000 B.C., sedentary communities and domestic plants and animals had appeared at several places along the flanks of the Zagros. At Ali Kosh, situated on the arid steppe of western Iran, at about 7500 B.C. people hunted gazelles, onagers (wild asses), and pigs, fished in the Meh-meh River, collected shellfish, and snared wild fowl. They also collected vetch and other plants, and between 8000 and 6500 B.C. they began growing domestic, two-rowed, hulled barley and emmer wheat. These early farmers lived in crude clay huts furnished with reed mats, and had stone bowls and a few other small household goods, but this settlement was neither rich nor impressive. Possibly the people came here only in the winter, since summers are unearthly hot and the cooler mountains would have provided many plant and animal products. Wild wheat is not native to the Ali Kosh area, but wild barley is available within a few kilometers, and the people here may have been growing grains that had been domesticated elsewhere.[29]

By 6000 B.C. there is evidence of domestic sheep and goats at sites all over Southwest Asia and even into Greece and southern Europe, and it appears that once domestication was well advanced, the spread of sheep and goat raising was very rapid. Most farming villages have hedgerows, patches of weeds and thorny plants, clippings, and stubble that are perfectly acceptable to the rather undiscriminating sheep and goats, and these animals, with their heavy fleece, are well protected against the sun and heat of the Middle East.

Domestic cattle were herded on the Anatolian Plateau (central Turkey) by about 5800 B.C. and were probably present in the Balkans by 6500 B.C. As with sheep and goats, cattle domestication seems to have been a widespread phenomenon, probably beginning sometime after 9000 B.C. and occurring in many areas from China to western Europe. Across this vast area, ancient farmers seem to have bred cattle for reduced size, increased docility and milk production, and increased tolerance of climatic conditions. An important step in the evolution of civilizations, as we will see in the next chapter, was the process by which cattle were adapted to the hot lowlands of the river valleys. The cattle found there today are thin, small animals that seem woefully scrawny by comparison to European cattle, but these Mesopotamia varieties are extraordinarily hardy. The late Shah of Iran tried to "improve" the cattle in the hot lowland of Iran by importing Dutch and Danish cows that were about four times the size of local varieties and produced vastly more milk and meat—until they strolled out into the hot plains (which we were surveying for archaeological remains), where most of them were felled by heat stroke. These European cattle had to be kept in air-conditioned barns to be productive, whereas the local varieties could do perfectly well on the hottest day, eating poor quality foods that would have killed the larger cattle.

Cattle were probably especially important to the first settlers on the southern Mesopotamian Alluvium. During the dry, hot summers in this region, few reliable protein sources are available to primitive agriculturalists, and cow meat and milk apparently provided a crucial nutritional component. Oxen (castrated bulls) may have been used to pull plows and carts. In many areas of Southwest Asia where rainfall is sufficient for cereal cultivation, plowing is essential because natural vegetation is thick. Later, the horse, donkey, and mule were also used as draught animals.

Another important domesticated animal was the pig, whose bones have been recovered from sites all over Southwest Asia. By 6000 B.C. and even as late as 2700 B.C., pig bones represented 20 to 30% of all mammal remains at many large sites. As discussed below, however, sometime after about 2400 B.C., pork apparently was religiously proscribed in most Mesopotamian cities, as well as in Egypt and elsewhere in Southwest Asia.

By 6000 B.C. agricultural villages spread over much of Southwest Asia, most of them comprising just a few score mud huts, a few hundred people, and the same essential economic functions. From Greece to Afghanistan, these villages looked very much alike and, taken on the basis of their archaeological remains, they were not impressive in their material wealth. Even sites like Jericho and Çatal Hüyük (see Chapter 8) which were larger than most and had relatively impressive art and architecture, were self-sufficient communities without many economic, political, or social ties outside their region. And the vast majority of these villagers were simple farmers without great aesthetic, religious, or social diversions.

As unremarkable as these villages were, however, their inhabitants were the first to focus their lives on agricultural products and to live in the sedentary communities that even today are the basic component of Middle Eastern settlement patterns. And as early as 6000 B.C., the processes that were to transmute these villages into states and empires were already underway.

Other Old World Domesticates

The domesticates and allied agricultural economies developed in Southwest Asia proved so successful that within centuries of their appearance they had spread far outside the Fertile Crescent. By 7000 B.C. farmers at Argissa-Maghula in Greek Thessaly were subsisting on cultivated emmer wheat and barley as well as domestic cattle and pigs. Recent recalibration of carbon-14 dates for scores of early European sites reveals that the basic wheat-barley/cattle-pigs-sheep complex diffused at the rate of about a mile a year, reaching Bulgaria about 5500 B.C., southern Italy about 5000 B.C., and Britain and Scandinavia between 4000 and 3000 B.C.[30] To the east, domestic wheat and barley reached the Indus Valley by at least 5000 B.C.

(and probably much earlier), and by the later first millennium B.C., domestic wheat was in cultivation in northeastern China.

The processes by which these domesticates and their associated agricultural techniques replaced hunting and gathering economies in much of the Old World are not well known but appear to have involved both the replacement of hunters and gatherers by agriculturalists and the conversion of hunters and gatherers to agricultural ways of life (see Chapter 12).

The grasslands and forests of temperate Europe and Eurasia contrast sharply with the steppes and arid plains of the Fertile Crescent, and the spread of agriculture northward and eastward required new strains of plants and animals and different social and technological adaptations. Methods had to be developed to clear the dense northern forests; and in some areas the rich hunting, gathering, and fishing resources formed such a productive food base that there was considerable "resistance" to the introduction of agriculture, with its unpredictability and heavy labor expenses.

Ammerman and Cavalli-Sforza formulated a "wave-of-advance" model for the spread of agricultural economies into Europe, postulating that the rate of spread could be predicted with some accuracy simply on the basis of the distance of any area from the centers of domestication in the Mediterranean world.[31] Their model is simple, elegant, and powerful, and the correlation of radiocarbon dates of agricultural evidence and human genetic evidence is impressive.[32] But their model—as all models are—is an oversimplification. Graeme Barker interprets the evidence to show that the spread of agriculture into Europe was not as predictable and independent of local factors as the "Wave of Advance" model suggests.[33] Colin Renfrew has suggested that Indo-European languages and culture spread into Europe through Turkey and came with agriculture, which seems to have been established in much of Turkey at least 8000 years ago.

Egypt's initial transition to agriculture was based on introductions of wheat, barley, sheep, and goats from outside the Nile Valley, probably both from the Middle East and from North African desert and coastal areas.[34]

In the Sudanese part of the Nile Valley, sorghum seems to have been the basis of initial agriculture—although the evidence for its use is scanty.[35]

Many herbs, fruits, and other plants were domesticated in India and Southeast Asia, but it was rice and millet that provided the majority of the food energy for the great Asian civilizations. Imprints from rice-grain husks have been found in potsherds from the site of Non Nok Tha in central Thailand in levels dating to between 6800 and 4000 B.C., but it is difficult to determine whether these represent domesticated rice. The use of the husks as tempering in ceramic manufacture may suggest domestication, but they may have been from collected wild species of rice. Bellwood identifies China's lower Yangtze drainage as a center of early rice cultivation, but he suspects that there were independent centers of domestication in India and, possibly, Thailand.[36]

Vegeculture of tubers and other crops probably has a long history in the moist lowlands of Asia. In vegeculture plants are propagated not from seeds, but from cuttings taken from leaves, stems, or tubers of plants like manioc, yams, potatoes, and taro. Bellwood disputes the notion that vegeculture generally preceded cereal agriculture in much of tropical Southeast Asia, but root-crop cultivation may have been the earliest kind of farming in areas like New Guinea.[37]

One of the world's most important cereals—millet—was apparently domesticated and first cultivated on the great Yellow River flood plain in North China (see Figure 6.3). The alluvial soils in this area are extremely fertile and sufficiently arid so that there was little vegetation to clear for agriculture in many areas. By about 4000 b.c. scores of villages existed in North China, most of them subsisting on millet and a few other domesticates and considerable hunted and gathered food. These villages usually contained about two or three hundred inhabitants, who lived in wattle-and-daub houses that looked very much like the circular houses and compounds marking the evolution of agricultural communities in Southwest Asia some 3000 years earlier.

Another crop of considerable importance in North China was soybeans; several wild varieties are native to this area. The little evidence available suggests soybeans were in cultivation by at least 1600 b.c. Soybeans are a remarkably versatile and nutritious crop. They are also good "green manure," enriching the soils on which they grow through nitrogen fixation. The substitution of soybeans for milk and meat in early Chinese diets explains why many Asian populations never evolved the enzymes necessary to digest milk products, causing them to suffer intestinal upsets if they eat these foods. Similar intolerances are found in Africa and South America.

The establishment of agricultural economies in Japan had some similarities to the pattern in Europe. In both areas, rich northern forested environments supported dense populations of hunter-foragers, and, when agriculture came, it supplanted the ancient hunting-foraging economies at different rates in different areas, slowly in areas heavily dependent on marine resources, more quickly in others.[38]

Agricultural Origins in the New World

In the New World, the most important domesticates were maize, beans, squash, and turkey (some people would add to this list cacao, from which chocolate is derived). For reasons not clearly understood, New World peoples did not domesticate any large draft animals, although the llama of South America certainly was an important transport animal. No animals equivalent to the pig or sheep were domesticated in the New World, perhaps because there were no suitable native species (although the Rocky Mountain sheep would seem to have possibilities). Like their Old World relatives, New World peoples complemented their staple domesticates with a wide range of herbs and spices, as well as "recreational" plants like coca, from which cocaine is derived, and tobacco.

The background of New World domestication, like that of the Old World, is many millennia of hunting, gathering, and foraging. Kent Flannery has suggested, for example, that the foods the aboriginal foragers of Mexico relied on, such as maguey cactus, prickly pear, and rabbit, were species that could withstand intensive human consumption for many millennia without changing much in their physical characteristics as genera, and that because of the nature of their seasonal movements, these foragers rarely exerted great pressure on these resources for any length of time.[39] For example, cactus-fruit collecting had to be done during a two- to three-week period, and during this time rabbits would not be hunted. This seasonal concentration on resources kept groups small and dispersed for most of the year, and one wonders why the Spanish found Mesoamerica (roughly from northern Mexico to southern Guatemala) a world of cities and farms instead of still the domain of hunters and gatherers.

Despite effective mechanisms that maintained the hunting and gathering way of life for so long in Mesoamerica, sometime after 8000 B.C. these people—probably unintentionally—began to domesticate maize, beans, squash, peppers, and other plant species. Why and how this happened is not clear. Climate changes, increasing population densities, and other factors may have been involved, and recent ecological analyses suggest that the shift to agriculture in the beginning may have been only a slight change in the way people in Mesoamerica hunted and foraged.[40]

Domesticated maize was the most important food through much of later Mesoamerican prehistory, and at present it is the only domesticated plant from this area whose evolutionary history we know in any detail. A few fragments of early forms of beans, squash, and peppers have been found, but not enough to reconstruct recent changes in their morphological characteristics.

Occasionally there are claims that maize was either brought to or from the Old World at a very early date, usually based on ancient sculptures or other art,[41] but no trace of ancient corn plants has ever been found in the Old World.

Until about 1970 the most widely accepted view of maize domestication was that advanced by a geneticist, Paul Mangelsdorf, who argued that domesticated maize evolved from a "wild maize," now extinct, with small

6.11   The evolution of maize cob size at Tehuacán. Smallest cob *(left)* dates to about 5000 B.C. Dates, successively *(left to right)*, are 4000 B.C., 3000 B.C., 1000 B.C., and *(far right)* an entirely modern variety dating to about B.C./A.D.

cobs topped by small tassels. This would have been a "pod-corn"—that is, the individual kernels would have been enclosed in chaff rather than the cup-like fruit-case of domesticated varieties. Mangelsdorf explains the extinction of this wild maize as being caused by overgrazing by European-introduced cattle and having been genetically "swamped" by continual hybridization with emerging species of domesticated maize.[42]

In the early 1960s, Richard MacNeish excavated several sites near Tehuacán, Mexico, and his findings seemed to confirm Mangelsdorf's hypothesis: the earliest corn cobs found (dating to about 5000 B.C.) were very small, and the tassels did indeed emerge from the tops of the fruits. But the early maize found by MacNeish seemed to have many morphological similarities to another wild perennial grass common in the semi-arid, subtemperate regions of Mesoamerica, a grass called teosinte (*Zea mays*), and some botanists questioned Mangelsdorf's reconstruction, partially because the supposed placement of kernels on a cob such as wild and domesticated maize would seem a very inefficient mechanism for seed dispersal.

In 1972 plant geneticist George Beadle reasserted his argument of some decades previous: that there had never been a "wild maize," that domesticated corn instead was a descendant of teosinte.[43] Teosinte is a tall (up to 2 meters) annual grass found throughout the semiarid and subtropical zones of Mexico and Guatemala, where it thrives in disturbed areas and rapidly invades open areas such as abandoned cornfields. Teosinte can be found growing in fields that also include wild beans and squash, with the beans twining around the teosinte stalk. Thus, the three staffs of life for Mesoamerica—maize, beans, and squash—are a "natural" association.[44] Teosinte looks like maize, except for the seed structure, where instead of maize's heavy husk-covered cobs of many rows of kernels, teosinte has just two rows of six to twelve triangular kernels enclosed in a very hard covering (Beadle produced reasonably good popcorn and mock "Fritos" from teosinte kernels).

Chemical analyses of teosinte and maize and studies of their genetic characteristics seem to support Beadle's view of teosinte as the major ancestor of maize—although Mangelsdorf mounted a spirited counterattack.[45]

It teosinte is the ancestor of domesticated maize, the primary changes in the domestication process were: (1) the development of a less brittle rachis, followed by the evolution of the cob; (2) the development of a soft fruit-case, so that the kernels could be shelled free of the cob; and (3) the evolution of larger cobs and more rows of kernels. A single gene—the so-called tunicate allele—controls to some extent the brittleness of the rachis and the toughness of the fruit case, and thus these features could easily have been produced by direct selection of mutants with these characteristics. The third change, increased cob size, was very gradual and probably differed sharply from area to area. The cobs from Tehuacán dating to about 5000 B.C. averaged a little less than 2 centimeters in length and were remarkably uniform. Cobs dated to between 3400 and 2300 B.C. averaged 4.3 centimeters, and by A.D. 700 the average size was still less than 4.4 centimeters. Between A.D. 700 and 1536, however, the maize cobs from Tehuacán reached an average of about 13 centimeters in length.

Stephen Gould accepts Beadle's claim for the derivation of maize from teosinte, but he rejects Beadle's hypothesis of a slow process of gradual genetic change.[46] Gould notes that the Tehuacán maize cobs are smaller than modern varieties, but at least they are cobs and we do not find intermediate stages between the teosinte "ears," which botanically are radically different from the maize ears. Also, as Gould notes, maize and teosinte are extremely similar genetically, and one would expect that a slow process of domestication toward maize would have resulted in some major genetic differences.

Gould champions Hugh Iltis' hypotheses about maize domestication. Iltis' argument is a complex botanical and genetic scenario that cannot be summarized here, but it basically involves a small genetic change that produces what Iltis called a "catastrophic sexual transmutation."[47] Some minor genetic mutations could transmute male teosinte tassel spikes to female corn ears. This kind of change accords well with Gould's conception of evolution as a series of sudden changes, not the accumulation of gradual changes. But Iltis' scheme requires that people recognize the potential of the "hopeful monsters" a mutant teosinte plant would represent, for it is unlikely this mutant would be able to reproduce easily. In this case, human intentions are an important part of the overall equation of domestication.

While maize is an excellent food source, it is deficient in a number of important proteins and vitamins, and the evolution of agricultural economies in Mesoamerica derived considerable impetus from the domestication of other species, the most important of which were beans and squash. Three species of beans (common beans, runner beans, and tepary beans) have wild ancestors in Mesoamerica, and changes in their morphological characteristics began to appear at about the same time as those of maize. Wild bean remains recovered in caves in Tamaulipas date to 7000 to 5500 B.C., and in Oaxaca from 8700 to 6700 B.C., but the earliest known domesticated beans did not make their appearance in these areas until between 4000 and 3000 B.C.

The domestication of beans seems to have involved: (1) increased seed permeability, so that the beans need not be soaked so long in water before being processed; (2) a change from a corkscrew-shaped, brittle pod that shatters easily to a straight, limp, nonshattering pod; and (3) in some cases, a shift from perennial to annual growth patterns.[48] The primary importance of beans is that they are rich in lysine, which maize is deficient in; thus the two are nutritionally complementary.

The domestication of squash and pumpkins, members of the genus *Cucurbita,* seems to have been aimed at improving the seeds (rather than the flesh) since wild cucurbita have flesh so bitter or thin that they have little food value. The earliest cucurbit seeds are found in cave deposits in Oaxaca and Tamaulipas dating to 8000 and 7000 B.C.

Some scholars think that maize, beans, squash, and perhaps other plants were domesticated over a wide area of highland, and perhaps lowland, Mesoamerica, but that the stimulus for the adoption of the agricultural and village way of life came from coastal areas. There, rich resources in the littoral/terrestrial interface may have provided sufficient food for non-

6.12 Some areas of early American domestication and agriculture. Other areas may be buried under coastal waters or undiscovered in dense coastal vegetation zones.

agricultural but sedentary communities.[49] In time, these sedentary communities could have incorporated the new domesticated plants into their "preadapted" economies and social organizations, forming the first agricultural communities. From that point on, the superior productivity of village-based agriculture would have ensured its rapid spread over all the areas where these crops could be grown successfully. Currently we do not have sufficient data to evaluate this reconstruction rigorously, but recent evidence indicates that it may be correct. Large mounds of shellfish remains and cracked rock and stone tools have been found on both the Gulf and Pacific coasts of Mesoamerica, but few have been excavated or firmly dated. Many groups of people lived in the rich terrestrial/marine niches along Mexico's coasts after 3000 B.C., and some of them may have lived in villages and farmed some maize and other crops, but the evidence for substantial farming and the spread of villages in Mexico is scanty until shortly after 2000 B.C.[50]

Some interesting evidence has come from Central and South America, however. Maize seems to have been cultivated in Peru, Argentina, and Chile as early as 3000 B.C.,[51] and there are even reports of maize phytoliths in the Las Vegas site, in Ecuador, in levels dated to 6000 to 7000 B.C.[52] Maize kernels dating to between 1400 and 900 B.C. have been found at the La Ponga site in Ecuador.[53] Maize may have been grown in this area already by 3000 B.C., by villagers of what archaeologists call the Valdivia and Early Cerro Narrio cultures, both of which have been radiocarbon dated to between 4000 B.C. and 2000 B.C. These sites appear to have been situated so as to exploit both coastal and floodplain environments, but some are in the highlands, At sites in lower elevations, large middens often contain shellfish remains, fish bones, and other refuse, and at several sites near the coast, investigators have found what they believe to be the impressions of corn kernels in the surfaces of ceramics. Other ceramics appear to have been decorated with applied clay effigies of maize cobs. If these artifacts are really between 2000 and 4000 years old, maize domestication here would seem to have been far more advanced than in the Tehuacán Valley, a thousand years later. Numerous grinding stones were found at these sites, as were many large storage pits. The humid conditions make it difficult to recover botanical remains, but at least one corn cob has been tentatively identified. Some archaeologists do not consider the evidence for early Valdivia maize agriculture to be persuasive, however.

Maize was just one of the important food crops in Andean South America, and in some areas potatoes, beans, and quinoa (pronounced "keen-wa") were the staffs of life. There is some evidence that beans may have been domesticated before 8000 B.C., based on a few remains from Guitarrero Cave, and there are fairly secure finds of beans there dating to about 5700 B.C.,[54] but the evidence of the earliest stages of the domestication of beans and many other plants is probably lost in the moist lowlands of the Amazon Basin.

It is an interesting footnote to Peruvian plant domestication that quinoa is becoming increasingly popular in North America. The seeds of this plant can be milled into a flour that has one of the highest protein contents of any plant. Some personal experiments with quinoa flour in the manufacture of chocolate chip cookies were bitter failures, but health food enthusiasts are using this plant in growing numbers.

The role of animal domestication in early Peru is unclear, but llamas and guinea pigs were certainly domesticated in central Peru by 3500 B.C. As in Mexico, however, hunting continued to play an important role in many areas until quite late.

The relationship of plants and animals, agriculture, and sedentary communities in northwestern South America in general suggests that domestication and sedentary communities may have preceded specialized agricultural economies in some areas by many centuries, particularly on the coast, where small sedentary communities of fishers, foragers, and part-time bean and squash cultivators were established before maize cultivation was of any importance.

Eventually, domesticated plants and animals and agriculture became the bases for human settlement in most of the prehistoric New World as maize, beans, squash, and other plants were adapted to environments as far north as Canada and as far south as the tip of South America.

Hypotheses About the Origins of Domestication, Agriculture, and Sedentary Communities

Cursed is the ground for thy sake;
In sorrow shalt thou eat of it all the days of thy life;
Thorns also and thistles shall it bring forth to thee; . . .
In the sweat of thy face shalt thou eat bread.

(Genesis 3:17–18)

We have described the botanical and cultural changes involved in the domestication of several plants and animals, noting in the process some of the relevant archaeological evidence. At this point let us consider some ideas about how and why domestication, agriculture, and sedentary communities appeared.

To begin with, we can dismiss the idea that people domesticated plants and animals because someone came from outer space and taught them how to do it, or because someone in ancient Syria had a brilliant idea and it spread around the world, or because people simply got sick of chasing animals and wanted an easier way to live.

The roughly contemporary appearance of a vast variety of domesticates, from palm trees to potatoes, across the same approximate latitudes around the world, and the several millennia it took to domesticate most plant and animal types, make any of the above scenarios improbable. And although there is great variability, hunters and gatherers tend to have more leisure time than primitive agriculturalists, and hunters and gatherers rarely spend this time analyzing their economy, designing cathedrals, or improving their standard of living; they spend it either talking or sleeping—skills, as Sharp notes, they have already thoroughly perfected.[55]

Even so, the idea that people became farmers because it's an easier way of life than hunting-foraging seems so plausible to most Westerners that it is hard to discredit this myth.

Anyone considering the problem of agricultural origins has to be impressed by three major categories of evidence: first, we know that world

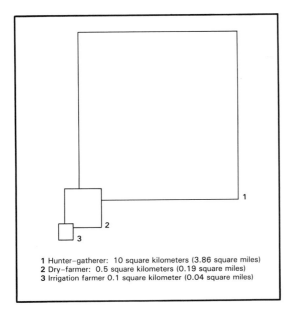

1 Hunter–gatherer: 10 square kilometers (3.86 square miles)
2 Dry–farmer: 0.5 square kilometers (0.19 square miles)
3 Irrigation farmer 0.1 square kilometer (0.04 square miles)

6.13 These three squares show the relative amount of land needed to feed individuals obtaining food in different ways.

climates changed greatly just before domestication and agriculture first appeared in the Old World; second, world population densities had also been increasing for hundreds of thousands of years before agricultural origins; third, human technology seems to have been improving in many ways over the whole history of our genus. Not surprisingly, climate change and population growth are included in most models of agricultural origins, and there is some suggestion in many hypotheses that the accumulation of knowledge about tool-making was an important element in the transition to agriculture.

Early Explanations of Agricultural Origins

Among the first hypotheses about agricultural origins was the so-called Oasis Hypothesis, also known as the propinquity hypothesis, which was an attempt to explain the origins of agriculture in terms of the climate changes associated with the end of the Pleistocene some 10,000 years ago. Various scholars suggested that as the world became warmer and drier, people, animals, and plants would be forced into close conjunction and out of this would arise domestication.[56] The Oasis Hypothesis was accepted in whole or in part for many years, but eventually it was demonstrated that the wild ancestors of wheat and barley did not grow in the areas where post-Pleistocene oases were thought to have been concentrated; and by the

1950s there was also evidence that deserts and oases had not formed in most of Southwest and Central Asia at the time domestication began—a blow to the Oasis Hypothesis.

Recent work in Egypt suggests that a change in rainfall patterns may have forced people out of oases and into the Nile Valley, where they may have become farmers of wheat and barley species that were first domesticated in Southwest Asia,[57] but, generally, the oasis hypothesis does not seem to explain much about the initial domestication process.

Another early idea about agricultural origins has been termed the "Natural Habitat Hypothesis." Harold Peake and Herbert Fleure suggested in 1926 that the first domesticates and agriculturalists would have appeared in the upper valley of the Euphrates River, because they knew that this is the "natural habitat" of wild species of wheat and barley. In the early 1950s Robert Braidwood of the University of Chicago organized a series of excavations to evaluate post-Pleistocene climatic changes and to look for early farming communities in northern Mesopotamia. His expeditions were among the first to include specialists in botany, geology, and zoology as well as archaeology, and this multidisciplinary approach has proven to be a highly successful research strategy.

Braidwood's excavations at Jarmo in the hill country of northern Iraq revealed an agricultural settlement dating to about 6500 B.C.—much earlier than had been found elsewhere.[58] About 6500 B.C. Jarmo was a settlement of a few dozen mud-walled huts inhabited by about 150 people who relied partly on wild plants and animals, such as snails, pistachios, and acorns, but who also seem to have been herding domesticated goats and, perhaps, sheep. But Braidwood also found at Jarmo the remains of partially domesticated wheat in association with grinding stones, sickle blades, and storage pits. Braidwood suggested that the cumulative effects of generations of interaction between people, plants, and animals in these natural habitat zones led to agriculture.

Braidwood's research into agricultural origins was one of the few systematic investigations into this problem at the time, and his work has had substantial and positive influences on subsequent investigations in this area. But as L. Binford pointed out, Braidwood's account of agricultural origins did not specify any testable causal factors.[59]

There were also empirical problems with Braidwood's hypothesis. Frank Hole, Kent Flannery, and James Neely conducted excavations at several sites on the Deh Luran Plain in Southwestern Iran and found evidence that by 6700 B.C. domestication and agriculture were already evolving in this area, which is just outside the natural habitat of wheat and barley (wild barley grows quite close by, but very sparsely).[60]

Also damaging to the basic premise of the natural habitat hypothesis was a series of experiments performed by J. Harlan in eastern Anatolia in 1966, in which, using a crude sickle made with flint blades set in a wooden handle, he was able to harvest wild emmer wheat at the rate of about 6.25 pounds per hour.[61] A family of four or five could probably have collected a year's supply of grain with only a few weeks' labor, and this would seem to suggest that the people who lived in the natural habitat of wheat and barley had perhaps the least incentive to domesticate and farm it, because they could collect more than enough from wild stands.

Contemporary Models of Agricultural Origins

Most contemporary attempts to explain agricultural origins concentrate on trying to understand why agriculture first appeared when and where it did. As such, most of these models, or explanations, are much alike, because they incorporate the same factors: (1) post-Pleistocene climate changes; (2) human population growth and dispersion; and (3) evolving technologies.

A particularly influential synthesis of these factors was Lewis Binford's "edge-zone hypothesis," which although widely regarded as flawed and to some extent obsolete, is still important because most other models have been written in reaction to it. Binford assumed that prior to the first farmers, hunting and gathering bands were in equilibrium with their natural environment, using wild plants and animals but not altering them in ways we recognize as domestication.

What could have upset this ancient equilibrium? Binford argues that this was essentially a case where changes in the demographic structure of one region resulted in the impingement of one group on the territory of another, upsetting the equilibrium and increasing the population density in some areas to the point that manipulation of the natural environment in new ways to increase productivity would be favored. He notes that Southwest Asia is an ecological mosaic, with close juxtapositioning of very different climates and plant and animal communities, and suggests that late Pleistocene hunters and gatherers used budding-off as the primary mechanism for population-size control. This phenomenon, noted in many parts of the world, involves a group slowly increasing its size up to a certain number, at which point, because of squabbling or insufficient resources, the group splits, with half the people emigrating to another, usually adjacent, territory.

Budding-off can also become the primary population-control mechanism if a new resource or subsistence technique—such as domestication and agriculture—opens a new niche.[62]

Given varying population growth rates, Binford maintains that budding-off would have certain consequences that could have led to domestication and agriculture.

> From the standpoint of the populations already in the recipient zone, the intrusion of immigrant groups would disturb the existing density equilibrium system and might raise the population density to the level at which we would expect diminishing food resources. This situation would serve to increase markedly for the recipient groups the pressures favoring means for increased productivity. The intrusive group, on the other hand, would be forced to make adaptive adjustments to their new environment. . . . There would be strong selective pressures favoring the development of more efficient subsistence techniques by both groups.[63]

Thus, Binford is particularly concerned with describing situations that would select for, or reward, a cultural innovation like domestication or agriculture. Once the population-to-resources balance was disturbed, there was a premium on every resource, and domestication might have come about in a number of ways. Perhaps the immigrant groups, in an attempt to regain the resources of their former habitats, would have tried to introduce wild wheat and barley into these marginal, "edge" zones. This would have exposed these plants to different selective environments, and domestication might have occurred as people manipulated these plant communities in these new environments.

Binford's edge-zone hypothesis provoked much discussion and research but it has some weaknesses. For one thing, as Flannery notes, it makes climate changes and population growth into "prime movers," and it has not been demonstrated how these forces would produce agriculture or that they even existed.[64] Most of the earliest agricultural communities in Southwest Asia, for example, have been found in areas like the Negev Desert, where it is difficult to imagine any form of "population pressure." In any case, in most instances of early domestication and agriculture, domesticates seems to represent only about 5% of the total diet for centuries after they begin to be domesticated—a slow process in which it's difficult to imagine much daily stress.[65]

On the other hand, aspects of the shift to cereals and other resources may help explain how agriculture—once it was in its initial stages—changed human demographic patterns. In hunting-gathering societies, fertility rates are suppressed significantly simply by maternal mobility: a pregnant woman's chances of spontaneous abortion go up considerably if she walks a lot and works heavily. Thus, those late Pleistocene groups that became less mobile, perhaps because they began to concentrate on salmon runs or wild cereal patches, might have experienced a rise in fertility rates. Also, a direct correlation exists between the amount of carbohydrates in the diet and fertility rates. Studies have shown that it is almost impossible for a woman to become pregnant until she has about 27,000 calories, or 20 to 25% of her body weight, stored as fat (the author assumes no liability for the accuracy of this statistic).[66] Nursing a child requires about 1,000 calories a day, and in many hunter-gatherer societies, the rigors of mobility and their high-protein diet can mean that nursing itself prevents sufficient fat buildup for a successful pregnancy for about three years. But with the change to a high-carbohydrate, cereal-based diet and restricted mobility of sedentary life, fertility rates may well have risen rapidly.[67]

Population growth, thus, has been an important element in most accounts of agricultural origins. Mark Cohen, for example, argues that "the nearly simultanous adoption of agricultural economies throughout the world could only be accounted for by assuming that hunting and gathering populations had saturated the world approximately 10,000 years ago and had exhausted all possible (or palatable) strategies for increasing their food supply within the constraints of the hunting-gathering life-style. The only possible reaction to further growth in population, worldwide, was to begin artificial augmentation of the food supply."[68]

Differences in the rate and timing of the adoption and spread of agriculture, Cohen suggests, may be found in local environmental variation,

6.14   The distribution of tooth size among modern peoples before European influences affected this characteristic. The largest teeth occur among groups used to eating tough foods raw; the smallest teeth occur among groups that have had agriculture the longest.

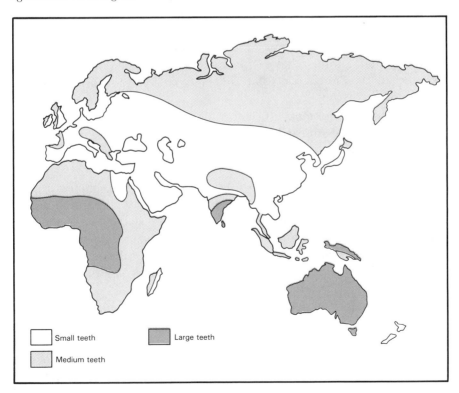

Small teeth

Large teeth

Medium teeth

such as the mobility that resources of a given area require of a hunter-gatherer for their exploitation.[69]

A major problem in assessing any hypotheses based on ancient population densities is the difficulty of estimating changes in these densities with any precision using archaeological data. One has to locate all the relevant archaeological sites (of which many are certain to have been destroyed or not found), then estimate the numbers of people at each site, and then develop a chronology for these sites that is fine enough to reconstruct changes over fairly short periods of time. Consequently, most population estimates based on archaeological data must be considered extremely speculative, and the relationship of agriculture and demographic change is complex and not well understood.

Some support for Cohen's arguments seems to exist in the paleopathological data from around the world. Agriculture allows more people to be fed, but compared to hunting-foraging in general agriculture is associated with shorter life expectancies, higher disease rates (especially among infants), smaller body size, and chronic malnutrition.[70]

Marvin Harris has formulated an explanation of agricultural origins that resembles Cohen's in its stress on increasing human populations, declining worldwide availability of large game animals, and the competition among groups for survival.[71] In addition, Harris considers a fascinating topic—food taboos. In trying to explain food taboos he has illuminated some of the central problems of trying to understand human economic history. Consider, for example, the humble pig. Eventually Islam, Judaism, and some Hindu religions instituted complete bans on pork to the extent that, even today, riots flare in some parts of the world if pigs invade religious areas.

Pigs are one of the most efficient converters of garbage, animal wastes, and other materials discarded by farmers—better in many respects than goats, sheep, or cows. In preindustrial economies and today around the world in Europe, America, and China the pig is used to convert grains that are not made into food or whisky into portable, storable food.

Why should the sturdy farmers of the Middle East reject such a good food source after millennia in which the pig was a staple part of the diet?

This question has set into conflict people who believe that all such religious prohibitions, and almost every other aspect of cultural behavior, can be understood directly in terms of technological, ecological, and demographic variables, and those who think it is impossible to explain adequately such things as religious food prohibitions and religious and social behavior generally in terms of the conditions of technology, ecology, and demography.

Harris dismisses the notion that this prohibition has anything to do with the common infection of pigs with trichinae, parasites that can kill people.[72] Recent studies have shown that pigs raised in hot climates seldom transmit this disease; moreover, Southwest Asian farmers ate cattle, sheep, and goats, which carry anthrax, brucellosis, and other diseases as dangerous or more so than anything pigs can transmit.

Instead, Harris explains prohibitions of pork in terms of the cost-benefit ratio of the animal in early subsistence agricultural systems. Pigs, unlike sheep, goats, and cattle, cannot subsist on husks, stalks, or other high-cellulose foods; their natural diet is tubers, roots, nuts, and fruits. Also, pigs are native to woodlands and swamps and do not tolerate direct sun and open country well. Thus, the clearing of land coincident with the spread of agriculture in Southwest Asia greatly reduced the habitat and natural foods available to pigs, and, increasingly, pigs had to be fed on grain, which brought them into competition with people; they also had to be provided with artificial shade and considerable water. Moreover, in contrast to other domesticates, as Harris points out, pigs cannot be milked, sheared, ridden, or used to pull a plow. In short, they lost their cost effectiveness relative to sheep, goats, and cattle, and their eventual proscription made excellent economic "sense."

And, as in all instances, it is not important that people ever recognize their motives in cases of such prohibitions: natural selection "fixes," or perpetuates, economically sound behaviors, regardless of what people think their motives are. Harris extends his argument to many other animals, in particular the sacred cow of India and the animals listed in the Old Testament. In every case he tries to show that when one considers the economic conditions in which these prohibitions arose, one can see that there was at least some minor overall advantage to be gained. Thus, he notes that the Old Testament forbids eating any insects other than locusts, which just happen to be the only ones that have much caloric value in relation to the cost of capturing them (and they are crop pests as well).

Against Harris and his supporters are many people who believe that Harris is propagating a bastardized version of a misconception of Marxian theory.[73] Sahlins, for example, points to the United States, where millions of cattle are slaughtered every year but where virtually every horse lives out its full life in relative luxury and, usually, enjoying considerable affection. Why, Sahlins asks, do Americans arbitrarily proscribe horsemeat and accept beef? Sahlins answers his own question, arguing that it is because every culture creates its own belief systems from its unique blend of economy, society, and ideology. The essential point, Sahlins says, "is that material rationality exists for men not as a fact of nature but as a construct of culture. . . . The natural conditions of viability (selective forces) comprise merely negative constraints, limits of functional possibilities, which remain indeterminate with respect to the generation of particular cultural forms."[74] Moreover, he says, one can never read directly from material circumstances to cultural order as from cause to effect. "It is not that these techno-environmental variables have no effect, rather that everything depends on the way these properties are culturally mediated—given meaningful organization by a mode of cultural organization."[75]

All this is rather far from the poor abandoned pigs of Southwest Asia, but the point is an important one. For if archaeologists must seek the causes of changes in the archaeological record (like the succession of pig bones by sheep and goat bones in the levels of a site) in the cultural mediation of "the mode of cultural organization," they will have to apply techniques of archaeological analysis currently not known or, possibly, not within the capacities of any other than a Divine Intellect.

Harris argues that one can show by the relative susceptibility to disease, dietary requirements, and other economic and historical factors that it makes economic "sense" to eat cattle and not horses in contemporary America. He does not insist all human behavior has such a delicate economic tuning or that there can be no whimsical aspects of culture. He says only that in the long run, if one is trying to understand something like these different uses of animals, one should look first at the economics of the situation.

Another general explanation of agricultural origins has been proposed by Brian Hayden. He begins with the assumption that ancient hunters and gatherers were, like most people, trying to maintain the reliability and stability with which they could get food.[76] He suggests that for the last several million years our ancestors diversified their "income" by eating a wide range of plant and animal species and developing digging sticks, hand-axes, nets, and all the other implements of pre-agricultural peoples. They also became in many areas sophisticated predators of large gregarious herbivores like deer.

Rather than a constant drive from population growth or "pressure" which forced these innovations and broadening of diets, Hayden envisions great stability in the population-to-resources balance. Unlike Cohen, he expects that if we had complete human skeletal evidence from the Pleistocene, we would see no more evidence on dietary stress in the late Pleistocene than in the early Pleistocene.[77]

But by the end of the Pleistocene 10,000 years ago, all the larger, more obvious animal species had been exploited, and people faced with the same kinds of cyclical food shortages that always had been humankind's lot could only turn to grasses, seeds, fish, and mice—in other words, the smaller, in some ways less desirable species. Hayden uses here a division of the animal world according to reproductive "strategy": "k"-selected species, like deer, are long-lived, tend to have long maturation periods, have only one or a few offspring each reproductive cycle, and are susceptible to overexploitation; *"r"*-selected species, on the other hand, like grasses, which reproduce in incalculably large numbers, are genetically plastic, live only a year, are extremely resistant to overexploitation, and quickly reestablish themselves after floods, fires, and droughts.

Switching to a diet of mainly *"r"*-selected species, Hayden argues, would have encouraged a sedentary way of life, stimulated the development of ground-stone tools, wooden implements, and other agricultural devices, stabilized the food supply, and lessened the need for group cooperation and common food storage. In Hayden's view, the rise of agriculture is just a "natural and logical extension of the trends" that led to the late Pleistocene refocusing of exploitation strategies from big-game animals to smaller animals and plants.

As for the timing and location of the first agricultural economies:

> It is especially notable that domestication did not first occur where the environments were rich enough to support sedentary, hunting-gathering-based ranked societies, with wealth competition and primitive valuables, such as those found in California, the Northwest Coast, Florida, and Palestine. According to my model, even though these areas had the highest population densities, their resource bases would have been much more stable and they would have experienced resource stress relatively infrequently. Because sedentism, wealth competition, and ranking in rich environments did not result in domestication in these areas, it is reasonable to conclude that such developments were not sufficient, or perhaps even necessary, conditions for domestication. Instead, domestication can be more usefully linked to the same Paleolithic processes which gave rise to the Mesolithic and Archaic: the effort to increase resource reliability in areas of frequent stress.[78]

Hayden's model, as well as the others discussed in this chapter, are to some extent incomplete and untested. We must presume that more powerful explanations will be formulated as the theory of archaeology develops and new evidence becomes available.

David Rindos's model (discussed earlier in this chapter) explains a great deal about what domestication and agriculture *are*. What remains to be explained is the timing and distribution of early agricultural communities.

Domestication, Agriculture, and Sedentary Communities: Summary and Conclusions

Generally, we have no evidence that people of the immediate post-Pleistocene era experienced recurrent periods of starvation, or that they "invented" domesticates and agriculture as a way of addressing their immediate food supply problems. Instead, the situation may well have been one in which, as population densities slowly rose, people gravitated into various niches where exploitation of wild wheat and barley, teosinte, and other third-choice plants was marginally increased. Even if they were not forced into these areas by expanding population densities and were not under the dire threat of imminent starvation so that they radically increased their consumption of these third-choice foods, the plasticity of some of these species was such that minimal changes in selective pressures might have quickly and directly rewarded this increased exploitation.

In this sort of reconstruction it does not seem likely that the timing of wheat and barley domestication, for example, was the result mainly of "lucky" mutations. All over the post-Pleistocene world so many diverse animal and plant species were domesticated that explaining them in terms of "lucky" mutations would require coincidences of a highly unlikely nature. Nor was this a technological revolution: the first intensive cereal collectors required only minimal tools—implements certainly no more complex or imaginative than the fish traps, bows, and arrows of the late Pleistocene.

Finally, for the individual with too few things to worry about, we might note that there is a certain danger in the fact that almost all the world's population now depends for most of its subsistence on only about twenty genera of plants. As hybrid strains are replacing local varieties at an ever-increasing rate, the chances for truly catastrophic mutations of new crop diseases are growing, particularly with expected worldwide changes in climate. The Irish potato famine was largely the result of a whole country becoming dependent on a single crop that was cloned from a small genetic population and therefore subject to almost complete destruction by disease. Eventually the world may be at the same risk. The transition to economies based on domesticated plants and animals thus is a process still in operation and one whose future direction is unclear.

Against this uncertain future seed "banks" have been established, such as the one at Washington State University where seeds of more than 1700 plant species are kept in special conditions, against the day when their genetic resources might be useful. Someone observed that weeds are just plants whose uses to humankind have not been fully realized, and there is great truth to this statement. Every species is a unique and complex product of evolution, whose members are in a sense chemical factories able to synthesize the most complex compounds from only air, water, and soil. Not only food but medicines and many other products are mainly made from plant materials, and we have every reason to believe that the continuing destruction and extinction of plant communities in the Amazon Basin and elsewhere in the developing world will eventually be recognized as irremedial catastrophes. As we have noted previously, in the evolutionary game in which all of earth's lifeforms participate, there is usually great advantage to the maintainance of genetic *variability*. We are so mutually interdependent with plants and animals that loss of variety in the nonhuman biological world will almost certainly come back to haunt us.

Notes

1. Rindos, *The Origins of Agriculture*, p. 143.
2. Rindos, "Symbiosis, Instability, and the Origins and Spread of Agriculture: A New Model," p. 754.
3. Rindos, *The Origins of Agriculture*, p. 6.
4. Tauber, "13C Evidence for Dietary Habits of Prehistoric Man in Denmark."
5. Pianka, *Evolutionary Ecology*, p. 12; Winterhalder and Smith, *Hunter-Gatherer Foraging Strategies*, pp. 14–15.
6. Keene, "Optimal Foraging in a Nonmarginal Environment: A Model of Prehistoric Subsistence Strategies in Michigan."
7. Frison, *Prehistoric Hunters on the High Plains*.
8. Barnard, "Contemporary Hunter-Gatherers."
9. Ellen, *Environment, Subsistence and System*.
10. Testart, "The Significance of Food Storage among Hunter-Gatherers," p. 530.
11. Flannery, "The Origins of Agriculture," p. 307.
12. Wright, "The Environmental Setting for Plant Domestication in the Near East."
13. Spuhler, "Anthropology, Evolution, and 'Scientific Creationalism,'" p. 115.
14. Kislev and Bar-Yosef, "The Legumes: The Earliest Domesticated Plants in the Near East?"
15. For this discussion of early Southwest Asian domestication I have relied heavily on Donald O. Henry's *The Levant at the End of the Ice Age*.
16. Kaufman, "A Reconsideration of Adaptive Change in the Levantine Epipaleolithic."
17. Binford, "Post-Pleistocene Adaptations"; Flannery, "The Origins of Agriculture"; Hayden, "Research and Development in the Stone Age: Technological Transitions among Hunter-Gatherers."
18. Solecki and Solecki, "Paleoecology of the Negev."
19. Henry, *From Foraging to Agriculture. The Levant at the End of the Ice Age*, pp. 29–30.
20. Legge and Rowley-Conwy, "Gazelle Hunting in Stone Age Syria."
21. Perkins and Daly, "A Hunter's Village in Neolithic Turkey."
22. Wright, "Origins of Food Production in Southwestern Asia: A Survey of Ideas," p. 463; see also Solecki, "An Early Village Site at Zawi Chemi Shanidar."
23. Gowlett, "The Archaeology of Radiocarbon Acclerator Dating."
24. Flannery, "The Origins of the Village as a Settlement Type in Mesoamerica and the Near East: A Comparative Study."
25. Ibid., p. 48.
26. Mellaart, *The Neolithic of the Near East*, p. 50.
27. Rosen, "Notes on the Origins of Pastoral Nomadism: A Case Study from the Negev and Sinai," p. 504.
28. Mellaart, *The Neolithic of the Near East*, p. 46.
29. Hole, "Comment on 'Origins of Food Production in Southwestern Asia' by G. Wright," p. 473.
30. Ammerman and Cavalli-Sforza, "A Population Model for the Diffusion of Early Farming in Europe."
31. Ammerman and Cavalli-Sforza, *The Neolithic Transition and the Genetics of Populations in Europe*.

32. Ibid.
33. Barker, *Prehistoric Farming in Europe*; also see Price, T. "The Mesolithic of Western Europe."
34. Butzer, *Early Hydraulic Civilization in Egypt*; Wenke et al., "Epipaleolithic and Neolithic Subsistence and Settlement in the Fayyum Oasis of Egypt."
35. Haaland, *Socio-Economic Differentiation in Neolithic Sudan.*
36. Bellwood, *The Prehistory of the Indo-Malaysian Archipelago.*
37. Ibid.
38. Akazawa and Aikens, *Prehistoric Hunter-Gatherers in Japan.*
39. Flannery, "Archeological Systems Theory and Early Mesoamerica."
40. Webster, "Optimization Theory and Pre-Columbian Hunting in the Tehuacan Valley."
41. Johannessen, "Indian Maize in the Twelfth Century B.C."
42. Mangelsdorf, *Corn: Its Origin, Evolution, and Improvement.*
43. Beadle, "The Mystery of Maize"; Beadle, "The Ancestry of Corn."
44. Flannery, "The Origins of Agriculture."
45. See, e.g., Galinat, "The Origin of Maize"; but cf. Mangelsdorf, "The Mystery of Corn: New Perspectives."
46. Gould, *The Flamingo's Smile.*
47. Iltis, "From Teosinte to Maize: The Catastrophic Sexual Mutation."
48. Flannery, "The Origins of Agriculture," p. 300.
49. Flannery and Coe, "Social and Economic Systems in Formative Mesoamerica"; Stark and Voorhies, *Prehistorical Coastal Adaptations: The Economy and Ecology of Maritime Middle America.*
50. Stark and Voorhies, *Prehistoric Coastal Adaptations: The Economy and Ecology of Maritime Middle America*; Fiedel, *Prehistory of the Americas*, pp. 160–84.
51. Lynch, "The South American Paleo-Indians."
52. Stothert, "The Preceramic Las Vegas Culture of Coastal Ecuador."
53. Lippi, Bird, and Stemper, "Maize Recovered at La Ponga, an Early Ecuadorian Site."
54. Lynch, *Guitarrero Cave: Early Man in the Andes.* Flannery, "The Origins of Agriculture," p. 303.
55. Quoted in Just, "Time and Leisure in the Elaboration of Culture."
56. Pumpelly, *Explorations in Turkey, the Expedition of 1904: Prehistoric Civilization of Anau*, pp. 65–66; Childe, *New Light on the Most Ancient East.*
57. See, e.g., Hassan, *Demographic Archaeology.*
58. Braidwood, "The Agriculture Revolution."
59. Binford, "Post-Pleistocene Adaptations."
60. See Hole, Flannery, and Neely, *Prehistory and Human Ecology of the Deh Luran Plain*; see also Hole, "Comment on 'Origins of Food Production in Southwestern Asia' by G. Wright."
61. Harlan and Zohary, "Distribution of Wild Wheats and Barley."
62. Binford, "Post-Pleistocene Adaptations," p. 334.
63. Ibid., p. 331.
64. Flannery, "The Origins of Agriculture."
65. Anderson, "On the Social Context of Early Food Production."
66. See, e.g. Frisch and McArthur, "Menstrual Cycles: Fatness as a Determinant of Minimum Weight for Height Necessary for Their Maintenance or Onset."
67. Harris, *Cannibals and Kings.*
68. Cohen, *The Food Crisis in Prehistory*, p. 279.
69. Ibid.
70. Cohen and Armelagos, *Paleopathology at the Origins of Culture.*
71. Harris, *Cannibals and Kings*; Harris, *Cultural Materialism.*
72. Harris, *Cows, Pigs, Wars, and Witches.*

73. See, e.g., Friedman, "Marxism, Structuralism, and Vulgar Materialism"; Sahlins, *Culture and Practical Reason.*
74. Sahlins, "Comment on A. H. Berger's *Structural and Eclectic Revisions of Marxist Strategy. A Cultural Materialist Critique.*"
75. Ibid.
76. Hayden, "Research and Development in the Stone Age: Technological Transitions among Hunter-Gatherers."
77. Cohen, *The Food Crisis in Prehistory.*
78. Hayden, "Research and Development in the Stone Age: Technological Transitions among Hunter-Gatherers," p. 530.

Bibliography

Adams, R. E. W. 1977. *Prehistoric Mesoamerica.* Boston: Little, Brown.

Aitkens, C. M., K. M. Ames, and D. Sanger. 1986. "Affluent Collectors at the Edges of Eurasia and North America: Some Comparisons and Observations on the Evolution of Society among North-Temperate Coastal Hunter-Gatherers." In *Prehistoric Hunter-Gatherers in Japan,* Akazawa, T. and C. M. Aikens, eds., Tokyo: University of Tokyo Press.

Akazawa, T. and C. M. Aikens, eds. 1986. *Prehistoric Hunter-Gatherers in Japan.* Tokyo: University of Tokyo Press.

Ammerman, A. J. and L. L. Cavalli-Sforza. 1972. "A Population Model for the Diffusion of Early Farming in Europe." In *The Explanation of Culture Change: Models in Prehistory,* ed. C. Renfrew. London: Duckworth.

———. 1984. *The Neolithic Transition and the Genetics of Populations in Europe.* Princeton: Princeton University Press.

Anderson, E. N., Jr. 1986. "On the Social Context of Early Food Production." *Current Anthropology* 27(3):262–63.

Athens, J. S. 1977. "Theory Building and the Study of Evolutionary Process in Complex Societies." In *For Theory Building in Archaeology,* ed. L. R. Binford. New York: Academic Press.

Barker, G. 1985. *Prehistoric Farming in Europe.* Cambridge, England: Cambridge University Press.

Barnard, A. 1983. "Contemporary Hunter-Gatherers: Current Theoretical Issues in Ecology and Social Organization." *Annual Review of Anthropology* 12:193–214.

Beadle, G. W. 1972. "The Mystery of Maize." *Field Museum of Natural History Bulletin* 43:2–11.

———. 1980. "The Ancestry of Corn." *Scientific American* 242:112–19.

Bellwood, P. 1986. *The Prehistory of the Indo-Malaysian Archipelago.* Orlando, Fla.: Academic Press.

Belyaev, D. K. 1969. "Domestication of Animals." *Science Journal* 5:47–52.

Binford, L. R. 1968. "Post-Pleistocene Adaptations." In *New Perspectives in Archaeology,* eds. S. R. Binford and L. R. Binford. Chicago: Aldine.

———. 1972. *An Archaeological Perspective.* New York: Seminar Press.

Boserup, E. 1965. *The Conditions of Agricultural Growth.* Chicago: Aldine.

———. 1981. *Population and Technology.* Chicago: University of Chicago Press.

Braidwood, R. J. 1960. "The Agricultural Revolution." *Scientific American* 203:130–41.

———. 1973. "The Early Village in Southwestern Asia." *Journal of Near Eastern Studies* 32:34–39.

Braidwood, R. J. et al. 1960. *Prehistoric Investigations in Iraqi Kurdistan.* Chicago: University of Chicago Press.

Braidwood, R. J., H. Cambel, and P. J. Watson. 1969. "Prehistoric Investigations in Southeastern Turkey." *Science* 164:1275–76.

Brush, C. 1965. "Pox Pottery: Earliest Identified Mexican Ceramic." *Science* 149:194–95.

Butzer, K. W. 1971. "The Significance of Agricultural Dispersal into Europe and Northern Africa." In *Prehistoric Agriculture,* ed. S. Struever. Garden City, N.Y.: Natural History Press.

———. 1976. *Early Hydraulic Civilization in Egypt.* Chicago: University of Chicago Press.

Carneiro, R. 1970. "A Theory of the Origin of the State." *Science* 169:733–38.

Carneiro, R. and D. Hilse. 1966. "On Determining the Probable Rate of Population Growth During the Neolithic." *American Anthropologist* 68:177–81.

Caton-Thompson, G., and E. W. Gardner. 1934. *The Desert Fayum.* London: Royal Anthropological Institute.

Cauvin, J. 1972. "Nouvelles Fouilles Tell Mureybet (Syria) 1971–1972. Rapport Preliminaire." *Annales Archeologiques de Syrie* 22:105–15.

Cavalli-Sforza, L. L. and M. W. Feldman. 1981. *Cultural Transmission and Evolution: A Quantitative Approach.* Princeton: Princeton University Press.

Chang, K. C. 1976. *Early Chinese Civilization. Anthropological Perspectives.* Cambridge, Mass.: Harvard University Press.

Childe, V. 1952. *New Light on the Most Ancient East.* 4th Ed. London: Routledge and Kegan Paul.

Clark, J. G. D. 1952. *Prehistoric Europe: The Economic Basis.* London: Methuen.

———. 1980. *Mesolithic Prelude. The Paleolithic-Neolithic Transition in Old World Prehistory.* Edinburgh: Edinburgh University Press.

Coe, M. D. 1960. "Archaeological Linkages with North and South America at La Victoria, Guatemala." *American Anthropologist* 62:363–93.

Cohen, N. M. 1977. *The Food Crisis in Prehistory.* New Haven and London: Yale University Press.

Cohen, N. M. and G. J. Armelagos. 1984. *Paleopathology at the Origins of Culture.* New York: Academic Press.

Diener, P. and E. E. Robkin. 1978. "Ecology, Evolution, and the Search for Cultural Origins: The Question of Islamic Pig Prohibition." *Current Anthropology* 19:493–540.

Ekholm, G. F. 1964. "Transpacific Contacts." In *Prehistoric Man in the New World,* eds. J. D. Jennings and E. Norbeck. Chicago: University of Chicago Press.

Ellen, R. 1982. *Environment, Subsistence and System.* Cambridge, England: Cambridge University Press.

Feldman, M. and E. R. Sears. 1981. "The Wild Gene Resources of Wheat." *Scientific American* 244:102–13.

Fieldel, S. J. 1987. *Prehistory of the Americas.* Cambridge, England: Cambridge University Press.

Flannery, K.V. 1965. "The Ecology of Early Food Production in Mesopotamia." *Science* 147:1247–56.

———. 1968. "Archeological Systems Theory and Early Mesoamerica." In *Anthropological Archeology in the Americas,* ed. B. J. Meggers. Washington, D.C.: The Anthropological Society of Washington.

———. 1969. "Origins and Ecological Effects of Early Domestication in Animals." In *The Domestication and Exploitation of Plants and Animals,* eds. P. J. Ucko and G. W. Dimbleby. London: Duckworth.

———. 1971. "Origins and Ecological Effects of Early Domestication in Iran and the Near East." In *Prehistoric Agriculture,* ed. S. Struever. Garden City, N.Y.: Natural History Press.

———. 1972. "The Origins of the Village as a Settlement Type in Mesoamerica and the Near East: A Comparative Study." In *Man, Settlement and Urbanism,* eds. P. J. Ucko, R. Tringham, and G. W. Dimbleby. London: Duckworth.

———. 1973. "The Origins of Agriculture." *Annual Review of Anthropology* 2:271–310.

Flannery, K. V. and M. D. Coe. 1968. "Social and Economic Systems in Formative

Mesoamerica." In *New Perspectives in Archeology,* eds., S. R. Binford and L. R. Binford. Chicago; Aldine.

Friedman, J. 1974. "Marxism, Structuralism, and Vulgar Materialism." *Man* 9;444–69.

Frisch, R. and J. McArthur. 1974. "Menstrual Cycles: Fatness as a Determinant of Minimum Weight for Height Necessary for Their Maintenance or Onset." *Science* 185:949–51.

Frison, G. C. 1978. *Prehistoric Hunters on the High Plains.* New York: Academic Press.

Galinat, W. C. 1971. "The Origin of Maize." *Annual Review of Genetics* 5:447–78.

Garrod, D. 1957. "The Natufian Culture. The Life and Economy of a Mesolithic People in the Near East." *Proceedings of the British Academy* 43:211–17.

Gould, S. J. 1985. *The Flamingo's Smile.* New York: W. W. Norton.

Gowlett, J. A. J. 1987. "The Archaeology of Radiocarbon Acclerator Dating." *Journal of World Prehistory* 1(2):127–70.

Haaland, R. 1987. *Socio-Economic Differentiation in Neolithic Sudan.* Oxford: British Archaeological Reports, International Series 350.

Halperin, R. 1980. "Ecology and Mode of Production: Seasonal Variation and the Division of Labor by Sex among Hunter-Gatherers." *Journal of Anthropological Research* 36:379–99.

Harlan, J. and D. Zohary. 1966. "Distribution of Wild Wheats and Barley." *Science* 153:1074–80.

Harlan, J., J. M. J. De Wet, and A. B. L. Stemler, eds. 1976. *Origins of African Plant Domestication.* The Hague: Mouton.

Harner, M. 1970. "Population Pressure and the Social Evolution of Agriculturalists." *Southwestern Journal of Anthropology* 26:67–86.

Harpending, H. and H. Davis. 1976. "Some Implications for Hunter-Gatherer Ecology Derived from the Spatial Structure of Resources." *World Archaeology* 8:275–86.

Harris, M. 1974. *Cows, Pigs, Wars, and Witches.* New York: Random House.

———. 1977. *Cannibals and Kings.* New York: Random House.

———. 1979. *Cultural Materialism.* New York: Random House.

Hassan, F. A. 1980. *Demographic Archaeology.* New York: Academic Press.

Hayden, B. 1981. "Research and Development in the Stone Age: Technological Transitions among Hunter-Gatherers." *Current Anthropology* 22:519–48.

Helback, H. 1964. "First Impressions of the Çatal Hüyük Plant Husbandry." *Anatolian Studies* 14:121–23.

———. 1969. "Plant Collecting, Dry-farming, and Irrigation Agriculture in Prehistoric Deh Luran." In *Prehistory and Human Ecology of the Deh Luran Plain,* eds. F. Hole, K. V. Flannery, and J. A. Neely. Ann Arbor: Memoirs of the Museum of Anthropology, University of Michigan, no. 1.

Henry, D. 1983. "Adaptive Evolution Within the Epipaleolithic of the Near East." In *Advances in World Archaeology, Vol. II,* eds. F. Wendorf and A. Close. New York: Academic Press.

———. 1989. *From Foraging to Agriculture. The Levant at the End of the Ice Age.* Philadelphia: University of Pennsylvania Press.

Higgs, E. S. and M. R. Jarman. 1969. "The Origins of Agriculture: A Reconsideration." *Antiquity* 43:31–41.

Ho, P. 1969. "The Loess and the Origin of Chinese Agriculture." *American Historical Review* 75:1–36.

Hole, F. 1962. "Archeological Survey and Excavation in Iran, 1961." *Science* 137:524–26.

———. 1971. "Comment on 'Origins of Food Production in Southwestern Asia' by G. Wright." *Current Anthropology* 12:472–73.

Hole, F., K. V. Flannery, and J. A. Neely. 1969. *Prehistory and Human Ecology of the Deh Luran Plain.* Ann Arbor: Memoirs of the Museum of Anthropology, University of Michigan, no 1.

Iltis, H. 1983. "From Teosinte to Maize: The Catastrophic Sexual Mutation." *Science* 222:886–94.

Johannessen, C. L. 1988. "Indian Maize in the Twelfth Century B.C." *Nature* 332:587.

Johnson, F., ed. 1972. *The Prehistory of the Tehuacan Valley*, Vol. 4. Austin: University of Texas Press.

Just, P. 1980. "Time and Leisure in the Elaboration of Culture." *Journal of Anthropological Research* 36:105–15.

Kaufman, D. 1986. "A Reconsideration of Adaptive Change in the Levantine Epipaleolithic." In *The End of the Paleolithic in the Old World*, ed. L. G. Straus. Oxford: British Archaeological Reports International Series 284.

Keene, A. S. 1981. "Optimal Foraging in a Nonmarginal Environment: A Model of Prehistoric Subsistence Strategies in Michigan." In *Hunter-Gatherer Foraging Strategies*, eds. B. Winterhalder and E. A. Smith. Chicago: University of Chicago Press.

Kirkbride, D. 1968. "Beidha: Early Neolithic Village Life South of the Dead Sea." *Antiquity* 42:263–74.

Kirkby, A. 1973. *The Use of Land and Water Resources in the Past and Present Valley of Oaxaca, Mexico*. Memoirs of the Museum of Anthropology, University of Michigan, no. 5.

Kislev, M. E. and O. Bar-Yosef. 1988. "The Legumes: The Earliest Domesticated Plants in the Near East?" *Current Anthropology* 29(1):175–78.

Kovar, A. 1970. "The Physical and Biological Environment of the Basin of Mexico." In *The Teotihuacan Valley Project. Final Report, vol. 1*, eds. W. Sanders et al. Occasional Papers in Anthropology, Pennsylvania State University.

Lange, F. W. 1971. "Marine Resources: A Viable Subsistence Alternative for the Prehistoric Lowland Maya." *American Antiquity* 73:619–39.

Lee, R. B. 1969. "!Kung Bushman Subsistence: An Input-Output Analysis." In *Environment and Cultural Behavior*, ed. A. P. Vayda. Garden City, N.Y.: Natural History Press.

Legge, A. J. and P. A. Rowley-Conwy. 1987. "Gazelle Killing in Stone Age Syria." *Scientific American* (August) 88–95.

Lippe, R. N., R. M. Bird, and D. M. Stemper. 1984. "Maize Recovered at La Ponga, an Early Ecuadorian Site." *American Antiquity* 49(1):118–24.

Lynch, T. F. 1980. *Guitarrero Cave: Early Man in the Andes*. New York: Academic Press.

———. 1983. "The South American Paleo-Indians." In *Ancient Native Americans*, ed. J. D. Jennings. San Francisco: Freeman.

MacNeish, R. S. 1964. "Ancient Mesoamerican Civilization." *Science* 143:531–37.

———. 1966. "Speculations about the Beginnings of Village Agriculture in Mesoamerica." *Actas y Memorials del 35a Congreso Internacional de Americanistas* 1:181–85.

———, gen. ed. 1972. *The Prehistory of the Tehuacan Valley: Chronology and Irrigation*, Vol. 4. Austin: University of Texas Press.

Mangelsdorf, P. 1974. *Corn: Its Origin, Evolution, and Improvement*. Cambridge, Mass.: Harvard University Press.

———. 1983. "The Mystery of Corn: New Perspectives." *Proceedings of the American Philosophical Society* 127(4):215–47.

Martin, P. S. and P. J. Mehringer, Jr. 1965. "Pleistocene Pollen Analysis and Biogeography of the Southwest." In *The Quaternary of the United States, Biogeography: Phytogeography and Palynology, part 2*, eds. H. E. Wright, Jr. and D. G. Frey. Princeton: Princeton University Press.

Matheny, R. T. and D. Gurr. 1983. "Variation in Prehistoric Agricultural Systems of the New World." *Annual Review of Anthropology* 12:79–103.

Meggers, B. 1975. "The Transpacific Origins of Mesoamerican Civilization: A Preliminary Review of the Evidence and Its Theoretical Implications." *American Anthropologist* 77:1–27.

Mellaart, J. 1966. *The Chalcolithic Early Bronze Ages of the Near East and Anatolia*. Beirut: Khayats.

———. 1975. *The Neolithic of the Near East*. London: Thames and Hudson.

Mortensen, P. 1972. "Seasonal Cacips and Early Villages in the Zagros." In *Man, Settlement and Urbanism,* eds. P. Ucko, R. Tringham, and G. W. Dimbleby. London: Duckworth.

Munchaev, R. M. and N. Y. Merpert. 1971. *New Studies of Early Agricultural Settlements in the Sinjar Valley.* VIII Congress International des Sciences Prehistoriques et Protohistoriques, Belgrade.

Palerm, A. and E. Wolf. 1960. "Ecological Potential and Cultural Development in Mesoamerica." *Social Science Monographs* 3:1–38.

Perkins, D., Jr. 1973. "The Beginnings of Animal Domestication in the Near East." *American Journal of Archaeology* 77:279–82.

Perkins, D., Jr. and P. Daly. 1968. "A Hunter's Village in Neolithic Turkey." *Scientific American* 210:94–105.

Pianka, E. R. 1978. *Evolutionary Ecology.* 2nd Ed. New York: Harper & Row.

Price, T. D. 1987. "The Mesolithic of Western Europe." *Journal of World Prehistory* 1(3):225–305.

Pryor, F. L. 1986. "The Adoption of Agriculture: Some Theoretical and Empirical Evidence." *American Anthropologist* 88(4):879–97.

Pumpelly, R. 1908. *Explorations in Turkey, the Expedition of 1904: Prehistoric Civilization of Anau.* Vol. 1. Washington, D.C.: Publications of the Carnegie Institution, no 73.

Reed, Charles A., ed. 1977. *Origins of Agriculture.* The Hague: Mouton.

Reichel-Dolmatoff, G. 1965. *Columbia.* New York: Praeger, Ancient Peoples and Places Series, no 44.

Renfrew, C., J. E. Dixon, and J. R. Cann. 1966. "Obsidian and Early Cultural Contacts in the Near East." *Proceedings of the Prehistoric Society* 32:30–72.

Renfrew, C. 1987. *Archaeology and Language.* London: Jonathan Cape.

Rindos, D. 1980. "Symbiosis, Instability, and the Origins and Spread of Agriculture: A New Model." *Current Anthropology* 21:751–72.

———. 1984. *The Origins of Agriculture: An Evolutionary Perspective.* New York: Academic Press.

Rosen, S. A. 1988. "Notes on the Origins of Pastoral Nomadism: A Case Study from the Negev and Sinai." *Current Anthropology* 29(3):498–506.

Sahlins, M. 1968. "Notes on the Original Affluent Society." In *Man the Hunter,* eds. R. B. Lee and I. DeVore. Chicago: Aldine.

———. 1976. *Culture and Practical Reason.* Chicago: University of Chicago Press.

———. 1978. "Comment on A. H. Berger's *Structural and Eclectic Revisions of Marxist Strategy.* A Cultural Materialist Critique." *Current Anthropology* 17:298–300.

Sanders, W. T. and B. J. Price. 1968. *Mesoamerica: The Evolution of a Civilization.* New York: Random House.

Smith, P. E. L. 1967. "New Investigations in the Late Pleistocene Archaeology of the Kom Ombo Plain (Upper Egypt)." *Quaternaria* 9:141–52.

———. 1972. "Ganj Dareh Tepe." *Iran* 10:165–68.

Solecki, R. L. 1981. "An Early Village Site at Zawi Chemi Shanidar." *Bibliotheca Mesopotamica.* Vol. 13.

Solecki, R. S. 1964. "Shanidar Cave, a Late Pleistocene Site in Northern Iraq." *VI International Congress on the Quaternary, Reports* 4:413–23.

Solecki, R. S. and R. L. Solecki. 1980. "Paleoecology of the Negev." *Quarterly Review of Archaeology* 1:8, 12.

Spuhler, J. 1985. "Anthropology, Evolution, and 'Scientific Creationism'." *Annual Review of Anthropology* 14:103–33.

Stark, B. and B. Voorhies, eds. 1978. *Prehistoric Coastal Adaptations: The Economy and Ecology of Maritime Middle America.* New York: Academic Press.

Stothert, K. E. 1985. "The Preceramic Las Vegas Culture of Coastal Ecuador." *American Antiquity* 50(3):613–37.

Tauber, H. 1981. "13C Evidence for Dietary Habits of Prehistoric Man in Denmark." *Nature* 292:332–33.

Testart, A. 1982. "The Significance of Food Storage among Hunter-Gatherers: Residence Patterns, Population Densities, and Social Inequalities." *Current Anthropology* 23(5):523–27.

Tringham, R. 1971. *Hunters, Fishers, and Farmers of Eastern Europe 6000–3000* B.C. London: Hutchinson University Library.

Turnball, P. F. and C. A. Reed. 1974. "The Fauna from the Terminal Pleistocene of Palegawra Cave." *Fieldiana* (Chicago Field Museum of Natural History) 63.

Van Loon, M. 1968. "The Oriental Institute Excavations at Mureybit, Syria: Preliminary Report on the 1965 Campaign." *Journal of Near Eastern Studies* 27:265–90.

Vavilov, N. I. 1949–50. "The Origin, Variation, Immunity, and Breeding of Cultivated Plants." *Chronica Botanica* 13.

Webster, G. S. 1986. "Optimization Theory and Pre-Columbian Hunting in the Tehuacan Valley." *Human Ecology* 14(4):415–35.

Wendorf, F. and A. E. Marks, eds. 1975. *Problems in Prehistory. North Africa and the Levant.* Dallas: SMU Press.

Wenke, R. J., J. E. Long, and P. E. Buck. 1988. "Epipaleolitic and Neolithic Subsistence and Settlement in the Fayyum Oasis of Egypt." *Journal of Field Archaeology* 15(1):29–51.

Western, C. 1971. "The Ecological Interpretation of Ancient Charcoals from Jericho." *Levant* 3:31–40.

Winterhalder, B. and E. A. Smith, eds. 1981. *Hunter-Gatherer Foraging Strategies.* Chicago: University of Chicago Press.

Wright, G. 1971. "Origins of Food Production in Southwestern Asia: A Survey of Ideas." *Current Anthropology* 12:447–77.

Wright, H. E., Jr. 1968. "Natural Environment of Early Food Production North of Mesopotamia." *Science* 161:334–39.

———. 1976. "The Environmental Setting for Plant Domestication in the Near East." *Science* 194:385–89.

Zeist, W. van. 1970. "The Paleobotany (Mureybit)." *Journal of Near Eastern Studies* 29:167–76.

Zeist, W. van and W. A. Casparie. 1968. "Wild Einkorn Wheat and Barley from Tell Mureybit in Northern Syria." *Acta Botanica Nederlandica* 17:44–53.

Zevallos, M. C. et al. 1977. "The San Pablo Corn Kernel and Its Friends." *Science* 196:385–89.

The Evolution of
Socially Complex Cultures

I should like to see, and this will be the last and
most ardent of my desires, I should like to see
the last king strangled with the guts of the last
priest.
        J. Messelier (clause in a will, Paris, 1733)

Although many people have echoed Messelier's sentiments, societies governed and exploited by kings and priests have existed for thousands of
years and persist in various places today. Kings and priests, and the social
and economic systems associated with them, are part of one of the most
profound cultural transformations of all time: the origins of *complex societies.*

Scholars, predictably, disagree on exactly what social or cultural "complexity" is, particularly when the term is applied to the bones and stones
and bricks of the archaeological record. But when archaeologists speak of
culture and social "complexity" most have in mind the kinds of changes
that we think happened for the first time long ago on the broad hot Tigris-
Euphrates alluvial plains of what is now Iraq. Based on well over a century of systematic archaeology, we know that if you were a citizen of an
ordinary community here at about 6000 B.C., you would have lived in a
village of a few hundred people, most of whom were your blood relatives;
you and almost everyone else would have worked in the fields to produce
the grain held in common stores for the whole settlement; and if you
were an older adult male you would have made most of the decisions for
you and your family about every aspect of your life. People just like you
and communities very like yours would be found in all directions from
your home, but your only contacts with them would have been minor
trade in obsidian, flint, semi-precious stone, and a few other commodities
and—perhaps—the occasional fight. For all practical purposes, you and

7.1 South American natives of the Monou-teri group shooting at a wooden dummy as practice for a raid. The day after this picture was taken these men raided a neighboring group.

your fellow community members were on your own in terms of religion, manufacturing tools, defense, and food production; probably you and every other adult would know most of the skills necessary for survival; and your extended family probably could have replicated the entire range of economic activities necessary for community survival.

But if you lived in this same area—perhaps the same town—3000 years later, at about 3000 B.C., you would have led a very different life. You could have been a slave or a king, depending on accidents of birth. Unlike your ancestors, who were almost all full-time farmers, you may have been a farmer, fisherman, potter, weaver, priest, or some other specialist; but unless you were among the elite, many of the decisions you made about your job would have been out of your hands—the responsibility of royal administrators. You would have been taxed and expected to fight with the army in any of the numerous wars and revolutions you would have seen in your lifetime. As a farmer or any other semiskilled or skilled worker, you would have been dependent for your continued existence and way of life on people with skills you yourself did not possess, such as potters, warriors, herdsmen, scribes, doctors, metalsmiths, sailors, and priests. Instead of a village of a few hundred people, you might have lived in a city of tens of thousands, you would probably have been a fervent believer in the national religion, and you would probably have been acutely aware of your social class, whether high or low. In these and many other ways, your society would have been "complex," at least in comparison to the simple, unworldly, communistic peasants of your ancient ancestry.

These same kinds of changes that occurred in Southwest Asia also happened in Egypt, the Indus Valley, China, Peru, Mesoamerica, and a few other places, and by now the reader will not be at all surprised to learn that the question that has fascinated archaeologists for centuries is, Why? Why in these specific areas did simple agriculturalists give way to socially more complex forms? Why did our ancestors not remain simple farmers, or revert to the ancient pleasures of the hunting-foraging way of life? Moreover, why was this transition so rapid? Although complex societies have existed for only the last five or six millennia, they have almost completely replaced the simpler cultural forms in which our ancestors had lived for a million years or more. Today, in the Arctic, the Kalahari Desert, and a few other places, hunting and gathering bands still follow the ancient ways, but soon they will be extinct and the "victory" of complex societies complete.

To answer all these various questions about the evolution of complex societies we will have to depend primarily on archaeological data. All independent cases of the evolution of cultural complexity happened before any written languages had been developed; and although some of the historically documented cases of evolving complexity, such as in West Africa, are instructive, none really developed in the same kind of isolation that Mesopotamia, Egypt, and a few others did. Today and in the recent past the cultures of the world truly constitute a "World System," in which the ancient linkages among peoples have been so multiplied and elaborated that we are all connected in the same social organism.[1] And thus it is no longer possible to study directly the kinds of independent changes in social structures that occurred thousands of years ago in the first states and empires.

But we also have some resources in the diversity of ancient complex cultures, particularly in the fact that, as in early agricultural communities, complex societies evolved independently in the Old and New Worlds, and this gives us the opportunity for *comparative analyses*.

The Psychological Effects of Social Complexity

Because of the nature of archaeological data, we will have to limit ourselves primarily to the artifacts of these extinct civilizations, the bones and
stones and houses that have survived them. But it is worth reflecting on
the tremendous impact that the evolution of cultural complexity has had
on the way people view themselves and the world. Immanuel Kant said
that the essence of immorality is to treat other people as objects, and in a
way this is the Original Sin of cultural complexity. If recent band societies
resemble Pleistocene band societies, most Pleistocene people were deeply
embedded in social and family relationships and had a clear role in society. Marshall Sahlins observed that our hunting and gathering ancestors
took the "Zen road to affluence": people living in complex sedentary communities seem to live in the eternal economic dilemma of unlimited wants
and limited means, but simpler societies have adjusted to their limited
means by having few wants.[2] Hunters and gatherers cannot accumulate
and therefore do not covet air-conditioners and trash-compactors, because they are frequently moving, and they live in such small and scattered groups that social hierarchies are of little use or relevance. Ethnologist Richard Lee described how one Christmas he supplied a group of
Kalahari Bushmen with a 1200-pound ox for a great feast—the ox represented far more meat than the group could eat.[3] Everyone in the group
complained, however, about how scrawny the animal was and how poor
his gift was. Lee realized eventually that the cool reception to his gener-
osity was the Bushmen's way of maintaining an egalitarian spirit: any particularly valuable or productive act or service met the same response, because group dynamics worked best if no one could take great personal
and public prestige from his accomplishments.

But in complex societies, from ancient times to the present, people often
have felt themselves to be minor replaceable cogs in a machine that operates mainly for the benefit of others. From the very first written documents, those from Mesopotamia of the fourth and third millennia B.C.,
we hear the age-old complaints about poverty, taxes, oppressive rulers,
governmental harassment, and other ills of cultural complexity. Little
wonder that from the earliest records we also see the beginnings of Uto-
pian movements, made up of those who yearn for a return to a simpler
place and time, when political, religious, and economic hierarchies did not
exist, when all people were considered of equal worth, where all had an
equal share, and where no one had power over any one else.

Ethnographic studies of hunter-gatherers have stripped away illusions
that these people are entirely warm, generous, nonviolent, pure expressions of humanity; they fight, lie, steal, and argue like the rest of us. But
cultural complexity seems to produce its own social pathologies; hunter-

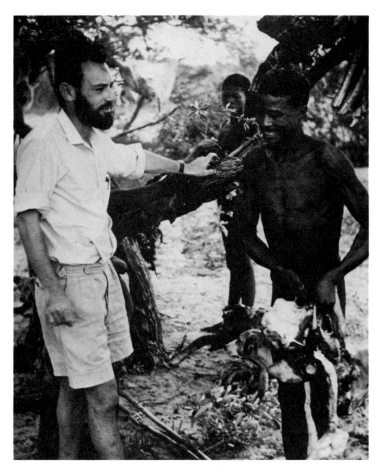

7.2   Richard Lee at a Christmas ox roast for a group of Kalahari Bushmen, southern Africa (see text for discussion of egalitarian principles of band organization demonstrated at this event).

foragers, including presumably our ancestors over the last three million years, seem to live with comparatively little of the crime, class consciousness, alienation, anxiety, and other ills of modern societies.

One of the oldest and commonest human errors has been to confuse cultural complexity and cultural worth. Already by 2500 B.C., a haughty citizen of a city-state in Iraq disparagingly described his nomadic neighbors as "[barbarians], who know no house or town, the boor of the mountains . . . who does not bend his knees [to cultivate the land] . . . who is not buried after his death."[4] Even in our own age, it is difficult to avoid the notions that civilizations have emerged because of the special gifts and vitality of their populaces, that simpler societies are incompletely developed, and that all the world's cultures are at various points along a gradient whose apex is the modern Western industrial community. This attitude was a great advantage in the age of European colonialism because it allowed the Spanish, British, Dutch, and others to treat other peoples as hardly human, and therefore undeserving of the full protection of laws and morality.

To classify, as archaeologists do, history's thousands of societies in terms of their inferred social stratification, size, and the complexity of information, matter, and energy exchanges is a research tactic that has as its goal the elucidation of the processes that produce these forms of complexity; we should not consider these measures as ultimate criteria. If we categorized human societies in terms of piety, social cohesiveness, "justice," or other abstract but important concepts, an ordering of societies very different from the following, having to do with "cultural complexity," would result.

Social and Cultural Typologies

Archaeologists have been heavily influenced in their conceptions of cultural complexity by anthropologists Julian Steward, Elman Service, and Morton Fried.[5] Although their schemes were formulated principally to classify the diversity of extant and recent cultures, their ideas have been widely applied by archaeologists to prehistoric societies. The use of ethnographic data to categorize ancient societies that exist only as rubble and discarded artifacts can, of course, be misleading, particularly since "it . . . seems possible that every hunter-gatherer or tribal society in the world was influenced to some degree by contact with technologically more advanced societies prior to ethnographic study."[6]

In some ways social typologies and taxonomies are a hindrance to good archaeological analysis because they inexactly lump together societies on the basis of unstated and sometimes irrelevant criteria.[7]

But almost all archaeologists continue to think of the archaeological record in terms of three or four basic *types* of societies. These types are abstractions that may fit precisely no particular case, especially with regard to extinct societies, but they are convenient summaries of the kinds of social differences archaeologists think were involved in the evolution of complex societies.

Bands

As we noted in Chapters 3 and 4, the archaeological evidence is undeniable that for the first three million years of our history as a genus, we lived in what anthropologists generally—if inexactly—call band societies.

Today's Eskimos who use snowmobiles to hunt bears, and the African Kalahari Bushmen who steal their neighbors' cows, are distorted remnants of a way of life and social form that is our "natural" lifeway, in that we have spent more than 99% of our history as a genus in it.

Archaeologists have developed a sense of what a "band" society is mainly by distilling and combining ethnographic studies of societies in many different parts of the world. The Copper Eskimo, African pygmies, Kalahari Bushmen, and Australian Aborigines are usually cited as examples of this social form. In contemporary and recent band societies, the most salient characteristic seems to be the only minor differences among members of the group in terms of prestige; no one has any greater claims to material resources than anyone else. In most of these societies, older males who are good providers gain the most respect, but they have little or no power to coerce other band members. This lack of social differentiation is tied to their economy: band members spend most of their lives in groups of fifteen to forty people, moving often as they exploit wild plants and animals. Steward concluded that almost all bands were patrilineal and patrilocal—that is, people reckoned their descent primarily through their father and newlyweds lived with the husband's band.[8] Whether or not this is true is still a major issue in ethnography, and many ethnologists think that bank societies can be extremely flexible in where newlyweds live, for example, based on available resources and other factors.[9] In any case, it is the kind of thing archaeologists can never be sure about when applying the model of a band society to a scatter of stones and bones. The division of labor in bands is generally along basic age and sex lines, and the economic structure is a sort of practical communism: money is not used, and exchange usually takes place between people who consider themselves friends or relatives. This gift giving is usually done very casually, and relationships are frequently cemented by offers of reciprocal hospitality.

This is the notion of band societies that archaeologists are accustomed to thinking in, but recently other typologies have been developed based on food storage and other characteristics.[10]

Tribes

"Tribe" is the rather ambiguous term that anthropologists have used to label social groupings that are larger than band societies but generally not particularly complex in terms of economy, social hierarchies, law, etc. People living in tribes are often subsistence farmers, such as the Pueblo Indian maize farmers of the American Southwest or New Guinea yam cultivators. Tribes often have a nominal leader who acts to redistribute food and perform a few minor ceremonial activities, but, as in band societies, he has no privileged access to wealth or power. He can lead only by example and serves at the pleasure of the tribe. Exchange in such societies is still usually accomplished through reciprocal trading within a kinship structure. Typically, tribal societies are larger, more territorial, have more elaborate ceremonialism and kinship systems, and make more distinctions in terms of prestige than band societies.[11]

In many cases, tribes may have been transitional forms, appearing where tribal farmers were in the transition to state societies, and in some cases tribal societies may have been direct outgrowths of the influences of state societies.[12]

Chiefdoms

Timothy Earle defines chiefdoms as "regionally organized societies with a centralized decision-making hierarchy coordinating activities among several village communities."[13] For many authorities, "chiefdoms" are different from bands and tribes both in degree and in kind. Chiefdoms are based on the concept of hereditary inequality: in a chiefdom, if you are the first son of a chief, chances are you will become chief no matter how unsuitable you may be, and if you are able but a "commoner" your options in life will be narrowly circumscribed. These differences in prestige usually correlate with preferential access to wealth; chiefs and their families can claim the best farmlands or fishing places as well as more food and more exotic and expensive items than "commoners." They are often regarded as divine and typically marry within noble families. The economies of these societies typically show a greater degree of specialization and diversification than those of tribes or bands. Craftsmen exist, but they are usually also farmers, and there is no permanent class of artisans as there is in states. Chiefdoms are much larger than tribes, often involving thousands of people.

Earle stresses the ideological bases of chiefdoms. He notes that chiefs typically create sacred places, such as pyramids; they also use symbols of individual power and position in burial cults, and their mortuary symbolism often includes expressions of military might. To some extent, recent archaeological research on chiefdoms has focused on the political and ideological activities of chiefs, rather than just the simple deterministic factors of agriculture and economy that permit the evolution of chiefdoms.[14]

Examples of chiefdoms include the precontact Nootka of British Columbia and early Hawaiian societies. The archaeology of chiefdoms is a particularly valuable endeavor, as Dick Drennan points out, because in some cases they seem to have been a transitional phase leading to states and elsewhere they were "terminal" in the sense that the indigenous developmental factors seem to have produced chiefdoms but not states.[15]

States

What a "state" is depends to some extent on what an analyst sees as the causal factors in producing cultural complexity. Generally, states are assumed to have centralized governments composed of political and religious elites who exercise economic and political control. In addition to being larger in population and territory than other societal forms, states are characterized by having full-time craftsmen and other specialists. The state codifies and enforces laws, drafts soldiers, levies taxes, and exacts tribute. States have powerful economic structures, often centered on market systems, and they have a diversity of settlement sizes, such as villages, towns, and sometimes cities.

Wright and Johnson, for example, define the state in terms of a political polity that has at least three levels in the decision-making hierarchy, such as village headmen, provincial governor, and national leader.[16]

Early states formed essentially independently in at east six areas of the ancient world: Mesopotamia, Egypt, the Indus Valley, China, Mesoamerica, and Peru; ethnographic and historical accounts of state formation include cases in Africa, Madagascar, and various other places.

Empires

Empires have been distinguished from states mainly on the basis of quantitative differences, not functional ones. Empires have more people, control more territory, exploit more environments, and have more levels of social, economic, and political stratification than early states.[17] Many of the early states seem to have been involved in competitive relationships with adjacent states, and for long periods this factor apparently limited their size and power. Eventually, however, in all the early centers of state formation these competitive relationships broke down and one state was able to increase its size and influence drastically—usually so rapidly that it had few competitors. In fact, its ultimate size seems to have been limited only by the level of its communications technology and its administrative efficiency.

Empires of this type first appeared in Mesopotamia toward the end of the second millennium B.C. and within a thousand years thereafter in Egypt, the Indus Valley, and China. The Inca state of Peru and the Aztec state of Mexico also seem to have achieved imperial dimensions just before the arrival of the Europeans, in the sixteenth century A.D.

Rather than think of the archaeological record primarily in terms of bands, tribes, chiefdoms, etc., it is probably more accurate to see the record as a great continuum of variability, reflecting the unique flexibility that humans bring to any situation. We can be certain the the Australopithecines had no Wagner festivals or international labor federations, and no great ancient state was founded principally on caribou hunting. But most other generalizations about these social typologies are likely to be imprecise.

Recent efforts to explain the evolution of cultural and social complexity have focused not so much on the categories of bands, tribes, chiefdoms, states, etc., but on the concepts of *specialization* and *exchange*. Specialization in this sense refers to occupational specialization, in which people produce various goods or services that are then circulated among the larger community. Brumfiel and Earle note that this sense of occupational specialization is a complex concept that varies in at least four dimensions: (1) specialists can be *independent* or *attached*, in the sense that a potter, for example, can supply pots for an unspecified, changing market, perhaps just for members of his or her own village, or for the larger region; or the potter can be attached to a particular patron, and make ceramics only for a given noble family, or governing institution; (2) specialists differ in the *nature of their products*, in that they can produce either *subsistence* items, such as grain, or *wealth*, such as ritual feather head-dresses or gold bracelets—or *services*, such as military duty or religious ceremonies; (3) specialists differ in the *intensity of production*, in that they could be part-time hunters, who contribute occasional deer to a village economy, or a full-time weaver, who does little else but produce textiles for trade; (4) specialists also differ in the *scale* of production, in the sense that in some economies a large group of people or a village, or even a region, may specialize in something, such as salt production, while in other cases, one or two people may produce a particular good or service.[18] Brumfiel and Earle distinguish between various kinds of exchange, between an elite and its supporting citizenry, among different elites, and the many kinds of political and social arrangements in which subsistence goods and nonsubsistence "wealth" can be circulated within and among communities, regions, and larger political entities.

The Archaeology of Complex Societies

Archaeologists regularly talk about the band societies of the Middle Pleistocene and the early chiefdoms of prehistoric Mesopotamia, but we do not have these societies trapped in amber; we have only the bones and stones and other artifacts, and we use the words bands, tribes, and states to describe them only with some license. To deal with this problem, archaeologists typically equate these terms with specific categories of physical evidence.

At the heart of many conceptions of cultural complexity is the idea of changing forms and levels of matter, energy, and information exchange. Each person and each society exist because it is able to divert energy from the natural world, through food sources and technology, and some of these changes are selective advantages. The greater the amount of energy a culture can capture and efficiently utilize, the better its competitive chances. And we can measure this in part by measuring such variables as population density and agricultural and commodity productivity.

Leslie White's equation of cultural evolution with the amount of energy captured per person per year has been widely criticized, but it remains popular among archaeologists.[19] Years of work amid peasant agriculturalists seems to make archaeologists somewhat blind to the psychological richness and ethnic charm of the village agricultural way of life; also, the fundamental data of archaeology are usually the stones and pots and other implements with which people converted energy to their purposes—the very appliances of energy use, in other words, that White stressed. So archaeologists necessarily take a very materialistic view of the past.

The particular artifacts that one considers most significant for studying cultural evolution depend entirely on one's theoretical perspective: those who believe in the impetus of population pressure organize their research to find evidence of this type, while Marxian theorists may look for the extravagant tombs and houses that bespeak social classes.

Architecture

Perhaps the most obvious differences between the archaeological record of the Pleistocene and that of the last five or six millennia is the presence in the latter period of massive amounts of residential and public architecture. All early states built palaces and tombs; hunter-gatherers occasionally build monuments, but rarely on the same scale as agriculturalists.

The appearance of substantial houses and other buildings is mainly a reflection of economic productivity: if a group produces or gathers sufficient resources within a small enough area, it can become sedentary, and in most climates shelter is worth the cost and effort required to build it.

Soon after permanent communities appeared in both the Old and New Worlds, the architecture of these settlements began to reflect changing levels of cultural and social complexity. Whereas the first houses in all communities were probably built very much alike and had the same contents, later communities incorporated residences that varied considerably in expense of construction and furnishing. Ethnographic evidence leaves little doubt that this architectural variability reflects economic, social, and political differentiation within the community, but the essential point is that, relative to earlier societies, there was a change in patterns of investment of societal energy and resources.

Similarly, once residential architectural variability appeared in many of these early communities, "monumental" architecture also appeared. Pyramids, earthen or brick platforms, "temples," "palaces," and other constructions protrude from the ruins of ancient settlements from North China to the high mountain valleys of Peru, and here, too, the important thing is that the ability and incentive to make these investments are radically different from the capacities of Pleistocene bands, in that they imply the ability of some members of the society to control and organize others.

Mortuary Evidence

For much of its history archaeology has been almost synonymous with grave-robbing. Its early practitioners were primarily concerned with finding ancient burials so that they could loot the beautiful goods that people so often have lavished on their departed.

The preservation of items carefully enclosed in burials is usually much better than those in houses or tool-making sites. Also, death for our ancestors, as for ourselves, was invested with more ritual than any other cultural aspect, and in many burials we have, so to speak, the crystallization of complex religious and social forces, as well as reflections of social status. It is in their great *variability* that mortuary customs are so informative: corpses can be buried, burned, ritually exposed, or entombed; they can be laid out flat, on their sides, flexed, or oriented to the cardinal points of the compass; they can be placed in earth, in caves, in crypts, in trees, or on refuse heaps. Burial contents can range from nothing to enormous quantities of jewelry and furnishings and scores of sacrificed human attendants and animals.

It is a fundamental archaeological assumption that a correlation exists between the level of social complexity of a people and the way they treat their dead.[20] Ethnographic studies show that the correlation between subsistence strategy, social organization, and mortuary practices is strong: bands and tribes differed comparatively little in mortuary practices, while sedentary agriculturalists vary their practices according to a wide range of age, sex, and status distinctions.[21]

The presence of juveniles buried with rich grave goods has been given considerable importance in defining the cultural complexity of ancient societies, because such burials are considered indications of ascribed status: it is assumed that young individuals could not have earned these goods on their own. Similarly, some ancient cemeteries have three or four distinct classes of burials. Some types are well constructed of stone, have rich grave goods, and are centrally located, while others are simple graves with little in them except the corpse. And it is a reasonable inference that these divisions correspond to different economic and social classes.[22]

Functional Differentiation and Interdependence

In Chapter 15 we will see that the aboriginal Americans who built the huge mounds that dot the river valleys of eastern North America had great trade systems, intensive maize agriculture, thousands of inhabitants, and a mortuary cult that involved human sacrifice and great expenditures of wealth on dead leaders, yet most archaeologists don't consider them to have constituted a "state," because these Native Americans were almost all full-time maize farmers; there were only a few specialists, in religion, hunting, warfare, etc. In some definitions of cultural complexity the essential component is the division of a community into functionally interdependent entities of such complexity that no small group of people can maintain all that community's activities. This is important because it produces a situation in which societies survive or die out as *societies,* as groups; in contrast, among hunter-gatherers the focus of selection is usually the *individual,* or at most the twelve or fifteen people with whom each individual spends most of the year. In other words, the unit of "selection" changes with cultural complexity. The inhabitants of contemporary North America, for example, are extremely interdependent, in the sense that we are all reliant for continued existence as both physical individuals and as social organizations on the 3 or 4% of the population that produces nearly all the food.

To translate this sense of functional interdependence into archaeological terms, we must look for concentrations and distributions of artifacts indicating a certain level of activity specialization. In early agricultural villages, each house and each group of houses had approximately the same contents in terms of numbers and types of ceramics, stone tools, figurines, and garbage. But in later, more complex societies we find concentrations of artifacts that clearly represent such things as pottery workshops and stone tool-manufacturing workshops, indicating that people specialized in these activities. Again, we infer that they were specialized, but the significant point archaeologically is that certain classes of artifacts are found in places, volumes, and diversities far different from what would be produced by, say, a hunting and gathering group. Certain differences will also be evident if we compare the contents of settlements. Some settlements might specialize in saltmaking, or barley agriculture, or pottery manufacture. This variability in the artifacts found within discrete but contemporary sites is a key element in our identification of cultural complexity.

Settlement Patterns

In addition to measuring social complexity by looking at things excavated at specific sites, we can also look at how settlements are distributed spa-

Communication

tially. First, we can examine variability in settlement size and configuration. Early agricultural villages in Southwest Asia were almost all approximately the same size, but settlements in the region several thousand years later were of many different sizes, ranging from a few hundred square meters to several square kilometers. Similarly, the basic shape of the settlements changed; some were apparently fortified rectangular compounds, while others were just five or six mudbrick houses. Thus, any archaeological analysis of cultural complexity will involve measuring the variability in site size and shape in a large sample of contemporary sites.

Second, we can look at the placement of settlements relative to the environment and to each other. A major part of the cost of exploiting any resource is the distance it must be transported. This applies equally to the deer hunted by Paleolithic bands and the irrigated rice of ancient China. It also applies to the cost of making decisions about resource production, movement, and storage. With primitive communications systems, for example, an official in one settlement cannot make many timely decisions about the agriculture or craft production of thirty or forty other settlements many kilometers away, because the cost of gathering the relevant information and accurately and rapidly acting on it is too high: he would need teams of observers and relay runners.

As a consequence, some arrangements of settlements are more common than others under certain conditions, and we can tell something about the relationships between settlements by analyzing their respective locations. On a relatively broad agricultural plain, towns and villages that exchange goods and services tend to be placed so as to form a pattern of interlocking hexagons (Figure 7.3) because this arrangement is especially efficient if there is a high level of movement of goods and people among the various settlements.[23] We will see several instances in our discussion of archaeological evidence relevant to the origins of cultural complexity where ancient settlements are arranged in an hexagonal pattern or some other form. Here too it is relatively unimportant whether or not the distribution of ancient settlements corresponds exactly to the patterns observed among present ones. What is important is that we know a major change in settlement patterning has occurred over time. Paleolithic hunters and gatherers and early agriculturalists lived in locations determined largely by the availability of material resources. But later in some areas of the ancient world, settlements began to be located with less regard for natural resources and more concern for trade routes, political frontiers, and administrative networks. Again, these changes occurred in settlements that were also building monumental structures, achieving denser population concentrations, and evolving some or all of the other elements of cultural complexity.

For these reasons, archaeological settlement pattern analyses have pro-

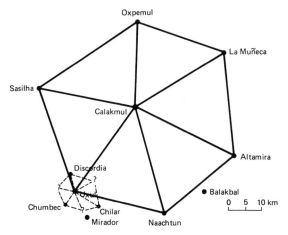

7.3 One of the initial archaeological signs of changing levels of complexity in regional economic and political systems is the appearance of settlement patterns in which settlements are arranged spatially in recognizable patterns. The *Central Place* model suggests that on a flat plain where all resources are uniformly distributed, the settlements will be arranged in different patterns depending on the strength of marketing, transport, administrative, and other factors. No perfectly uniformly endowed plains exist, of course, but actual maps of archaeological settlements can be matched against the theoretical ideal and similarities noted to these ideal models. The hexagonal areas surrounding the settlements in the above models are produced by assuming that on a featureless plain each settlement would have a circular area around it, the inhabitants of which would go to the center for goods and services. But since a plain packed with such circular "demand zones" would leave some areas outside these circles, the theoretical model assumes that the circles are slightly overlapped and that the overlapping areas are bisected. The resulting figures are hexagons. Modern settlement maps as well as archaeological site maps for a given area often roughly approximate the variations of hexagonal distributions illustrated above. Here some of the major settlements of the Maya in lowland Mexico were spaced in an hexagonal arrangement.

vided significant insights into the nature of ancient social, political, and economic changes.[24]

To summarize archaeological approaches to complex cultures, we have several specific lines of evidence: we can look for changes in architecture, technology, settlement size and location, and mortuary complexes, and we can attempt to link these changes to different levels and forms of energy and information usage and overall thermodynamic capture.

Most of the rest of this book is a summary of how different cultures in various parts of the world made the transition to complexity, and this summary is based on these forms of evidence of evolving complexity. But before considering the specifics, let us look first at the general problem of explaining these evolutionary patterns. As in the case of the origins of cultural behavior itself, and the appearance of agricultural economies, we would like some kind of theoretical constructs that will help tie all these separate early states together as members of a class and then will account for the time and location of their appearance.

Explaining the Origins of
Cultural Complexity: The Search for Causes

The earliest scholars believed the rise of cities and states and other elements of evolving cultural complexity required no explanation, because they assumed these developments to be mainly or entirely the work of the gods. The scholars of the Enlightenment and subsequent centuries usually explained the origins of cultural complexity in evolutionary terms. Drawing a parallel with the biological world, Europeans felt that competition between human societies was inevitable, and that to a large extent they had already "won."

Darwin thought that

> civilized nations are everywhere supplanting barbarous nations, excepting where the climate opposes a deadly barrier; and they succeed mainly, though not exclusively, through their arts, which are the products of the intellect. It is, therefore, highly probable that with mankind the intellectual facilities have been gradually perfected through natural selection.[25]

Even in the early part of this century, many scholars believed that the cultural evolution of the whole world was to some extent a result of the rise of the West. The ancient states of China, India, and even the Americas were thought to have been prodded to higher achievements by contact with the European/Near Eastern core areas. Thor Heyerdahl's expeditions, for example, perpetuate this notion that all civilizations are derived from a Near Eastern, or at least Old World, source.

The coincidence of the village farming way of life with the rise of cultural complexity represented to some scholars a sufficient explanation of the evolution of cultural complexity. People then finally had enough leisure time and sedentary habits, it was argued, to develop architecture, art, writing, cities, and the rest of "civilization." The problems with this explanation are apparent even with a superficial examination. Many agricultural groups apparently never developed into "states," while at least one early complex culture (in Peru) may have evolved without a primary agricultural economy. In any case, many hunters and gatherers have more leisure time than primitive agriculturalists.

If you knew nothing about the archaeology of early complex cultures and began to research them, chances are you would be struck by the same facts that so forcefully impressed Julian Steward, V. G. Childe, and other early scholars who studied the problem of cultural complexity origins:

most developed in similar physical environments and were the creations of similar economies.[26] Early states and empires arose for the most part in arid or semiarid environments where agricultural production could be easily intensified, either by canals, terracing, building up fields in lake beds, or some other method; and in zones where many different products were available to everyone, if everyone were linked in exchange networks. As a beginning student of early civilizations you would doubtless also be impressed by how all these civilizations followed similar developmental trajectories, and yet had negligible contact with each other.

Thus, you might look for one or two key factors that had operated in each of these early societies to *cause* civilization in a rather mechanical fashion, in which these civilizations were the outcome of the proper mix of population growth, agricultural intensification, trade, and the peculiarities of the human mind.

In some ways it is easy to imagine societies becoming more complex once the first elements of complexity are in place. A powerful chief, for example, could become a national king, if he were successful in battle and conquered a large, exploitable hinterland like the Mesopotamian Alluvium. And once a city had organized its production and distribution of pottery and grain, it seems an easy step to extend these administrative institutions to fish, beer, plows, and tax revenues; and once the first royal tombs filled with grave gifts are constructed, the building of great mortuary pyramids seems a fairly simple extension of the basic idea.

But what could produce the first irrevocable breaks with the strong traditions of practical communism and social equality of the first agricultural communities?

To begin with the most obvious factor, as noted above, even though intensive agriculture is the foundation of almost all early complex cultures, it is not a sufficient explanation in and of itself. So most attempts to understand the origins of complexity try to link specific agricultural patterns with some other factors. Anthropologist J. Athens notes that agriculture is an effort to maintain an artificial ecosystem and, in some climates, such as arid or temperate environments, the plowing, irrigating, and other efforts needed to maintain agricultural ecosystems are so great that it is doubtful that "the more intense forms of agricultural production would be developed or become adopted unless there was a compelling reason to do so."[27] Athens maintains (as does Ester Boserup)[28] that the only reason sufficient to account for the enormous efforts required to maintain agricultural systems would be an imbalance between the population and available food supply.

In arid and temperate environments, annual agricultural production can vary greatly because of crop disease, weather, and other factors, and there is some incentive to try to stabilize production in these areas by augmenting the irrigation system, intensive weeding, land leveling, and other tasks that require a lot of work. In arid, semiarid, and temperate regions, the growing season is often sharply restricted by the weather, and thus "cultivation . . . does not permit cycling of plantings in such a way as to equalize the labor requirement throughout the year."[29] Each spring, for example, many different activities might have to be performed to avoid poor harvests, and under these conditions, according to Athens, there is a strong selection for certain kinds of cultural complexity. Increasing the territorial size of the cultural system would help meet crises brought on by a flood or some other disaster striking a single village; individuals and villages might also become specialized in trades and crafts to make production more efficient; and, perhaps most important, it would be advantageous to have a hierarchical administrative organization, so that work and production could be closely and efficiently administered.

Many explanations of the evolution of complex societies combine certain forms of agricultural intensification with *population growth*. What causes human population growth? When I framed that question in a graduate school paper, the instructor's marginal comment was brief and rude. The question, rather, is what causes variations in human reproductive rates and population densities.

George Cowgill remarked that many analysts of cultural evolution have assumed

> that a pervasive and powerful factor in human history has been the strong tendency of human populations to increase up to the point where serious shortages of important resources are in the offing; and that experience or anticipation of such shortages has been a major factor, or even the dominant factor, in stimulating intensification of agricultural production and other technical and social innovations. In extreme versions, the entire history of complex societies and civilizations is seen as hardly more than the outcome of measures that began as ways of coping with problems posed by relentless human fertility—what might be called the "strictly from hunger" point of view of developmental processes.[30]

It is easy to see the attractiveness of these ideas, for if one examines history, a strong positive statistical correlation between population growth and cultural complexity is evident. The relationship between human population growth and cultural complexity may not be one of direct cause and effect, however, for correlation does not necessarily demonstrate causation. Moreover, even if the relationship is in some sense causal, it may be that the revolution of cultural complexity leads to rising population densities, rather than the reverse. Empirically, too, there seem to be some problems with the idea that human population growth somehow caused the evolution of cultural complexity. All societies have evolved mechanisms like migration, abortion, infanticide, marriage rules, and contraceptive techniques to control population growth, and thus we might expect people faced with stresses because of overpopulation to impose population controls, rather than "invent" cultural complexity.

It is worth noting that the greatest recent falls in fertility rates come not as the result of food shortages or technical advances in contraception, but as correlates of increasing educational levels for women, greater social mobility, increasing urbanization, and the expanding role of women in the work force.

The actual causal mechanisms by which these factors are translated into reduced fertility are not clear, however. In any case, there is no evidence human populations have ever increased at anything approaching the biologically feasible rate. If the world's population 5570 years ago were only one thousand people and their annual rate of increase since then were four per thousand people—a relatively moderate growth rate—the world's present population would be between 7 and 8 trillion. Obviously, human populations in the past have been under fairly stringent natural and cultural controls, and if we are to link population growth to increasing cultural complexity, we must specify additional factors or principles in causal "models" of how these variables were related.

It should be stressed, however, that the following "models" of cultural evolution are relatively simple and many authorities consider them incomplete as explanations of early complexity. But the ideas they embody are so basic to the analysis of cultural complexity that they are reviewed in detail here, and most contemporary explanatory models have been written in reaction to these earlier formulations. More contemporary explanations of cultural complexity are illustrated in Chapters 8 through 15.

Irrigation Agriculture and the Evolution of Cultural Complexity

Perhaps the most obvious common denominator of ancient complex societies was extensive irrigation systems. Even today aerial photographs of Mesopotamia, Peru, and most other areas of early state formation clearly show the massive remnants of these ancient structures, and similar constructions were built by early "chiefdoms" in such places as Hawaii and southwestern North America. This led some scholars to conclude that the construction and operation of complex irrigation systems were at the heart of the origins of complex societies. A particularly influential proponent of this view is Karl Wittfogel, whose *Oriental Despotism* is a detailed excursion into comparative history and sociological analysis.[31]

Wittfogel notes that the limiting factors on agriculture are soil conditions, temperature, and the availability of water. Of these, water is the

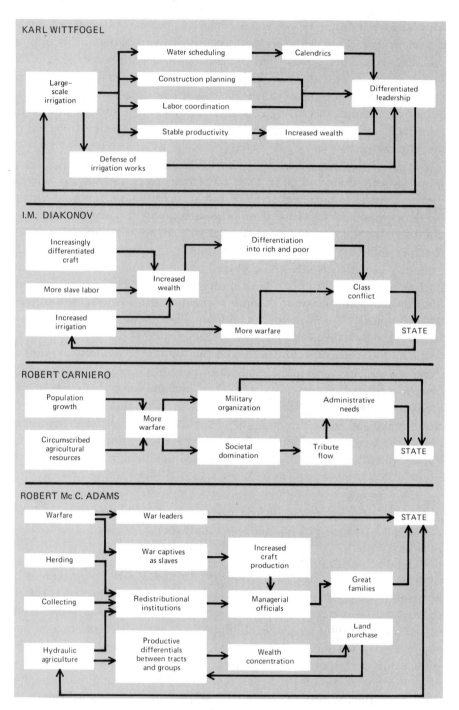

7.4 Several models of the evolution of cultural complexity. In these diagrams Henry Wright (1977) has depicted the most important hypothetical cause-and-effect relationships in several models of the origins of states and urbanism.

most easily manipulated, but its weight and physical characteristics impose limitations on this manipulation. To divert water to agricultural fields requires canal systems, dams, and drainage constructions that can only be built efficiently with organized mass labor; and, once built, irrigation systems require enormous investments of labor and resources annually to clean and maintain them. In addition, these systems necessitate complex administration and communication, because crucial decisions have to be made about construction and repairs, water allocation, and crop harvesting and storage. Thus, a complex irrigation system under ancient conditions required cooperation and centralized hierarchical decision-making institutions.

Irrigation systems also have the intrinsic capacity to create another element in the process of the evolution of complex societies: wealth and status differentials. Fields closer to main rivers are better drained, more easily irrigated, and possess a higher natural fertility, and thus control of such lands would create immediate wealth differentials. Correspondingly, wealth and status would most likely accrue to the elites of the decision-making hierarchies. Wittfogel concludes that irrigation-based agriculture has many other effects on a society. It encourages the development of writing and calendrical systems, so that records can be kept of periods of annual flooding, agricultural production statistics, the amounts of products in storage, and the allocation of water. Construction of roads, palaces, and temples would also be encouraged, because the mobilization of labor for the canal works could be generalized to these other endeavors very easily, and roads would contribute to the movement of agricultural produce and to the communication required for efficient operation of the systems. The construction of temples and palaces would also serve to reinforce the position of the hierarchy. The creation of standing armies and defensive works would also likely follow, because irrigation systems are extremely valuable but not very portable, and they are easily damaged by neglect or intentional destruction.

Wittfogel's hydraulic hypothesis still has considerable currency, as we will see, but there seem to be logical and empirical problems with his ideas as a general model of the origins of cultural complexity. Simple societies in several parts of the world have been observed operating extensive irrigation works with no perceptible despotic administrative systems or rapid increases in social complexity.[32] More damaging to Wittfogel's hypothesis is the scarcity of archaeological evidence of complex irrigation systems dating to before, or to the same time as, the appearance of monumental architecture, urbanism, and other reflections of increasing cultural complexity in Southwest Asia and perhaps other areas where complex societies appeared independently and early.

Nonetheless, difficulties of dating irrigation canals, inadequate archae-

ological samples, and other deficiencies of evidence are such that we cannot conclude irrigation was unimportant in virtually any case of early cultural complexity.

Warfare, Population Growth, and Environment

"War is the father of all things" said Heraclitus, and, given its frequency
in human affairs, we should not be surprised that many scholars see warfare as a natural adjunct of population growth in driving cultural evolution. In the film *The Third Man,* Orson Welles—justifying his profiteering
in postwar Vienna to Joseph Cotton—says

> in Italy for thirty years under the Borgias they had warfare, terror, mur-
> der, bloodshed but they produced Michelangelo, Leonardo da Vinci, and
> the Renaissance. In Switzerland, they had brotherly love, they had five
> hundred years of democracy and peace. And what did that produce? The
> cuckoo-clock.

Welles' character was stating what for many is an uncomfortable truth:
human competition seems to be a powerful engine of cultural evolution.
No early state, not Mexico, or China, or Sumer, or Peru, was without a
background level of organized violence that occasionally erupted into great
wars spanning decades.

Why should this be so? Why doesn't "natural selection" favor those cultures that develop institutions to resolve conflicts? Why doesn't cultural
selection seem to give the edge to communities that pool their resources
to build irrigation canals, establish universities, and work for world peace?

Desmond Morris, Robert Ardrey, and others say it is because we have
a propensity for competition or violence woven deeply into our chromosomes. But for most scholars, warfare is not genetically expressed; it is
simply a cultural behavior that is elicited by environmental and cultural
conditions.

Anthropologist Robert Carneiro argues that warfare was the primary
mechanism for the evolution of social complexity in ancient Peru, Mesopotamia, Egypt, Rome, northern Europe, central Africa, Polynesia, Mesoamerica, Colombia, and elsewhere. He believes, however, that

> warfare cannot be the only factor. After all, wars have been fought in
> many parts of the world where the state never emerged. Thus while war-
> fare may be a necessary condition for the rise of the state, it is not a
> sufficient one. Or, to put it another way, while we can identify war as the
> mechanism of state formation, we need also to specify the conditions un-
> der which it gave rise to the state.[33]

Carneiro sees two such conditions as essential to the formation of complex societies in concert with warfare: population growth and environ-

mental circumscription. He notes that human population densities have been increasing in many areas for millennia, but that only a certain environmental zones can population growth join with warfare to produce highly complex early civilizations. These environmental zones are exceptionally fertile areas "circumscribed," or surrounded, by areas of lesser productivity such as deserts, mountains, or oceans. As an example, Carneiro points to the coast of Peru, where approximately seventy-eight rivers run from the Andes to the ocean through an 80-kilometer stretch of some of the driest deserts on earth. Here, he says, are fertile, easily irrigated strips of land along the rivers, but in any direction one soon encounters desert, mountains, or the ocean. Similar conditions, he asserts, prevailed in Mesopotamia, Egypt, and the other centers of early civilizations.

Again using Peru as an example, Carneiro suggests that shortly after the appearance of the village farming way of life, these fertile riverine areas were sparsely occupied by small autonomous villages. He assumes that in such conditions populations grew and, as these populations increased, villages tended to divide because of internal conflicts and pressure on agricultural lands. Some of the inhabitants would then establish a new community some distance away. Such movements were easily accomplished in this early period because there was no shortage of land and little investment in terracing or irrigation systems. As a consequence, the number of villages increased faster than village size, and all communities remained essentially the same in political and social organization.

Eventually, however, given this constant population growth and the proliferation of villages, all the land that could be irrigated and exploited easily became occupied, and the expanding population rapidly began to outrun the available food supplies. Since they could not move into the sea or deserts or easily colonize the mountains, early Peruvian farmers chose agricultural intensification. They built terraces and irrigation canals and tried to keep pace with their population growth rates, but they were caught in the Malthusian dilemma: food supplies can be increased, but not nearly as quickly as population increases. At this point, Carneiro concludes, people turned to warfare as the only alternative. The village under the most stress would attack the weakest adjacent village, and the victor would expropriate the land and harvests of the loser. The conquered people not killed in the fighting could not simply move away and reestablish their villages, and they could not emigrate to the highlands because their whole culture was based on the village farming way of life. They were either taken back to the victors' village, where they became slaves or artisans, or they were left as serfs who were taxed so heavily that they had to reduce their own consumption and intensify their production still further.

These developments encouraged the formation of an institutionalized bureaucracy to administer the taxes and slaves, and the establishment of the bureaucracy in turn intensified wealth and status differentials, as the most successful military men were given the administrative posts. In addition, the defeated peoples came to constitute a lower class, and thus the stratification of society increased as the level of warfare rose. Carneiro believes that warfare continued in Peru until all of each river valley was under the control of one integrated authority, a development he terms a state. Subsequently, again because of the never-ending pressure of population, these states contended with each other until a whole series of river valleys was controlled by a single dominant center.

Carneiro uses ancient Peru as an example of this developmental pattern but he argues that it applies almost point-for-point to the other major centers of early development.

Carneiro's ideas are diagrammed in Figure 7.4, where it can be seen that the whole structure rests on two "causal" factors: (1) the assumption that constant population growth among early village agriculturalists would inexorably demand increases in food production; and (2) the assumption that warfare is the most likely response to these conditions.

Since many primitive societies had remarkably precise control of their population-to-resources balance, population growth cannot be regarded as automatic. There is no demonstrated and inevitable reason why these populations could not have maintained their size below the stress level rather than resorting to agricultural intensification or warfare. Thus, to strengthen Carneiro's hypotheses we must stipulate other factors that encouraged or allowed these presumed growth rates.

In a reconsideration of Carneiro's model, David Webster maintains that warfare's principal importance in the evolution of the first states was the role it played in breaking down the kinship ties that organized early chiefdoms.[34] He notes that chiefdoms apparently are kept from evolving into states partially because the chief's power and prestige are tied to his role as a redistributive head, and if he begins to hoard wealth or exploit people, the chief begins to lose the support of his kinsmen and deputy rulers. Webster proposes that warfare produces a potent environment for evolutionary change to state-level societies by rendering ineffective many of the internal constraints that keep chiefdoms in a stable sociopolitical status. Continued warfare between chiefdoms would place great adaptive value on a stable military leadership, thereby dampening the constant petty squabbles between rival rulers. A chief who is successful in warfare can also claim more wealth in the form of booty than he could on the basis of his redistribution of his own society's production.

It is difficult to test archaeologically the many hypotheses that Carneiro's model has stimulated. Analyzing the possible interrelationship of, for example, population growth, warfare, and cultural change, would require that we find evidence of conflict and demonstrate that it is linked to pressures exerted on resources by increasing population densities. If Carneiro is correct, complex societies appeared only after a long period of population growth in circumscribed environments, and monumental architecture, irrigation systems, urbanism, land terracing, and other aspects of "civilization" emerged only at population density peaks and are associated in time and space with defensive constructions, mass burials, burned settlements, concentrations of weapons, and other evidences of conflict.

In a recent reevaluation of Carneiro's ideas, Robert Schacht suggests that Carneiro's model is potentially of great usefulness but requires redefinition.[35] He says that environmental circumscription, for example, should be expressed in terms of quantified degrees of certain variables, and population growth must be analyzed in terms of specific responses of a given culture to demographic changes; warfare too, Schacht says, could be better expressed in terms of general conflict, or competition. Patrick Kirch has redefined some of Carneiro's terms and has found that the model has considerable applicability to Polynesia where the ocean circumscribed ancient cultures on Hawaii, Tonga, and other islands.[36] Other archaeological evaluations of Carneiro's model are considered in subsequent chapters.

Marxian Explanations of Early Cultural Complexity

For many of today's social scientists, there is really no doubt about the answer to the problem of the evolution of social complexity. Friedrich Engels' remarks at Karl Marx's grave expressed this certainty: "As Darwin discovered the law of evolution in organic nature so Marx discovered the law of evolution in human history."[37]

Despite attempts by many to discard Marx's ideas on this point as simplistic and wrong, and despite the interweaving of Marx's economic analysis with dubious political polemic, there is no denying the tremendous influence Marx's contributions have had on the analysis of social systems.

Almost any attempt to sum up Marx's theories about the origins of cultural complexity and the dynamics of history necessarily involves great oversimplification and arguable interpretations. Marx's famous statement of his basic ideas is worth quoting at length and studying carefully because it is still one of the most astute and revolutionary observations on the nature of cultural evolutionary history:

> In the social production of their subsistence men enter into determined and necessary relations with each other which are independent of their wills—production-relations which correspond to a definite stage of development of their material productive forces. The sum total of these production-relations forms the economic structure of society, the real basis,

upon which a juridical and political superstructure arises, and to which definite forms of social consciousness correspond. The mode of production of material subsistence conditions the social, political and spiritual life-process in general. It is not the consciousness of men which determines their existence but on the contrary it is their social existence that determines their consciousness. At a certain stage of their development, the material productive forces of society come in conflict with the existing production relations, or what is merely a juridical expression for the same thing, the property relations within which they operated before. From being forms of development of the productive forces, these relations turn into fetters upon their development. Then comes an epoch of social revolution. With the change in the economic foundation the whole immense superstructure is slowly or rapidly transformed. In studying such a transformation one must always distinguish between the material transformation in the economic conditions essential to production—which can be established with the precision of the natural science—and the juridical, political, religious, artistic, or philosophic, in short ideological forms, in which men become conscious of this conflict and fight it out. As little as one judges what an individual is by what he thinks of himself, so little can one judge such an epoch of transformation by its consciousness; one must rather explain this consciousness by the contradictions in the material life, the conflict at hand between the social forces of production and the relations in which production is carried on.[38]

Marx continues:

No social formation ever perishes before all the productive forces for which there is room in it have been developed; and new, higher relations of production never appear before the material conditions of their existence have matured in the womb of the old society itself. . . . In broad outline Asiatic, ancient, feudal, and modern bourgeois modes of production can be designated as progressive epochs in the economic formation of society. The bourgeois relations of production are the last antagonistic form of the social process of production. . . . This social form brings, therefore, the prehistory of human society to a close.[39]

What all this means in terms of the whole history of complex societies is the point of enough books to fill a large library, and Marx's basic premises are being constantly reinterpreted. Still, many accept as valid the basic tenents of Marxian analyses of history, and there is certainly great power to some of his ideas.[40]

Marx himself, however, had very little to say specifically about the origins of complex societies. His primary attention was given to detailing the problems of capitalism and the dynamics of the transition from feudalism to capitalist societies. Much more attention was paid to the problem by later followers of Marx, particularly Engels and Lenin, and, recently, V. V. Streuve and I. Diakonov.

Until the origins of agriculture, these scholars suggest, all societies were classless, all goods were shared, no one really owned anything, and all were treated equally. But gradually, after the achievement of domestication and the agricultural way of life, some people managed to control more than their fair share of the land, which is of course the basic source of wealth in an agricultural community. By controlling land, these elites were able to enslave others and force these people to work the land for them. In time the ruling classes developed the state, laws, and the church to justify, protect, and perpetuate their economic and political privileges. The state is then seen as an exploitative mechanism created by the elites to control and oppress the workers. Marx promulgated a "labor theory of value," according to which capitalists steal much of the value that a worker produces, by paying the worker only a fraction of the value his labor confers on goods and produce, and appropriating the rest.

According to Marxian theory, every economic system based on the division of society into socioeconomic classes and on exploitation carries within itself the seeds of its own destruction, because generally the means of producing wealth constantly improve, technologically and otherwise, and at a certain stage outgrow the social system constructed on them. Thus, slave societies would eventually give way to feudal societies, and eventually, Communist societies will replace capitalist societies.

Diakonov's model of early Mesopotamian state formation (Figure 7.4) rests on the assumption that if wealth differentials can arise, they will, and that once these differentials exist, antagonism between socioeconomic classes will follow and eventually the state will form to promote and protect the vested interests of the ruling class.[41]

Economists and others have challenged many of Marx's ideas, but evidence from early Mesopotamian and Egyptian societies does support some aspects of the Marxian reconstruction: wealth differentials developed early and were impressive, and slavery existed, as did communal labor pools, warfare, irrigation systems, trade networks, and other elements integral to the Marxian scheme. But that is not to say that this scheme is correct or complete. Much of it was constructed on the basis of evidence from early documents, and some forms of the Mesopotamian states evolved several hundred years prior to writing; thus the textual evidence might be of only limited relevance to the origins of social complexity. More important, it is very difficult to test the Marxian reconstruction with archaeological data. We cannot conclusively demonstrate class conflict or slavery in the absence of written records, and many of the crucial elements of the Marxian paradigm, such as the labor theory of value and the contradictions between economies and their social correlates, seem utterly beyond the reach of archaeological research per se. Zagarell uses documents from Mesopotamia to show that by the time the first extensive texts were available, some correspondences exist between Marxian ideas and what we know from these texts about property, the rights of women, etc.[42]

Other scholars have focused on the importance of trade in a Marxian context. Kipp and Schortman, for example, argue that in some cases of early state formation an important factor was the destabilizing effects markets in luxury goods had on early chiefdoms. In chiefdoms personal relationships between elites and the populace are important, but, as market economies evolved, chiefs and other elites had an increasingly difficult time trying to control these markets: "When leadership is undermined by a market blind to everything except profits, policies of systematic impoverishment are as essential to leaders as armies. Economic exploitation joins tyranny, and so states are born."[43]

A recent and detailed attempt to understand the origins of complex societies from a perspective at once Marxian and archaeological is that of Friedman and Rowlands.[44] Like most other contemporary scholars, they have interpreted Marx in a way such that the focus of the analysis is the social relations of production that economies embody, rather than just blunt forces of technology, environment, and agriculture. This study defies easy translation and summary, but contains at least the following elements. Friedman and Rowlands assert that we must "reconstruct the structures of reproduction of particular social forms," which they define as the "social structures that dominate the process of production and circulation and which therefore constitute the socially determined form by which populations reproduce themselves as economic entities."

Friedman and Rowlands attempt to explain the evolution of the "asiatic state" in North China, Mesopotamia, Mesoamerica, and Peru in terms of a single, generalized model, which is illustrated in Figure 7.5.

Although Friedman and Rowlands' complex approach includes much of interest, it involves various problems. The primary one is that so much of it seems untestable with the archaeological record. Testability with archaeological data cannot, of course, be taken as the ultimate criterion of social theory, but archaeologists must try to measure the "fit" of their ideas against the reality of the bones and stones they excavate. In Friedman and Rowlands' approach, as in so many Marxist and Structuralist endeavors, causation is expressed in terms of dominances and constraints (Figure 7.5). Friedman and Rowlands say that the physical environment, technology, and general economic forces impose a system of constraints that make certain kinds of things unlikely to evolve (like a drive-through bank set up and frequented by hunter-gatherers); but, working in the opposite direction, the relations of production dominate the entire functioning of the system, determining its characteristics and developmental pattern. To find "relations of production" archaeologically, however, and then to demonstrate how they were the causes of other cultural phenomena is difficult. In Friedman and Rowlands' approach, something like an organized system of ancestor worship can in a given society and instance

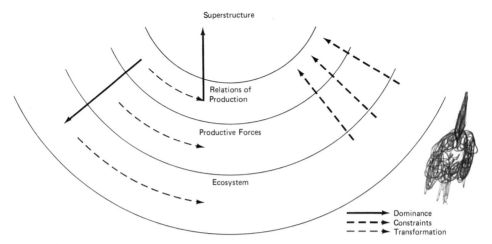

Superstructure

Relations of
Production

Productive Forces

Ecosystem

Dominance
Constraints
Transformation

7.5 Friedman and Rowlands' view of cultural evolution is a modified Marxian model in which the four levels of social formation depicted above have structurally autonomous properties that cannot be derived from those of another level. That is, one can never be certain about the properties of one level, for example, just because one knows that the society involved is a hunting-gathering one, or that they have a national religious cult. But some levels can impose "constraints" on other levels (rice farmers cannot be truly nomadic). The "relations of production" do, however, organize and dominate the whole process of social development. An appreciation of what relations of production are and how these levels of social formation inevitably become incompatible—and why the cultural system breaks them down—are explained in the works of Marx and his successors.

organized system of ancestor worship can in a given society and instance be the *result* of a complex of economic factors, and in another society and time organized ancestor worship can be a *cause* of economic behavior.

Marxian analyses seem to leave the poor archaeologist sitting amidst his or her pile of bones, stones, and pots, without much hope of ever understanding much about the causes of more than the most obvious cultural attributes of the societies whose remains are being studied. Still, Marxist analyses of archaeological data have shown potential, and additional examples of them are presented in subsequent chapters.

Marvin Harris' Explanation of State Origins

Marvin Harris has used elements of the ideas of Wittfogel, Marx, Carneiro, and others to formulate a model that, although many reject it, at least has the advantages of being stipulated and partially testable with the archaeological data.[45]

Harris attributes the slower rate of the agricultural revolution in the New World to the lack of a large domesticable animal, so the basic conditions from which societies in the New and Old Worlds are transformed into complex societies are somewhat different. But in both areas he thinks warfare was a common and early solution to the requirements of life as low-density cultivators. Warfare, he says, affects the demography of early cultivators in various ways. Warfare, with its extreme valuation of and emphasis on males, promotes female infanticide and nutritional neglect of young girls, thereby suppressing population growth rates; fights between members of villages and among villages fosters the dispersal of villages, as people move away and create no-man's lands between them (which preserves ecological systems and hunting opportunities).

But the big step in Harris' view is the rise of chiefdoms, which he ties to the appearance of "big men," particularly influential older men whose advice and guidance the community seeks. Village big men act as the "nodes" of three important institutional complexes: they intensify production, carry out redistribution of harvest surpluses and trade goods, and use their prestige and position to lead the way in fighting or trading with neighboring villages.

Harris says that simple

> chiefdoms come into being with the first intensifier-redistributor-warrior complexes. The more production is intensified, the more there is to redistribute and to trade, the larger the population, the more intense the warfare, the more complex and powerful the chiefly sector. Other things being equal, all such systems tend to move from symmetric forms of redistribution (in which the primary producers get back everything they produce) to asymmetric forms (in which the redistributor gets more of what is produced for longer and longer periods). Eventually the retained portion of the harvest surplus provides the chief with the material means for coercing his followers into further intensifications.[46]

If all this is a standard developmental trajectory, why did chiefdoms never appear among the Eskimo, while appearing in Melanesia but stabilizing at that level of development, and while in Mesopotamia and many other areas chiefdoms quickly became international empires?

Harris turns to basic factors of demography, economy, environment, and technology to answer this question. He suggests that for chiefdoms to appear and then to become states and empires, the appropriate "energy gates" had to be available: yams and tubers are poor energy gates because they do not store well and have no clearly defined harvest period, so a big man or chief cannot easily shut off the flow of proteins and calories produced by the farmers; but grains store well and have defined periods of harvest, so a chief with command of community grain stores absolutely controls the lives of his associates.

Harris notes that the "paleotechnic infrastructures most amenable to intensification, redistribution, and expansion of managerial functions were those based on the grain and ruminant [e.g., cattle] complexes of the Near and Middle East, southern Europe, northern China, and northern India."[47]

Harris thinks that most advanced chiefdoms did not evolve directly into states because they were based on incipient social stratification and other characteristics that made them relatively unstable and prone to factional disputes, revolts, and frequent migrations by people seeking to leave the sphere of influence of an exploitative chief.

Thus he expects the first "pristine" states to arise in areas of sharp ecotones, where there is good potential for intensifying agriculture but where away from the agricultural lands the environment is such that a family would suffer a sharp drop in standard of living if they moved away.[48]

Drawing on the work of Robert Carneiro (see above), Harris identifies the sharply circumscribed farmlands of Egypt, Mesopotamia, northern India, the Yellow River Basin, central highland Mexico, the Peruvian coast, and the Andes highlands as areas where pristine states could be expected.[49] And once pristine states form in an area, their expansionist tendencies and economic impact act as a "single gigantic amplifier," as the whole region is caught in cycles of agricultural intensification, population growth, warfare, etc.

Harris explains the difference between early states principally in terms of their environments. The collapse of the Mayan state in lowland central America he links to the limits on agricultural intensification in an area of relatively poor soil and prolonged drought. To explain Egypt, Peru, Mexico, and other cultural centers, he relies heavily on Karl Wittfogel's ideas about "oriental despotism" (see above). In those areas where a single river was the source of much of the agricultural wealth, highly stratified, centralized despotic states would emerge because of the needs for control, administration, and repair of irrigation works, and the ability of a single monarch to control such systems. But rainfall "agriculture leads to dispersed, multicentered forms of production. Hence it is doubtful that any pristine state ever developed on a rainfall base. Most rainfall states were probably secondary formations brought into existence to take advantage of the opportunity for trade and plunder created by the expansion of hydraulic empires."[50] Harris links the rise of feudalism and capitalism in the West to the fact that rainfall agriculture is not as easily controlled as the produce of a single great river system, such as in Egypt.

Other Models of Cultural Evolution

The explanations of cultural evolution presented above are some of the earlier and simpler efforts to deal with this issue, and in contemporary archaeology these explanations have been combined with many other ideas. Some of these will be reviewed in subsequent chapters. Brumfiel and Earle have categorized these "models" of explanation into three categories: (1)

commercial development models, in which increasing occupational specialization and exchange are seen as the "natural" outgrowth of economic growth; (2) *adaptationist* models, in which political leaders are assumed to have intervened directly in the economy, to redistribute goods, for example, or to manage irrigation systems; and (3) *political models,* in which local rulers intervene in economies but, unlike their assumed role in adaptationist models, these leaders are assumed to be the primary beneficiaries of their efforts.[51] Brumfiel and Earle propose that "political elites consciously and strategically employ specialization and exchange to create and maintain social inequity, strengthen political coalitions, and fund new institutions of control, often in the face of substantial opposition from those whose well-being is reduced by such actions."[52]

It is worth noting here that most of these many kinds of explanations of cultural complexity are *functional arguments.* Functional arguments attempt to explain the origins of something (e.g., the human heart) in terms of the functions it performs (e.g., blood circulation). To the question, why do people have hearts, the answer that some device is needed to circulate blood is an explanation of sorts, but it does not explain why some other kind of life support system did not evolve, nor does it explain the evolutionary history of the heart or the selective pressures that shaped this history.

Similarly, for example, to assert that complex societies developed because a leader was needed to coordinate irrigation and redistribute agricultural production is a functional explanation and does not explain why the society did not remain egalitarian, or develop a capitalist economic system, or go off in some other developmental direction.

Yet functional explanations seem to work so well for certain phenomena. The national religious cults that all early states developed, for example, seem so transparently a device for social control. Montaigne said that "Man is certainly stark mad. He cannot make a worm, and yet he will be making gods by the dozen." But for an early state few things are as useful as gods in whom everyone believes. Then one can despise and kill (and take possession of the property of) all nonbelievers, foreign and domestic, without qualms; one is willing to sacrifice one's self in battles, or participate in pyramid building, or accept a social hierarchy, simply because the gods have so decreed. And the best part is that one does all these things without much cost to the state—people fight in wars, work for the common good, or accept life as a disenfranchised slave often on the premise that in the afterlife things will be greatly improved.

Thus, functionalist arguments have the strong heuristic value of suggesting hypotheses about a specific development and indicating crucial variable relationships.

In part because of the limitations of functionalist explanations, in part because of the great power of modern evolutionary theory, there has been a revival of interest in applying the principles of biological evolution to cultural phenomena.[53] Many scholars argue that the modern theory of evolution has never really been applied to archaeological data. A key difficulty has always been that the objects of the archaeological record do not reproduce as people do, and thus the rules of genetics cannot be applied; also, whereas change in the biological world is through the relatively slow processes of genetic mutation, drift, selection, and so on, cultural changes can be conveyed quickly and pervasively from one group to another (as in the spread of agriculture).

So how might we apply evolutionary principles to archaeological problems? The answer is not at all clear, but there are a few interesting ideas. As Robert Dunnell has observed, what matters in evolutionary theory is not so much how a characteristic is transmitted—whether by genes or culture—as it is the mechanisms by which traits are perpetuated in an individual.[54] Thus, whether a person gets eye color through genetic inheritance or religious beliefs through parental instruction is irrelevant in the sense that both traits have been transmitted. In short, we don't have to concern ourselves overmuch with the fact that the behaviors at the base of cultural complexity are not transmitted genetically.

Cavalli-Sforza and Feldman, for example, have tried to derive mathematical models that describe the propagation of whole sets of cultural behavior (such as agriculture) through space and time, based on their putative selective costs and benefits.[55] Ammerman and Cavalli-Sforza have produced an evolutionary model of the spread of agriculture across prehistoric Europe using evolutionary ideas.

One issue in evolutionary models of cultural complexity involves the *scale* of selection. In the biological universe, the transmission of traits takes place at the level of the individual. The individual genes of the plant or animal do or do not get perpetuated, not the species as a whole. Therefore, the most productive point at which to analyze a given evolutionary problem is the transmission of traits from individual to individual. But in cultural situations, many individual traits, specifically behavioral ones, are the products of instruction by the complete community of parents, teachers, and friends. And people act in corporate groups in ways that make these the functional units of the society. For example, in the next chapter we will see that religions appeared early in all great civilizations and that these formed effective ways to get people to act in concert for the corporate good, such as in fighting wars, clearing irrigation canals, and building pyramids. To summarize, in the production and transmission of cultural characteristics, cultural selection can act on groups as well as individuals.

Figure 7.6, a speculative reconstruction of change in selected variables in Egypt during its period of state-formation, is an attempt to illustrate

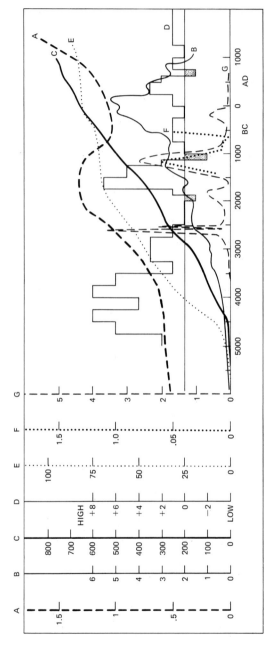

7.6 A major theoretical problem in modern archaeology is how to study the causes and effects of cultural changes on the basis of archaeological data. In this chart, a sample of variables from ancient Egypt has been plotted on comparative scales. One might look for intervals where many important variables changed significantly (such as between 3000 and 2500 B.C. here) to try to explain the origins of complex societies in Egypt. But the archaeological data are meager, correlation does not prove causal connection, and the shape of these plots will depend on how the data are statistically transformed. The variables here are (A) coefficient of rank-size distribution of settlement size, (B) population in millions, (C) population density per square kilometer, (D) lake volumes and stream discharge in east Africa, (E) percentage of domesticated animals in total faunal assemblage (estimated from reports on excavated materials), (F) price of farmland per unit (in silver), (G) monumental architecture in cubic meters of worked stone. Many other variables could be plotted (e.g., average transport distance of craft items).

some of these questions. This graph shows that had we enough data, we could try to find a point in time where intersecting patterns of rapid change occurred, which would suggest that in seeking the explanation of this cultural history we might concentrate our attention on that period. It also illustrates that not all variables—once adjusted for increasing population size—are significant in cultural evolution.

Many ambiguities and complexities are associated with applying evolutionary ideas to culture. To take a simple example, we assume that much of the human physiology is the product of long-term selection. Ageing, for example, seems to make evolutionary sense, in that the lifespan of a human being is probably the result of complex selection for organisms that can live long enough to reach reproductive age in a condition where they will make effective parents—that is, they know how to feed and secure their offspring; selection would not favor infinite lifespans, of course, because change is essential: what works is what is adaptable, and each new generation offers new chances at evolution. Are social forms in the same way selected for lifespan? Could the collapse of the Mayan societies be the result of selection producing social forms that live long enough to be minimally effective but that do not last forever to the exclusion or retardation of novelty, of mutation and change?

Even if we had the data to establish very accurately this sort of graphical cultural history, however, we would not have explained this history. These graphs serve only to draw our interest to the time, place, and variables that may be most important in the developmental pattern we are trying to understand.

Summary

In these analyses of cultural and social complexity, archaeologists must come to terms with a fundamental problem: most of the origins and development of ancient societies cannot be fully explained on the basis of the archaeologically retrievable facts of climate, technology, economy, and demography. Cultural evolution instead must be analyzed at some level above these basic conditions—at the higher level of the social, economic, and political relationships of peoples and social entities. But how do we get at these higher-level interactions through the data and methods of archaeology?

The approaches we will review in subsequent chapters are not solutions to this problem,[56] but they may offer various possibilities.

It is also important to note that the cases of cultural evolution detailed in the following chapters repeatedly show that cultural evolution is not a continuous, cumulative, gradual change, in most places. "Fits and starts" better describes it. In Mesopotamia, the Indus Valley, and elsewhere, there is clear evidence of communities that seem on the verge of cultural complexity, either to die out without further development or to be overtaken by developments in some core area to which they have become the periphery.[57]

Notes

1. Shennan, "Trends in the Study of Later European Prehistory," pp. 373–76; Wallerstein, *The Modern World System.*
2. Sahlins, "Notes on the Original Affluent Society."
3. Lee, "!Kung Bushmen Subsistence."
4. Roux, *Ancient Iraq,* p. 161.
5. See, e.g., Steward, "The Economic and Social Basis of Primitive Bands"; Service, *Primitive Social Organization;* and Fried, *The Evolution of Political Society.*
6. Trigger, "Archaeology at the Crossroads: What's New?" p. 287.
7. Wenke, "Explaining the Evolution of Cultural Complexity: A Review."
8. Steward, "The Economic and Social Basis of Primitive Bands."
9. Barnard, "Contemporary Hunter-Gatherers: Current Theoretical Issues in Ecology and Social Organization," pp. 195–7; Ember, "Residential Variation among Hunter-Gatherers."
10. Summarized in Barnard, "Contemporary Hunter-Gatherers: Current Theoretical Issues in Ecology and Social Organization," pp. 204–10. See also Testart, "The Significance of Food Storage among Hunter-gatherers: Residence Patterns, Population Densities, and Social Inequalities," p. 530.
11. Flannery, "The Cultural Evolution of Civilizations."
12. Service, *Origins of the State and Civilization.*
13. Earle, "Chiefdoms in Archaeological and Ethnohistorical Perspective," p. 288.
14. Ibid.
15. Drennan and Uribe, *Chiefdoms in the Americas.*
16. Wright and Johnson, "Population, Exchange, and Early State Formation in Southwestern Iran."
17. Eisenstadt, *The Political System of Empires.*
18. Brumfiel and Earle, "Specialization, Exchange, and Complex Societies: An Introduction," p. 5.
19. White, *The Science of Culture.*
20. Binford, "Mortuary Practices. Their Study and Their Potential"; Rothschild, "Mortuary Behavior and Social Organization at Indian Knoll and Dickson Mounds."
21. See also O'Shea, *Mortuary Variability—An Archaeological Investigation;* Peebles, "Moundville and Surrounding Sites"; Bartel, "A Historical Review of Ethnological and Archaeological Analyses of Mortuary Practice."
22. O'Shea, *Mortuary Variability—An Archaeological Investigation.*
23. Berry, *Geography of Market Centers and Retail Distribution.*
24. Johnson, "Aspects of Regional Analysis in Archaeology"; Johnson, "Monitoring Complex System Integration and Boundary Phenomena with Settlement Size Data"; Evans and Gould, "Settlement Models in Archaeology"; Crumley and Marquardt, *Regional Dynamics;* Madsen, "Settlement Systems of Early Agricultural Societies in East Jutand, Denmark."
25. Darwin, *The Descent of Man and Selection in Relation to Sex,* p. 154.
26. See, e.g., Steward, "Cultural Causality and Law: A Trial Formulation of the Development of Early Civilizations."
27. Athens, "Theory Building and the Study of Evolutionary Process in Complex Societies," p. 375.
28. Boserup, *The Conditions of Agricultural Growth.*

29. Athens, "Theory Building and the Study of Evolutionary Process in Complex Societies," p. 366.
30. Cowgill, "On the Causes and Consequences of Ancient and Modern Population Changes," p. 505.
31. Wittfogel, *Oriental Despotism.*
32. Woodbury, "A Reappraisal of Hohokam Irrigation."
33. Carniero, "A Theory of the Origin of the State," p. 734.
34. Webster, "Warfare and the Evolution of the State: A Reconsideration."
35. Schacht, "Circumscription Theory."
36. Kirch, "Circumscription Theory and Sociopolitical Evolution in Polynesia."
37. Quoted in Harris, *The Rise of Anthropological Theory*, p. 217.
38. Marx (1932:10–11; original 1859), *Capital.*
39. Quoted in Cohen, *Karl Marx's Theory of History. A Defense*, pp. viii–ix.
40. See, e.g., G. A. Cohen's excellent *Karl Marx's Theory of History: A Defence.*
41. Diakanov, *Ancient Mesopotamia.*
42. Zagarell, "Trade, Women, Class, and Society in Ancient Western Asia."
43. For example, Kipp and Schortman, "The Political Impact of Trade in Chiefdoms."
44. Friedman and Rowlands, *The Evolution of Social Systems;* also see Patterson and Gailey, *Power Relations and State Formation.*
45. Harris, *Cannibals and Kings;* idem., *Cultural Materialism.*
46. Harris, *Cultural Materialism*, p. 92.
47. Ibid., p. 94.
48. Webster, "Warfare and the Evolution of the State: A Reconsideration"; Webb, "The Flag Follows Trade."
49. See Carniero, "A Theory of the Origin of the State."
50. Harris, *Cultural Materialism*, p. 105.
51. Brumfiel and Earle, "Specialization, Exchange, and Complex Societies: An Introduction."
52. Ibid., p. 3.
53. See, e.g., Adams, "Natural Selection, Energetics, and Cultural Materialism"; Alexander, "The Search for a General Theory of Behavior"; Alexander, "Evolution and Culture"; Cohen, "Evolutionary Epistemology and Human Values"; Dunnell, "Style and Function: A Fundamental Dichotomy"; Dunnell and Wenke, "An Evolutionary Model of the Development of Complex Societies"; Kirch, "Circumscription Theory and Sociopolitical Evolution in Polynesia"; Boyd and Richerson, *Culture and the Evolutionary Process.*
54. Dunnell, "Evolutionary Theory and Archaeology."
55. Cavalli-Sforza and Feldman, *Cultural Transmission and Evolution. A Quantitative Approach.*
56. Lumsden and Wilson, *Genes, Mind, and Culture;* Johnson, "Rank-size Convexity and System Integration: A View from Archaeology"; Johnson, "Organizational Structure Scalar Stress."
57. Miller, "Ideology and the Harappan Civilization," p. 38.

Bibliography

Adams, R. N. 1981. "Natural Selection, Energetics, and Cultural Materialism." *Current Anthropology* 22:603–24.
———. 1988. *The Eighth Day.* Austin: University of Texas Press.
Alexander, R. D. 1975. "The Search for a General Theory of Behavior." *Behavioral Science* 20:77–100.
———. 1979. "Evolution and Culture." In *Evolutionary Biology and Human Social Behavior. An Anthropological Perspective*, eds. N. Chagnon and W. Irons. North Scituate, Mass.: Duxbury Press. Pp. 59–78.

Athens, J. S. 1977. "Theory Building and the Study of Evolutionary Process in Complex Societies." In *For Theory Building in Archaeology.* New York: Academic Press.

Barnard, A. 1983. "Contemporary Hunter-Gatherers: Current Theoretical Issues in Ecology and Social Organization." *Annual Review of Anthropology* 12:193–214.

Bartel, B. 1982. "A Historical Review of Ethnological and Archeological Analyses of Mortuary Practice." *Journal of Anthropological Archaeology* 1(1):32–58.

Berlinski, D. 1976. *On Systems Analysis.* Cambridge, Mass.: MIT Press.

Berry, B. 1967. *Geography of Market Centers and Retail Distribution.* Englewood Cliffs, N.J.: Prentice-Hall.

Binford, L. R. 1971. "Mortuary Practices. Their Study and Their Potential." In *Approaches to the Social Dimensions of Mortuary Practices,* ed. J. A. Brown. Memoirs of the Society for American Archaeology 25:6–29.

Boserup, E. 1965. *The Conditions of Agricultural Growth.* Chicago: Aldine.

Boyd, R. and P. J. Richerson. *Culture and the Evolutionary Process.* Chicago: U. of Chicago Press.

Brumfiel, E. M. and T. K. Earle, eds. 1987. *Specialization, Exchange, and Complex Societies.* Cambridge, England: Cambridge University Press.

———. 1987. "Specialization, Exchange, and Complex Societies: An Introduction." In *Specialization, Exchange, and Complex Societies,* eds. E. M. Brumfiel and T. K. Earle. Cambridge, England: Cambridge University Press.

Buck, R. C. 1956. "On the Logic of General Behavior Systems Theory." In *The Foundations of Science and the Concept of Psychology and Psychoanalysis,* eds. H. Feigl and M. Scriven. Minnesota Studies in the Philosophy of Science 1:223–28. Minneapolis: University of Minnesota Press.

Carniero, R. 1970. "A Theory of the Origin of the State." *Science* 169:733–38.

Cavalli-Sforza, L. L. and M. W. Feldman. 1981. *Cultural Transmission and Evolution. A Quantitative Approach.* Princeton: Princeton University Press.

Cohen, G. A. 1978. *Karl Marx's Theory of History. A Defense.* Princeton: Princeton University Press.

Cohen, R. 1981. "Evolutionary Epistemology and Human Values." *Current Anthropology* 22:201–18.

Cowgill, G. 1975. "On the Causes and Consequences of Ancient and Modern Population Changes." *American Anthropologist* 77:505–25.

Crumley, C. L. 1979. "Three Locational Models. An Epistemological Assessment for Anthropology and Archaeology." In *Advances in Archaeological Method and Theory,* Vol. 2, ed. M. B. Shiffer. New York: Academic Press.

Crumley, C. L. and W. H. Marquardt, eds. 1987. *Regional Dynamics. Burgundian Landscapes in Historical Perspective.* Orlando, Fla.: Academic Press.

Darwin, C. 1871. *The Descent of Man and Selection in Relation to Sex.* New York: Appleton.

Diakonov, I., ed. 1969. *Ancient Mesopotamia.* Moscow: Nauka.

Drennan, R. D. and C. A. Uribe, eds. 1987. *Chiefdoms in the Americas.* Lanham, Md.: University Press of America.

Dunnell, R. C. 1978. "Style and Function: A Fundamental Dichotomy." *American Antiquity* 43:192–202.

———. 1980. "Evolutionary Theory and Archaeology." In *Advances in Archaeological Method and Theory,* Vol. 3, ed. M. B. Schiffer. New York: Academic Press.

Dunnell, R. C. and R. J. Wenke. 1980. "An Evolutionary Model of the Development of Complex Societies." Paper presented at the Annual Meeting of the American Association for the Advancement of Science, San Francisco.

Earle, T. K. 1987. "Chiefdoms in Archaeological and Ethnohistorical Perspective." *Annual Review of Anthropology* 16:279–308.

Eisenstadt, S. N. 1963. *The Political System of Empires.* New York: Free Press of Glencoe.

Ember, C. R. 1975. "Residential Variation among Hunter-Gatherers." *Behavioral Science Research* 10:199–227.

Evans, S. and P. Gould, 1982. 'Settlement Models in Archaeology." *Journal of Anthropological Archaeology* 1:275–304.

Flannery, K. V. 1972. "The Cultural Evolution of Civilizations." *Annual Review of Ecology and Systematics* 3:399–426.

———. 1973. "Archeology with a Capital S." In *Research and Theory in Current Archeology*, ed. C. L. Redman. New York: Wiley.

Fried, M. H. 1960. "On the Evolution of Social Stratification and the State." In *Culture in History*, ed. S. Diamond. New York: Columbia University Press.

———. 1967. *The Evolution of Political Society*. New York: Random House.

———. 1975. Public lecture at the University of California at Santa Barbara, Spring 1975.

Friedman, J. and M. J. Rowlands. 1977. *The Evolution of Social Systems*. Pittsburgh: University of Pittsburgh Press.

Gellner, E., ed. 1980. *Soviet and Western Anthropology*. London: Duckworth.

Harris, M. 1968. *The Rise of Anthropological Theory*. New York: Crowell.

———. 1977. *Cannibals and Kings*. New York: Random House.

———. 1979. *Cultural Materialism: The Struggle for a Science of Culture*. New York: Vintage Press.

Hill, J. N. 1977. *Explanation of Prehistoric Change*. Albuquerque: University of New Mexico Press.

Johnson, G. A. 1977. "Aspects of Regional Analysis in Archaeology." *Annual Review of Anthropology* 6:479–508.

———. 1980. "Rank-size Convexity and System Integration: A View from Archaeology." *Economic Geography* 56:234–47.

———. 1982. "Organizational Structure Scalar Stress." In *Theory and Explanation in Archaeology*. New York: Academic Press.

———. 1981. "Monitoring Complex System Integration and Boundary Phenomena with Settlement Size Data." In *Archaeological Approaches to Complexity*, ed. S. E. van der Leeuw. Amsterdam: University of Amsterdam.

Kipp, R. S. and E. M. Schortman. 1989. "The Political Impact of Trade in Chiefdoms." *American Anthropologist* 91(2):370–85.

Kirch, P. V. 1988. "Circumscription Theory and Sociopolitical Evolution in Polynesia." *American Behavioral Scientist* 31(4):416–27.

LeBlanc, S. 1973. "Two Points of Logic Concerning Data, Hypotheses, General Laws, and Systems." In *Research and Theory in Current Archeology*, ed. C. L. Redman. New York: Wiley.

Lee, R. E. 1969. "!Kung Bushmen Subsistence: An Input-Output Analysis." In *Environment and Cultural Behavior*, ed. A. P. Vayda. Garden City, N.Y.: Natural History Press.

Legros, D. 1977. "Chance, Necessity and Mode of Production: A Marxist Critique of Cultural Evolutionism." *American Anthropologist* 79:26–41.

Lewarch, D. E. 1977. "Locational Models and the Archaeological Study of Complex Societies: A Dilemma in Data Requirements and Research Design." Paper presented at the 76th Annual Meeting of the American Anthropological Association, Houston, Texas.

Lumsden, C. J. and E. O. Wilson. 1981. *Genes, Mind, and Culture*. Cambridge, Mass.: Harvard University Press.

Madsen, T. 1982. "Settlement Systems of Early Agricultural Societies in East Jutand, Denmark: A Regional Study of Change." *Journal of Anthropological Archaeology* 1:197–236.

Marx, K. 1932. (orig. 1859). *Capital and Other Writings*. New York: The Modern Library.

———. 1973. *Grundrisse. Foundations of the Critique of Political Economy*. New York: Vintage Press. Original manuscript 1857–1858.

Marx, K. and F. Engels. 1970. *Selected Works*, in 3 vols. Moscow: Progress Publishers.

May, D. A. and D. M. Heer. 1968. "Son Survivorship, Motivation and Family Size in India: A Computer Simultation." *Population Studies* 22:199–210.

Miller, D. 1985. "Ideology and the Harappan Civilization." *Journal of Anthropological Archaeology* 4(1):34–71.

O'Shea, J. M. 1984. *Mortuary Variability—An Archaeological Investigation.* New York: Academic Press.

Patterson, T. C. and C. W. Gailey, eds. 1987. *Power Relations and State Formation.* Washington, D.C.: American Anthropological Association.

Peebles, C. S 1971. "Moundville and Surrounding Sites: Some Structural Considerations of Mortuary Practices, II." In *Approaches to the Social Dimensions of Mortuary Practices,* ed. J. A. Brown. Society for American Archaeology Memoir no. 25.

Redman, C. L. 1978. *The Rise of Civilization.* San Francisco: Freeman.

Renfrew, C. 1972. *The Emergence of Civilization.* London: Methuen.

Rothschild, N. A. 1979. "Mortuary Behavior and Social Organization at Indian Knoll and Dickson Mounds." *American Antiquity* 44(4):658–75.

Roux, G. 1964. *Ancient Iraq.* Baltimore: Penguin.

Sahlins, M. 1968. "Notes on the Original Affluent Society." In *Man the Hunter,* eds. R. Lee and I. DeVore. Chicago: Aldine.

Salmon, M. H. and W. C. Salmon. 1979. "Alternative Models of Scientific Explanation." *American Anthropologist* 81:61–74.

Sanders, W. T. and B. J. Price. 1968. *Mesoamerica.* New York: Random House.

Schacht, R. M. 1988. "Circumscription Theory." *American Behavioral Scientist* 31(4):438–48.

Service, E. 1962. *Primitive Social Organization.* New York: Random House.

———. 1975. *Origins of the State and Civilization.* New York: Norton.

Shennan, S. J. 1987. "Trends in the Study of Later European Prehistory." *Annual Review of Anthropology* 16:365–82.

Smith, M. E. 1977. "State Systems of Settlement: Response to Crumley." *American Anthropologist* 79:903–6.

Spooner, B., ed. 1972. *Population Growth: Anthropological Implications.* Cambridge, Mass.: MIT Press.

Steward, J. 1949. "Cultural Causality and Law: A Trial Formulation of the Development of Early Civilizations." American Anthropologist 51:1–27.

———. 1936. "The Economic and Social Basis of Primitive Bands." In *Essays in Anthropology in Honor of Alfred Louis Kroeber.* Berkeley: University of California Press.

Streuve, V. V. 1969. "The Problem of the Genesis, Development and Disintegration of the Slave Societies in the Ancient Orient." Trans. I. Levit. In *Ancient Mesopotamia,* ed. I. M. Diakonov. Moscow: Nauka.

Tainter, J. A. 1987. *The Collapse of Complex Societies.* Cambridge, England: Cambridge University Press.

Testart, A. 1982. "The Significance of Food Storage among Hunter-gatherers: Residence Patterns, Population Densities, and Social Inequalities." *Current Anthropology* 23(5):523–37.

Trigger, B. 1972. "Determinants of Urban Growth in Pre-Industrial Societics." In *Man, Settlement and Urbanism,* eds. P. J. Ucko, R. Tringham, and G. W. Dimbleby. London: Duckworth.

———. 1984. "Archaeology at the Crossroads: What's New?" *Annual Review of Anthropology* 13:275–300.

Wallerstein, I. 1974. *The Modern World System.* New York: Academic Press.

Webb, M. 1975. "The Flag Follows Trade: An Essay on the Necessary Integration of Military and Commercial Factors in State Formation." In *Ancient Civilization and Trade,* eds. J. Sabloff and C. C. Lamberg-Karlovsky. Albuquerque: University of New Mexico Press.

Webster, D. 1975. "Warfare and the Evolution of the State: A Reconsideration." *American Antiquity* 40:471–75.

Weiss, R. M. 1976. "Demographic Theory and Anthropological Inference." *Annual Review of Anthropology* 5:351–81.

Wenke, R. J. 1981. "Explaining the Evolution of Cultural Complexity: A Review." In *Advances in Archaeological Method and Theory*, Vol. 4, ed. M. B. Schiffer. New York: Academic Press.

White. L. 1949. *The Science of Culture*. New York: Grove Press.

Wittfogel, K. A. 1957. *Oriental Despotism: A Comparative Study of Total Power*. New Haven: Yale University Press.

Wolf, E. R. 1966. *Peasants*. Englewood Cliffs, N.J.: Prentice-Hall.

Woodburn, J. 1982. "Egalitarian Societies." *Man* (NS) 17:431–51.

Woodbury, R. B. 1961. "A Reappraisal of Hohokam Irrigation." *American Anthropologist* 63(3):550–60.

Wright, H. 1977. "Recent Research on the Origin of the State." *Annual Review of Anthropology* 6:379–97.

Wright, H. and G. A. Johnson. 1975. "Population, Exchange, and Early State Formation in Southwestern Iran." *American Anthropologist* 77:267–89.

Yellen, J. E. 1976. "Settlement Patterns of the !Kung: An Archaeological Perspective." In *Kalahari Hunter-Gatherers: Studies of the !Kung San and Their Neighbors*, eds. R. Lee and I. DeVore. Cambridge, Mass.: Harvard University Press.

Zagarell, A. 1986. "Trade, Women, Class, and Society in Ancient Western Asia." *Current Anthropology* 27(5):415–30.

8

The Origins of Complex Societies in Southwest Asia

And Babylon, the glory of kingdoms, the
beauty of the Chaldees' excellency, shall be as
when God overthrew Sodom and Gomorrah.

It shall never be inhabited, neither shall it be
dwelt in from generation to generation: neither shall the Arabian pitch tent there; neither
shall the shepherds make their fold there.

But wild beasts of the desert shall lie there;
and their houses shall be full of doleful creatures; and owls shall dwell there, and satyrs shall
dance there.

And the wild beasts of the islands cry in their
desolate houses, and dragons in their pleasant
places: and her time is near to come, and her
days shall not be prolonged.

Isaiah 13:19–22

Isaiah was right. Today the city of Babylon—once the brightest star in a
galaxy of brilliant ancient cities—is a great rubbish heap, picked over by
such "doleful creatures" as archaeologists and the infrequent tourist.

But five thousand years ago, when most of the world's people were dirt-
poor illiterate farmers or hunters and gatherers, and when the peoples of
the New World were still thousands of years from village life, Babylon
and its surroundings were a cosmopolitan world of cities, libraries, schools,
shops, international trade, roads, taxes, temples, and many of the other
elements we identify with "civilization." Indeed, many of us today are still
living in the basic *urban* way of life that first evolved in these ancient
Southwest Asian cities.

Southwest Asian culture history is so rich, so ancient, that it almost defies general interpretation. Before the Iran-Iraq war it was possible to
drive from the eastern outskirts of Babylonia, near Baghdad, Iraq, and
the heartland of the oldest cities in the world, then along the road north

and east to Susa, one of the world's first true metropolises, where the Biblical Daniel and Esther once lived, then north through some of the first farms in the world, and still further north, a route, Xenephon tells us, Greek soldiers once took, as homesick for Greece and the sea they fought their way to the Black Sea to ships to take them home.

But amid this romantic historical pageant, scholars have long sought some general sense of why and how these societies developed as they did, why this part of the world was the first to produce complex cultures, and why the basic pattern of cultural development in ancient Southwest Asia was repeated in most of its essentials in Egypt, the Indus Valley, China, Mesoamerica, Peru, and perhaps elsewhere.

The Ecological Setting

Complex societies first developed in a relatively small area of southern Iraq and Iran, but the factors that produced them involved most of the Fertile Crescent and the surrounding lowlands it encompasses, or "Greater Mesopotamia," as it is known (Figure 8.2).

The lower elevations of the great arc of mountians of the Fertile Crescent were the natural habitats of wild wheat, barley, sheep, and goats. In ancient times this area was covered with vast grasslands and oak and pistachio forests, and, even as late as the nineteenth century, sheep grazing these verdant uplands were brought to lowland markets with their wool stained scarlet by wildflowers in their range. Today the forests are almost entirely gone, and the whole area is severely overgrazed.

From the very beginnings of life on the alluvial lowlands, people there fought and traded for the gold, silver, copper, stone, wheat, sheep, and goats of the mountain areas. Today there is still no better way to integrate economically the highlands and lowlands than the ancient pattern in which nomads and herders take sheep and goats to these highlands in the lush summer months, using the animals to convert mountian vegetation into milk, meat, and hides, which can then be exchanged for lowland products.

Below the high mountains, the foothills and rolling plains of the piedmont were the location of some of the earliest agricultural villages. Ancient Neolithic farms are found in almost every place where rainfall is sufficient to grow wheat and barley with a reasonable reliability.

Coming down from the piedmont, one finds north and northwest of Baghdad the dry, undulating northern Mesopotamian Plains. The Tigris and Euphrates have cut deeply into the land surface, so that irrigation here requires sophisticated damming and canalization. But with irrigation, the high productivity of this area made the northern plains the "breadbasket" of some of the greatest empires of antiquity.

	MESOPOTAMIA AND IRAN	SYRIA–PALESTINE	ANATOLIA	AEGEAN AND CYPRUS
IRON AGE — 0 / 500	PERSIAN EMPIRE (539–330) ACHAEMENID DYNASTY Ecbatana, Susa, Pasargadae, Persepolis *Imperial Rule* Zoroastrianism *War with Macedon and Alexander the Great*	PERSIAN RULE	PERSIAN RULE	
	NEO–ASSYRIAN EMPIRE (900–626) SARGONID DYNASTY Assur, Dur-Shurruken, Khalku, Nineveh NEO–BABYLONIAN STATE (626–539) CHALDEAN DYNASTY Babylon	IRON AGE (1200–539) *Mediterranean trade* PHOENICIANS: Sidon Syre Byblos ——— HEBREWS: Jerusalem United Monarchy Kingdoms of Israel and Judah Old Testament Babylonian captivity	IRON AGE Neo–HITTITE STATES Carchemish URARTU Tushpa (Van)	
MIDDLE BRONZE AGE — 1000 / MILLENNIUM BC			SEA PEOPLES ——————————————→	
	KASSITE DYNASTY (1500–1150) *Trade/diplomacy* (*Amarna Letters*) Dur Kurigalzu HURRIANS (Mitanni) Northwest Mesopotamia Jezirah MIDDLE ASSYRIAN STATE	LATE BRONZE AGE (1500–1200) INTERNATIONAL ERA (*Amarna Letters*) *Hapiru–Hebrew Mediterranean trade ports* Berytus, Byblos, Ugarit, Tyre *Alphabetic origins Peoples of the sea*	LATE BRONZE AGE (1500–1200) HITTITE NEW KINGDOM *Agriculture and metallurgy Expansion into Syria* Hattusha *Western Anatolia* AHHIJAWA (ACHAEANS?) Arzawa	LATE MINOAN NEW PALACE PERIOD *Minoan settlements abroad* Knossos, Phalstos, Mallia, Zakro, Khania ——— LATE HELLADIC *Mycenaean palaces–commerce* Mycenae, Pylos, Tiryns ——— LATE CYPRIOT II–III *Commercial centers–trade* Enkomi, Kition, Hala Sultan Tekke, Aylos Dhimitrios
— 1500	EARLY OLD BABYLONIAN Isin, Larsa OLD ASSYRIAN *Anatolian trading colonies* Assur, Kanesh ——— OLD BABYLONIAN Shamshi–Adad and Hammurabi Assur, Shubat–Enill, Babylon, Marl	MIDDLE BRONZE AGE *Strong Fortified Cities* (Palestine) Aphek Hazor Megiddo *Palaces and Emporia* (Syria) Alalakh, Ebia, Mari, Yamkhad, Qatna HURRIANS Jezirah	MIDDLE BRONZE AGE OLD ASSYRIAN COLONIES *Karum* Kanesh HITTITE OLD KINGDOM INDO–EUROPEANS Hattusha	CRETE: MIDDLE MINOAN *First palaces–commerce* Knossos, Mallia, Phalstos ——— CYPRUS: MID–LATE CYPRIOT *Urbanism–metallurgy* Enkomi ——— GREECE: MIDDLE HELLADIC *Shaft graves* Mycenae
— 2000	(2350–2200) AKKADIAN NATION STATE *Secular power and private land ownership* Agade (Akkad) Ur, Kish ——— (2112–2004) UR III DYNASTY (Neo–Sumerian) *State ownership of land* Ur, Umma, Sellush–Dagan	EARLY BRONZE AGE *Urbanism, Literacy, Commerce* (Syria) Ebia Mari	EARLY BRONZE AGE Troy	EARLY BRONZE AGE (Aegean) *Development toward palatial centers* Knossos ——— EARLY BRONZE AGE (Cyprus) Philla Sotira Kaminoudhia
— 2500				

8.1   A cultural chronology of Southwest Asia and the Mediterranean world.

For the very first cities, however, the heartland is the lower alluvial plains of Mesopotamia. Historian Arnold Toynbee argued that civilizations first evolved in physical circumstances that challenged the inhabitants to overcome severe problems of climate and resources, and he was thinking in part about southern Mesopotamia. Nothing in this region appears to account for its pivotal role in human history. The unprepossessing hot plains have no usable stone or metal, few trees, and a climate whose extremes of

MILLENNIUM BC	MESOPOTAMIA AND IRAN	SYRIA–PALESTINE	ANATOLIA	AEGEAN AND CYPRUS
2500	EARLY BRONZE AGE *Use of bronze tools/weapons* SUMERIAN CITY STATES *Temple Elites Palatial Power Widespread use of writing* Eridu, Kish, Lagash, Nippur, Ur, Uruk	EARLY BRONZE AGE *Urban Centers (Palestine)* Al Arad Beth–Shan Megiddo	EARLY BRONZE AGE Troy	EARLY BRONZE AGE *(Aegean) Urban Centers* Lerna Tiryns Crete Cyclades ——— CHALCOLITHIC *(Cyprus)* Erimi
3000	*Urban beginnings Ceremonial architecture Long–distance trade* Eridu, Uruk, Khafaje, Gawra, Brak, Nineveh *Writing* Uruk, Jemdet Nasr, SUMERIANS	CHALCOLITHIC *Walled urban center Long–distance trade* Habuba Kabira	NEOLITHIC Kum Tepe (Troad)	FINAL NEOLITHIC Kea Lemnos Knossos Phalstos (Crete) ——— CERAMIC NEOLITHIC Aylos Epiktitos Vrysi Sotira Teppes
4000	*Agricultural Settlements* (SOUTH MESOPOTAMIA) Eridu Ubaid	NEOLITHIC (PNB) Byblos Ramad Ras Shamra	NEOLITHIC Beycesultan Can Hassan	MIDDLE/LATE NEOLITHIC *Large Aegean islands settled* Euboea, Chios, Rhodes, Samos ——— *Occupation gap (Failed colonization?)* Dhali Agridi
5000	*Rainfed agriculture and pottery* (NORTH MESOPOTAMIA) Hassuna, Halaf *Irrigation farming and pottery* (CENTRAL MESOPOTAMIA) Baghouz, Samarra, Choga Mami, Matarrah, Tell es–Sawwan *Specialized production* Umm Dabaglyah	*Settlement on Mediterranean coast and in northern Levant* NEOLITHIC (PNA) *Desiccation and widespread settlement abandonment* Continuity at: Jericho Munhata	NEOLITHIC *Early village sites* Can Hassan III Hacilar	EARLY NEOLITHIC (AEGEAN) *Early village sites* Nea Nikomedeia Franchthi Cave, Lerna Knossos (Crete) ——— ACERAMIC NEOLITHIC (CYPRUS) Khirokitia Tenta Cape Andreas Kastros
6000	NEOLITHIC *Spread of farming in Zagros* Jarmo Sarab (Iran) Guran (Iran)	NEOLITHIC (PPNB) *Craft specialization* Beidha *Increasing use of pottery* Ramad Tell Abu Hureyra	NEOLITHIC *Early village sites* Çatal Hüyük Hacilar *Obsidian production/trade Wall paintings and molded relief sculpture* Çatal Hüyük	EPIPALEOLITHIC *First island settlement* Crete Cyprus?
7000	NEOLITHIC *Small agricultural settlements* Ali Kosh (Iran) Ganj Dareh (Iran)	NEOLITHIC (PPNA) *Walled community* Jericho *Earliest use of pottery* Mureybet *Small agricultural settlements* Beidha Tell Abu Hureyra Mureybet	NEOLITHIC *Small agricultural settlements Early use of metals* Çayönü	EPIPALEOLITHIC
8000	EPIPALEOLITHIC *Advanced hunter–gatherers* Asiab (Iran) Karim Shahir Zawi Shemi *Sheep herding in Zagros?*	EPIPALEOLITHIC *Earliest settled villages* Ain Mallaha Beidha Jericho	UPPER PALEOLITHIC	UPPER PALEOLITHIC (AEGEAN) *Sea travel and resource exploitation* Franchthi Cave Melos (Cyclades)
9000				

heat and humidity have been a main topic of conversation for generations of archaeologists.

But most early civilizations are the "gift" of some great river system, and this is particularly true of Mesopotamian cultures. The Tigris and Euphrates rivers created the Alluvium with annual deposits of flood-borne fertile silt and clay, and they provide the irrigation water that makes agriculture possible here. The swamps and wetlands formed by the rivers

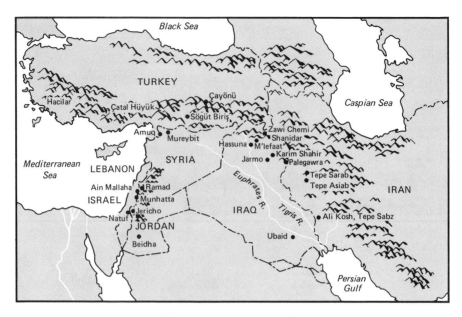

8.2 Southwest Asia. The world's first complex societies evolved in the alluvial plains of the Tigris and Euphrates Rivers and their tributaries, but they had their origins in these early agricultural communities.

support a variety of usable wild plants, such as flax for textiles and rushes for basketry, but in addition to irrigation water the major gift of the rivers is fish. The scrub forests along the rivers cannot support many game animals, and thus fish—and later, domestic cattle—furnished the protein indispensable to survival on the Alluvium. Vegetarianism was not an option for ancient agriculturalists here, because the blend of vegetable proteins necessary to substitute for animal protein was not always available. Fish fill this lack admirably, however, and many clues point to their importance for the early inhabitants: "altars" in buildings dating to the fifth millennium B.C. have been found covered with layer upon layer of fish bones.

From May to October, the average daily high temperature on the Alluvium is over 40° C (104° F), and hot winds dry the soil to over a meter and darken the sky with choking clouds of dust. Anyone who has spent a summer in Mesopotamia immediately understands why several Middle Eastern religions have envisioned heaven as an eternity of sitting by shaded streams in cool palm groves, eating fruit. But the fiery summers are balanced by temperate, lovely, autumns, winters, and springs; and with adequate irrigation, the Mesopotamian Alluvium yields rich harvests of wheat, barley, dates, olives, lentils, oranges, onions, and other crops.

Southwest Asian peoples have lived in villages for nine millennia, and there is hardly a square meter that does not contain a few sherds, stone tools, bones, or old irrigation canal banks. All over the Alluvium one finds the mounds ("tell" in Arabic or "teppeh" in Persian) that result from the construction and continual reconstruction of buildings on the same spot. For thousands of years, people here have used clay bricks as their basic building material, and their settlements have taken the form of closely packed small rectangular structures. Although ideally suited to the climate and resources of the area, such buildings become so dilapidated after fifty or a hundred years that it is easier to rebuild than to repair them, and because there are incentives to rebuild on the same spot (less land is lost to cultivation and higher elevation gives better drainage and protection against floods and attack), settlements become mound-shaped as they are constantly reconstructed on the debris of the previous ones.

Hundreds of thousands of such mounds dot the landscape of Southwest Asia, some of them rising 50 meters or more above plain level. The mounds are littered with stone tools, bones, broken pottery, broken clay bricks, collapsed walls, eroding ovens and pottery kilns, and corroding metal. Burrowing animals, well and terrace construction, and erosion often mix the layers of these mounds so that usually one finds the remnants of every phase of a site's occupation on its surface. This allows archaeologists to estimate a site's periods of occupation simply by inspecting the surface artifacts.

Besides mounds, the most obvious archaeological features of Southwest Asia are irrigation canals. Seven millennia of irrigation agriculture have resulted in a landscape criss-crossed with canals, and it is not unusual to find abandoned irrigation canals several thousand years old with banks still 2 or 3 meters high. Aerial photographs are particularly useful in charting these ancient waterworks.

Because of the time and expense of archaeological excavation and the destruction of sites, far fewer than 5% of all sites dating from 8000 to 2350 B.C., for example, have been, or will ever be, properly excavated, and thus our analyses must be tempered with some tentativeness. On the other hand, the rich archaeological remains provide an opportunity to use sampling procedures and test hypotheses; and by using aerial reconnaissance, surface surveys, and regional analyses, archaeologists can determine when and for how long unexcavated sites were occupied. In the case of Mesopotamia, a generation of archaeologists has followed Robert McC. Adams in performing regional surveys.[1]

In my own surveys in Iranian Mesopotamia, as I visited the various mounds evident on my aerial photographs, I found on top of every mound a small pile of about twenty to thirty pottery sherds.[2] These had been left there ten years previously by Robert McC. Adams, who had used them to assess the periods of occupation of each site. Although my samples were somewhat larger and much more brutalized by computerized statistical analysis, Adams' summary culture history was remarkably in agreement with my own.[3]

If we had had millions of research dollars and years of time, both Adams and I should have made a detailed topographic map of each site, collected artifacts in a stratified random sampling design of hundreds of 2 × 2 meter squares, then test-excavated each site to confirm its occupational periods. In the real world, each of us spent many months just making sketch maps of the sites, picking up informal samples, and inferring the site's cultural history from these evidences. Archaeologists' time and resources are usually so limited that it is on this kind of research that the culture history of Mesopotamia has been reconstructed.

Early Agriculture (c. 7000–6000 B.C.)

When we left the first farmers of Southwest Asia in Chapter 6, they were living in thousands of small villages, from Afghanistan to western Turkey, in almost every place with sufficient rainfall to support wheat and barley without extensive irrigation.

In most areas the simple village farming way of life continued largely

8.3 Millennia of rebuilding at Jericho formed many superimposed layers containing stone walls.

8.4 The people of Neolithic Jericho removed the skulls from some bodies and formed human features on them in plaster.

unchanged into the last few millennia B.C., but in other areas already by 7000 B.C. changes were underway.

At Jericho, in Jordan, even though the populace was still hunting and gathering wild resources to complement their primitive agriculture, soon after 7000 B.C. they had built a stone wall some 1.5 meters thick and more than 3.5 meters high around a complex of stone houses (Figure 8.3). If an essential element in cultural complexity is activity specialization to the point that large populations are socially and economically interdependent, Jericho was not a complex community; but the construction of the wall indicates that its people were beginning to direct their energies in a way quite different from that of most hunters and gatherers.

The disposal of the dead, usually an excellent reflection of changing cultural complexity, suggests that people at Jericho and perhaps elsewhere were not much different socially from their predecessors. Forty headless adult bodies were found buried beneath one room at Jericho, and further excavations revealed a cache of skulls that had been reconstructed with plaster, painted, and then decorated with "eyes" made from seashells (Figure 8.4). Whether these represent ancestor worship, war trophies, or some other ritual will probably never be known.

The importance of the bulldozer as an archaeological tool was demonstrated in the late 1970s, when one cut a roadway through a hill in Amman, Jordan, exposing a huge archaeological site. In 1982 archaeologists began excavating this site, now called 'Ain Ghazal ("spring of the gazelles"), one of the largest known Neolithic sites in the Middle East.[4] The first major period of occupation at 'Ain Ghazal began at about 7200 B.C., and it was probably occupied most or all of the time until about 5000 B.C. Covering approximately 30 acres, 'Ain Ghazal is about three times larger than Jericho, but it is unclear how much of the site was occupied at any one time. It probably had at one point at least several hundred inhabitants, who ate wheat, barley, lentils, sheep, and goats, and lived in mudbrick buildings of various sizes and shapes. For its early age, the community of 'Ain Ghazal has impressive art. Large plaster figures of people were found under house floors. Numerous figures of people and animals have been unearthed, and in one cache two clay figures of cattle had flint blades stuck into their heads, necks, and chests.

The diet of the early inhabitants of 'Ain Ghazal is impressive in its diversity. Goats were probably both hunted and herded, and pigs, gazelles, birds, and many other species were exploited, along with some wheat and barley, peas, lentils, and many wild plants. But toward the end of its occupation, the people of 'Ain Ghazal seem to have fallen on hard times. Late in its history, it seems to have been occupied only seasonally by pastoral peoples.

After 6000 B.C., the developmental focus of Southwest Asia shifted somewhat, as villages in the eastern Levant, in what is now Israel and Jordan were abandoned—perhaps because of climate changes.[5] James Mellaart, in fact, claimed that

> It may be said without undue exaggeration that Anatolia, long regarded as the barbarous fringe to the fertile crescent, has now been established as the most advanced center of neolithic culture in the Near East. The neolithic civilization revealed at Çatal Hüyük shines like a supernova among the rather dim galaxies of contemporary peasant culture.[6]

Çatal Hüyük (pronounced rather like "Chatal Huooyook") may not merit all the superlatives its excavators have showered on it, but it is one of the most interesting sites in Southwest Asia. Located in south central Anatolia (Turkey) and first occupied at about 6250 B.C., Çatal Hüyük was probably inhabited continuously until its abandonment at about 5400 B.C. During some of this time it may have extended over 13 hectares and had a population of about 4000 to 6000—several times larger than any other known site in this period.

A great fire swept through Çatal Hüyük in the middle of its history, nicely preserving the earliest levels, but unfortunately only less than a

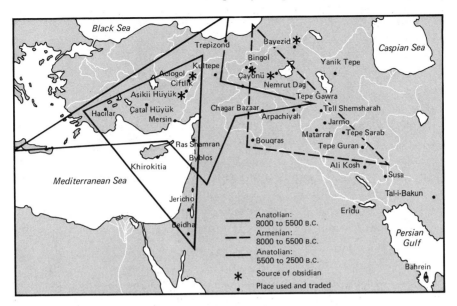

8.5 The circulation of obsidian throughout Southwest Asia was an important economic process that may have contributed to the evolution of political and economic institutions.

hectare of the site was excavated before the project was ended by disputes between Mellaart and the Turkish government and other problems.

Nestled against the first ranges of the Taurus Mountains, Çatal Hüyük controls access to a critical resource: the obsidian sources at the Hasan Dag volcano. Because each obsidian source is chemically distinctive, we know that after 7000 B.C. great quantities of obsidian from the Çatal Hüyük area were distributed throughout Anatolia, the Levant, and Cyprus (Figure 8.5). Beautiful obsidian artifacts were found in Çatal Hüyük itself, but so far no obsidian storage or workshop areas have been discovered in the tiny fraction of the site that has been excavated.

Most of the 158 structures uncovered at Çatal Hüyük are little different from their contemporaries elsewhere in Southwest Asia: each is built of shaped mud and composed of rectangular rooms with plastered walls and floors, and most houses are about 25 meters square, one story high, and abut one another, except where occasionally separated by an open courtyard. Inside most rooms are two raised platforms, probably for sleeping, and an occasional rectangular bench. Unlike any other community of this period and area, access to the rooms of Çatal Hüyük was only by ladder through the roof—there are no front doors—and the close packing of structures is such that a lot of the movement among the houses must have been on the roofs. The roof access may reflect a need for defense, for once the inhabitants had pulled up the ladders on the outside walls, the settlement would have been difficult to attack. Other Southwest Asian sites had walls at this time, but most settlements were small unfortified hamlets.

The forty or so "shrines" at Çatal Hüyük have essentially the same floor plan as the other structures, but their walls are richly decorated with paintings, reliefs, and engravings expressing many naturalistic themes, most concentrating on the two favorites of ancient art—fertility and death. Vultures are portrayed ripping apart headless human corpses, women give birth to bulls and ride leopards, and other symbols such as breasts, vultures, bulls, and rams abound. Some of the rooms have intricate arrangements of cattle skulls and horns (Figure 8.6). Walls were plastered and

8.6   These reconstructions of "cult-centers" at Çatal Hüyük show the importance of the cattle motif. All early complex societies evolved rituals and religions that functioned as organizing and controlling institutions.

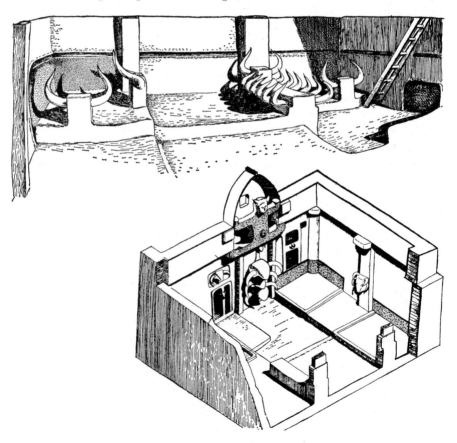

then painted, and periodically the walls were replastered and new designs painted or engraved on them. One wall was plastered and painted forty times. These shrines do not seem to have required the expenditure of vast amounts of labor and resources for essentially noneconomic purposes, as did the monumental construction of later periods. Nor do they enclose radically different amounts of expensive goods, indicating great wealth disparities. Little about them, in fact, conflicts with the interpretation that they were kinship-cult centers in a simple ranked society.

Such a conclusion would be strengthened if we found only minor variability in mortuary practices, and indeed this seems to be the case. Many corpses appear to have been taken outside the settlement and exposed to the vultures and the elements until the flesh was gone, after which the bones were interred in the house floors. Men, women, and children were buried in much the same way, either in baskets or simply in holes. Most of the graves contained no goods, but some women and children were accompanied by shell and stone necklaces, armlets and anklets, and, occasionally, obsidian mirrors and bone cosmetic implements. Some men were buried with mace heads, flint daggers, obsidian projectile points, clay seals, and other items.

In one complex of rooms, the so-called Vulture Shrine, six individuals were buried in the floor with significantly richer grave goods than were found in the residential burials.[7] However, we do not find any infants at Çatal Hüyük buried with disproportionately rich or numerous grave goods, nor is there any significant variation in construction expense of the graves themselves; there are no stone coffins, tombs, or the like.

The people of Çatal Hüyük subsisted on the typical late Neolithic combination of agriculture, hunting, and gathering. Emmer, einkorn, a bread wheat, barley, pea vetch, and other crops were grown in quantity, but with simple techniques. Domestic cattle supplied meat, hides, milk, and perhaps traction. There was considerable trade but mainly in small quantities of exotic items. Shells from the Mediterranean (160 kilometers distant) and Syrian flint were found here, perhaps taken in trade for obsidian artifacts (Figure 8.5). But there is no evidence of voluminous trade in agricultural products, or even in large amounts of obsidian, and there are no obvious workshops for goods, no stores of obsidian, no complex technologies.

Settlements dating to between 7500 and 6000 B.C. have also been found elsewhere in Anatolia, but all are very simple, undifferentiated farming communities with little discernible public architecture, intense occupational specialization, or elaborate mortuary practices.

Jarmo, in Iraqi Kurdestan, is an example of the drab peasant culture that Mellaart invidiously compared to Çatal Hüyük. Jarmo was first settled sometime before 6750 B.C. and was occupied intermittently to 5000

B.C., and thus it overlaps with Çatal Hüyük for perhaps as much as 1000 years. But unlike Çatal Hüyük, Jarmo was probably home to no more than 200 people, and for most of its existence it consisted of only about twenty small mud houses. Burials at Jarmo are quite uniform, as are the contents of the houses. There is not nearly the diversity of aesthetic expression found at Çatal Hüyük nor wall paintings, finely worked obsidian, or the like; there are just a few clay figurines of pregnant women and animals. Jarmo has no fortified walls or large, nonresidential buildings, and the technology seems to have been mainly devoted to the processing of plant foods. Perhaps the most significant difference between the two sites is that Çatal Hüyük controlled a localized and important resource (obsidian), whereas Jarmo did not.

Elsewhere in the Zagros, northern Mesopotamia, and the Iranian Plateau, most communities established between 8000 and 6000 B.C. were also farming villages, with little public architecture, elaborate mortuary cults, or occupational specialization.

Charles Redman has pointed out that:

> Çatal Hüyük exemplifies those communities that may have made many of the crucial advances leading to civilization, but for one reason or another did not become urban societies. In some ways, the town of Çatal Hüyük and its inhabitants should be considered a premature flash of brilliance and complexity that was a thousand years before its time.[8]

The various excavations reviewed here at Jericho, Çatal Hüyük, etc., have provided a picture of the kinds of communities that immediately preceded "civilization" in Southwest Asia, but we still lack many elements of this picture. In a rightly ordered world, the combined international archaeology faculties and their students—funded by the United Nations— would begin surveys in western Anatolia and work their ways east and south, into Afghanistan, mapping and excavating sites along the way. Instead, our evidence for the world's first socially complex societies is mainly from fewer than a hundred well-excavated settlements and perhaps twenty or thirty regional site surveys in this huge area.

Initial Cultural Complexity (c. 6000–4000 B.C.)

An early indication of social change in Southwest Asia is evident in the distribution of pottery styles. As with agriculture, the invention of pottery seems to have occurred independently in many areas of Southwest Asia, where clay had previously been used for figurines and storage pits for centuries. The multiple origins and rapid spread of pottery after about 6500 B.C. no doubt reflect the increasing importance of containers in these agricultural economies—probably for carrying water and for storing, cooking, and serving food.

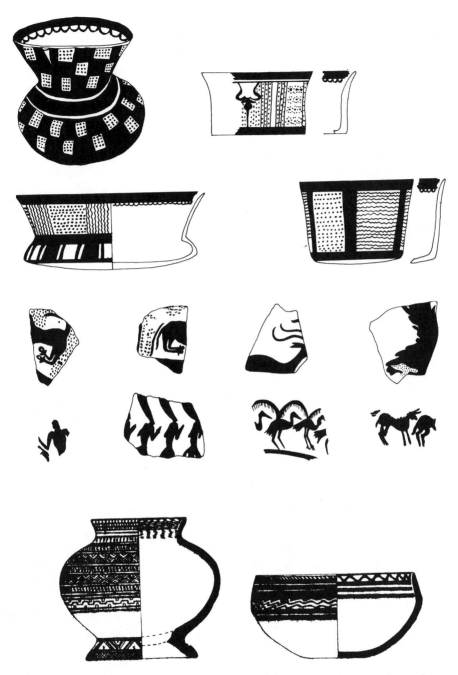

8.7 Halafian and Samarran (bottom two vessels) pottery styles spread over large areas of sixth millennium B.C. Mesopotamia.

Soon after ceramic vessels came into general use in Southwest Asia, the first sophisticated pottery styles appeared, and by about 5500 B.C. two distinctive styles, the Samarran and the Halafian, had achieved wide distribution (Figure 8.7). Samarran pottery tends to be in simpler shapes than the Halafian and is painted in matte colors, whereas the Halafian colors are glossy. Halafian and Samarran ceramics are important beyond their aesthetics: they are the first complex *styles* of artifacts to be so widely distributed. These highly stylized ceramics were first made with a "tournette," wooden slab affixed to a peg that could be set in a hole in the ground and act as a pivot around which the slab could be turned by hand. Simple as this sounds, it was a major advance over simply hand-molding a pot; the tournette allowed the potter to turn the pot so as to shape and paint it.[9] So intricate and delicate are the Halafian and Samarran designs that one might assume that specialists are producing them. But Frank Hole cautions us not to underestimate the skill of villagers—"Persian carpets" being just one example of an extraordinarily complex and beautiful product of simple villagers.[10] And despite the wide geographical distribution of the Halafian and Samarran designs, there is little evidence that

8.8 Distribution of Samarran and Halafian pottery. One of the first steps in the evolution of complex societies seems to be a rapid expansion of an art style over a large area.

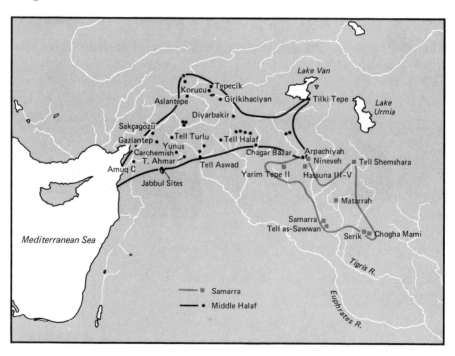

ceramic production was centrally organized and administered. Exactly what the spread of these styles reflects is obscure. Their distributions tend to overlap a great deal and there is no convincing proof that they belong to different time periods or social groups.[11]

The distribution of Halafian and Samarran ceramics coincides with the extension of cultivation to the arid lowlands and the consequent accelerating pace of cultural evolution. A significant site of this period is Tell as-Sawwan, located in north central Iraq at the juncture of the northern arid steppe country and the alluvial plains.[12]

Preliminary excavations indicate that from 6000 to 5000 B.C., Tell as-Sawwan was home to a few hundred people, most of whom were engaged in simple irrigation farming of wheat, barley, and linseed. They also kept domestic goats, sheep, and cattle, hunted onagers, antelopes, and other animals, and gathered fish and mussels from the Tigris, which ran close to the site.

In most ways, this settlement resembled other early agricultural villages; granaries, kilns, ovens, and small mudbrick houses are arranged around open courtyards. In addition, a large ditch or moat was constructed around the site by cutting into the natural conglomerate rock on which the site was located. The discovery of many hardened clay balls ("sling missiles") in the ditch led its excavators to conclude it was a defensive structure, reminiscent of the fortifications at Jericho.

The residential architecture of Tell as-Sawwan contrasts somewhat with that of other settlements of the period, consisting mainly of huts with stone foundations and the more common rectangular clay structures, although there is still little variability in building size or apparent construction cost.

At least 128 burials at Tell as-Sawwan date to approximately 5500 B.C., and they contain the richest assortment of grave goods for this period of any site known in Southwest Asia. Of the classifiable skeletons, fifty-five were infants, sixteen were adolescents, and thirteen adults. The graves differ little in orientation, location, or construction. Most bodies were placed in a contracted position, facing west, in simple shallow oval pits dug into house floors. Although most rooms had many burials (one had up to twenty-three), one room contained only an adult male's burial, whose relatively rich grave goods may reflect emerging status ascription.

Most graves had at least one craft item, mainly carved alabaster, beads in exotic stones, or pottery. But the Tell as-Sawwan burials do not seem to reflect great social differences: burials of adults and juveniles have, on the average, more numerous and varied goods than infants, and no obvious signs indicate inherited wealth, power, and prestige. Yet these graves are very rich, compared to those of contemporary sites, and the disparity among them may indicate emergent ranking and status ascription.

Tell as-Sawwan is just one of many sites of this period, and soon after Halafian, Samarran, and other pottery styles spread through the highlands, the developmental focus shifted to the Mesopotamian Alluvium.

The earliest known sites on the Alluvium are near the estuary of the Persian Gulf, and there is some evidence that with the end of the Pleistocene, the Gulf shoreline moved inland as much as 180 kilometers.[13] Thus, the key settlements in the transition to life on the Alluvium—and the origins of cultural complexity here—may be under seawater. Between about 5500 and 3500 B.C. Mesopotamian climates were comparatively humid, with a change to cooler and drier conditions after about 3500 B.C. Subtle changes in climate like this might have had major effects on the settlement history of the area, because the zones in which dry-field rainfall-based grain cultivation was possible may have shifted. Also, changes in sea levels at this time may have meant that lands that previously were flooded were opened to settlement.[14]

The few sites on the Alluvium that may antedate 5800 B.C. are poorly known—most are probably covered by much larger and later sites.

Subsistence in most parts of the Alluvium is more complex than in the highlands, requiring many timely decisions in which floods must be anticipated and controlled, land irrigated and drained, and cattle pastured, tended, and milked. Fishing adds to this complexity because it is seasonal to some extent and requires coordination to be maximally effective. Even getting sufficient stone for simple agricultural implements necessitates considerable organization in the lowlands, for such stone often had to be obtained in the mountains, far from lowland settlements.

If Ali Kosh in Iranian Mesopotamia is typical, the movement onto the Alluvium required a very generalized diet, in which many small plant seeds eventually were discarded in favor of the wheat-barley-lentil group augmented by sheep-goat-cattle proteins.[15] There is some evidence that the cuisine centered on a variant of the one-pot approach, a stone or clay vessel filled with water, cereals, and other plants, and bits of whatever animals were about. Like the meals of ancient Sparta (described as consisting of two courses, the first a kind of porridge, and the second, which was a kind of porridge), the diets of these Mesopotamian villagers probably relied on blending as many plant and animal nutrients as possible in soups and stews. What breakfast must have been does not bear thinking about.

In the ancient Sumerian account of the creation of the universe, the city of Eridu was the first to have emerged from the primeval sea that covered the world before humans. Eridu is in fact one of the earliest known settlements on the southern Mesopotamian Alluvium, having been established at about 5400 B.C. Few contemporary sites have been found here,

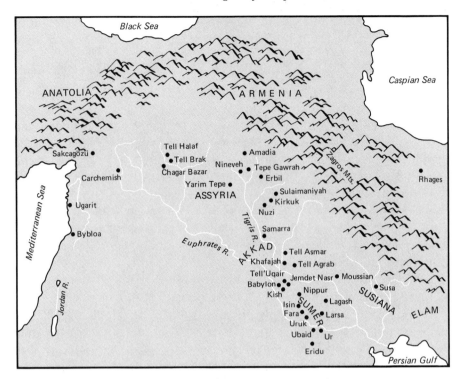

8.9   The first socially complex societies were concentrated in areas of Mesopotamia where rivers allowed great agricultural productivity.

but the lowest levels of many large ancient mounds have not been excavated. At Eridu, the earliest known structure is thought to have been a temple. Archaeologists have been accused of bestowing this term on virtually any structure large enough to stand upright in, but the earliest building at Eridu is very similar to others known through texts and other evidence to have been temples. A single small room (3.5 by 4.5 meters) with an "altar" faces the entrance, and a pedestal is in the center; in these specifics the building is nearly identical to the temples of later periods. The ceramics found here resemble those of Halaf and Samarra, but few other artifacts were discovered. This "temple" may reflect a low level of social complexity, but we can say little on the basis of this single occupation.

But by 4500 B.C. some signs of cultural changes toward complexity appeared in a large region of the lower Alluvium, in the foothills of the Zagros, and in some of the larger valleys in the Zagros. Henry Wright notes that in these areas there were large centers with populations of 1000 to 3000, which dominated networks of smaller settlements. Excavation on some of these larger centers revealed central platforms, supporting ritual buildings, segregated elite residences with large storage structures, and indications of socially segregated cemeteries.[16]

As Wright notes, there is evidence that there had been a *previous* period in which larger centers had emerged, but here too, as in the case described above for Çatal Hüyük, no uniform gradual pattern of increasing complexity exists. The whole history of early complexity, in fact, seems to be a messy "boom-or-bust" cycle, with only a very general long-term overall trend toward complexity.

In addition to the central Mesopotamian plains in Iraq, the Susiana Plain (Figure 8.9) in southwestern Iran was also a major developmental center for early complexity—probably because one could do both rainfall agriculture and irrigation farming here, and because trade routes from the Gulf to the Iranian plateau ran across the Susiana. By about 4500 B.C.—perhaps much earlier—several large centers may have dominated surrounding areas economically and politically. The people of this era made beautiful hand-painted pottery in such large quantities and with such great similarity that, even given the virtuosity of peasant artisans, some occupational specialization may be reflected.[17]

In summarizing this and other evidence from the Susiana, Gregory Johnson observed that

> In sharp contrast to the pervasive smaller villages of the [earlier] periods, Choga Mish on the Susiana Plain covered an area of some 15 hectares by 4300 B.C. Most architecture consisted of residences and associated ceramic kilns. The community also contained at least one monumental building, perhaps more. This 10 × 15 meter structure had walls of between 1 and 2 meters thick and contained several interior rooms. One of these was stacked with storage jars, while another was apparently used in working flint. A substantial mud-brick platform stood nearby. This period in Susiana and elsewhere saw the introduction of formalized closure of containers using clay sealings that carried the impression of a decorated seal. This practice is ordinarily associated with problems of security in materials storage or shipment.[18]

Johnson goes on to note that the high degree of similarity in ceramics across the plain indicates that not everyone was making these—that their production might already have been somewhat centralized. All this evidence, he says,

> strikes me as indicative of a hierarchically organized social system with some degree of influence over labor and resource allocation. The fact that the known major architecture at Choga Mish was destroyed by fire and the community abandoned . . . suggests that elite influence over the population in general was less than elites might have wished.[19]

Frank Hole's interpretation of Susiana culture history differs somewhat from Johnson's.[20] He suspects that even toward the end of the fifth millennium B.C., the Susiana was loosely tied together by small-scale trade and a regional religious cult, centered at Susa, in which priests and people were in a voluntary mutually supportive relationship dedicated to the continued success of agriculture.[21]

To the west of the Susiana, in central Mesopotamia, the 'Ubaid Period (c. 5300–3600 B.C.) saw the gradual settlement of the plain by villagers. By 4350 B.C. the 'Ubaid culture was quite uniform over most of the Alluvium: all the settlements seem to have been located on reliable water courses and almost all were less than ten hectares in size (most of them only one or two).

The spread of the 'Ubaid culture is remarkable for many reasons, particularly its great extent. 'Ubaid-style ceramics are found far into central Turkey, to the southwest in the Arabian Plateau, and in highland Iran—an area much greater than that encompassed within the Halafian and Samarran stylistic zones. The potter's wheel was probably invented by the end of the 'Ubaid Period, when the pivoted working surface, or tournette, was altered by setting the wheel's axle in bearings and weighting the wheel in such a way that high speeds could be produced. This allowed delicate vessels to be made and painted decorations like bands could be easily applied simply by holding a paint brush steady against a rotating ceramic vessel.[22]

Virtually every 'Ubaid settlement had a large nonresidential building, probably a temple, built of mudbrick on platforms of clay or imported stone. Access typically was by a flight of stairs, to a room about ten meters in length, with a broad platform at one end and a table or small "altar" at the other. Smaller rooms were built on both sides of the main room, and ladders in these would sometimes give access to a second story. The exteriors of the buildings were often decorated with projections and recesses, where light and shadows created pleasing effects. In later periods, mosaics of colored ceramic cones and bitumen were used as decorations. At Eridu, seventeen such "temples" were found superimposed, giving the later ones considerable elevation. Such structures are found all over Greater Mesopotamia soon after their appearance on the southern Alluvium.

Until about 5000 B.C., settlements seem to have been located primarily with regard to the availability of resources and the land's agricultural potential, not on the basis of political and economic relationships. People tend to organize their territories in patterns reflecting changing social and economic conditions, and thus, when the movement of people and goods between settlements becomes important, and the area is agricultural and relatively flat, these settlements often are quite regularly spaced, as is evident in both the Warka area and the Susiana Plain.

By 4000 B.C., the number of small settlements had increased dramatically in many areas, and there was increasing variability in their arrangement and composition.

EARLY STATES: THE URUK PERIOD (3600–3100 B.C.)

The Uruk period is regarded as the era of primary state formation in Southwest Asia. Cultural forces and processes probably in operation for thousands of years resulted in the appearance in this interval of the complete checklist of "civilization": cities, warfare, writing, social hierarchies, etc.

The central Mesopotamian Alluvium and the Susiana Plain offer our earliest and best known examples of the changes that occurred in Greater Mesopotamia in the Uruk Period. As many as 10,000 people may have lived in the city of Uruk by 3800 B.C., and around the town were many smaller villages and towns whose sizes and distribution suggest they may not have been tightly integrated into Uruk's political and economic systems.[23] Then, at about 3000 B.C., the city of Uruk apparently grew rapidly to about 50,000 people, who lived behind substantial defensive walls. There is also evidence of widespread simultaneous abandonment of almost all the rural settlements surrounding Uruk—leaving little doubt, Robert McC. Adams suggests, that the growth of the city was a result of the immigration or forcible transference of the population from the hinterlands into the city.[24]

Archaeologists who have excavated Uruk occupations usually have found themselves ankle deep in the remains of millions of bevel-rimmed bowls, surely one of the ugliest ceramic types ever made outside a kindergarten, but also one of the most significant. Gregory Johnson has shown that various measurements of these bowls change over time, so that the Late, Middle, and Early Uruk periods can be defined in part by changes in their form.[25] Also, Hans Nissen has argued that these bowls probably were ration bowls and were used to "pay" workers in grain for their labors.[26] Most have a capacity that corresponds to what a laborer's daily ration was asserted to be in ancient texts, and the symbol for "to eat" in early Mesopotamian writing seems to be that of a person eating from a bevel-rimmed bowl. Others have suggested that these bowls were used to bake bread. These bowls were mass produced, probably by simply molding, and they represent one of the first craft items mass produced on a massive scale in what must have been government-controlled workshops.

The urbanization of this period is difficult to explain. Toward the end of the 'Ubaid period, almost all major settlements were fortified, and documents written much later, at c. 2600 B.C., speak at length of conflict between the people of Ur, Uruk, Umma, and the other city-states. In

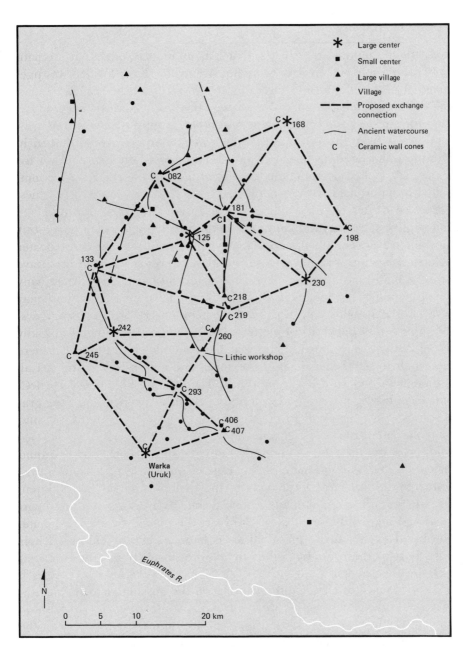

8.10 In simple societies the location of settlements with respect to each other is dictated mainly by ecological factors like water availability. But as societies become more complex, political and economic relationships begin to influence where settlements are located and how large they become. In Gregory Johnson's reconstruction (1975) of exchange networks among late Uruk Period (c. 3200 B.C.) settlements in southern Iraq, the spatial arrangement of the settlements with respect to one another is quite regular (the wall cones were used to decorate administrative centers, and their distribution here may indicate lines of political authority).

339

subsequent chapters, we will see that urbanization in other parts of the world also seems to be related to defensive needs: Egypt, which was protected from hostile outside forces by the deserts and the sea, developed urban societies only comparatively late; while in the Indus Valley, much more of a crossroads for nomadic and other groups, urbanism was present almost from the very first. In this context, Adams argues that early Mesopotamian urbanization may have been imposed on a rural populace by a small, politically conscious superstratum that was motivated principally by military and economic interests.[27]

Because of defensive considerations and the cost of transporting labor and products, agricultural land nearest the urban areas would have been most intensively exploited, and this may have stimulated the construction of large irrigation systems. Another important possible effect of urbanism is that it might have created—and to some extent have been created by—the large nomadic populations thought to have been present in Southwest Asia as early as 6000 B.C. Most of these people were probably similar to the present-day Bakhtiari of western Iran, who herd sheep and goats in the uplands during much of the year but come down to the lowlands in the winter to sell their animals' wool, meat, and milk products and to buy craft items, food, and other products. The relationships between these upland nomadic and lowland sedentary groups are varied and complex. Historically, when central governments have weakened in Southwest Asia, some marginal cultivators, in times of war or poor harvests, revert to nomadic pastoralism, while some pastoralists become laborers or marginal farmers if they lose too many sheep. One of the effects of urbanism suggested by the archaeological record—the depopulation of the countryside—may have given nomads a new niche. By working the mountain areas and highlands for most of the year, the nomads could have come down and exploited the marginal areas between urban centers—peacefully most of the time, but other times with hostility—and then traded with the townspeople.

The sequence of Uruk culture history on the Susiana Plain parallels in many ways developments on the central alluvium. Gregory Johnson and Henry Wright have analyzed Susiana cultural history by focusing on the evolution of administrative institutions.[28] They have been principally concerned with the origins of the "state," which they define as a society with at least three levels of hierarchically arranged, specialized administrators. For example, in a simple agricultural village, many decisions are made about what crops to plant, how much of the harvest is to be stored, who gets what share of the land, who marries whom, and so forth. Many of these decisions are made by individuals, but some of those that directly affect the whole community are made by a village headman. We might say that this individual represents the first level of the decision-making hierarchy—he directs the activities of others who do the work. A second level of administrative hierarchy would exist if there were people who coordinate the activities of these village headmen and correct or approve their decisions—perhaps government agents charged with taxing and administering local affairs. Such agents would be under a third administrative level, and additional levels may exist above this.

Wright and Johnson suggest that the ancient Mesopotamian state can be defined as a society with at least three such levels of specialized administrators and that the effectiveness of such societies and their dominance over other societal forms is tied to their ability to store and process information and make correct decisions at specific points along the control hierarchy.

Wright and Johnson examined some areas where the first Mesopotamian states developed, looking for evidence for this change. One type of evidence was the actual administrative "documents" themselves. In most of Mesopotamia at this time, the administration of people and goods was facilitated by using pieces of inscribed stone to impress clay with signs of authorization. These stones were usually either in the form of "stamps," used like the rubber stamps of today, or cylinders that were rolled across the clay to make an impression, and they varied in size and in the complexity of the symbols incised on them. Fortunately, the clay impressed with these seals and the seals themselves are often preserved in archaeological sites, and they can be used to infer the levels of administration in these extinct societies. The impressed clay can be divided into two classes. *Commodity sealings* were used to certify the contents of a container such as a vessel, basket, bale, or storeroom. They were made by placing a lump of clay over a knot that had to be untied to gain access to the container, so that unauthorized entry could be detected. Discarded commodity sealings indicate the receipt of stored or redistributed goods. Other seals are termed *message sealings* and convey or store facts about goods or people. Some are plain counters whose shape indicates a numerical unit, others (bullae) are small spheroidal jackets that were once wrapped around sets of counters, and still others are flat rectangular "tablets" stamped with numerical symbols. Writing, as we know it, did not exist until after the state, as defined by Wright and Johnson, emerged, but these stamps and seals obviously conveyed a great deal of information.

Wright and Johnson used commodity and message sealings found at sites on the Susiana to reconstruct some aspects of the production, transport, and administration of goods. They concluded that they could determine when the change was made from one- and two-level decision-making institutions to the three levels that define the state.

Wright and Johnson also analyzed the locational arrangement of the settlements in southwestern Iran and found that after about 3600 B.C.

there were trends toward more regular spacing of settlements and the emergence of distinctive site size groupings—both consistent with the change from a two-level to a three-level control hierarchy. This pattern of developing settlement arrangements correlating with changes in the technology of administration is apparently not unique to southwestern Iran.

> The transition seems to occur in several adjacent regions in Iraq between [2700 B.C. and 3250 B.C.] around the ancient centers of Nippur, Nineveh, and Uruk. . . . Thus rather than one case of state emergence, there was a series of emergences of individual states in a network of politics.[29]

Johnson concludes that although a state society formed on the Susiana by about 3500 B.C., by 3200 B.C.

> Conflict broke out between elites of Susa . . . [and] Choga Mish. These apparently found their source in an attempt of Choga Mish to sever its dependent relationship with longer established elites in western Susiana. Scenes of violence appeared on cylinder seals, many villages and small centers were abandoned, and a no-man's land separated the contending parties east and west. Settled population declined rapidly, a process that would continue into the third millennium.[30]

Wright and Johnson suggest that decisions involving problems posed by drought, overpopulation, conflict, or some other single factor could probably be handled by a two-level hierarchy—something on the order of a chiefdom, perhaps—but a combination of problems would require additional decision-making capacity. They suspect that one important factor in early Southwest Asian state formation may have been the economic interactions between nomadic pastoralists and lowland farmers, probably in the form of the exchange of cheese, rugs, meat, minerals, and other highland resources for pottery, grain, and other lowland commodities. Economic models suggest that fluctuating demand for certain craft products in relatively simple economies often stimulates the centralization of workshops and economic administration, and Wright and Johnson suggest that in prehistoric Mesopotamia the economic demands of nomads on lowland economies may have fluctuated sufficiently to produce a similar effect, with the eventual emergence of a three-level hierarchy.

Johnson has applied various concepts from what is known as "hierarchy theory" and the so-called rank-size rule to the Susiana data.[31] Hierarchy theory is essentially a search for explanations of how everything, from molecular structures to world governments, seems to involve hierarchical orderings. The assumption here is that there are principles common to all these forms of hierarchies and that the processes by which hierarchies change and appear can be mathematically analyzed.

The rank-size rule is an empirical observation that many settlement patterns are similar in terms of the relative sizes of their component settlements. Thus, in any region, state, or country, one can expect that the distribution of people will be somewhere between the extreme distributions in which everyone lives in one settlement, or in which the populace is divided into settlements of exactly equal population size. A rank-size plot is produced by ranking each settlement in terms of its population, assigning the number one to the largest, two to the second largest, and so on, and then graphing the actual population of each settlement against the rank of that settlement on a full logarithmic scale. Geographers have noted that for many developing countries, such as Egypt, the rank-size plot will be a *primate* one. For example, Cairo is much larger than any other settlement and forms such a large proportion of the total Egyptian population that a rank-size plot of Egypt is sharply concave, in contrast to a linear distribution in which the second-ranked city is only slightly smaller than the largest, the third is slightly smaller than the second, and so on. It is sufficient here to note that major shifts in the political and economic organization of an area are usually reflected in its rank-size plot. To use Egypt as an example again, the great growth of Cairo to over twelve million people during the last fifteen years has correlated with the growing power of the national government and the rapidly industrializing economy. Prior to 1965, a much greater percentage of Egyptians lived in smaller towns and villages than do now.

The archaeological applications of this are obvious. If we know when a site was occupied and how big it was, we can then estimate population sizes, construct rank-size plots, and perhaps detect when great political and economic changes were taking place. The application of this idea to various areas of Mesopotamia yields mixed results. Clearly there were different rank-size patterns at different times, which might signal the origins of state-level societies.[32] This would not *explain* their appearance, but it would help identify them. But it is difficult to estimate the populations of sites in different time periods with precision. Also, no one really knows what the different kinds of rank-size plots mean. They are essentially empirical generalizations without any theory to explain them.

Nonetheless, the fundamental problem of archaeology is to define patterns in the distribution of artifacts through space and time, and these rank-size plots at least have the virtue of showing us some possible patterns.

The Rise of City-States (c. 3200–2350 B.C.)

The residential architecture of Uruk reflects a diversity of occupational, economic, and social classes. All the buildings were mudbrick, but some were larger, better built, and more elaborately decorated. Many of the people lived in small rectangular buildings built along narrow winding streets through which ran both above- and below-ground drainage canals. Apparently, houses were one story high for the poor and two stories for the wealthier, but both types were similar. Built of mudbrick and white-washed, they represent an ideal architectural adaptation to the climate. Similar houses today are comfortable and attractive, with their white-washed walls contrasting effectively with beautiful dyed rugs and textiles.

By the middle of the fourth millennium B.C., the population of Uruk included specialists in scores of arts and crafts. Potters, using molds and mass-production techniques, turned out enormous quantities of pottery. In earlier periods great numbers of beautifully painted vessels were made at most larger settlements, but by the middle and late fourth miillennium, pottery manufacture had become a centralized, administered activity at Uruk and many other settlements. Other specialists included stonecutters, metalsmiths, bricklayers, farmers, fishermen, shepherds, and sailors. Writing did not come into general use in Mesopotamia until about 3000 B.C., but at Uruk and other sites, in levels dating from the fifth millennium on-ward, archaeologists frequently find stamp and cylinder seals, which were used to impress clay sealings for containers, bales of commodities, and documents. Some of these seals convey in picture form the economic specialization of the community. Boats, domestic animals, grain, deities, and many other motifs are portrayed. Clay tokens have been found in about ten geometric shapes at many early Mesopotamian sites, and Denise Schmandt-Besserat argues that these tokens, which were usually enclosed in clay envelopes, represented such commodities as pots of grain, numbers of animals, and areas of land.[33]

The ancient settlement of Uruk (known as Warka in Arabic and Erech in the Bible), located in the heart of the southern Alluvium, is one of the oldest cities in the world and may have been the center of one of the earliest states.

Initially settled before 4000 B.C., by the late fourth millennium B.C. Uruk was already an impressive settlement of perhaps 10,000 people and several large temples. Between about 3200 and 3100 B.C., the "White Temple" at Uruk was built on a ziggurat (stepped pyramid) some 12 meters above ground level. Constructed of whitewashed mudbrick and decorated with elaborate recesses, columns, and buttresses, it must have been an impressive sight—especially to peasants coming into the city on market days. Inside the temple were tables and altars, all arranged according to the same ritual plan evident at Eridu some 2500 years earlier.

The temple architecture and ziggurats of Mesopotamian antiquity raise an important point. As we will see in surveying other early complex cultures around the world, one of the most consistent signs of evolving cultural complexity is the diversion of enormous amounts of energy and materials into pyramids, ziggurats, palaces, platforms, and other huge constructions. Why would people who could just as easily, and certainly more profitably, build irrigation canals, terrace slopes, weave wool, or do other productive tasks instead "waste" their labors on these mammoth construction projects?

The answer to this question may lie precisely in the fact that these constructions are, in a limited sense, "wasteful." They may, for example, have helped prevent high and unstable rates of population growth, perhaps by deflecting investments in food production. As Leslie White suggested, "it would appear that the ruling class was frequently confronted with the problems of over-production and the threat of technological unemployment or a surplus of population among the lower classes. Their great public works programs, the wholesale disposition of wealth in mortuary customs, etc., enabled them to solve both these problems with one stroke."[34]

Just as important, monumental construction programs could reinforce social and political hierarchies. Frankfort speculates that

> the huge building, raised to establish a bond with the power on which the city depended, proclaimed not only the ineffable majesty of the gods but also the might of the community which had been capable of such an effort. The great temples were witnesses to piety, but also objects of civic pride. Built to ensure divine protection for the city, they also enhanced the significance of citizenship. Outlasting the generation of their builders, they were true monuments of the cities' greatness.[35]

Toward the end of the fourth millennium B.C., the Mesopotamian alluvium and adjacent regions were the heartland of a developed sociopolitical and economic organization that most would call a civilization. Writing had been developed, great temples soared up from the centers of large cities, and the economy was both productive and complexly interdependent. Also, the Uruk polity spread far beyond the central Alluvium. At Habuba Kabira South, in Syria, there was a large later Uruk settlement that is so similar to the great Uruk cities to the south in styles of ceramics and architecture that it may well have been built by people who came directly from the south to found this city.[36]

SUMERIAN CIVILIZATION

To the modern urbanite, it may seem inappropriate to label as a "state" a collection of twelve or fourteen cities with a combined population of perhaps 100,000 and an area of only a few thousand square kilometers, but, as someone once observed, it is the small states, like Sumer, Athens, Florence, and Elizabethan England, that have put posterity most in their debt.

Approximately thirteen city-states made up Sumerian "civilization" between about 3000 and 2350 B.C. Through most of their history, these city-states were politically autonomous, but they belonged to the same cultural tradition and by 3000 B.C. had collectively developed many of the classical

elements of Southwest Asian civilization, including ziggurats, brick platforms, the potter's wheel, wheeled carts, metal-working, sailboats, and
writing.

The fact that the Sumerian language was essentially unlike those of the
contemporary but less-developed Semitic cultures that surrounded the
Sumerians has led some to place Sumerian origins in Turkey, Bahrain, or
even outer space. The Sumerians, of course, thought of themselves as a
distinct and superior culture, and their myths speak of origins in some
distant land; but we will probably never determine their ethnic origins,
and in a sense this is an unimportant problem. The achievements of cultures cannot be explained in terms of the special characteristics and mental gifts of the people of these cultures: people are constants in the equations of cultural evolution, and it is their circumstances and position in
place and time that determine their cultural "achievements."

Sumerian Writing. We have a remarkably detailed picture of life in these
Sumerian city-states, because shortly before 3000 B.C. they began to develop a written language. What we know about the Sumerian language is
derived from the thousands of clay tablets on which they wrote (Figure
8.11). Their script is known as cuneiform, from the Latin for wedge-shaped,

8.11 A cuneiform text on
clay.

EARLIEST PICTOGRAPHS (3000 B.C.)	DENOTATION OF PICTOGRAPHS	PICTOGRAPHS IN ROTATED POSITION	CUNEIFORM SIGNS CA. 1900 B.C.	BASIC LOGOGRAPHIC VALUES		ADDITIONAL LOGOGRAPHIC VALUES		SYLLABARY (PHONETIC VALUES)
				READING	MEANING	READING	MEANING	
	HEAD AND BODY OF A MAN			LÚ	MAN			
	HEAD WITH MOUTH INDICATED			KA	MOUTH	KIRI₃ ZÚ GÙ DUG₄ INIM	NOSE TEETH VOICE TO SPEAK WORD	KA ZÚ
	BOWL OF FOOD			NINDA	FOOD, BREAD	NÍG GAR	THING TO PLACE	
	MOUTH + FOOD			KÚ	TO EAT	ŠAGAR	HUNGER	
	STREAM OF WATER			A	WATER	DURU₅	MOIST	A
	MOUTH + WATER			NAG	TO DRINK	EMMEN	THIRST	
	FISH			KUA	FISH			KU₆ HA
	BIRD			MUŠEN	BIRD			HU PAG
	HEAD OF AN ASS			ANŠE	ASS			
	EAR OF BARLEY			ŠE	BARLEY			ŠE

8.12 The evolution of Sumerian writing. Fourth millennium B.C. tablets were inscribed vertically with pictographs, but in the early third millennium, the direction of the writing and the pictographs were rotated to the horizontal. In succeeding millennia the symbols were stylized and given phonetic meanings.

a reference to the fact that Sumerian was written by impressing wet clay with the end of a reed, leaving wedge-shaped marks. When baked, these clay tablets can survive thousands of years.

The earliest known written documents may be clay tablets and sealings from early occupations at Uruk (c. 3400 B.C.). There are signs for carpenter, donkey, boat, copper, and many other things, totaling fifteen hundred symbols in all (Figure 8.12). Some signs seem to mean to buy, others refer to *en,* the title of a lord, and to *unken,* which may have been a people's assembly.[37] The hierarchical nature of Mesopotamian society by the end of the Uruk Period is vividly illustrated by one of the oldest documents known, the "Standard Professions List," which gives the titles of officials and names of professions, all arranged according to what is apparently a composite sense of power and prestige.[38]

The ability of this cumbersome writing to convey abstract concepts or the spoken language was initially quite limited, but in the centuries after 2900 B.C. the Sumerians improved it. Phoneticization, by which some signs came to represent distinct words and syllables of the spoken language, was most important. Thus, the sign of an arrow came to mean both arrow and the Sumerian word for life, the connecting link being that the spoken word for both was *ti*.[39] Eventually much of spoken Sumerian was represented by written symbols, and the pictographic elements slowly lost their representational character as the scribes stylized them and reduced the number of directions in which the stylus had to move to write them.[40] Unique signs were developed for most Sumerian vowels and syllables, but the language was never reduced to an alphabetic system where every distinct sound is represented by a unique sign. Instead, it remained a welter of signs that represented pictographs of concrete objects, signs that represented syllables of speech, and signs that represented ideas. This made reading the script a complicated process, requiring the memorization of hundreds of different characters. One sign, for example, which ultimately derived from a pictograph representing a mountain, acquired a total of ten possible phonetic values and four ideographic values as well.[41] At first the reader had to infer the exact meaning of the word by considering its context, but eventually the Sumerians devised a system of determinatives—signs placed before or after a word to indicate the general category to which the word belonged, such as birds, male proper nouns, or deities.

Over the centuries the Sumerians' successors reduced the complexity of the written language, but even as late as about 1900 B.C., it had between six and seven hundred unique elements. At this stage it was similar to Chinese and a few other modern languages, which faithfully represent the spoken language and are adequate for most purposes, but which, compared to alphabetic systems, are very cumbersome. It is difficult to construct typewriters or computers for languages with hundreds of unique elements, and even minimal literacy in such languages is the product of long and arduous training. (To become literate in modern Chinese, for example, one must memorize several thousand characters.) Most Semitic languages were—and still are—written without the short vowels, but this is no barrier to effective communication if one is accstmd t rdng th scrpt.

The first truly alphabetic written languages appear to have developed toward the end of the second millennium B.C. among Semitic-speaking peoples in Palestine and northern Syria. In the tenth or ninth centuries B.C., the Greeks adapted the Syrian or Phoenician variant of these early alphabets to their own language, reducing the number of signs to fewer than twenty-five and making several major refinements in the process.

The Greek alphabet was the basis for all modern European writing systems, including the Cyrillic alphabet of eastern Europe.

The role of writing in early Mesopotamian societies seems largely economic. Simple pictographs and the spoken language cannot efficiently meet the requirements of a society that has surpluses to be stored and redistributed, water to be allocated, land rights to be assigned and adjudicated, ritual prayers to be said, and all the other tasks we find in complex cultures. In fact, only the Inca of Peru managed to develop states and empires without a written language, but they had a fairly efficient substitute in the form of a vast bureaucracy and the *quipu*, a system of knotted strings in which the length of strings and placement of knots was used as a device to assist the memory of the recordkeeper.

The economic element in ancient Mesopotamian scripts is evident in their mathematical notations. Georges Ifrah notes that the Sumerians used a counting system with a base of 60, instead of the base-10 system we have.[42] It is difficult for most people to accept that there is nothing inherently necessary about a base-10 numbering system. But modern computers employ a base-2 system, for example, using just 1s and 0s to represent all numbers. We probably use a base-10 system because we have 10 fingers—there are many base-20 systems that probably reflect our 20 fingers and toes. Whatever base is used, people usually have given a unique name to *each* of the numbers, so a base-2 system is the simplest and a base-60 system is rather unwieldy. Mathematicians have often proposed that we scrap our base-10 system and use something more convenient, like a base-12 system, where numbers would have a large number of divisors, or a base-11 system, which, as a prime number, would eliminate the ambiguities of a quantity like "0.36," which can reduce to 36/100, 18/50, 9/25, etc.[43] The Sumerians got around the problem of having to learn and use a unique name for each number between zero and 60 by giving a name to each multiple of 10 lower than or equal to 60. They may have developed a base-60 system because six times 60 is close to the number of days in the year, but the origins of the numbering system probably go far back beyond the first written evidences we have of it. As H. W. F. Saggs notes, the ancient Sumerians and their successors used their mathematical systems not only for economic accounting but also for land surveying, calendrical calculations, and astronomical science.[44] By about 1800 B.C. the people of Mesopotamia had calculated the square root of 2 to a value correct to 1 in two million, they had formulated massive tables of reciprocals of sexagesimal numbers, they knew how to calculate the length of a hypotenuse from the lengths of the sides of right-angled triangles (1200 years before Pythagoras), they could calculate cube roots, and they knew some algebraic operations.[45]

The way of life described by the Sumerian texts, and by texts of other

Southwest Asian cultures of the third millennium B.C., is still recognizable to anyone who has traveled in these areas. Sheep, goats, and cattle are tabulated, taxed, and exchanged; children are shepherded to school—as always, much against their will; a council of elders meets to consider grievances against the inhabitants of an adjoining city-state; and Sumerian proverbs express ideas recognizable in many societies:

> Upon my escaping from the wild ox,
> The wild cow confronted me.
>
> When a poor man dies,
> do not try to revive him.[46]

But when they turn to mythological and eschatological themes, the ancient Mesopotamians are less accessible—perhaps even a bit bizarre—by our standards, as in this section of the myth of the revival of the goddess Inanna.

> Go to the underworld.
> Enter the door like flies.
> Ereshkigal, the Queen of the Underworld, is moaning
> With the cries of a woman about to give birth.
> No linen is spread over her body.
> Her breasts are uncovered.
> Her hair swirls about her head like leeks.
>
> When she cries, "Oh! Oh! My inside!"
> Cry also, "Oh! Oh! Your Inside!"
> When she cries, "Oh! Oh! My outside!"
> Cry also, "Oh! Oh! Your outside!"
> The queen will be pleased.
> She will offer you a gift.
> Ask her only for the corpse that hangs from the hook on the wall.
> One of you will sprinkle the food of life on it.
> The other will sprinkle the water of life.
> Inanna will arise.[47]

Western philosophy and theology are deeply influenced by the philosophies first propounded in ancient Greece, philosophies very different from those of ancient Sumer. Thus, most Westerners see the world in terms of beginnings and ends, causes and effects, and the importance and "will" of the individual. We cannot completely reconstruct Sumerian philosophy on the basis of fragmentary texts, but it seems evident that the Sumerians saw a much more static and magical world than we do. Although their technology and complex organizations demonstrate that they were shrewd, rational people, there seems to have been little emphasis or analysis of human motivation or the physical world. They viewed the earth as a flat disk under a vaulted heaven and believed that various gods guided history according to well-laid-out plans and that the world continues without end and with little change. Each god was in charge of something—the movements of the planets, irrigation, or brickmaking, for example—and each was immortal and inflexible. As with humans, the deities were hierarchically arranged in power and authority and were given to power struggles and many vices.

Like most people, the Sumerians thought that the gods had a special interest in them. Just as every American dollar bill has on the back of it the assertion that 'ANNUIT COEPTIS' ("[God] has favored our undertaking"), almost every Sumerian official pronouncement and inscription called on the gods known to favor the city.

Sumerian Economy. In Sumer, wheat, barley, vegetables, and dates were the major crops, while cattle raising and fishing were of almost equal importance. Cattle were raised for draft power, hides, and milk and meat. Fish was a staple, as were pigs.

Out of each measure of wheat or barley, the Mesopotamian farmer probably fed about 16% to his animals, reserved about 10% for the next year's sowing, lost about 25% in storage, and ate the rest.[48] If ethnographic studies are to be believed, in the absence of modern transport people did not cultivate fields much more than 4 kilometers from their houses. At any given time, much of the land around a village in ancient times would have been fallow, while other land would have been unirrigatable or of marginal productivity. All in all, Adams estimates that each person would need for subsistence about one hectare of barley and wheat fields, along with at least some pastures and orchards.[49]

One of the less desirable "firsts" of Sumerian civilization was probably in the field of epidemic diseases. Just as there are certain disastrous things a hunter and gatherer can do (e.g., presume on too slight evidence that a cave bear is not at home), one of the worst things a villager can do is contaminate drinking water with sewage, and this is hard to avoid in a primitive town. Typhoid, cholera, and many other diseases require certain levels of population density to evolve, to maintain a reservoir of infected individuals, and to perpetuate themselves. These levels were probably reached for the first time in Sumer. Once people started digging wells and irrigation canals in areas with many people and animals, disease and epidemics quickly followed.

Few economies in history or prehistory have been as organized as the Sumerian. Tablet after tablet records endless lists of commodities produced, stored, and allotted. Ration lists, work forces, guild members—all are recorded in numbing detail. Even the city's snakecharmers were organized.

The state and church controlled economic matters through much of Mesopotamian history, but individual merchants and capitalist elements also seem to have been significant fairly early on. For the Marxian this is important because once you have individual wealth and capital, the possibilities for class conflict and the rest of the Marxian paradigm exist. Although Sumerian society was organized on the basis of kinship, people also belonged to and acted through occupational and social classes. In the event of war, for example, members of different "guilds," such as silversmiths or potters, would be under the command of their "guild president." One of the major trends in the evolution of complex societies generally was the change from a kinship-based society to one based on divisions along occupational, social, and economic class lines, and by 3000 B.C. there is evidence that this trend was developing in Mesopotamian societies. Even so, throughout the history of Southwest Asia kinship ties have been powerful social forces. At the pinnacle of Sumerian society was a god-king, assumed to be a descendant of and in contact with the gods. Beneath him was a leisured class of nobles. There was also a class of wealthy businessmen who lived in the larger, better houses of the city; of lesser wealth and prestige were the many artisans and farmers, including smiths, leatherworkers, fishermen, bricklayers, weavers, and potters. Scribes apparently held fairly important positions and literacy was an admired accomplishment. At the bottom of society were the slaves, often war captives or dispossessed farmers.

Money, as we know it, did not exist in ancient Sumer; most exchange was "in kind," the trading of products for other products. Local and long-distance trade was voluminous, however, and ships sailed up the rivers from the gulf carrying shell, carnelian, lapis lazuli, silver, gold, onyx, alabaster, textiles, and food and other produce.

One of the most spectacular differences between Sumerian societies after 3000 B.C. and their predecessors is in mortuary practices. At the end of the 'Ubaid period (3800 B.C.), graves varied little, even at the largest settlements; but after 3000 B.C. a radical shift occurred. The famous death pit at Ur is an impressive display of wealth and pomp. Excavating here in 1927–1928, Sir Leonard Woolley came upon five bodies lying side by side, each with a copper dagger and a few other items.[50] Beneath them was a layer of matting on which the bodies of ten women were encountered, lying in two rows, each richly ornamented with gold, lapis lazuli, and carnelian jewelry. Nearby was a gold- and jewel-encrusted harp, across which were the bones of the gold-crowned harpist. The bodies were lying on a ramp, and as the excavators continued down this they encountered a heavily jeweled chariot, complete with oxen and grooms. Then the investigators began unearthing masses of gold, silver, stone, and copper vessels, as well as additional human bodies, weapons, and other items. Nearby another set of six male skeletons equipped with copper knives and helmets was found, as well as the remains of two four-wheeled wooden wagons—also decorated with harnesses of gold and silver and accompanied by the skel-

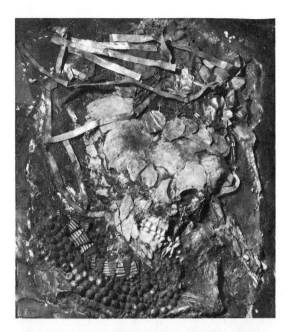

8.13 Crushed skull of a female attendant in the death pit at Ur. Note the gold jewelry and precious stones. This Sumerian model of a goat and a tree is made of wood, lapis, and gold, and is about 51 cm high. It was found at Ur and dates to about 2600 B.C.

etons of grooms and drivers. Other arrangements of human skeletons, harps, wagons, and model boats appeared as the excavations continued. At the end of the tomb was a wooden bier containing the remains of the queen. The entire upper part of her body was hidden by a mass of beads of gold, silver, lapis lazuli, carnelian, agate, and chalcedony. Her head-dress and other furnishings were lavishly ornamented with gold, silver, and precious jewels. Liberally strewn about the chamber were human bodies, jewelry, vessels of precious metals, silver figurines, silver tables, cosmetics, seashell ornaments, and a number of other treasures.

All together some sixteen "royal" burials were found at Ur, all of them distinguished from the myriad common graves by the fact that each was not merely a coffin but a structure of stone, or stone and mudbrick, and by the inclusion of human sacrifices—up to eighty in one case. At least three categories of burials seemed evident, ranging from the sixteen royal graves to less elaborate but still richly furnished graves in which presumably the common people were placed.

The Biblical Flood? The Biblical account of the flood is clearly related to Sumerian stories that predated the Old Testament by thousands of years. Finegan has identified some of the similarities[51]:

BABYLONIAN CUNEIFORM	GENESIS
1. The gods decide to make a flood	1. The Lord decides to destroy wicked mankind
2. The God Ea warns Artarhasis to build a ship	2. The Lord warns Noah to build an ark
3. He is to take his family and animals aboard	3. Noah is to take his family and animals aboard
4. The flood turns mankind into clay	4. The flood destroys all flesh
5. The ship grounds on Mount Nisir	5. The ark comes to rest on the Ararat Mountains
6. Artarhasis learns when the waters have subsided by sending out a dove, a swallow, and a raven	6. Noah learns when the waters have subsided by sending out a dove, a swallow, and a raven.
7. He offers sacrifice to the gods	7. He offers sacrifice to the Lord
8. The gods smell the sweet savor	8. The Lord smells the pleasing odor
9. The god Enlil blesses Artarhasis and his wife	9. God blesses Noah and his sons

In 1929 S. Langdon, excavating the Sumerian city of Kish, and Sir L. Woolley, excavating at Ur, both announced that they had found thick levels of clean sand that they presumed was evidence of the biblical flood. Some members of these expeditions are even said to have sold vials of sand labeled "Samples from the Great Flood."

But Spuhler notes that many layers thought to be flood deposits are really sand dunes produced by winds, while others are ordinary river deposits.[52] In any case, radiocarbon dates for levels thought to be of the biblical flood range from 4500 B.C. to 2700 B.C., and the absence of sand

8.14 Gypsum statuettes of an aged couple, residents at about 2500 B.C. of Nippur, one of the largest cities in southern Mesopotamia during this period.

strata at many other nearby sites, and the evidence of life as usual all over Mesopotamia at this time, rule out fairly conclusively any scientific evidence for the biblical flood—except in the opinion of the people who keep climbing Mount Ararat looking for pieces of the ark. Popular press reports of a Russian pilot who saw the outline of a boat far above the tree line on this mountain have fueled several expeditions to recover the ark, but so far only a few questionable bits of wood have been found. Only the drearily unimaginative archaeologist would be impressed by the fact that all the samples of wood were radiocarbon-dated to about 3500 years ago—a time for which there is no evidence that the rest of the world was flooded.[53]

Brilliant as Sumerian civilization was, one must realize that it was just one of the states of this era. In the southwestern highlands of Iran, for example, between 3300 and 2800 B.C. a great "Proto-Elamite" trading state arose, perhaps on the basis of its ability to dominate trade over the rich routes between Mesopotamia and the Iranian Plateau.[54]

Southwest Asia after 2350 B.C.

For centuries after 3000 B.C., the Sumerian city-states engaged in almost constant warfare, with first one and then another gaining temporary ascendancy. With the rise to power of Sargon of Akkad at about 2350 B.C., however, the political fabric of ancient Southwest Asia was forever changed. Sargon and his several immediate successors used the city of Akkad as a military base from which they mounted spectacularly successful attacks in all directions. Akkadian historical documents recount thirty-four battles fought by Sargon against the southern city-states, during which he moved down the Alluvium, capturing many kings, smashing city walls, and finally "cleansing his weapons in the sea." Sargon appointed Akkadians to administrative posts in the conquered city-states and then began expanding his other frontiers, invading Syria, Lebanon, and western Iran.

After 2200 B.C. quarrels arose among rival claimants to the Akkadian throne, and the "empire" fragmented under the onslaught of peoples moving in from the highlands on the margins of the empire. This "dark age" lasted until about 2100 B.C., when one or more Sumerian kings were able to evict the invaders and reestablish political control over much of the southern Alluvium under what is known as the Ur III dynasty. One ruler, Ur-Nammu, based at the ancient city of Ur, aggressively extended his influence into much of the area formerly encompassed by the Akkadians. Great volumes of obsidian, lapis lazuli, and copper are thought to have passed into central Mesopotamia from as far away as India and the Aegean. Legal texts of the late third millennium describe in detail problems of land use, irrigation rights, compensation for bodily injury, penalties for adultery, and many other elements of daily life.

Despite its apparent stability, the Ur III political system of the late third millennium was constantly under pressure from internal political rivalries, as well as from the incursions of nomads and rival groups along the empire's frontiers. The coup de grace was administered at about 2004 B.C. with the invasion from western Iran of the Elamites, who led the king of Ur away in captivity.

From about 2000 to 1800 B.C., Greater Mesopotamia was politically fragmented as kings at Isin, Larsa, Susa, and elsewhere established contending states.[55] Eventually the ancient city of Babylon became the most powerful political entity and by 1792 B.C. Hammurabi established the Babylonian Empire, based mainly on the southern Alluvium. The many documents of his reign reflect a skillful politician adept at bureaucratic, military, and political uses of power: his famous law code, although harsh by modern standards, reflects efficient administration. Both Hammurabi and his successors encountered opposition from southern city-states and from a rival state in Assyria, to the north. As always, the nomads and other peoples on the empire's periphery, in this case the Kassites and Hurrians, made inroads as soon as the central government weakened, and eventually they overran much of the Babylonian Empire.

After about 1600 B.C., the political history of Southwest Asia becomes

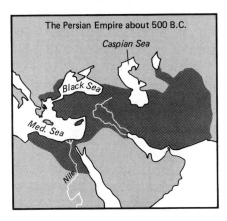

The Assyrian Empire about 700 B.C.

The Persian Empire about 500 B.C.

8.15   The extent of the Assyrian and Persian (Achaemenid) empires.

extremely complicated, with frequent political realignments and, overall, the gradual extension of imperial power (Figure 8.15). Assyrians, Elamites, Achaemenids, and other cultural groups established empires, and eventually the political and military scale became distinctly international as empires centered in Egypt, Anatolia, and Iran met and, more often than not, came into conflict.

The Origins of Cultural Complexity in Southwest Asia: Summary and Conclusions

It is possible that some archaeologist will one day be sweating over a Mesopotamian mudbrick building foundation and discover a library of cuneiform tablets that record in great detail a history of the origins of the Mesopotamian state, complete with population statistics, war casualties, changes in the weather, average fertility rates, and a demographic analysis of the whole critical two millennia of state origins.

In fact, recent discoveries in Syria of previously unknown archives of thousands of texts do provide some information of this kind. But for the most part only archaeological data can really give us the evidence of change over thousands of years and great areas that could constitute the basis for an analysis of Southwest Asian cultural complexity. Now that the Iran-Iraq war has ended, we may eventually be able to add to the critical data on this problem. Unfortunately, Thank to Saddam, this is no longer true

With the information now available, the evolution of complex societies in Mesopotamia seems to have been the result of multiple factors, operating in complex patterns that differed from place to place. The Alluvium has the potential to produce tremendous quantities of food, more than is required in most years to sustain millions of people and their domestic animals. Such surpluses do not necessarily create great differences in wealth, but they certainly allow the formation of rich and poor social classes, and it seems to be a pattern that wherever one finds economic systems that produce great surpluses, one also discovers elaborate social hierarchies of administrators to organize, store, distribute, and exploit these surpluses.

What factors stimulate the production of surpluses and thus create the conditions for the establishment of social classes and administrative hierarchies? Our review of the evidence from Southwest Asia suggests that irrigation cannot be said to be the "primary cause" of early Southwest Asian cultural complexity. Some elements of the very first stages of complexity seem to correlate with initial attempts to irrigate on the Susiana Plain,[56] but activity specialization, monumental architecture, changes in settlement spacing and size hierarchies, architectural variability, mortuary stratification—in short, the whole range of physical evidence of cultural complexity—appear before evidence of significant extension of irrigation systems. This is true for both the heartland of Sumer and for the Susiana Plain in southwestern Iran—the two areas for which we have the best archaeological data.

8.16 At about 500 B.C. the capital of the Achaemenid empire was Persepolis, in central Iran. This grand stairway ascends to the Palace, and along it are reliefs that illustrate real life processions to pay homage to the king. Some of these sculpted figures carry flowers and other gifts for the ruler.

In much of Southwest Asia, in fact, increased investments in irrigation systems appear to be the result of urbanism.[57] Population agglomeration (urbanism) requires that the surrounding areas be particularly productive, because it is not feasible for the city dwellers either to farm or defend areas more than several kilometers from the city. Irrigation was absolutely necessary for any sedentary existence on the Alluvium, but if the spread, integration, and construction of irrigation systems were the mechanisms whereby complex societies first evolved and developed, we would expect to see a direct and positive correlation between the size and complexity of irrigation systems and the complexity of cultures based on them.

Instead, we find that 'Ubaid, Uruk, and early dynastic settlements subsisted on the produce of fields that could be irrigated by relatively simple, autonomous canal works. Only many centuries after the appearance of the first complex societies were there complex, integrated irrigation systems.

Some scholars have tried to explain the origins of cultural complexity in Southwest Asia and elsewhere as mainly the result of human population growth. The persistence of this view must be seen in large part as a result of the fact that if one graphs the approximate population of Greater Mesopotamia from about 8000 B.C. to about 3000 B.C. against evidence of activity specialization, monumental architecture, agricultural productivity, and the other evidences of cultural complexity, then an impressively close correlation is apparent. Also, it is difficult not to be impressed by the evidence for warfare as a potent factor in Mesopotamian cultural evolution. Great brick walls ring cities and villages alike in some periods, and there are caches of weapons, chariots, documents, and representational art to tell us that warfare formed the very weft and warp of Mesopotamian history. Nor is this sustained level of warfare merely a historical curiosity: if some adult male citizen of ancient Mesopotamia of almost any century were transported through time to our own age, the thing he would find most familiar, perhaps, is the battle that raged through the 1980s between Iran and Iraq. Armies have met here in this same general area in almost every century for more than 5000 years.

If we apply the principles of evolutionary biology, warfare is simply an expression of competition, which is common to all life forms. To give it a name, however, is not to explain it, nor is the importance of warfare in *initial* Mesopotamian cultural complexity manifestly evident. One might expect that with the cheapness of clay construction there would be many more walled settlements in the early civilizations, if warfare were a frequent curse. The walling of towns does occur very soon after the emergence of urbanism, and warfare may have played a significant role in the major increases of cultural complexity that followed the earliest "states," but extensive circumvallation of sites is common on the southern Allu-

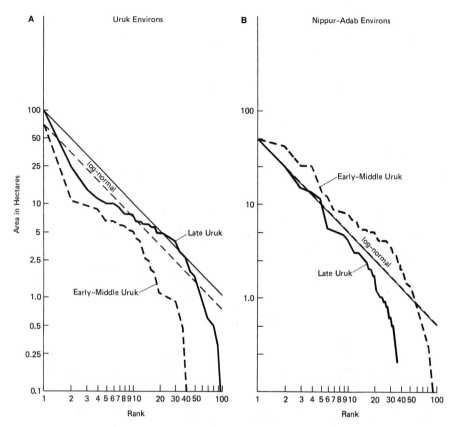

8.17 Settlement rank-size plots produced by Robert McC. Adams (1981) for two areas of Mesopotamia during the early Middle Uruk period (c. 3400 B.C.) and the Late Uruk period (c. 3200 B.C.), the periods when the first "states" were appearing in this part of Mesopotamia. Rank-size plots like these are formed by ranking the settlements from largest to smallest and then plotting a settlement's rank against its size on a log-log scale. Geographers have found that different shapes of such plots correlate with certain kinds of economic and political situations. The "flat-top" plot of the Early Uruk in the Nippur-Adab region (b) often is associated with very loose economic integration, whereas a log-normal distribution (the dotted lines in the graphs) is associated with considerable political and economic integration. The fact that both the plots for the Late Uruk period—when the first developed complex societies here were in operation—changed over time in the same direction from markedly different original plots, may reflect the growing economic and political interdependence of settlements in the late Uruk "state."

vium mainly after about 2900 B.C.—some centuries after the appearance of other evidence of cultural complexity. After 2900 B.C. the historical records leave no doubt that warfare was almost continuous throughout Southwest Asia.

But can population growth and warfare be shown to have been antecedent conditions to the rise of Mesopotamian states? In Iranian Mesopotamia, on the Susiana Plain, Wright and Johnson conclude that

> The available data show that there was a period of population decline prior to state formation. States emerged perhaps during a period of unsettled conditions as population climbed back toward its former level. As Carneiro suggested, warfare may have a role in state formation, but in this case, increasing population in a circumscribed area cannot be the sole or direct cause of such warfare. If the hypothesis that population increase was the primary cause of state formation were correct, the state should have emerged in Susiana times [before 4000 B.C.] because population in that period seems to have been as high as in early Uruk times [3700 B.C.].[58]

The Susiana Plain, however, is an area where rainfall agriculture is possible, and there are some questions about the reliability of Wright and Johnson's population estimates.[59] What about the southern Mesopotamian Alluvium, the Sumerian heartland? Robert McC. Adams concludes that

> possibly the attainment of some minimal population level was necessary to set the process [of urbanization] into motion. But such evidence as there is suggests that appreciable population increases generally followed, rather than preceded, the core processes of the Urban Revolution. Particularly in Mesopotamia, where the sedentary village pattern seems to have been stabilized for several millennia between the establishment of effective food production and the "take-off" into urbanism, it may be noted that there is simply no evidence for gradual population increases that might have helped to precipitate the Urban Revolution.[60]

Adams here is principally concerned with the phenomenon of urbanism rather than complex societies in general, but his "Urban Revolution" includes many of the essential transitions we have defined as the basis for the evolution of complex societies. What about agricultural intensification? In Carneiro's scheme this is a direct result of population pressure; thus we should find evidence of increasingly intensive irrigation and of farming of marginal lands just before, and along with, the evolution of cultural complexity. But major investments in irrigation and land reclamation seem to have occurred after the emergence of urbanism and other evidences of cultural complexity and, as Adams notes, urbanization appears to have involved the widespread abandonment of large areas of formerly intensively farmed lands. Furthermore, there is no evidence that any of the areas of early state formation ever approached their agricultural limits: that is, with only minor investments in additional irrigation systems, they could have enormously increased the numbers of people who could be supported; yet the population size remained quite stable. In short, agricultural intensification seems to have been more a result than a cause of the emergence of complex societies.

What about the possibilities of a Marxian analysis of Mesopotamian state origins? As noted in Chapter 7, the Marxian view of state origins can be a complex, multilevel one, not just a dubious assertion that by and large, economic and technological factors are important determinants of socio-political arrangements. But despite elaborate attempts to link the "things" of the archaeological record with the notions of modes of production, historical dialectics, class structures, and the like,[61] the links inferred are tenuous and the available data are simply not adequate to the task of testing these ideas in any meaningful manner.

Zagarell has argued that various evidence supports the idea that already by the Late Uruk period, Mesopotamian communities were in many ways kin-based, production and social arrangements were quickly breaking down, and personal debt, the exploitation of women, and an emerging merchant class were common.[62] But the data simply are not adequate to put a Marxian analysis to an acceptable test.

Adams' general interpretations of the whole Mesopotamian sequence have led him to conclude that this pattern of urbanization and economic development

> was not generated by any unique propensities of the landscape, and that we must look instead to the human forces that were harnessed in the building of the cities themselves.[63]

But Adams sees environmental factors as important. While rejecting Wittfogel's major premise that irrigation is the primary cause of cultural complexity, he nonetheless concludes that

> In the largest sense, Mesopotamian cities can be viewed as an adaptation to [the] perennial problem of periodic, unpredictable shortages. They provided concentration points for the storage of surpluses, necessarily soon walled to assure their defensibility. The initial distribution of smaller communities around them suggests primarily localized exploitation of land, with much of the producing population being persuaded or compelled to take up residence within individual walled centers rather than remaining in villages closer to their fields. Tending to contradict a narrowly determinist view of urban genesis as merely the formation of walled storage depots, the drawing together of significantly larger settlements than had existed previously not only created an essentially new basis for cul-

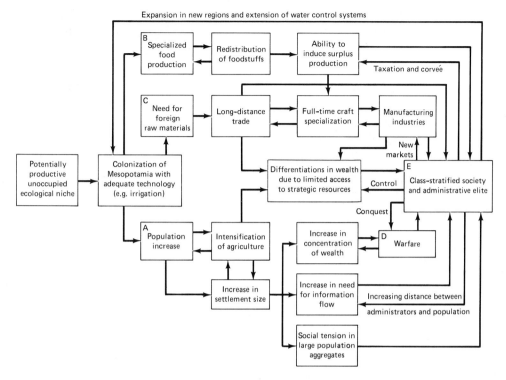

8.18 A model (by C. Redman [1978]) illustrating hypothetical cause-and-effect relationships among variables that led to the formation of class-structured "states" in Mesopotamia.

tural and organizational growth but could hardly have been brought about without the development of powerful new means for unifying what originally were socially and culturally heterogeneous groups.[64]

Adams argues "that the primary basis for [political-economic] organization was . . . religious allegiance to deities or cults identified with particular localities, political superordination resting ultimately on the possibility of military coercion, or a fluid mixture of both."[65]

Charles Redman concluded that

City-states were the pervasive political units that were only temporarily unified into larger national states. The causes for these periods of unification and centralized authority are diverse, but economic factors seem to be a recurrent theme. Lowland Mesopotamia is devoid of some of the most basic raw materials. . . . The importation of these goods may have been crucial in the emergence of urbanism in Mesopotamia. . . . Two major forces were called upon to forge nation-states out of the heterogeneous Mesopotamian cities: militarism and complex administration. These two factors were the primary unifying elements, although in different proportions, in all Mesopotamian state governments and in subsequent ones.[66]

Notes

1. Adams, *Land Behind Baghdad;* Adams, *Heartland of Cities;* Adams and Nissen, *The Uruk Countryside;* also see Young, et al., *The Hilly Flanks and Beyond.*
2. Wenke, "Western Iran in the Partho-Sasanian Period."
3. Wenke, Imperial Investments and Agricultural Developments in Parthian and Sasanian Khuzestan: 150 B.C. to A.D. 640"; Wenke, "Explaining the Evolution of Cultural Complexity. A Review."
4. Simmons et al., "'Ain Ghazal: A Major Neolithic Settlement in Central Jordan."
5. Redman, *The Rise of Civilization,* p. 180.
6. Mellaart, *Earliest Civilizations of the Near East,* p. 77.
7. Mellaart, *The Neolithic of the Near East,* pp. 101–5.
8. Redman, *The Rise of Civilization,* p. 187.
9. Nissen, *The Early History of the Ancient Near East, 9000–2000 B.C.,* p. 46.
10. Hole, "Settlement and Society in the Village Period," p. 95, citing Childe, *New Light on the Most Ancient East,* p. 111.
11. Nissen, *The Early History of the Ancient Near East, 9000–2000 B.C.,* p. 57.
12. Abu es-Soof, "Tell Es-Sawwan: Excavations of the Fourth Season (Spring 1967) Interim Report"; Yasin, "Excavation at Tell Es-Sawwan, 1969 (6th Season)."
13. Larsen and Evans, "The Holocene History of the Tigris-Euphrates-Karun Delta."
14. Nutzel, "The Climatic Changes of Mesopotamia and Bordering Areas."
15. Hole, Flannery, and Neely, *Prehistory and Human Ecology of the Deh Luran Plain.*
16. Wright, "The Evolution of Civilizations."
17. Cf. Hole, "Settlement and Society in the Village Period," p. 95.
18. Johnson, "Nine Thousand Years of Social Change in Western Iran," p. 284.
19. Ibid., p. 285.
20. Hole, "Settlement and Society in the Village Period."
21. Ibid., p. 96.
22. Nissen, *The Early History of the Ancient Near East, 9000–2000 B.C.,* pp. 47–48.
23. Adams, "Patterns of Urbanization in Early Southern Mesopotamia."
24. Ibid., p. 739.
25. Johnson, *Local Exchange and Early State Development in Southwestern Iran;* Johnson, "Organizational Structure Scalar Stress."
26. Nissen, *The Early History of the Ancient Near East, 9000–2000 B.C.,* pp. 84–85.
27. Adams, "Patterns of Urbanization in Early Southern Mesopotamia," p. 743.
28. Johnson, *Local Exchange and Early State Development in Southwestern Iran;* Johnson, "The Changing Organization of Uruk Administration on the Susiana Plain"; Wright, "The Evolution of Civilizations"; Wright, "The Susiana Hinterlands During the Era of Primary State Formation"; Wright and Johnson, "Population, Exchange, and Early State Formation in Southwestern Iran."
29. Wright and Johnson, "Population, Exchange, and Early State Formation in Southwestern Iran," pp. 273–74.
30. Johnson, "Nine Thousand Years of Social Change in Western Iran, p. 287.
31. Johnson, "Aspects of Regional Analysis in Archaeology"; Johnson, "Organizational Structure Scalar Stress." For a description of hierarchy theory, see Pattee, *Hierarchy Theory. The Challenge of Complex Systems.*
32. Crumley, "Three Locational Models: An Epistemological Assessment for Anthropology and Archaeology."
33. Schmandt-Besserat, "Decipherment of the Earliest Tablets."

34. White, *The Science of Culture*, p. 383.
35. Frankfort, *The Birth of Civilization in the Near East*, pp. 56–58.
36. Nissen, *The Early History of the Ancient Near East, 9000–2000 B.C.*, pp. 120–21.
37. Oates, "The Emergence of Cities in the Near East."
38. Nissen, *The Early History of the Ancient Near East, 9000–2000 B.C.*, p. 81.
39. Diringer, *Writing*.
40. Nissen, *The Early History of the Ancient Near East, 9000–2000 B.C.*, p. 136.
41. Ibid., p. 40.
42. Ifrah, *From One to Zero*.
43. Ibid., p. 35.
44. Saggs, *Civilization Before Greece and Rome*.
45. Ibid., pp. 225–27.
46. Hamblin et al., *The First Cities*, pp. 103–4.
47. Wolkstein and Kramer, *Inanna, Queen of Heaven and Earth*.
48. Adams, *Heartland of Cities*.
49. Ibid., p. 87.
50. Woolley, *Excavation at Ur*.
51. Finegan, *Handbook of Biblical Chronology: Principles of Time Reckoning in the Ancient World and Problems of Chronology in the Bible*, pp. 42–43; quoted in Spuhler, "Anthropology, Evolution, and 'Scientific Creationism,' " p. 116.
52. Spuhler, "Anthropology, Evolution, and 'Scientific Creationism.' "
53. Taylor and Berger, "The Date of Noah's Ark."
54. Alden, "Trade and Politics in Proto-Elamite Iran."
55. Schacht, "Early Historic Cultures."
56. Pollock, "Style and Information," p. 368.
57. Adams, "Patterns of Urbanization in Early Southern Mesopotamia."
58. Wright and Johnson, "Population, Exchange, and Early State Formation in Southwestern Iran," p. 276.
59. Weiss, "Periodization, Population, and Early State Formation in Khuzestan."
60. Adams, *The Evolution of Urban Society: Early Mesopotamia and Prehispanic Mexico*, pp. 44–45.
61. Friedman and Rowlands, *The Evolution of Social Systems*, pp. 201–76.
62. Zagarell, "Trade, Women, Class, and Society in Ancient Western Asia." See also Tosi, "The Notion of Craft Specialization."
63. Adams, *Heartland of Cities*, p. 252.
64. Ibid., p. 244.
65. Ibid., p. 78.
66. Redman, *The Rise of Civilization*, p. 319.

Bibliography

Abu es-Soof, B. 1968. "Tell Es-Sawwan: Excavations of the Fourth Season (Spring 1967) Interim Report." *Sumer* 24:3–16.

Adams, R. McC. 1955. "Developmental Stages in Ancient Mesopotamia." In *Irrigation Civilizations. A Comparative Study*, ed. J. H. Steward. Washington, D.C.: Pan-American Union, Social Science Monographs.

———. 1965. *Land Behind Baghdad*. Chicago: University of Chicago Press.

———. 1966. *The Evolution of Urban Society: Early Mesopotamia and Prehispanic Mexico*. Chicago: Aldine.

———. 1972. "Patterns of Urbanization in Early Southern Mesopotamia." In *Man, Settlement and Urbanism*, eds. P. G. Ucko, R. Tringham, and G. Dimbleby. London: Duckworth.

———. 1975. "The Mesopotamian Social Landscape: A View from the Frontier." In

Reconstructing Complex Societies. Supplement to the Bulletin of the American Schools of Oriental Research, no. 20.

———. 1981. *Heartland of Cities.* Chicago: Aldine.

Adams, R. McC. and H. Nissen. 1972. *The Uruk Countryside.* Chicago: University of Chicago Press.

Alden, J. R. 1982. "Trade and Politics in Proto-Elamite Iran." *Current Anthropology* 23(6):613–40.

Athens, J. S. 1977. "Theory Building and the Study of Evolutionary Process in Complex Societies." In *For Theory Building in Archaeology,* ed. L. R. Binford. New York: Academic Press.

Childe, V. G. 1952. *New Light on the Most Ancient East.* 4th Ed. London: Routledge and Kegan Paul.

Cowgill, G. 1975. "On Causes and Consequences of Ancient and Modern Population Changes." *American Anthropologist* 77:505–25.

Crumley, C. L. 1979. "Three Locational Models: An Epistemological Assessment for Anthropology and Archaeology." In *Advances in Archaeological Method and Theory,* Vol. 2, ed. M. B. Schiffer. New York: Academic Press.

Diakonoff, I. M., ed. 1969. *Ancient Mesopotamia.* Moscow: Nauka.

Diringer, D. 1962. *Writing.* New York: Praeger.

Finegan, J. 1964. *Handbook of Biblical Chronology: Principles of Time Reckoning in the Ancient World and Problems of Chronology in the Bible.* Princeton: Princeton University Press.

Flannery, K. V. 1972. "The Cultural Evolution of Civilizations." *Annual Review of Ecology and Systematics* 3:399–426.

Frankfort, H. 1956. *The Birth of Civilization in the Near East.* Garden City, N.Y.: Doubleday.

Friedman, J. and M. J. Rowlands. 1977. *The Evolution of Social Systems.* Pittsburgh: University of Pittsburgh Press.

Gelb, I. J. 1952. *A Study of Writing: The Foundations of Grammatology.* Chicago: University of Chicago Press.

———. 1969. "On the Alleged Temple and State Economics in Ancient Mesopotamia." *Estratto da studi in onore di Edouard Vtoltera* 4:139–54.

Gibson, M. 1972. "Population Shift and the Rise of Mesopotamian Civilization." In *The Explanation of Cultural Change: Models in Prehistory,* ed. C. Renfrew. London: Duckworth.

Hamblin, D. J. et al. 1973. *The First Cities.* New York: Time-Life.

Hole, F. 1977. "Pastoral Nomadism in Western Iran." In *Explorations in Ethnoarcheology,* ed. R. A. Gould. Albuquerque: University of New Mexico Press.

———. 1978. "The Prehistory of Herding: Some Suggestions from Ethnography." *Colloques international du CNRS,* no. 580.

———. 1987. "Settlement and Society in the Village Period." In *The Archaeology of Western Iran,* ed. F. Hole. Washington, D.C.: Smithsonian Institution Press.

Hole, F., K. V. Flannery, and J. A. Neely. 1969. *Prehistory and Human Ecology of the Deh Luran Plain.* Memoirs of the Museum of Anthropology, University of Michigan, no. 1.

Ibrahim, M., J. A. Sauer, and K. Yassine. 1976. "The East Jordan Valley Survey, 1975." *Bulletin of the School of Oriental Research* 22:42–64.

Ifrah, G. 1985. *From One to Zero.* Trans. L. Bair. New York: Viking Penguin.

Jacobsen, T. and R. Adams. 1958. "Salt and Silt in Mesopotamian Agriculture." *Science* 128:1251–58.

Jawad, A. J. 1965. *The Advent of the Era of Townships of Northern Mesopotamia.* Leiden: Brill.

Johnson, G. A. 1973. *Local Exchange and Early State Development in Southwestern Iran.* Museum of Anthropology, Anthropological Papers, University of Michigan, no. 51. Ann Arbor.

————. 1975a. "Locational Analysis and the Investigation of Uruk Local Exchange Systems." In *Ancient Civilization and Trade*, eds. J. Sabloff and C. Lamberg-Karlovsky. Albuquerque: University of New Mexico Press.

————. 1975b. "Early State Organization in Southwestern Iran: Preliminary Field Report." *Proceedings of the 4th Annual Symposium on Archaeological Research in Iran.* Teheran.

————. 1977. "Aspects of Regional Analysis in Archaeology." *Annual Review of Anthropology* 6:479–508.

————. 1981. "Monitoring Complex System Integration and Boundary Phenomena with Settlement Size Data." In *Archaeological Approaches to the Study of Complexity*, ed. S. E. van der Leeuw. Amsterdam: University of Amsterdam.

————. 1982. "Organizational Structure Scalar Stress." In *Theory and Explanation in Archaeology.* New York: Academic Press.

————. 1987. "The Changing Organization in Uruk Administration on the Susiana Plain." In *The Archaeology of Western Iran*, ed. F. Hole. Washington, D.C.: Smithsonian Institution Press.

————. 1987. "Nine Thousand Years of Social Change in Western Iran." In *The Archaeology of Western Iran*, ed. F. Hole. Washington, D.C.: Smithsonian Institution Press.

Just, P. 1980. "Time and Leisure in the Elaboration of Culture." *Journal of Anthropological Research* 36:105–15.

Kramer, S. N. 1959. *History Begins at Sumer.* Garden City, N.Y.: Doubleday.

Larsen, C. E. 1975. "The Mesopotamian Delta Region: A Reconsideration of Lees and Falcon." *Journal of the American Oriental Society* 95:43–57.

Larsen, C. E. and G. Evans. 1978. "The Holocene History of the Tigris-Euphrates-Karun Delta." In *The Environmental History of the Near and Middle East Since the Last Ice Age*, ed. W. C. Brice. Pp. 227–44. London: Academic Press.

Lees, S. H. and D. G. Bates. 1974. "The Origins of Specialized Nomadic Pastoralism: A Systematic Model." *American Antiquity* 30:187–93.

Loding, D. 1976. *Ur Excavations: Economic Texts from the Third Dynasty.* Philadelphia: Publication of the Joint Expedition of the British Museum and the University Museum, University of Pennsylvania.

Mallowan, Sir M. E. L. 1965. *Early Mesopotamia and Iran.* London: Thames and Hudson.

Melikishvili, G. A. 1978. "Some Aspects of the Question of the Socioeconomic Structure of Ancient Near Eastern Societies." *Soviet Anthropology and Archaeology* 7:25–72.

Mellaart, J. 1965. *Earliest Civilizations of the Near East.* London: Thames and Hudson.

————. 1975. *The Neolithic of the Near East.* London: Thames and Hudson.

Mitchell, W. 1973. "The Hydraulic Hypothesis: A Reappraisal." *Current Anthropology* 4:532–34.

Nissen, H. 1972. "The City Wall of Uruk." In *The Explanation of Cultural Change: Models in Prehistory*, ed. C. Renfrew. London: Duckworth.

————. 1988. *The Early History of the Ancient Near East, 9000–2000 B.C.* Chicago: University of Chicago Press.

Nutzel, W. 1976. "The Climatic Changes of Mesopotamia and Bordering Areas." *Sumer* 32.

Oates, J. 1980. "The Emergence of Cities in the Near East." *The Cambridge Encyclopedia of Archaeology*, pp. 112–19. New York: Crown Publishers/Cambridge University Press.

Oppenheim, A. L. 1964. *Ancient Mesopotamia: Portrait of a Dead Civilization.* Chicago: University of Chicago Press.

Pattee, H. H. 1973. *Hierarchy Theory. The Challenge of Complex Systems.* New York: Braziller.

Pollock, S. 1983. "Style and Information: An Analysis of Susiana Ceramics." *Journal of Anthropological Archaeology* 2(4):354–90.

Redman, C. L. 1978. *The Rise of Civilization*. San Francisco: W. H. Freeman.

Renfrew, C. 1975. *The Emergence of Civilization*. London: Methuen.

———, ed. 1972. *The Explanation of Cultural Change: Models in Prehistory*. London: Duckworth.

Roux, G. 1976. *Ancient Iraq*. Baltimore: Penguin.

Sabloff, J. and C. C. Lamberg-Karlovsky, eds. 1975. *Ancient Civilization and Trade*. Albuquerque: University of New Mexico Press.

Saggs, H. W. F. 1989. *Civilization Before Greece and Rome*. New Haven: Yale University Press.

Schacht, R. 1987. "Early Historic Cultures." In *The Archaeology of Western Iran*, ed. F. Hole. Washington, D.C.: Smithsonian Institution Press.

Schmandt-Besserat, D. 1981. "Decipherment of the Earliest Tablets." *Science* 211:283–84.

Service, E. R. 1962. *Primitive Social Organization*. New York: Random House.

———. 1975. *Origins of the State and Civilization*. New York: Norton.

Simmons, A. H., I. Kohler-Rollefson, G. O. Rollefson, R. Mandel, and Z. Kafafi. 1988. " 'Ain Ghazal: A Major Neolithic Settlement in Central Jordan." *Science* 240:35–39.

Smith, P. E. L. and T. C. Young, Jr. 1972. "The Evolution of Early Agriculture and Culture in Greater Mesopotamia. A Trial Model." In *Population Growth: Anthropological Implications*, ed. B. Spooner. Cambridge, Mass.: MIT Press.

Spooner, B., ed. 1972. *Population Growth, Anthropological Implications*. Cambridge, Mass.: MIT Press.

Spuhler, J. 1985. "Anthropology, Evolution, and 'Scientific Creationism.' " *Annual Review of Anthropology* 14:103–33.

Steward, J. H. 1949. "Cultural Causality and Law: A Trial Formulation of the Development of Early Civilizations." *American Anthropologist* 51:1–27.

Taylor, R. E. and R. Berger. 1980. "The Date of Noah's Ark." *Antiquity* 54:35–36.

Tosi, M. 1972. "The Early Urban Revolution and Settlement Pattern in the Indo-European Borderland." In *The Explanation of Cultural Change: Models in Prehistory*, ed. C. Renfrew. London: Duckworth.

———. 1984. "The Notion of Craft Specialization and Its Representation in the Archaeological Record of Early States in the Turanian Basin." In *Marxist Perspectives in Archaeology*, ed. M. Spriggs. Cambridge, England: Cambridge University Press.

Ucko, P., R. Tringham, and G. Dimbleby, eds. 1972. *Man, Settlement and Urbanism*. London: Duckworth.

Webb, M. 1975. "The Flag Follows Trade: An Essay on the Necessary Integration of Military and Commercial Factors in State Formation." In *Ancient Civilization and Trade*, eds. J. Sabloff and C. C. Lamberg-Karlovsky. Albuquerque: University of New Mexico Press.

Weiss, H. 1977. "Periodization, Population, and Early State Formation in Khuzestan." In *Mountains and Lowlands: Essays in the Archaeology of Greater Mesopotamia*, eds. L. D. Levine and T. C. Young, Jr. Malibu: Undena.

Wenke, R. J. 1975–76. "Imperial Investments and Agricultural Developments in Parthian and Sasanian Khuzestan: 150 B.C. to A.D. 640." *Mesopotamia* 10–11:31–217.

———. 1981. "Explaining the Evolution of Cultural Complexity. A Review." In *Advances in Archaeological Method and Theory*, Vol. 4, ed. M. B. Schiffer. New York: Academic Press.

———. 1987. "Western Iran in the Partho-Sasanian Period: The Imperial Transformation." In *The Archaeology of Western Iran*, ed. F. Hole. Washington, D.C.: Smithsonian Institution Press.

White, L. 1949. *The Science of Culture*. New York: Grove Press.

Wittfogel, K. A. 1957. *Oriental Despotism: A Comparative Study of Total Power*. New Haven: Yale University Press.

Wolkstein, D. and S. N. Kramer. 1983. *Inanna, Queen of Heaven and Earth*. New York: Harper & Row.

Woolley, Sir L. 1965. *Excavations at Ur*. New York: Crowell.

Wright, H. T. 1969. *The Administration of Rural Production in an Early Mesopotamian Town*. Ann Arbor: Museum of Anthropology, Anthropological Papers, University of Michigan, n. 38.

———. 1977. "Recent Research on the Origin of the State." *Annual Review of Anthropology* 6:379–97.

———. 1986. "The Evolution of Civilizations." In *American Archaeology. Past and Future*, eds. D. Meltzer, D. Fowler, J. Sabloff. Washington, D.C.: Smithsonian Institution Press.

———. 1987. "The Susiana Hinterlands During the Era of Primary State Formation." In *The Archaeology of Western Iran*, ed. F. Hole. Washington, D.C.: Smithsonian Institution Press.

Wright, H. T. and G. A. Johnson. 1975. "Population, Exchange, and Early State Formation in Southwestern Iran." *American Anthropologist* 77:267–89.

Yasin, W. 1970. "Excavation at Tell Es-Sawwan, 1969 (6th Season)." *Sumer* 26:4–11.

Young, T. C., Jr. 1972. "Population Densities and Early Mesopotamian Origins." In *Man, Settlement and Urbanism*, eds. P. J. Ucko, R. Tringham, and G. W. Dimbleby. London: Duckworth.

Young, C. T., Jr., P. E. L. Smith, and P. Mortensen, eds. 1983. *The Hilly Flanks and Beyond: Essays on the Prehistory of Southwestern Asia*. Chicago: Studies in Ancient Oriental Civilization, No. 36, University of Chicago.

Zagarell, A. 1986. "Trade, Women, Class, and Society in Ancient Western Asia." *Current Anthropology* 27(5):415–30.

9

The Origins of
Egyptian Civilization

Concerning Egypt I shall extend my remarks
to a great length, because there is no country
that possesses so many wonders, nor any that
has such a number of works that defy description.

Herodotus (c. 440 b.c.)

Ancient Egyptian civilization came into full flower a few centuries later
than Mesopotamian cultures, and it might have been influenced by them,
but Egypt's cultural evolution was largely an independent process so distinctive and brilliant that ancient Egypt remains the most widely renowned of all ancient civilizations.

For the archaeologist long trained in seeing the functional, the economic, in art, artifacts, and architecture, Egypt's similarities to other early
civilizations are obvious: Egypt is yet another variation on an ancient theme,
a developmental pattern in which people turned from hunting and foraging to agriculture, eventually built huge monuments, sent armies against
each other first and then against their neighbors, and eventually formed
great empires.

Yet the similarities among ancient civilizations should not blind us to
the unique genius of each. Egypt was a distinctive civilization, of an excellence in arts, letters, and science that can only be dimly glimpsed in its
artifacts that lie in museum cases all over the world.

Ancient Egypt's most important contributions to the comparative analyses of early civilizations may derive from its distinctive, almost contradic-
tory cultural characteristics: Egypt was one of the most centralized of early
political systems, yet it also seems to have been the least urban; its bureaucratic complexity was extraordinary, but the vast majority of people seem
to have lived in largely self-sufficient villages and towns; and though its
political cycles were closely related to a single environmental factor (Nile

370

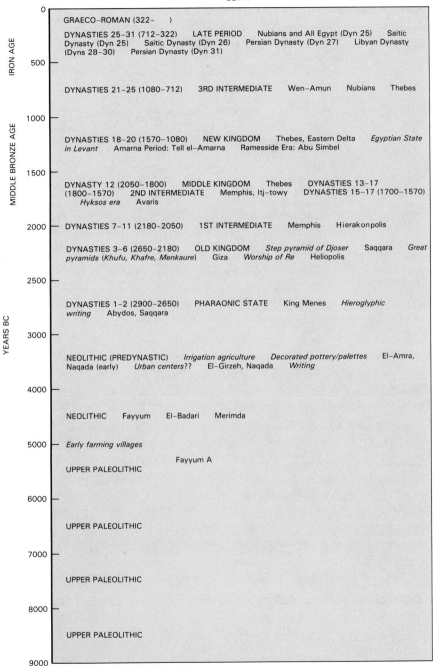

EGYPT

IRON AGE	
MIDDLE BRONZE AGE	
YEARS BC	

0

GRAECO—ROMAN (322—   )

DYNASTIES 25–31 (712–322)   LATE PERIOD   Nubians and All Egypt (Dyn 25)   Saitic
Dynasty (Dyn 25)   Saitic Dynasty (Dyn 26)   Persian Dynasty (Dyn 27)   Libyan Dynasty
(Dyns 28–30)   Persian Dynasty (Dyn 31)

500

DYNASTIES 21–25 (1080–712)   3RD INTERMEDIATE   Wen–Amun   Nubians   Thebes

1000

DYNASTIES 18–20 (1570–1080)   NEW KINGDOM   Thebes, Eastern Delta   *Egyptian State
in Levant*   Amarna Period: Tell el–Amarna   Ramesside Era: Abu Simbel

1500

DYNASTY 12 (2050–1800)   MIDDLE KINGDOM   Thebes   DYNASTIES 13–17
(1800–1570)   2ND INTERMEDIATE   Memphis, Itj–towy   DYNASTIES 15–17 (1700–1570)
    Hyksos era   Avaris

2000   DYNASTIES 7–11 (2180–2050)   1ST INTERMEDIATE   Memphis   Hierakonpolis

DYNASTIES 3–6 (2650–2180)   OLD KINGDOM   *Step pyramid of Djoser*   Saqqara   *Great
pyramids (Khufu, Khafre, Menkaure)*   Giza   *Worship of Re*   Heliopolis

2500

DYNASTIES 1–2 (2900–2650)   PHARAONIC STATE   King Menes   *Hieroglyphic
writing*   Abydos, Saqqara

3000

NEOLITHIC (PREDYNASTIC)   *Irrigation agriculture*   *Decorated pottery/palettes*   El–Amra,
Naqada (early)   *Urban centers??*   El–Girzeh, Naqada   *Writing*

4000

NEOLITHIC   Fayyum   El–Badari   Merimda

5000

Early farming villages

Fayyum A

UPPER PALEOLITHIC

6000

UPPER PALEOLITHIC

7000

UPPER PALEOLITHIC

8000

UPPER PALEOLITHIC

9000

9.1.   A cultural chronology of ancient Egypt.

371

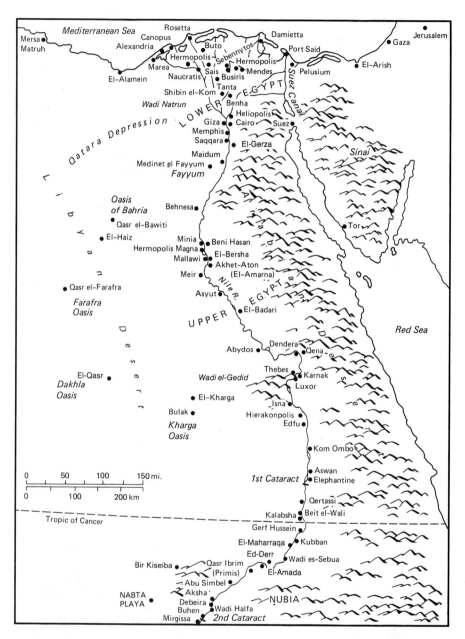

9.2 Some major Egyptian sites. Egypt's desert and mountain frontiers protected it for many centuries against outside influences, but already by 3000 B.C. its trade and military contacts extended into Palestine, Nubia, and the Mediterranean world.

flood levels), within these environmental limits Egypt's sociopolitical evolution was a baroque interweaving of factors, personalities, and events.

The Ecological Setting

Until this century, when dams were built in southern Egypt, the torrential spring rains of east and central Africa sent silt-choked floods pulsing down the Nile Valley, depositing along the way innumerable tons of rich soil and flowing finally, in autumn, into the Mediterranean. Along the river's course this natural alluviation has produced one of the world's richest agricultural niches, which with even the simplest tools supports as many as 450 people per square kilometer.

Although highly productive, the Nile Valley is really only an extremely elongated oasis (Figure 9.2) where agricultural lands are sharply circumscribed in most regions by rocky deserts. From the Sudanese border to Cairo, the cultivable strip along both banks of the river is only 3 kilometers wide in most places, and so sharp is the demarcation that, as the cliché has it, one may stand with one foot on the red desert sands and the other foot on the black, irrigated croplands. North of Cairo, the Nile divides into several main branches and many smaller streams, creating a delta of flat, well-watered, fertile land.

Thanks to the Nile, ancient Egyptians could traverse the length of their country in just a few days of pleasant sailing, providing a rapid, reliable channel for the movement of goods, people, and information. This was undoubtedly a key factor in the centralized governments that developed in later Egyptian antiquity, and it is not surprising that in the southern reaches of the Nile, where five cataracts (steep rapids) constitute impassable barriers to navigation, the power of the Egyptian states weakened.

Although there have been major fluctuations in the annual volume of the Nile—for example, catastrophically low water levels in some of the years between 2250 and 1950 B.C.—the Nile floodplain has existed in essentially its present form since about 3800 B.C.[1] The gradient, periodicity, and other attributes of the Nile are such that complex irrigation systems using dams and long feeder canals up river became important only fairly late in Egypt's history. For the most part the Nile forms small flood basins that a few villages could manage successfully.[2] One simply waited for the spring floods to recede, leaving behind their rich sediments, and then scattered seeds on the ground, herded pigs or sheep over the cropland to trample the seeds into the ground, weeded and hoed in a small way, relied on cats to keep down rats, birds, and other small crop pests, and then waited for the bountiful harvest.

But if spring floods were too high, then water stayed on the lands too long and crops could not be sown in time to harvest them before the summer heat destroyed them; if the floods were too low, not enough sediment and water were available for good harvests. Ancient Egyptians used tripods to dip water from the river to irrigate crops, and in some cases they trapped water behind dams during low flood years, so that they could use it later for irrigation, but in the long run—and with sensible storing of grains—the Nile Valley was rich enough to support high population densities.

Temperatures along the Nile regularly exceed 37°C (99°F), and rainfall is sparse: irrigation is required everywhere for agriculture. The heat can often be oppressive, but the climate allows three harvests a year of some crops and the cultivation of a wide range of plants. Since antiquity the staples have been wheat, barley, lentils, beans, onions, cucumbers, melons, and figs, and the pastures and gardens of the alluvium have long sustained sheep, goats, pigs, cattle, and fowl. Even without agriculture, the Nile itself is generous, supporting myriad fish, ducks, geese, turtles, crocodiles, hippopotami, and other game animals, dense stands of rushes and reeds for basketry, flax for linen and canvas, and papyrus for cordage and paper.

The deserts that parallel the Nile offer a few oases, but elsewhere they provide barely enough to support a few nomads. They are, however, rich in limestone, granite, and other building stone. Copper can be found in the Sinai Desert, while immense gold and silver reserves used to be available in the Eastern Desert.

Although there is evidence of contacts between Egypt and Palestine by at least 3700 B.C., Egypt's inhospitable deserts and the scarcity of natural harbors in the Delta isolated it to some extent from foreign influences and invasions, and probably did much to keep Egypt for thousands of years an insular country of small towns and villages, with few cities.[3]

In some ways the history of Egyptian archaeology is as interesting as the archaeology itself. Some of the exploits of Belzoni were recorded in Chapter 1, and he was just one of many early travelers to Egypt. Egyptian tombs and temples have been looted for thousands of years, but French, German, British, Italian, and American collectors have done the most damage. Today, there are more Egyptian obelisks in Italy and England than in Egypt, and Egyptian antiquities can still be bought in New York, Paris, Tokyo, and many other cities.

But European influences on the study of the Egyptian past have not been entirely malign. The British scholar Sir Flinders Petrie is a good example. A prodigy of sorts, he was a full professor at thirty-nine and spent years meticulously excavating and recording Egyptian sites. By the standards of his era he was an excellent technical archaeologist, he developed some brilliant applications of mathematics to archaeology, and he published the results of his research quickly and competently. In the field he was dedicated and never minded the rigors of camp life. He fed his staff on a diet that would have been proscribed for prisoners of war by the Geneva Convention, but he subsisted on it himself. His composure and discipline were impressive. Once, it is reported, when Petrie was excavating a brick wall a scorpion got into his long flowing white beard and Petrie nonchalantly laid his beard out on the wall and smashed it along its length with his trowel.

Much of what we know about the earliest periods in Egypt we owe to Petrie. But most early archaeologists in Egypt concentrated on tombs and temples, so that our knowledge of the Egyptian past is heavily biased toward the rich ceremonial centers. Also, life is so concentrated in the narrow valley that millennia of reoccupation and flooding have destroyed or obscured most of the older communities, in contrast to the extraordinary preservation of the mortuary centers along the desert margins.

Early Egyptian Agriculture (c. 5500–4000 B.C.)

Egypt beyond the green and vibrant Nile Valley is a starkly beautiful land, where clear blue sky and rocky reddish-yellow desert frame a simplified and, to all appearances, lifeless world. But at many times during past millennia, these deserts bloomed under seasonal rains that supported vast grasslands and cattle, gazelle, hares, antelopes, birds, and other animals.

Along the great erosional gulleys that lead into the river valley, one finds scatters of stone tools and bleached bones—all that remains of Paleolithic Egyptians and the animals they hunted. They began living here hundreds of thousands of years ago, perhaps earlier, and as the rain patterns shifted and made these areas deserts again, they probably moved back into the Nile Valley and its rich riverine resources.

An interesting glimpse into Paleolithic life in Egypt comes from Jebel Sahaba (in southern Egypt), where Fred Wendorf excavated a cemetery containing the skeletons of fifty-eight people who had died between about 12,000 and 10,000 B.C., many of them children under the age of three. Wendorf describes the burials:

> one of the unusual features of the burials was the direct association of 110 artifacts, almost all in positions which indicate they had penetrated the body either as points or barbs on projectiles of spears. They were not grave offerings. Many of the artifacts were found along the vertebral column, but other favored target areas were the chest cavity, lower abdomen, arms, and the skull. Several pieces were found inside the skull, and two of these were still embedded in the sphenoid bones in positions which indicate that the pieces entered from the lower jaw.[4]

One must be a bit wary of cemetery data, for they may give us a biased view of death. For all we know, 99% of the people contemporary with those in the Jebel Sahaba cemetery died in peace, full of years and in the bosom of their families. But the Jebel Sahaba corpses are not reassuring.

With the end of the last ice age at about 8000 B.C., the Nile Valley and its adjacent deserts took on much of the form and climate that they have today. The way of life in the period just after 8000 B.C., at least as we know it from excavations in the southwestern Egyptian desert, the Fayyum, and elsewhere, was apparently a slow seasonal round of hunting the desert margins for wild cattle, gazelle, birds, and many other animals, and exploiting the Nile for fish, fowl, and other animals and plants.

There is some suggestion of intensified plant use at late Pleistocene northeast African sites in areas like Kom Ombo and Wadi Kubbaniyya, in the form of mortars, sickle blades, and other implements, but the adaptation appears to have been a mobile one, based on small groups pursuing a diversified hunting-foraging economy.[5]

The earliest evidence of forms of subsistence, settlement, and technology in northeast Africa that differed significantly from those of the late Pleistocene comes from the desert areas of Bir Kiseiba and Nabta in southwest Egypt (Figure 9.2). On the basis of evidence from this area, Wendorf, Schild, and Close note that both "cattle and pottery seem to have been known in the Sahara as early as anywhere else in the world."[6] Moreover, they observe that the people of about 6200–5900 B.C. lived in communities in which

> The houses and pits indicate long term or, at least, recurrent settlement. At the very least, they must have been occupied for most of the year, and it seems likely that the Nile no longer played an important role in the settlement system, suggesting that a different kind of exploitation was now being employed in the desert. Instead of sites representing small family units or task-groups, there are now medium-sized villages, composed of perhaps as many as 14 family units . . . where there was at least sufficient social control to determine the arrangement of the community.[7]

We really don't know what these early hunter-foragers of the Egyptian deserts based their diets on—there are numerous ground stone tools but few plant remains from these sites.[8] They may have been cattle herders,[9] but the evidence is not conclusive.[10]

Apparently these desert peoples did not make the transition to a fully agricultural, sedentary way of life in these arid zones. Fekri Hassan has argued that eventually climate changes induced people to abandon the desert areas and migrate into the richer Nile Valley.[11]

Sometime between 6000 and 5000 B.C., domesticated wheat, barley, sheep, goats, and cattle were introduced from outside Egypt and became the basis for the evolution of Egyptian civilization. Whether these domesticates were introduced for the first time in this millennium or were just successfully farmed for the first time is unknown, as are their ultimate origins. Some of these plants and animals probably came from Southwest Asia, where they had already been in agricultural use for 2000 years, but others may have come from oases in the Libyan Desert.[12]

Our most systematic evidence about these early agriculturalists is from the Fayyum Oasis. British archaeologist Gertrude Caton-Thompson did excellent excavations along the lake's north shore in the 1930s, and Fred Wendorf's work there in the 1970s greatly augmented Caton-Thompson's findings. As noted in Chapter 2, our own work there involved six months of excavations and surface-collection in the deserts on the southern side of the lake.

On the basis of this combined evidence, the Fayyum appears to have been a rather unstable, only partly agricultural adaptation. Agriculture has such great inertia in that once you make large grinding stones, big grain storage silos, sickles, and all the rest of the primitive farmer's tool kit, you have real incentives not to move. Also, cereal grains ripen over a short interval, and a farmer who is out hunting the week they ripen can expect to come back to a field stripped bare by rats and birds.

Thus it is puzzling that the ancient peoples of the Fayyum left no traces of houses or prolonged occupation (Figure 9.2). Caton-Thompson found complete sickles in silos full of wheat and barley in Fayyum sites dating to about 5000 B.C., but people here seem to have combined some minor cereal farming with the ancient hunting-foraging ways. Partly for this reason, some scholars see a resistance to the spread of agriculture in Egypt because of the richness of the lake and river environments.

By 4000 B.C. agriculture had spread over much of Egypt, including the southern areas that had taken a few tentative steps toward agriculture in the late Pleistocene. Some people still depended on fish and wild plants for much of their food, others were locked into the wheat-barley, sheep-goat combination that underlies so much of Middle Eastern cultural evolution.

The Predynastic Period (c. 4000–3100 B.C.)

One of the most impressive and puzzling things about ancient Egypt is the apparent rapidity and comprehensiveness with which hundreds of unconnected and functionally similar villages were transmuted into an organized social, economic, and political unity—the first Egyptian state. This transmutation began at about 4000 B.C. in the south and rapidly spread to the north, encompassing most of Egypt by 3100 B.C.

The rise of the Egyptian "state" after 3100 B.C. is really just a bold inference based on: the spread over much of Egypt of pottery and architectural styles that suggest close, continuing contacts among people over large areas of the country; the "waste" of massive amounts of resources in tombs and monumental buildings in such a way as to imply an unequal distribution of wealth, power, and prestige; and some equivocal signs, like the Narmer Palette (Figure 9.4), that seem to indicate a potentate in the process of exercising kingly authority.

In 1894 Sir Flanders Petrie excavated more than 2000 graves at the site of Naqada, just north of Luxor, and defined the Predynastic cultural sequence (Figure 9.1). These graves were filled with pottery in different

9.3 As in other early civilizations, the rise of complex societies in Egypt was reflected in graves and tombs. This grave, from Minshat Abu Omar, in the eastern Egyptian Delta, dates to about 3050 B.C. By this time in Egypt some people such as this individual were buried simply and with just a few pots, while others were interred in rich tombs.

9.4 The Narmer Palette, a thin sheet of stone engraved with symbols that may reflect the political unification of Egypt. King Narmer is shown wearing different crowns on the different sides of the palette, possibly symbolizing the political unification of the Delta and Nile Valley, under the representation of the hawk-headed god Horus. Scenes of battle and what appears to be a beheading suggest a military unification of the country.

styles, as well as slate palettes, beautiful flint tools, jewelry made of bone and other materials, untold beads, figurines, and other riches. These graves and other Predynastic sites reveal a society quickly changing, in the direction of the class-stratified, religious unity that we know from documents of less than a thousand years later.[13]

By far the largest and most complex cluster of Predynastic settlements in Egypt are those at Hierakonpolis, the "City of the Hawk," where excavations have spanned almost a century (until recently under the direction of the late Michael Hoffman).[14]

The Predynastic people of Hierakonpolis lived in rectangular, semisubterranean houses of mudbrick and thatch, they apparently worshiped in small, perhaps wooden shrines, they made and distributed regionally several kinds of pottery, some if it very beautiful, they hunted, herded, fished, and farmed the by now traditional array of Egyptian plants and animals, and they buried their dead in rock and mudbrick tombs of a size and content to reflect the social power and prestige of the individual.

Much of the settlement of Hierakonpolis lies under the water-table. Slow, painstaking excavations aided by pumps were begun in the 1980s, and we eventually should know a great deal more about the structure and function of this capital of southern Egypt. The fact that the Narmer Palette (Figure 9.4) was found here and other evidence suggests that Hierakonpolis may have been the place where the unification of Egypt, north and south, Upper and Lower, was celebrated, and where, in a sense, "Egypt" may have first formed.

Excavations by Fekri Hassan at Naqada suggest that already by the middle of fourth millennium B.C. this site was a large town that already was fully dependent on agriculture. Naqada and Hierakonpolis may have been by far the largest communities in all of southern Egypt at this time, with much smaller communities scattered down the valley.

In the north, by about 3650 B.C. there was a large and comparatively wealthy community at Maadi, just south of what is now Cairo. Stone tools, pottery, and other artifacts show that the people of Maadi exchanged products with towns in Palestine, and they may have been part of a large trade network that integrated commodities from the Levant, the Nile Valley, and even the deserts and oases. The bones of donkeys have been found at Maadi, and one can imagine these people sending caravans off across the desert to Palestine, while their boats plied the Nile River trade.

Other Predynastic communities have been found in northern Egypt, most of them dating to just before 3000 B.C. At Minshat Abu Omar, hundreds of graves have been found containing the distinctive pottery, alabaster jars, slate palettes, and other riches of the late Predynastic. Recent excavations at Buto, in the north Delta, have revealed clay cones identical to those used to decorate temples in the Uruk state centered in Mesopotamia at about 3200 B.C. Buto may have been one of the most important ports on maritime trade routes over which vast quantities of timber, oil, wine, minerals, pottery, and other commodities passed.[15]

Fekri Hassan has shown that the evolution of complex societies in Predynastic Egypt was a long, involved process, in which climate changes, warfare, evolving agricultural efficiency, and the development of religions and political institutions were all subtly interrelated. He stresses that no simple ecological model can account for these changes, and that Egypt is not likely to have been unified by any single battle or ruler.[16]

The Archaic, Old Kingdom, and First Intermediate Periods (c. 3100–2040 B.C.)

The Predynastic period saw the first complex Egyptian societies, in trade patterns, occupational specialization, and settlement patterns. But Archaic and Old Kingdom periods were the great formative era of Egyptian civilization, the time when Upper and Lower Egypt were first united politically and Egyptian forms of writing, architecture, administration, and ideology emerged.

Traditional sources suggest that Menes (also known as Narmer [see the Narmer Palette, Figure 9.4]), a minor official from Upper Egypt, rose to power and conquered Lower Egypt at about 3100 B.C., and that he and his successors established a theocratic political system over the entire navigable length of the Nile. Menes is recorded as having built a capital at Memphis, diverting the stream of the Nile to create a strategic position at the junction of Upper and Lower Egypt. His next several successors were also powerful kings, but there is some evidence of internal dissension at about 2900 B.C. Later, peace appears to have been restored, and major construction projects were undertaken in the centuries before 2700 B.C. We know from tombs and other archaeological evidence that already by the early Old Kingdom (c. 2700 B.C.), Egypt was a complexly organized nation-state, with monumental architecture, a multi-tiered economy, and a centralized and hierarchically arranged bureaucracy.

Most of our evidence about these early centuries comes from a few badly looted tombs and sites. From these have been recovered beautiful artifacts of diorite and other varieties of hard stone, as well as skillfully fashioned copper implements and a few items of gold. Such distinctive Egyptian materials as faience (a glassy substance made from molded and fired crushed quartz) and papyrus paper were in use at this time, and contemporary documents written on this paper show that the Egyptians were already skilled in astronomy, geometry, accounting, surgery, and architecture.

By 2700 B.C. the economic sphere was already quite complex, involving long-distance trade to Syria and beyond and considerable local exchange of craft goods and foodstuffs; but most Egyptians of the Old Kingdom period continued to live in unwalled, largely self-sufficient villages. Apart from Memphis there were few towns or cities, a situation that may have contributed to the political integration of the country, since there were no urban power centers to resist incorporation.[17] Large areas of the middle Nile Valley were only sparsely settled, and population growth was quite slow, with little competition for agricultural land or irrigation water (although many new settlements appear to have been founded in the Delta). Apparently, the slow population growth during the Old Kingdom period (Butzer estimates an annual rate of 0.8 per 1000) eventually did begin to exert some pressure on available resources toward the end of the period, since large game almost disappeared from the Alluvium and contemporary documents describe a shift away from pastoralism to a greater reliance on grain agriculture.[18]

9.5 The Pyramids at Giza are the focal points in a vast mortuary complex of tombs and temples. But outside these mortuary centers Egypt was primarily a rural, agricultural society, with few cities.

The middle of the third millennium B.C. was for Egypt a marvelous age in which many of the greatest pyramids and palaces were built, an integrated royal bureaucracy formed, and arts and crafts were brilliantly executed. Because of the relatively comprehensive documents from this period, we know many of its political and social details.[19]

Djoser, the second king of the Old Kingdom period, was able to organize the people and economy of Egypt to the extent that he, or his grand vizier, Imhotep, could arrange construction of the great step pyramid at

Saqqara as his tomb. The actual crypt was built inside the pyramid, whose six levels rose over 60 meters and were surrounded by large buildings and a stone wall more than 9 meters wide with a perimeter of more than 1.6 kilometers. The pyramid complex at Saqqara was the world's first large-scale stone building and one of the most beautiful and, in terms of the efforts and materials required for its construction, it dwarfs the monumental architecture of all other early complex societies. It truly must have been an impressive sight, forty-five centuries ago, with its crisp white limestone facing contrasting with the cobalt blue sky, green palm groves, and desert sands.

Djoser's successors, particularly those of the Fourth Dynasty (c. 2613–2494 B.C.), also built massive pyramids and experimented with designs and constructions until the "perfect" pyramid form was achieved by King Khufu—as exemplified by the pyramids at Giza (Figure 9.5). It is not just the massive size of this and other pyramids of this era that is so impressive, but also the complex engineering, the deft execution of stone sculpture, and the precise planning such projects would have required.

So much has been said and imagined about the pyramids, yet we can hope to know so little. The minds that designed them and invested them with meaning are these many centuries gone to dust. Cairo sprawls out and around the pyramids today in such clamorous ubiquity that the only way to see the pyramids in a manner anything like that of the ancient Egyptian is to come in from the west, in the early morning, in the quiet of the desert.

The pyramids were all located on the west bank of the Nile, an equation no doubt with death and the setting sun. All are situated on the stone outcrop near the river, to facilitate the transfer of millions of blocks of limestone and alabaster, some of which had to be shipped down the Nile and then transported up the bank to the construction area.

Exactly how the blocks were quarried and transported remains unknown, but we have some clues. Quarrying probably involved a combination of hammering with hard stones, chopping with copper adzes, and fracturing by heating the rock with fires and then splashing cold water on it. The surveying and construction methods used to build the pyramids are the subject of an extended treatise by Mark Lehner, where the various possible combinations of ramps, rollers, and so on are checked against the physical remains.[20] The Great Pyramid of Giza required the quarrying, transport, preparation, and laying of 2,300,000 stone blocks, each with an average weight of 2.5 tons, and it is estimated to have required a labor force equivalent to about 84,000 people employed for eighty days a year for twenty years. It is not known how these people were mobilized and administered, but many think the construction was done by the peasantry during seasons when little agricultural activity was required. The administration, feeding, direction, and planning required to control such a work force, which included many highly trained craftsmen as well as laborers, would obviously argue a high degree of political and bureaucratic centralization. The king was apparently able to call on all the resources of the country, and direct them and the people to virtually any end, and at times the entire national economy was probably focused on these projects. The absolute control of the monarch is directly reflected in the texts and in the mortuary complexes of the various levels of high-ranking administrators who served him, many of whose tombs are laid out around the king's, reflecting the king's control over them even into eternity.

With the construction techniques available to Old Kingdom craftsmen, a pyramid is the only architectural form that could support its own weight when built to the scale that the Old Kingdom pyramids were. But we also have to account for the dimensions and angles used. Edwards makes the interesting suggestion that the upward angle of the pyramids parallels that of the slant of light on winter afternoons in Egypt, and he says the texts hint that the pharaohs ascended into heaven by walking up the rays of light.[21] The pyramids thus would be a first step to the union with the sun god and eternity.

Limited as they were by the lack of electronic survey equipment and precisely engineered tools, we cannot expect Egyptian craftsmen to have been up to modern construction standards. The north corner of the Great Pyramid, for example, is almost a whole inch higher than the south corner.

In preparing the mortar to bind the blocks of the three pyramids at Giza, the Egyptians burned limestone, and some of the charcoal from these fires is mixed in the mortar. As mentioned in Chapter 2, in 1984 Mark Lehner and I began collecting these charcoal samples from these and other Old Kingdom pyramids with the idea of radiocarbon dating them and thereby establishing a chronology for the pyramids and associated buildings, and perhaps even getting an idea of how long it took to build individual monuments. We reasoned that if we could get hundreds of samples of carbon from each pyramid, then the random errors and imprecisions of any single date would be lost in the general pattern of dates.[22] Our average dates are three to four hundred years older than the historical chronology, and almost no one believes that our dates are an accurate estimate of the actual time the pyramids were built. We hope to collect more samples, however, and our research is a test both of the dates of the pyramids and the dating methods themselves.

The pyramids fascinate most people who see them, and on hot summer nights tourists continue experiments to determine if the pyramids retard aging, heighten sexual potency, cure diseases, or in other ways give off powerful emanations.

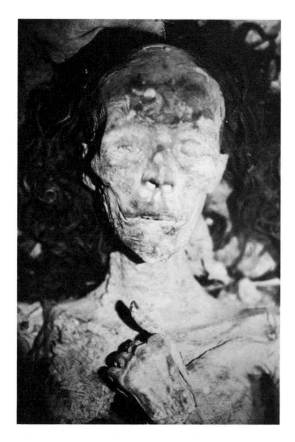

9.6 Mummy of Queen Tiye—her lovely hair still largely intact—who died about 1350 B.C. The position of the left hand is typical of Egyptian royalty.

Some of the most impressive features associated with the pyramids were the "solar boats," which are thought to have been buried at the time of the pharaoh's death, to facilitate his sailing into eternity. These boats were disassembled and buried in stone pits near the pyramids, and recently one was excavated and reconstructed—the wood, ropes, and other components almost perfectly preserved by the dark and aridity of the boat pits. In 1987 scientists introduced a camera through an airlock into another boat pit, in the hopes that the atmosphere would be unchanged since the time of burial, but the camera showed dung beetles walking around somewhat decayed segments of the boat.

Mark Lehner has shown that the Sphinx, the three main Giza pyramids, and various temples are part of a single unified design, in which the different strata of limestone on the Giza Plateau were carefully used to produce this magnificent ceremonial center.[23]

Shortly after 2495 B.C. there was a change in dynasties as well as in the religious and political texture of the Old Kingdom. The worship of the

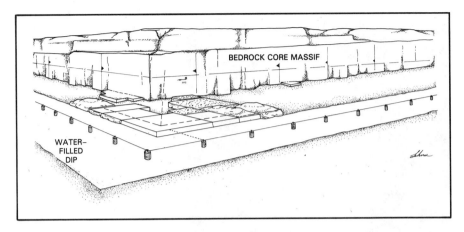

9.7 Mark Lehner (1987) has reconstructed one method whereby the ancient Egyptians may have laid out a nearly perfect square around a limestone mound over which the Great Pyramid of Khufu at Giza was built. Because of the mound, surveyors could not have used the diagonal of a square to calculate and lay out a perfect square. But by placing stakes in water in a channel around the pyramid's base they could have determined a truly level plane, and then used successive measurements from the stakes to the first row of stones to approximate very closely a square.

sun god, Re, emerged as the dominant religion, and the nobility and provincial authorities began to encroach on the king's authority.

What Egypt was like outside the main centers during the Old Kingdom is less well known. Since 1984, we have been excavating the remains of a small Old Kingdom town, at Kom el-Hisn, in the western Delta.[24] The composition and history of communities like Kom el-Hisn must be assumed to be partly determined by variability in the power of the pharaoh and elites to control rural sectors of the state. Wilson notes that the written language of Old Kingdom Egypt had no words for "government" or "state" as impersonal terms, conceived apart from the pharaoh; the Egyptian "theory of government was that the king was everywhere and did everything. . . . The fiction of direct delegation of duty and of a direct report to the king was impossible to maintain in practice; but in the theory of government it was no fiction, it was a working reality."[25] Pharaoh Wahkare' Khety III (2070–2040 B.C.), in his instructions to his son, Merykare, forcefully recommended building towns as a means to counteract political fragmentation and inefficient organization, especially in the eastern Delta, which, he lamented, was being subdivided into rival provinces and cities.[26] Kemp suggests that in Upper Egypt the control of local affairs by the pharaoh's overseer was gradually diluted during the late Old Kingdom, culminating in the appearance of provincial governors, or *nomarchs*.[27] Scholars differ on the extent of fluctuations of royal power during the 4th and 5th Dynasties—when Kom el-Hisn was occupied. Trigger raises the possibility that a slow but continuous expansion and elaboration of society and economy in the Old Kingdom may have been accompanied by growing complexity and power of provincial administrative institutions.[28] The apparent emergence of powerful *nomarchs* in the 6th Dynasty may reflect a reduction of pharaonic power, but the pharaohs of this period were still able to send expeditions to Nubia and Palestine and exert considerable internal control as well.[29] Unfortunately, except for Kom el-Hisn, only a few small areas of provincial early pharaonic communities have been excavated.[30] These communities seem small by Mesopotamian standards; some were walled and possessed modest public architecture. Recent surveys by Dutch archaeologists in the east Delta revealed scores of Old Kingdom villages, however, and their future excavations might be very informative as to the nature of provincial Old Kingdom Egypt.[31]

The breakup of central control came in the First Intermediate Period (c. 2160–2040 B.C.), a time of political and religious upheaval. The increasing prominence of the god Osiris cut away at the foundation of the old state religion of Re, in which the king was central and absolute.

Some characteristics distinguish Old Kingdom Egypt from Mesopotamia and other early complex societies. For example, there seem to have been no standing armies during most of the Old Kingdom and no economically significant slavery. In some ways the economic system—although highly administered—was a simple redistributive, almost chiefdom-like system, quite different from that of early Mesopotamian states.

Economic exchange was apparently controlled almost entirely through the king; there were no "merchants," in the capitalistic sense at least, until centuries after the end of the Old Kingdom. Craftsmen, scribes, peasants, and everyone else were required to perform some services in the name of the king and were liable for military and civil conscription, but there is a clear contrast here with the partially capitalistic, multi-tiered, highly differentiated economic system characteristic of the later Mesopotamian states.

The political breakdown of Egypt toward the end of the Old Kingdom may have been in part a result of dramatic climatic changes. Rainfall decreased in much of Egypt after about 2900 B.C., and Butzer notes that decreasing rainfall would have reduced the resources and numbers of desert nomads as well as eliminated much of the seasonal pastoral movements into the deserts.[32]

The Middle Kingdom and Second Intermediate Periods (2040–1570 B.C.)

Reconstructing the sociopolitical changes of the Middle Kingdom (c. 2080–1640 B.C.)—Egypt's first true imperial age—is a complex matter. In this period, successive rulers sought to increase national integration while directing defense and trade along increasingly active frontiers. For example, Ammenemes I (c. 1930 B.C.) seems to have been in a constant struggle for power with provincial governors during a period of Asiatic threat to the east Delta.[33]

In general, the history of the Middle Kingdom contains the same cycles of expansion and collapse that can be seen in all the great ancient empires. Periods of well-regulated trade, prosperity, and brilliance in art, architecture, and literature were punctuated by periods of revolution, poverty, and political fragmentation. The Middle Kingdom originated in the great civil unrest of the twenty-first century B.C., when, according to a contemporary account,

> Corn has perished everywhere. . . . People are stripped of clothing, perfume, and oil. . . . Everyone says, "There is no more." . . . Strangers have come into Egypt everywhere. . . . Men do not sail to Byblos today: What shall we do for fine wood? Princes and pious men everywhere as far as the land of Crete are embalmed with the resins of Lebanon, but now we have no supplies. . . . The dead are thrown in the river. . . . Laughter has perished. Grief walks the land.[34]

Conditions began to improve radically after 2040 B.C., when Mentuhotep II brought Upper and Lower Egypt once again under the rule of a single royal house. Mentuhotep and his next several successors reorganized the country with considerable energy, undertaking expeditions into Nubia, Libya, and Syria, reopening trade routes to the Red Sea, and commencing again the construction of monumental buildings.

In about 1999 B.C., Amenemhet came to power, and he and his successors inaugurated one of the most glorious epochs of Egyptian civilization. The capital was reestablished near Memphis, from which both Upper and Lower Egypt could be ruled effectively, trade routes were extended, fortresses were built along the country's frontiers, and the tradition of co-regency was established, in which sons were made co-rulers toward the end of their fathers' reigns, thereby eliminating some of the bloody battles for succession that had plagued previous dynasties.

There were also advances in Egyptian art and architecture during this period. Many literary classics were composed, and the cult of Osiris completed its replacement of the colder, sterner religion of Re and gave the common people some hope of the afterlife that in the past had been restricted to royalty.

From about 1786 to about 1720 B.C., various kings managed to remain in general control of most of Egypt, but gradually the power of Asiatic peoples in the eastern Delta increased. The origins of these foreigners, collectively referred to as the Hyksos, are uncertain, as is the manner in which they took over Egypt and the extent of their domination. But at about 1674 B.C. they captured Memphis, and the Hyksos king adopted the trappings of Egyptian royalty. Artifacts made in the manner of the Hyksos have been found all along the Nile Valley and as far south as Karnak, but it is not clear how directly they were able to control most of the population.

The New Kingdom and Third Intermediate Periods (1570–664 B.C.)

By the mid-fifteenth century B.C. Egypt probably had a population of many millions of people, whose governmental institutions, religion, language, economy, and most other aspects of life were probably remarkably like those of their ancestors during all of the preceding 2000 years.[35] This great weight of *tradition* must be considered when analyzing Egyptian history. Time and time again Egypt would fragment under revolt and invasion, but the ancient order would always persist and reform.

New Kingdom documents, such as the Wilbour Papyrus, describe a highly stratified society, with a king and elite who could draw freely on the country's wealth, and under them provincial nobles, who inherited their titles and wealth, lesser bureaucrats, priests, military officers, wealthy farmers, craftsmen, soldiers, tenant-farmers, and, supporting the whole structure, the peasants (Figure 9.8).[36]

The major cities of the New Kingdom were Memphis, near what is now Cairo, and Thebes, at Luxor. King Ahmose, a native of Thebes, began at about 1570 B.C. to expel the Hyksos from the Delta, and after several battles drove them beyond the eastern frontier. He even captured the rich city of Sharuhen, in Palestine, and he and his successors reformed the bureaucracy, modeling it after that of the Middle Kingdom.

Perhaps the greatest ruler of this period was Thutmosis III. He reigned for fifty-four years (1490–1463 B.C.), and his mummified body suggests that he died still looking quite young for his many years and accomplishments.[37] Thutmosis III established Egypt's Asiatic empire with his conquest of much of the eastern Mediterranean coastal areas. Even powerful Assyria paid material tribute to the Egyptian Empire, as did the Babylonians and the Hittites. His surprise attack on Megiddo and his amphibious operation against the Mitanni, a powerful kingdom in Southwest Asia, established Egypt as a world power. By about 1450 B.C. Egypt had commercial contacts on a large scale, exchanging products with Phoenecia, Crete, the Aegean Islands, and its traditional African trading partners. Military pacification programs were extended far into Nubia, and vast quantities of Nubian gold and building stone were shipped to the Nile Valley.

One of the most famous monarchs of the New Kingdom was Akhenaten (ruled 1364–1347 B.C.), who introduced a semimonotheistic religion

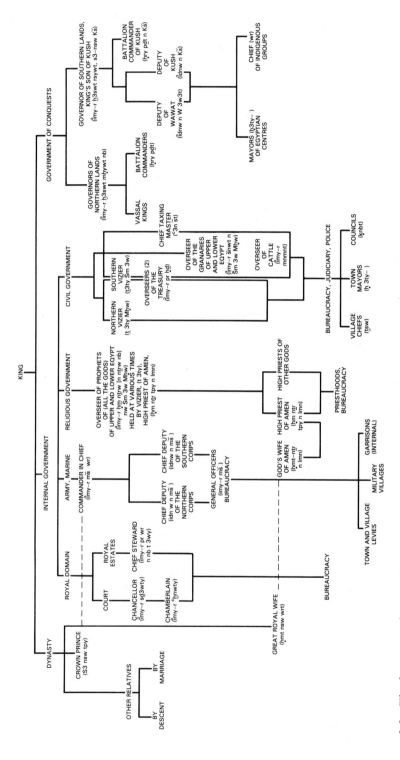

9.8 The bureaucratic organization of Egypt in the New Kingdom Period. This reconstruction by David O'Connor (1983) shows the numerous officials who organized most aspects of Egyptian society.

and tried to eradicate vestiges of older polytheistic cults. He built marvelous temples to Aton, the new god, and constructed a new capital at Tell el-Amarna in central Egypt, complete with magnificent religious and administrative buildings.

Akhenaten's physical appearance suggests he may have been a hermaphrodite, and some suspect that he had physical problems that produced some mental instabilities, but he was a reasonably effective civil and military leader.

> Akhenaten started off with certain disadvantages. From statues and portraits medical historians have diagnosed tuberculosis, hyperpituaritism, hypogonadism, and acromegaly. . . . Nor was his emotional life any healthier. His first wife appears to have been his mother, Tiy . . . they had one daughter. Then he married his maternal cousin, Nefretiti, and fathered three more daughters. His third and fourth wives were not blood relatives, and there was one son of each marriage, the second of whom was to rule as the boy pharaoh Tutankhamon. Akhenaten's fifth and last marriage was to the third of his own daughters by Nefretiti . . .[38]

Soon after Akhenaten's death the old religions were reestablished by Smenkhare and the famous Tutankhamen. Akhenaten's mummy was probably intentionally destroyed by bitter opponents of his attacks on the old religious traditions of Egypt.[39] The "Boy King," Tutankhamen, reigned as a teenager from about 1313 to 1300 B.C., and powerful members of his court tried to reestablish the power of the central court by reviving the state religion in the classical traditions of the cult of Amun.

When Howard Carter and his financial sponsor, Lord Carnarvon, discovered Tutankhamen's undisturbed tomb in 1922, their find captured the world's attention as no other archaeological discovery has. Lord Carnarvon's death from an infected mosquito bite in 1923 and the family's later run of bad luck inspired the notion of the "Curse of the Pharoahs" and made Lord Carnarvon's son so disenchanted with things Egyptian that he had some of his father's antiquities sealed away in the family castle. In 1988, with the help of a 75-year-old butler, these antiquities were rediscovered, but the most important finds from this tomb had long since been sent to New York's Metropolitan Museum and other institutions.

Between about 1225 B.C. and the accession of Rameses III (ruled 1184–1153 B.C.), Egypt's fortunes varied, with periods of foreign invasion and some internal breakdowns of the social order. And from Rameses III on, "[c]ontraction, not expansion, characterized the foreign policy while the disintegration of government became evident."[40] Even in an era of megalomaniacs, Rameses III is impressive. While invading Palestine, he incompetently allowed himself and his army to be ambushed by Hittites. He barely salvaged himself by a strategic retreat, and then preceded to cover almost every available flat stone surface in Egypt with heroic accounts of his mighty victory. Rameses III and his successors were also plagued by court intrigues, probably some assassination attempts, various economic problems, and threats along the frontiers.

It may have been his successor, Merneptah, who refused to let the Israelites leave and gave Charlton Heston such a difficult time. The pyramids were built thousands of years before the Israelites arrived, and it may be that the building projects described in the Bible as their work were small mudbrick complexes in the eastern Delta.

At about 1000 B.C., Egypt lost military control of Nubia, and the breakup of its Asiatic empire brought it into confrontation with the Israel of David and Solomon. The Egyptians captured a city on the border of Israel and agreed to peace with the marriage of the pharaoh's daughter to Solomon. But five years after Solomon's death Seshonk I invaded Israel, plundered Jerusalem, and reestablished Egypt's control.

During the first millennium B.C., Egypt had various periods of resurgence when various kings reasserted Egyptian influence in Palestine and Africa, and as late as 715 B.C. Shako conquered the Delta and forged a unified country once again. But continued military pressures from Kush and the disintegration of national political and economic systems ate away at the country's structure and stability and, at about 525 B.C., Cambyses, a Persian king, conquered Egypt and reduced it to a vassal kingdom, proclaiming himself pharaoh. In 332 B.C., Alexander the Great marched into Egypt, evicted the Persians, and built the city of Alexandria. Later, the Romans, Arabs, and British would complete the conquest of Egypt, submerging almost entirely this distinctive civilization that was for so many years the light of the ancient world. Not until A.D. 1952 was Egypt again ruled by Egyptians.

Ancient Egyptian Art and Thought

As we noted in Chapter 1, some archaeologists now argue that the archaeological record can only really be interpreted if we shift our attention away from the simplicities of technology, environment, and economy, to the ideas, concepts, and world-views of ancient peoples. Unlike many areas, in Egypt it is in some ways possible to do this, because of its rich and early literature. Through most of antiquity, Egyptians believed that their society and life in general had been established by the gods and that they and all their social, political, and economic relationships were part of a divinely designed, immutable world order. To some extent, the Egyptian past makes sense only when viewed in that way.

Certain themes are deeply embedded in ancient Egyptian culture. For one thing, Egyptians seem to have been a "God-intoxicated" people, "half in love with easeful death." Herodotus wrote that they were the most "religious" people he had encountered, given to incessant and elaborate religious rituals and an enormous priestly bureaucracy. Their concern with death, and the vast energies and richness they invested in preparing for it, are manifestly evident, but it is also a testament to a people so passionately alive that they tried everything to perpetuate life into death.

Mummification was an attempt to preserve the body for use in the afterlife, when it would be revived and rehabilitated. Burial in the arid desert sands must have been the first form of mummification, but chemical methods were already developed by the Old Kingdom period. After death, the corpse was placed on a board and washed. The brain was removed by a hooked wire passed through the nose, an incision was made in the abdomen, and all the internal organs except the kidneys and heart were removed. A Greek of the third century A.D. reported that at that time the stomach was placed in a box and offered to the sun god, with the incantation "If I have sinned in eating or drinking what was unlawful, the fault was not mine, but of this" (showing the box in which was the stomach).

The heart was left in—if the tomb paintings are realistic—so that on the final day it could be weighed against a feather, to see if its sins would tip the balance against eternal life. After the internal organs had been removed, the abdomen and chest cavity were washed in palm wine, and the viscera were placed in a container of natron (hydrated sodium carbonate) for forty days. The liver, lungs, stomach, and intestines were then placed separately in four "canopic" jars, to be guarded over by four different deities. The body was then stuffed with sand or straw and covered with natron to desiccate it for another forty days. Then it was washed once more and rubbed with wine, spices, and oils. The cheeks were restored to lifelike dimensions with rag stuffings, the incised abdomen was sewn up, and the hair and complexion were touched up with paint. The entire body was protected with a coating of resin, placed in a coffin or tomb, and carried across to the west bank of the Nile for interment. At the close of ceremonies, the priest would incant: "You live again, you revive always, you have become young again, you are young again, and forever . . ." It must have been a great comfort to the ancient Egyptians to hear these words and see the care of the dead.

To our own, essentially Greek, minds, the Egyptians seem to have been unable to distinguish between things and their substances. J. Wilson argues that the Egyptians saw no difference between supplying a dead king with real loaves of bread, wooden models of bread, or loaves painted on the walls; it was not the actual thing that mattered, it was the idea.[41] The physical man needed physical bread, but in the spirit world, "spiritual" bread was appropriate.[42]

Egyptian deities were conceived of as very human in their behavior, even to the extent that they could be intimidated. Egyptians recited prayers in which they ticked off the services rendered a god, demanding payment in the form of the prayer answered.

For the Westerner, the physical world is a rather neutral place, where a lightning bolt or virus may strike one down, but on the basis of chance, without malevolence. Rocks are rocks, the dark is peopled only by morbid human projections, and death is inevitable, final, complete. For the ancient Egyptian, however, the world teemed with unseen but animate, conscious forces; malignant spirits were everywhere, as were forces for good; and, with sufficient effort, some of the inconveniences of being dead could be mitigated.

Ancient Egyptians seem to have had a monophysitic perspective in that everything in the universe was thought to have been derived from one substance and was an expression of that substance. The god Amon, for example, might reside in a stone statue, but also in a well-formed ram or a duck, or in all three at the same time. Nor were these considered just different representations; rather, "the image was the god for all working purposes."[43] The symbolic, mythical element seemed to pervade even quite practical areas, such as medicine. One remedy for schistosomiasis (a disease characterized by bloody urine) was to shape some cake dough like a penis, then wrap it in meat, recite an incantation, and feed it to a cat.[44]

In all this supernatural and symbolic content, we should not lose sight of the practical, canny Egyptian. These were not people paralyzed by the Infinite: they built houses, boats, and beautiful buildings, and they enjoyed themselves in a world of color, play, and physical pleasures that still seems attractive and alive, even when viewed only in fragmentary 4000-year-old paintings on tomb walls. When contemporary tourists see Egyptian wall paintings they are often surprised at the riot of colors—even the great temple columns of Karnak were brightly colored. This seems strange to many because we see beauty in the monochrome simplicity of Egyptian stone reliefs and buildings. But it helps to understand their art if one experiences the qualities of Egyptian natural light, the sun rolling up redly over the horizon, illuminating the blues and greens of river, sky, and vegetation with a vividness rarely seen outside the Nile Valley, and the limestone absorbing and transforming the light into weights and colors and dimensions and textures that constantly change.

For much of its history, Egyptian society seems to have been elaborately hierarchical, but well ordered and even "fair." Central to the culture was the concept of *ma'at*, which is usually translated as "justice," "truth," or "right-dealing." It is interesting that English really has no word equivalent to the sense of *ma'at*, or to the Greek concept of "virtue," which has a little of the same sense of *ma'at*. For the Egyptian, *ma'at* was recognizing the order of the world and universe and the necessity of doing the right thing, which usually meant following religious and civil laws and customs. Justice

9.9 Members of a 1920s era Chicago Cubs baseball team play a road game.

EGYPTIAN SCRIPTS (Alphabet)

Hieroglyphic sign	Meaning	Tran-scription	Sound value	New Kingdom Hieratic	Demotic	Coptic
	vulture	3	glottal stop		2	omitted or єı
	flowering reed	ı	I			єı or є
	forearm & hand	ꜥ	ayin			omitted
	quail chick	w	W			oγ
	foot	b	B			π or в
	stool	p	P			π or в
	horned viper	f	F			ϥ
	owl	m	M	ꞋꞋ or)		м
	water	n	N	—	-	N
	mouth	r	R		о [√]	p or λ [є]
	reed shelter	h	H			8
	twisted flax	ḥ	slightly guttural			8 or omitted
	placenta (?)	ḫ	H as in "loch"			8 or ϩ
	animal's belly	ẖ	slightly softer than ḫ			8
	door bolt	s, z	S			c
	folded cloth	s, ś	S			c
	pool	š	SH		λ	ω
	hill	ḳ	Q			к, σ
	basket w. handle	k	K			к, σ
	jar stand	g	G			σ
	loaf	t	T			т. θ
	tethering rope	ṯ	TJ		ϳ (ϼ)	x. т
	hand	d	D		(◻—Δ)	т
	snake	ḏ	DJ		ϲ	x

EGYPTIAN ROYAL TITULARY

Horus Name	Nebty Name	Golden Horus Name	Prenomen	Nomen
(srḫ)	(nbty)	(Ḥr nbw)	(ny-sw bit)	(s3 Rꜥ)

9.10 The hieroglyphic signs in early Egyptian writing were representations of common objects but they eventually acquired phonetic meaning—sound values—that allowed Egyptians to express their entire spoken language in writing. In later periods the signs were stylized and simplified to facilitate writing. The names of royalty were written in ovals—cartouches—and each ruler used several different names.

tempered with mercy, giving to the widow and the orphan but encouraging self-reliance and planning, doing one's share—all these and more were part of *ma'at*. One could demand justice and respect as a moral right, based on *ma'at*.

EGYPTIAN WRITING

One of the many mysteries of Egypt is the origin of its written language, or hieroglyphs ("sacred carvings") (Figure 9.10). These were used from shortly before 3100 B.C. until about A.D. 40 (and still today by the crafty forgers in modern Cairo's tourist traps), and they went through a period of aesthetic and linguistic development. But hieroglyphic writing first appeared in such a developed form that we cannot see the full transition from what must be assumed to be a pictographic writing that may have been first expressed on papyrus. Some scholars think writing was introduced from Mesopotamia, where it may have been in use some centuries before it appeared in Egypt, but there are great differences between the characters and materials used.

Ancient Egyptian was a mixture of signs and symbols, some of them expressing sounds in the spoken language, others indicating to the reader how a written character with several possible meanings was to be read in that specific context (Figure 9.11). It was not used as a truly alphabetic system, but virtually everything in the spoken language could be efficiently conveyed in the written language.

To a much greater degree than in modern English, hieroglyphs were written in a manner that expressed both contemporary aesthetic styles and the subject matter. Thus, we get simple, grave characters in early religious writings and more overblown, showy texts in later military texts.

The story of how Egyptian hieroglyphs were deciphered is interesting and somewhat strange. The hero is Jean-François Champollion, born in France A.D. 1790. By the age of sixteen he was fluent in Latin, Greek, and at least six other Oriental languages, and from boyhood he was of such an Oriental appearance that eventually he was nicknamed "The Egyptian." To enter a Parisian school he was asked to write a paper on the subject of his choice, and he wrote on *Egypt under the Pharaohs* with such obvious intelligence that he was immediately asked to join the faculty.

Champollion applied his genius to the translation of the Rosetta Stone, which had been found by one of Napoleon's soldiers in 1799 in the northern Delta. We now know that its text is an unimportant tribute to the Pharaoh Ptolemy V, written at about 196 B.C., but it is significant because the same text was written in Greek as well as in two forms of ancient Egyptian—hieroglyphic and demotic. The first word recognized was *Ptolemaios,* the king's name, and Champollion extended his analysis to an inscription on an obelisk in which the name of some ruler was evident in

Hook signs and long-handled mace, phonograms for *s* and *hd*, spell the word for illumine, *shd*. The sun is a determinative, the basket a masculine suffix for "you." Literally, "May you illumine."

Face ideogram has value of *hr* and also means *face*. Vertical rod signals "here symbol means what it depicts."

Horned viper is masculine suffix -*f* and signifies "he," "him," or "his."

Desert hare stands for the sound *wm*—the word for "open"—reinforced by the wavy water symbol *n*. Two determinatives follow: Door on its side indicates "open," forearm holding stick adds the idea of "force" or "effort."

Basket: masculine suffix for "you." Coupled with preceding group, it makes the five signs read: "May you open."

One eye is an ideogram that can stand for "see." But two indicate the "eyes" themselves.

Horned viper: "his."

9.11   Reading hieroglyphic writing required recognizing symbols that convey sound values, symbols that are determinatives—which clarify meanings in ambiguous contexts—and symbols that are grammatical markers. The inscription outlined above is on the sarcophagus of Amenhotep II (died c. 1412 B.C.) and reads "Illumine his face, open his eyes."

the *cartouche*, with the oval always enclosing the ruler's name in Egyptian inscriptions. Correspondences between letters common both to Ptolemy and Cleopatra gave Champollion four letters, or sound values, and then by examining hundreds of cartouches he was able to determine the sound values and meanings of much of the written language.

The ancient written Egyptian language was different from ancient Mesopotamian scripts in a significant structural element. Sumerian nouns and verbs did not vary in their written forms depending on grammatical context, so that the word for mouth *(ka)*, for example, was always the same, whether it was the subject, direct object, or some other grammatical element; but in ancient Egyptian the first letter in the word for mouth, r, was followed by a different vowel, depending on the grammatical position of the word.[45] As H. W. F. Saggs notes, an important consequence of this is that the ancient Egyptians "already had an alphabet within their grasp." Ancient Egyptians could have written all or most of their spoken language with just 24 signs. Saggs points out, however, that the Egyptians never simplified their written language to this alphabetic format—probably, as he says, because it would have been against the interest of the scribes who had mastered the difficult and more cumbersome syllabic forms to use a system anyone could have mastered in just a few months—a reaction similar to that of print unions throughout history.[46]

The Origins of Complex Societies in Egypt: Conclusions

Many simplistic notions about the origins of cultural complexity have foundered on the evidence from Egypt. Population growth, for example, may have produced fairly dense concentrations of people in favorable agricultural areas, but the long-term pattern of population growth was probably one of very slow increase through most of antiquity. Karl Wittfogel's notion of large-scale irrigation as a powerful impetus to empire (Chapter 7) may have some application to Egypt, for irrigation no doubt was important early on, and some of the first Egyptian stone engravings apparently show royalty in the process of opening irrigation canals.[47] But irrigation in ancient Egypt was primarily through the passive blessings of the Nile flood, and such irrigation works as were constructed seem to have been small, local installations that did not require a lot of people or "paperwork" to run them. Similarly, the Egyptians were fond of murals showing their kings defeating hordes of foreigners, and people in the competing areas of pre-state Egypt may well have been given to bloody civil wars. But the disposition of settlements, the lack of walled towns and forts, and the art and literature do not support the notion that the critical factor in Egyptian state origins was Menes or someone else strapping on his sword, massing his troops, and marching to the Mediterranean, leaving a unified political state in his wake. Warfare was part of state formation processes, but probably as a mechanism, not a primary cause.

Nor is there much solace for the orthodox Marxian searching for the power of class conflict to produce states. Many scholars can and do see Egypt as an example of the validity of Marxian theory, but Egyptian lit-

9.12 This late second-millennium B.C. Egyptian sculpture of a noble woman has the vibrant, sensual quality typical of some later periods but in great contrast to archaic forms.

erature and iconography seem to exhibit more mutually supportive bonds of kinship and religion than naked class struggle.

The pressure of nomads and others on the Egyptian periphery seems to have been a factor occasionally, but there is not much of a parallel here with the situation in ancient China, where nomad-farmer relationships were a central theme in imperial developments.

A comparison of Mesopotamian and Egyptian settlement patterns underscores that urbanization is just one strategy, not an indispensable condition of cultural complexity. Whereas the people of Mesopotamia early and dramatically aggregated into fortified towns and cities, from which they conducted agricultural, industrial, religious, and administrative operations, the Egyptians did not even have a permanent capital until late in the second millennium, when Thebes emerged as a center—although recent excavations reveal a greater degree of urbanization than was previously thought to have existed. The comparatively slow development of Egyptian urbanism probably had many causes, including: (1) the absence of any powerful foreign peoples on Egyptian borders; (2) the uniformity of the environment all along the Nile, so that there was little to be gained from large-volume, interregional exchange of food or craft products; and (3) the pronounced political centralization, which inhibited development of secular, economic differentiation.

When the largest pyramids and other structures were built, population growth rates were slow, there was apparently little pressure on the country's resources, and large areas of uninhabited but fertile land existed. If we view these pyramids as mechanisms to mobilize and train a large work force, we must ask why such a work force would be an advantage, because when the first pyramids were built there were few large irrigation works and little demand for a standing army. If we view the vast expenditures of wealth in the funerary complexes as a means of "balancing" the economy by taking out of circulation inordinate amounts of gold, silver, or craft items, there is some difficulty in explaining why this would have been necessary in a society whose economic system and long-distance trade were strictly controlled by the monarchy and where there were few large markets and almost no free enterprise or capitalism of any kind.

In conclusion, it has proven remarkably difficult to prise apart the many causes and effects that make up Egyptian cultural history, but we are beginning to understand at least some aspects of this history. And, like other early civilizations, there is much about Egypt that is interesting and rewarding at a level beyond that of the mechanics of its history.

Notes

1. Butzer, *Early Hydraulic Civilization in Egypt*, p. 28.
2. Butzer, "Long-term Nile Flood Variation."
3. Butzer, *Early Hydraulic Civilization in Egypt*, p. 226.
4. Wendorf, *The Prehistory of Nubia*, p. 959.
5. Wendorf, Schild, and Close, *Cattle-Keepers of the Eastern Sahara. The Neolithic of Bir Kiseiba*; Wendorf, F., R. Schild, and A. E. Close. *The Prehistory of Wadi Kubbaniya, Vols. 2 and 3*; Close, *Prehistory of Arid North Africa. Essays in Honor of Fred Wendorf*; Lubell, Sheppard, and Jackes, "Continuity in the Epipaleolithic of Northern Africa with Emphasis on the Maghreb."
6. Wendorf, Schild, and Close, *Cattle-Keepers of the Eastern Sahara. The Neolithic of Bir Kiseiba*, p. 428.

7. Ibid., p. 425.
8. Hadidi, "Vegetation of the Nubian Desert (Nabta Region)," p. 347.
9. Gautier, "Archaeozoology of the Bir Kiseiba Region, Eastern Sahara," p. 70.
10. Gautier, "Prehistoric Men and Cattle in North Africa."
11. Hassan, "Desert Environment and Origins of Agriculture in Egypt"; Hassan, "The Predynastic of Egypt."
12. Butzer, *Early Hydraulic Civilization in Egypt;* Hoffman, *Egypt Before the Pharaohs: The Prehistoric Foundations of Egyptian Civilization.*
13. Hassan, "The Predynastic of Egypt."
14. Hoffman, *Egypt Before the Pharaohs;* Hoffman, *The Predynastic of Hierakonpolis—An Interim Report;* Hoffman et al., "A Model of Urban Development for the Hierakonpolis Region from Predynastic through Old Kingdom Times."
15. von der Way, "Investigations Concerning the Early Periods in the Northern Delta of Egypt;" van den Brink, ed. *The Archaeology of the Nile Delta: Problems and Priorities.*
16. Hassan, "The Predynastic of Egypt."
17. Service, *Origins of the State and Civilization,* p. 228.
18. Butzer, *Early Hydraulic Civilization in Egypt.*
19. Trigger, "The Rise of Civilization in Egypt."
20. Lehner, "Some Observations on the Layout of the Khufu and Khafre Pyramids."
21. Edwards, *The Pyramids of Egypt.*
22. Haas et al., "Radiocarbon Chronology and the Historical Calendar in Egypt."
23. Lehner, "Some Observations on the Layout of the Khufu and Khafre Pyramids."
24. Wenke et al., "Kom el-Hisn: Excavations of an Old Kingdom West Delta Community."
25. Wilson, *The Culture of Ancient Egypt,* p. 79. See also Kemp, "Old Kingdom, Middle Kingdom and Second Intermediate Period c. 2686–1552 B.C."; Trigger, "Egyptology and Anthropology," pp. 32–50.
26. Badawy, "The Civic Sense of the Pharaoh and Urban Development in Ancient Egypt," p. 105.
27. Kemp, "Old Kingdom, Middle Kingdom and Second Intermediate Period c. 2686–1552 B.C.," p. 108.
28. Trigger, "The Mainlines of Socioeconomic Development in Dynamic Egypt to the End of the Old Kingdom," p. 107. See also Kanawati, *The Egyptian Administration in the Old Kingdom. Evidence on its Economic Decline,* pp. 69–77; Goedicke, *Konigliche Dokumente aus dem Alten Reich.*
29. Butzer, "Pleistocene History of the Nile Valley in Egypt and Lower Nubia," p. 278; Kemp, "Old Kingdom, Middle Kingdom and Second Intermediate Period c. 2686–1552 B.C.," p. 113.
30. Reviewed in Kemp, "Old Kingdom, Middle Kingdom and Second Intermediate Period c. 2686–1552 B.C."; Bietak, "The Present State of Egyptian Archaeology"; Mills, "Research in the Dakhleh Oasis"; Krzyzaniak and Kobusiewicz, *Late Prehistory of the Nile Basin and Sahara;* Hoffman, *The Predynastic of Hierakonpolis—An Interim Report;* Fairservis, "Preliminary Report on the First Two Seasons at Hierakonpolis."
31. van den Brink, *The Archaeology of the Nile Delta.*
32. Butzer, *Early Hydraulic Civilization in Egypt,* pp. 26–27.
33. Kemp, "Old Kingdom, Middle Kingdom and Second Intermediate Period c. 2686–1552 B.C."
34. Aldred, *The Egyptians,* p. 102.
35. O'Connor, "New Kingdom and Third Intermediate Period, 1552–664 B.C.," p. 190.
36. Ibid., pp. 194–195.
37. Ibid., p. 219.
38. Tannahill, *Sex in History,* pp. 77–78.
39. O'Connor, "New Kingdom and Third Intermediate Period, 1552–664 B.C.," p. 222.

40. Ibid.
41. Wilson, "Egypt: The Nature of the Universe," p. 72.
42. Ibid.
43. Ibid., p. 73.
44. Farooq, "Historical Development," p. 2.
45. Saggs, *Civilization Before Greece and Rome*, pp. 73–74.
46. Ibid.
47. Wittfogel, *Oriental Despotism: A Comparative Study of Total Power;* Hoffman, *Egypt Before the Pharaohs: The Prehistoric Foundations of Egyptian Civilization*, p. 315.

Bibliography

Adams, R. McC. 1981. *Heartland of Cities*. Chicago: University of Chicago Press.

Aldred, C. 1961. *The Egyptians*. New York: Praeger.

Badawy, A. 1967. "The Civic Sense of the Pharaoh and Urban Development in Ancient Egypt. *Journal of the American Research Center in Egypt* 6:103–9.

Bietak, M. 1979. "The Present State of Egyptian Archaeology." *Journal of Egyptian Archaeology* 65:156–60.

Brothwell, D. R. and B. A. Chiarelli, eds. 1973. *Population Biology of the Ancient Egyptians*. London: Academic Press.

Butzer, K. W. 1976. *Early Hydraulic Civilization in Egypt*. Chicago: University of Chicago Press.

———. 1978. "Perspectives on Irrigation Civilization in Egypt." In *Immortal Egypt*, ed. D. Schmandt-Beserat. Mailbu: Undena.

———. 1980. "Pleistocene History of the Nile Valley in Egypt and Lower Nubia." In *The Sahara and The Nile*, ed. M. Williams and H. Faure. Rotterdam: Balkema.

———. 1984. "Long-Term Nile Flood Variation and Political Discontinuities in Pharaonic Egypt." In *From Hunters to Farmers*, eds. J. D. Clark and S. A. Brandt. Berkeley: University of California Press.

Butzer, K. W., G. Isaac, J. L. Richardson, and C. K. Washbourn-Ramau. 1972. "Radiocarbon Dating of East African Lake Levels." *Science* 175:1069–76.

Clark, J. D. 1971. "A Re-Examination of the Evidence for Agricultural Origins in the Nile Valley." *Proceedings of the Prehistoric Society* 37(2).

———. 1976. "Prehistoric Populations and Resources Favoring Plant Domestication in Africa." In *Origins of African Plant Domestication*, eds. J. R. Harlan et al. The Hague: Mouton.

Close, A. E. ed. 1987. *Prehistory of Arid North Africa. Essays in Honor of Fred Wendorf*. Dallas: Southern Methodist University Press.

Cruz-Uribe, E. 1985. *Saite and Persian Demotic Cattle Documents*. Chico, Calif.: Scholars Press.

Edwards, I. E. S. 1961. *The Pyramids of Egypt*. London: Parrish.

Fairservis, W. 1972. "Preliminary Report on the First Two Seasons at Hierakonpolis." *Journal of the American Research Center in Egypt* 9:7–27, 67–99.

Farooq, M. 1973. "Historical Development." In *Epidemiology and Control of Schistosomiasis (Bilharziasis)*, ed. N. Ansari. Baltimore: University Park Press.

Frankfort, H. 1956. *The Birth of Civilization in the Near East*. Garden City, N.Y.: Doubleday.

Frankfort, H., J. Wilson, and T. Jacobsen. 1949. *Before Philosophy*. Baltimore: Penguin.

Friedman, J. and M. J. Rowlands. 1977. *The Evolution of Social Systems*. Pittsburgh: University of Pittsburgh Press.

Gautier, A. 1984. "Archaeozoology of the Bir Kiseiba Region, Eastern Sahara." In *Cattle-*

Keepers of the Eastern Sahara. The Neolithic of Bir Kiseiba, ed. A. E. Close. New Delhi: Pauls Press. Pp. 49–72.

————. 1987. "Prehistoric Men and Cattle in North Africa: A Dearth of Data and a Surfeit of Models." In *Prehistory of Arid North Africa,* ed. A. E. Close. Dallas: Southern Methodist University.

Ghoneim, W. 1977. "Die Okonomische Bedeutung des Rindes im Alten Agypten." *Lexikon der Agyptologie* 5:259.

Goedicke, H. 1967. *Konigliche Dokumente aus dem alten Reich.* Wiesbaden: Harrasowitz.

Haaland, R. 1987. *Socio-Economic Differentiation in the Neolithic Sudan.* Oxford: British Archaeological Reports International Series 350.

Haas, H., J. Devine, R. J. Wenke, M. E. Lehner, W. Wolfli, G. Bonani. 1987. "Radiocarbon Chronology and the Historical Calendar in Egypt." In *Chronologies in the Near East,* eds. O. Avrenche, J. Evin, and P. Hours. *British Archaeological Reports* 379:585–606.

Hadidi, N. 1980. "Vegetation of the Nubian Desert (Nabta Region)." In *Prehistory of the Eastern Sahara,* eds. F. Wendorf and R. Schild. New York: Academic Press.

Hamilton, E. 1930. *The Greek Way.* New York: Norton.

Harlan, J. R. 1982. "The Origins of Indigenous African Agriculture." In *The Cambridge History of Africa,* Vol. 1, ed. J. D. Clark. Cambridge, England: Cambridge University Press.

Harlan, J. R., J. M. de Wet, and A. B. Stemler, eds. 1976. *Origins of African Plant Domestication.* The Hague: Mouton.

Harris, J. E. and K. R. Weeks. 1973. *X-Raying the Pharaohs.* New York: Scribner's.

Hassan, F. 1986. "Desert Environment and Origins of Agriculture in Egypt." *Norwegian Archaeological Review* 19:63–76.

————. 1988. "The Predynastic of Egypt." *Journal of World Prehistory* 2(2):135–85.

Helck, W. 1971. *Die Beziehungen Agyptens su Vorderasien im 3. und 2. Jahrtausend v. Chr.* Wiesbaden: Harrasowitz.

————. 1975. *Wirtschaftsgeschichte des alten Agyptens im 3. und 2. Jahrtausend vor Chr.* Leiden-Koln: Brill.

Hoffman, M. 1980. *Egypt Before the Pharaohs: The Prehistoric Foundations of Egyptian Civilization.* New York: Knopf.

————. 1982. *The Predynastic of Hierakonpolis—An Interim Report.* Egyptian Studies Association, Publication no. 1. Cairo: Cairo University Herbarium and the authors.

Hoffman, M. A., H. A. Hamroush, and R. O. Allen. 1986. "A Model of Urban Development for the Hierakonpolis Region from Predynastic Through Old Kingdom Times." *Journal of the American Research Center in Egypt* 23:175–87.

Jacquet-Gordon, H. 1962. *Les Noms des Domaines Funeraires sous l'Ancien Empire Egyptien.* Cairo: Institut Francais d'Archeologie Orientale.

Kanawati, N. 1977. *The Egyptian Administration in the Old Kingdom. Evidence on its Economic Decline.* London: Warminster, Aris & Phillips.

Kaster, J., trans. and ed. 1968. *Wings of the Falcon.* New York: Holt.

Kees. H. 1961. *Ancient Egypt. A Cultural Topography.* Chicago: University of Chicago Press.

Kemp, B. J. 1977. "The Early Development of Towns in Egypt." *Antiquity* 51:185–200.

————. 1982. "Old Kingdom, Middle Kingdom and Second Intermediate Period in Egypt." In *The Cambridge History of Africa,* Vol. 1, ed. J. D. Clark. Cambridge, England: Cambridge University Press.

Kitto, H. D. F. 1951. *The Greeks.* Harmondsworth: Pelican.

Krzyzaniak, L. and M. Kobusiewicz, eds. In press. *Late Prehistory of the Nile Basin and Sahara.* Poznan: Poznan Archaeological Museum.

Lehner, M. 1983. "Some Observations on the Layout of the Khufu and Khafre Pyramids." *Journal of the American Research Center in Egypt* 20:7–29.

————. 1985. *The Pyramid Tomb of Hetep-heres and the Satellite Pyramid of Khufu.* Mainz am Rhein: Philipp von Zabern.

Lubell, D., P. Sheppard, and M. Jackes. 1984. "Continuity in the Epipaleolithic of Northern Africa with emphasis on the Maghreb." *Advances in World Archaeology* 3:143–68.

Mills, A. 1984. "Research in the Dakhleh Oasis." In *Origin and Early Development of Food-Producing Cultures in North-Eastern Africa*, eds. L. Krzyzaniak and M. Kobusiewicz. Poznan: Poznan Archaeological Museum.

Moens, M. and W. Wetterstrom. 1989. "The Agricultural Economy of an Old Kingdom Town in Egypt's West Delta: Insights from the Plant Remains." *Journal of Near Eastern Studies* 3:159–73.

O'Connor, D. 1982. "New Kingdom and Third Intermediate Period, 1552–664 B.C." In *Ancient Egypt. A Social History*, eds., B. G. Trigger, B. J. Kemp, D. O'Connor, and A. B. Lloyd. Cambridge, England: Cambridge University Press.

———. 1983. "New Kingdom and Third Intermediate Period, 1552–664 B.C." In *Ancient Egypt. A Social History*, eds. B. Trigger, B. Kemp, D. O'Connor, and A. Lloyd. Cambridge, England: Cambridge University Press.

———. 1987. "The Old Kingdom Town at Buhen." Paper presented at the Annual Meeting of the American Research Center in Egypt.

O'Connor, P. 1972. "A Regional Population in Egypt to Circa 600 B.C." In *Population Growth—Anthropological Implications*, ed. B. Spooner. Cambridge, Mass.: MIT Press.

Posener-Krieger, P. 1976. *Les Archives du Temple Funeraire de Neferirkare-kaki.* 2 vols. Cairo: Institut Francais d'Archeologie Orientale du Caire.

Posnansky, M. and R. McIntosh. 1976. "New Radiocarbon Dates for Northern and Western Africa." *Journal of African History* 17:161–95.

Service, E. 1975. *Origins of the State and Civilization.* New York: Norton.

Tannahill, R. 1982. *Sex in History.* Briarcliff Manor, N.Y.: Scarborough Books.

Trigger, B. G. 1979. "Egyptology and Anthropology." In *Egyptology and the Social Sciences*, ed. K. Weeks. Cairo: American University in Cairo Press.

———. 1982. "The Rise of Civilization in Egypt." In *The Cambridge History of Africa*, Vol. 1, ed. J. D. Clark. Cambridge, England: Cambridge University Press.

———. 1984. "The Mainlines of Socioeconomic Development in Dynastic Egypt to the End of the Old Kingdom." In *Origin and Early Development of Food-Producing Cultures in North-Eastern Africa*, eds. L. Krzyzaniak and M. Kobusiewicz. Poznan: Poznan Archaeological Museum.

———. 1985. "The Evolution of Pre-industrial Cities: A Multilinear Perspective." *Melanges offerts à Jean Vercoutter.* Paris: Editions Recerce sur les Civilisations.

Van den Brink, E. 1988. *The Archaeology of the Nile Delta: Problems and Priorities.* Amsterdam: Amsterdam Foundation for Arch. Research in Egypt.

Wendorf, F. 1968. *The Prehistory of Nubia.* 2 vols. and atlas. Dallas: Fort Bergwin Research Center and Southern Methodist University Press.

———. 1976. "The Use of Ground Grain During the Late Paleolithic of the Lower Nile Valley, Egypt." In *Origins of African Plant Domestication*, eds. J. R. Harlan et al. The Hague: Mouton.

Wendorf, F., R. Said, and R. Schild. 1970. "Egyptian Prehistory: Some New Concepts." *Science* 169:1161.

Wendorf, F. and R. Schild. 1975. "The Paleolithic of the Lower Nile Valley." In *Problems in Prehistory: North Africa and the Levant*, eds. F. Wendorf and A. Marks. Dallas: Southern Methodist University Press.

———. 1980. *Prehistory of the Eastern Sahara.* New York: Academic Press.

———. ed. A. E. Close. 1980. *Loaves and Fishes: The Prehistory of Wadi Kubbaniya.* New Delhi: Pauls Press.

———. 1989. *The Prehistory of Wadi Kubbaniya, Vols. 2 and 3.* Dallas: Southern Methodist University Press.

———. 1984. *Cattle-Keepers of the Eastern Sahara. The Neolithic of Bir Kiseiba.* New Delhi: Pauls Press.

Wenke, R. J., R. Redding, P. Buck, H. A. Hamroush, M. Kobusiewicz, and K. Kroeper.

1988. "Kom el-Hisn: Exvations of an Old Kingdom West Delta Community." *Journal of the American Research Center in Egypt* xxv:5–34.

White, L. 1949. *The Science of Culture.* New York: Grove Press.

Wilson, J. A. 1946. "Egypt: The Nature of the Universe." In *Before Philosophy,* eds. H. A. Frankfort et al. Baltimore: Penguin.

———. 1951. *The Culture of Ancient Egypt.* Chicago: University of Chicago Press.

———. 1960. "Civilizations Without Cities." In *City Invincible,* eds. C. H. Kraeling and R. McC. Adams. Chicago: University of Chicago Press.

Wittfogel, K. A. 1957. *Oriental Despotism: A Comparative Study of Total Power.* New Haven: Yale University Press.

10

Indus Valley Civilization

> Aristobulus . . . says that when he was sent
> upon a certain mission [in Pakistan and India]
> he saw a country of more than a thousand cit-
> ies, together with villages, that had been de-
> serted because the Indus had abandoned its
> proper bed.
>
> Strabo (c. 63 B.C.–A.D. 24)[1]

Since the discovery of its archaeological remains in the 1920s, the Harappan civilization that flourished in the Indus Valley, in what is now Pakistan and India, in the third millennium B.C. has been unfairly considered
a sort of poor relation to the great states and empires that graced the
Mesopotamian Alluvium and the Nile Valley. Not only did the Indus Valley cultures mature thousands of years later than those in Egypt and
Southwest Asia, they also neglected to leave much in the way of the pyramids, tombs, and palaces so prized by archaeologists. Nor was the Harappan a particularly long-lived civilization, having appeared and disappeared within the space of about five centuries.

But the Indus is yet another example of a sociopolitical system whose
major evolutionary dynamics now appear to have been primarily self-
contained, with only minor aspects the results of outside influences.

In fact, the Indus Valley cultures are of great interest for the study of
the origins of cultural complexity: (1) they are now the only major early
civilization whose writing system has not yet been deciphered; (2) they
constructed massive cities with perhaps the ancient world's most advanced
municipal water and sewage system; (3) their area of cultural and political
influence and control extended over almost 1,300,000 square kilometers—considerably more territory than any other Old World civilization
of this period; and (4) the Indus cities seem to have a very equitable distribution of wealth, a primitive socialism in an era when other comparable
civilizations were highly stratified groupings of a few extraordinarily wealthy

407

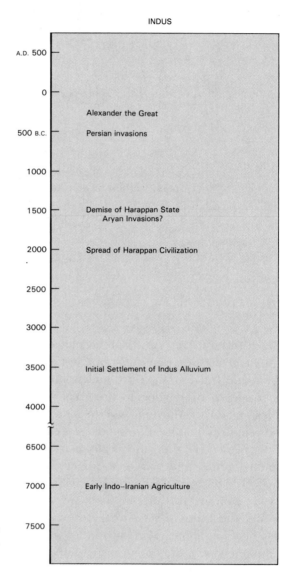

INDUS

A.D. 500

0

Alexander the Great

500 B.C.   Persian invasions

1000

1500   Demise of Harappan State
Aryan Invasions?

2000   Spread of Harappan Civilization

2500

3000

3500   Initial Settlement of Indus Alluvium

4000

6500

7000   Early Indo–Iranian Agriculture

7500

10.1 A simplified chronology of South Asian prehistory.

and powerful nobles and priests supported by tens of thousands of peasants.

It is not an uncommon archaeological irony that the Indus civilization, so interesting in various ways, is one of the most archaeologically inaccessible. Most critical phases of its archaeological record lie more than 10 meters under the Indus floodplain, far beneath the water table. Along with the pleasures of sloshing through mud against a background whine of pumps, the archaeologist working on early complex societies in the

Indus must be prepared for the collapse of huge walls of debris as he or she cuts down through the alluvium to the earliest levels.

The Ecological Setting

Millions of years ago, movements of the earth's crust forced the Indian subcontinent against the main Eurasian land mass with such pressure that the land in between was squeezed upward, creating the Himalayan Mountains. Every spring melting snow in these highlands sends floods down the mountains and across the lowlands to the sea.

So little rain falls on this plain during the average year that reliable agriculture is possible only through irrigation from the river. Thus, Indus civilization, like those in all other "pristine" cases except Mexico, is the gift of a huge river. But—unlike the Nile—the Indus is relatively unpredictable, with great annual fluctuations in its volume and course. Frequently the river has flowed across the countryside in devastating floods. Centuries of this have left the Indus floodplain a maze of old river channels and great deposits of silt, all smoothed down by the action of wind and water. The great fertility of these soils is complemented by the arid, hot climate, which, like that of Egypt and Mesopotamia, supports several crops a year and a great diversity of plant species.

Because the Indus is navigable over much of its length, ships could sail between Harappan cities with goods and, presumably, with the bureaucrats sent by the central government to collect taxes and administer the provinces.

To the west of the hot Indus plains, although the arid foothills and mountains of Pakistan, Afghanistan, and Iran sharply limit the extension of agriculture, they also provide valuable minerals, metals, animal products, and other goods, and are the homelands of pastoral and nomadic peoples who exerted great influence on lowland civilizations. The diffusion of new ideas, objects, and peoples into the Indus Valley was mainly along routes through these western borderlands, or along the thin coastal strip on the Arabian Sea, since the Himalayas to the north were a formidable barrier. The Great Indian Desert to the east of the Indus Valley reduced contacts with the rest of the subcontinent.

Thus, the Indus Valley was, like Egypt and Peru, an area of rich agricultural lands sharply bounded by highlands, desert, and ocean.

The Neolithic Background
to South Asian Cultural Complexity

The ancestors of the people who built the Indus civilizations spent thousands of years as small-time farmers and herders in the highlands above the Indus, during centuries when most of the plain was only lightly occupied. Domesticated wheat and the remains of domesticated sheep and goats have been found in levels dating to about 7000 B.C. in several sites in Afghanistan and Baluchistan, and the evidence suggests that thereafter the agricultural and pastoral ways of life spread gradually, from west to east, throughout highland areas where rainfall and streams provided enough water.

Many settlements in the highlands west of the Indus at about 3500 B.C. were probably based on simple wheat farming supplemented by sheep and goat raising and some hunting and gathering. These farmers made pottery and used a few copper tools, but some villages are so insubstantial that they suggest a relatively mobile population and perhaps only seasonal occupation.[2]

Rice may well have been in cultivation on the Ganges Plain by 4500 B.C.; all over South Asia at this time people were making intensive use of many cereals, most of which never became staples like wheat and rice.[3]

Pre-Harappan Cultures

The Indus Valley civilizations seem another classic demonstration of the discontinuous nature of cultural evolution. Rather than a smooth building process of Neolithic villages adding trait by trait the components of civilization, we see in the Indus area the same pattern evident in Mesopotamia, where initial moves toward complexity stall and disappear, while areas that were once on the periphery become the core developmental regions.

In Mesopotamia Çatal Hüyük and other highland settlements "are overtaken by the more sustained trajectory observable in Mesopotamia."[4] Daniel Miller sees the same pattern of core-periphery relationships between the edge of the highlands bordering the Indus Plain and the Plain itself. He notes that from 6000 B.C. sites on the Indus Plain show signs of greater size and architectural complexity than the farming villages of previous periods, and by 5000 B.C. some communities have what seem to be large nonresidential buildings that may be for communal activities. By 4000 B.C. communities all over what is now Pakistan were fortified, traded in marine shells, lapis lazuli, and other goods, had a degree of overall planning evident in the architecture, and produced pottery in regional styles that may reflect competitive political entities.[5]

In comparing the Indus Valley and Mesopotamia, Henry Wright notes that already by the late fifth and early fourth millennia B.C. in some areas of the Indus Plain there were settlements of such different sizes that we may suspect that they reflect hierarchically organized societies.[6] The distribution of their sizes and locations he interprets as evidence of a group of poorly integrated competing centers.[7]

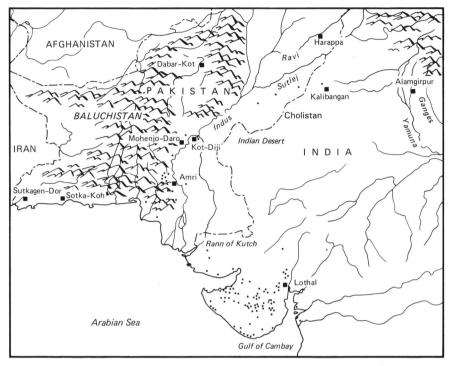

10.2 The heartland of the Harappan civilization as the Indus Valley. The larger cities are shown as squares in this map, the smaller communities as dots. Thousands of other communities are probably buried under the alluvium.

Miller observes that several regions on the Indus Plain and in Iran at this time were "moving toward a greater degree of social complexity and urbanization,"[8] although eventually the Indus Plain became the developmental center. And although much of the mature Harappan civilization can be found in these developments of the fifth and fourth millennia B.C., the cultural changes that occurred as part of the Harappan civilization after about 2600 B.C. were pervasive, profound, and in some ways revolutionary.[9]

Between 3000 and 2400 B.C., settlements appeared at Kot Diji, Harappa, Kalibangan, and elsewhere, perhaps founded by people moving in from the western highlands. The few excavated settlements of this period reveal simple mudbrick houses in small villages scattered in areas where no extensive irrigation would have been necessary. Some villages were walled, though there was certainly no shortage of land or pressure on other resources at this time. These various lowland settlements prior to 2600 B.C. show some stylistic uniformity and a great deal of economic and architectural similarity, but they appear to have been economically and politically independent and self-sufficient, and reflect none of the rigid planning typical of later settlements here.

Indus Valley artifacts found in Mesopotamian sites show that by 2500 B.C. Harappan culture was already a regional force. And within a few centuries after 2400 B.C., the simple, scattered agricultural societies of the Indus Valley were transformed into a large, complex, urban-based socio-political system that we might legitimately call a state society. At Kot Diji and Amri, thick layers of ash suggest the transformation was not a peaceful one, but this is not at all clear.

It may be significant that the emergence of Indus Valley urban cultures occurred at the same time that the first Near Eastern empires were forming, since there appears to have been increased trade and other contacts between these two areas at this time, but there is no evidence that these Mesopotamian contacts somehow induced the Indus Valley peoples to form states or aggregate in cities.

Harappan Civilization

The appearance of large cities shortly after 2600 B.C., and the associated spread of a distinctive constellation of artifact and architectural styles over much of the Indus Valley marked the emergence of Harappan civilization, a political system that survived only about five hundred years but which managed to integrate much of the Indus Plain in a political and cultural unit.

More than eight hundred "mature Harappan" sites are known, and these are spread over an area that at 1.25 million square kilometers is larger than the first Egyptian state and the total area of Sumer, Akkad, and Assyria.[10]

Miller notes that Harappan civilization is not really a strictly riverine civilization, because the area of Harappan settlements involves many ecological zones, and only after the main Harappan developments were well underway did settlements begin to cluster along the river. Most research on Indus sites has been devoted to the largest settlements, especially Mohenjo-daro, Harappa, Chanhu-daro, Pathiani Kot, Judeirjo-daro, Kalibangan, and Lothal. Doubtless many smaller Harappan settlements are buried beneath silt or have been washed away by floods.

The largest Harappan settlement, Mohenjo-daro (Figure 10.3), covers at least 2.5 square kilometers and may have had 40,000 inhabitants. Recent studies of this and other Harappan settlements show that these communities were probably not so regularly constructed as archaeologists once thought, but, compared to the jumbled anarchy of most Mesopotamian city plans, Mohenjo-daro was quite orderly.[11] Fentress has questioned the overall similarity of Mohenjo-daro and Harappa, but here too, by Mesopotamian standards, these settlements are quite similar.[12] Miller, in fact, notes that even tiny Harappan hamlets seem to be trying to emulate the grid-pattern and other designs of the major Harappan metropolises.[13]

Mohenjo-daro was bisected by a north–south running street some 9 meters wide that was flanked by drainage ditches. Public toilets and sewers and bathrooms in houses were connected to the main sewage lines. Most residences were made of fired brick, comprising several rooms arranged around an open courtyard, and the majority appear to have had private showers and toilets drained by municipal sewage systems. Some houses were two stories high and larger and more elaborate than others, but the overall impression is one of uniformity. In fact, Sarcina found that at Mohenjo-daro almost all houses were of two basic patterns: having a courtyard at one corner that was flanked by rooms on two sides, or with a courtyard in the center and rooms on three sides.[14] Also, almost all the houses were of the same approximate size and construction: 77% of the

10.3 Some elements of the "Citadel" at Mohenjo-daro are evident in this photograph (the large circular Buddhist shrine in the background was built many centuries after Mohenjo-daro was abandoned).

buildings seem to have been ordinary houses, with the rest being small shops and businesses distributed throughout the living areas.[15] If gross differences in wealth divided the inhabitants, these inequities are not reflected in residential architecture, at least not to the extent that they were in Mesopotamia.

At Mohenjo-daro and the other two largest Harappan sites, Harappa and Kalibangan, the carefully arranged residential areas were flanked on the west by a great fortified "citadel" mound. Mohenjo-daro's citadel is about 150 meters to the northwest of the main settlement, separated from the residences by land that seems to have been devoid of any settlement and perhaps was regularly flooded by a branch of the river. It has even been suggested that this flooding may have been deliberate, to create a pool of water for bathing, fishing, or some other activity. The northwest complex at Mohenjo-daro is dominated by a brick platform some 12 meters high.

One of the most interesting structures at the northwest complex is the "Great Bath," a swimming-pool-like building about 12 by 7 meters and 2.5 meters in depth, constructed of baked brick and lined with bitumen. Flanking the pool are what appear to have been dressing rooms. They were carefully staggered to give maximum privacy, and some were equipped with toilets.[16] The Great Bath may have figured in some religious activities, but it contained no obvious icons or other religious elements and may have been mainly just a public bathing facility.

Adjacent to the bath was a cluster of platforms and rooms, variously—and imaginatively—interpreted as granaries, assembly halls, and garrisons—but there is little evidence with which to infer their functions. Overall, the complex was about 450 meters long and 90 meters wide at its maximum extent—representing a major investment of labor and materials. Thus, while there is nothing at Mohenjo-daro or any other Indus city to compare with the Egyptian pyramids or the White Temple at Uruk, the Indus city dwellers nonetheless were also diverting considerable energy and resources to building projects. The major difference seems to have been that, in contrast to the largely ceremonial public architecture in Mesopotamia and Egypt, the Indus Valley constructions provided some return in the way of administrative buildings, better defenses, and more storage space.

There is at least a possibility that Harappan civilization was "cut off" in the midst of its development by invasion, flood, interrupted trade routes, or some other factor, and that more "wasteful" monumental architecture might eventually have appeared at the Indus cities had they been allowed to develop for a longer period. The absence of easily accessible building stone may also account for the comparatively utilitarian and drab Harappan monumental architecture.

The thousands of people who lived at Mohenjo-daro included farmers, herdsmen, goldsmiths, potters, weavers, brickmasons, architects, and many other specialists, and streets were lined with stores and shops. Wheat and barley were the basis of the economy, supplemented by dates, melons, sesame, peas, mustard, and other crops. Cattle, sheep, goats, pigs, and domestic fowl were the major animal foods, and buffalos, camels, asses, dogs, and cats were also kept. A few elephant bones have even been found. The horse was apparently rarely used until the very end of the Harappan period.

Most of the larger Harappan settlements were similar to Mohenjo-daro in architecture and economy, but the site of Lothal, although only about 300 to 250 meters in size, shows an impressive complexity for such a tiny settlement.[17] A large tank, which Rao and others have interpreted as a dock,[18] may not be;[19] its function is unclear. There was a factory for making beads from carnelian, crystal, jasper, and other stones, and facilities for making ornaments out of bronze, elephant tusks, and many other commodities. Possehl sees in Lothal's precise layout and concentrations of exotic commodities an entrepôt, a frontier settlement of entrepreneurs who were processing raw materials from the hinterlands and sending products on to the great Harappan cities.[20]

We know little about the smaller Harappan settlements. The few that have been examined seem to have brick walls around a district within the site, perhaps in imitation of the citadels at the larger cities, and the basic arts and crafts and subsistence practices also appear to have been patterned after those of the cities.

Relatively few Harappan cemeteries have been found, but the bodies so far excavated were buried with far fewer evidences of social differentiation than in Mesopotamia and Egypt.[21]

Harappan settlements, with their precisely administered character, would seem to be excellent subjects for the systematic analysis of settlement patterns, to try to discern the economic and political forces that dictated where people lived, but generally archaeologists have not competed with each other for a chance to survey systematically the intensely hot, heavily populated Indus Plain.

What we do know of Indus settlement patterns suggests at least four different size categories, comparable to, for example, those of Mesopotamia in the third millennium B.C., and if we consider three or more administrative levels to be evidence of state-level political organization, then the Harappan civilization was certainly a state. In fact, the Harappan population at the beginning of the second millennium B.C. was probably at least 200,000, and the tightly organized fabric of their lives suggests an empire-like political system.

But it is interesting that Harappan civilization seems to have "taken off" at a time of relatively low population densities and when there is some evidence in the form of burned buildings of regional warfare.[22]

The Indus Script

The great similarity of cities, towns, and villages of the Harappan civilization probably reflects a nation in which various settlements were in regular and diverse communication with each other. Clay and metal models of wheeled carts and river boats are all that remain of a transportation system that once probably moved large quantities of flint tools, pots, and other artifacts, some mass produced at a few locations. The considerable degree of occupational specialization evident in these artifacts suggests intensive local trade, but there is a strong possibility that this was a kind of administered, noncapitalistic redistribution, rather than a free-enterprise

10.4 Harappan seals were inscribed with many different symbols in complex arrangements, suggesting a form of written language, but this language has not yet been deciphered.

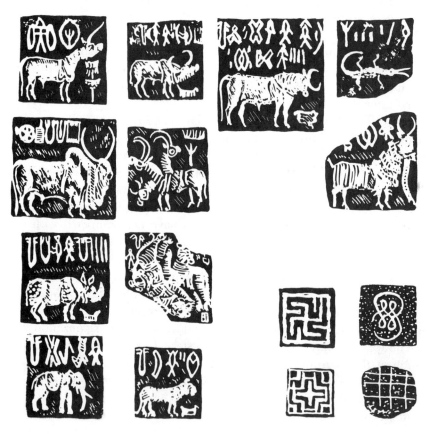

or peasant marketing system. One possible key to the Harappan economic system is the hundreds of Harappan stamps and seals (Figure 10.4), which may have been used to denote ownership or make records of transactions. The apparently ritual scenes depicted in so many of these might argue against this interpretation, but the question will remain unresolved as long as the Harappan script is undeciphered.

Almost all Harappan writing is in the form of inscriptions on these seals. The estimated number of unique symbols is between 350 and 425, which means that the language is almost certainly not alphabetic: alphabetic languages use one symbol for each unique sound in the spoken language, and no known language has more than about fifty alphabetic characters. Since the Indus script contains about four hundred unique characters, it is probably made up of symbols indicating syllables in the spoken language, plus some determinatives and other marks to aid interpretation. Many other documents could have been written on palm leaves or some other long-since decayed medium.

The Indus language may well be Dravidian, from southern India, or Indo-European—a relative of our own. If other early scripts are any guide, the majority of these seals and the lost documents were simply records of Farmer A selling eighteen cows to Farmer B and paying the appropriate taxes, or a ritual prayer or blessing on the ruler, followed by some record of events and transactions of ordinary daily life. Computers now hum away trying to find patterns of co-occurrence among the four hundred Indus symbols, and some day we may have a clear sense of what they mean. But the texts are so few and the characters so stylized that it will not be easy to decipher them.

The writing system was complemented by a standardized system of weights and measures. Small, precisely cut pieces of chert in both binary and decimal arrangements were used as counterweights in balances, and several measuring sticks marked off in units of about 33.5 centimeters have been found; apparently this unit was the common measure of length, much like the English "foot," for many of the buildings are precisely constructed to this scale.[23]

The writing, seals, and weights suggest substantial interregional—perhaps international—trade. There are some reports of Harappan colonies as far away as Afghanistan and Oman.[24] But Miller claims that "It can, however, by now be asserted that there is a remarkable lack of evidence for external trade in Harappan material."[25] Harappan peoples may have traded principally in raw materials and not the finished goods that would allow identification of trade patterns, but the overall evidence of foreign trade is slight, when compared with Mesopotamia at the same period of development.

The Harappan art and religious architecture so far discovered seem

10.5 This Harappan toy cart is one of the earliest known models of wheeled vehicles.

less varied, complex, and beautiful than that of the same period in Mesopotamia, but they have a certain affecting quality. The most popular art was in the form of terra cotta figurines, the majority of which were standing females, heavily adorned with jewelry. For reasons that will probably forever remain unclear, the Harappan people simply did not produce large amounts of "art," in sculpture, figurines, paintings, or even bodily adornment—or at least few such things have been preserved into our own time.

The Decline of Harappan Civilization

As we have noted, most civilizations do not die the same way that biological organisms do. Civilizations are transmuted by time and circumstance so that they change, but usually complexes of their constituent elements live on for millennia in languages, art, and other forms.

The Harappan civilization is part of the cultural stream that connects even the present-day inhabitants of the Indus to their long and complex ancestry, but archaeologists have long wondered if the Indus might really be that rare civilization that rather than being transformed was obliterated, perhaps, by flood, drought, invasion, or some other calamity. The principal archaeological evidences relating to the demise of Harappan civilization are: (1) the increasing heterogeneity of pottery and other artifact styles within the same area that in earlier centuries have been so uniform stylistically; (2) the "degradation" of art and architecture toward the end of the Harappan period—which has led some imaginative scholars to suppose that the Harappans had lost their sense of cultural unity and purpose; and (3) the discovery of about a score of human skeletons "sprawled" in the streets of Mohenjo-daro (Figure 10.7), thought by some to be perhaps the aftermath of an invasion.

Hydrologist Robert Raikes suggests that Harappan civilization was terminated by destruction of their fields and settlements through floods brought on by major shifts of the earth's crust near the mouth of the Indus River.[26] Raikes notes the lack of settlements in the area near the mouth of the river and also that the fossil beaches are many miles inland

10.6 This lovely bronze figure from Mohenjo-daro is one of the few examples of Harappan art.

from the present coast. This is what one would expect if the river's route to the sea were blocked by an uplift of land near the mouth, since water would have been backed up into a large lake that eventually could have inundated the Harappan area.

Alternatively, Possehl notes that several strains of domesticated African millet were introduced to South Asia at about the time of the Harappan decline, and suggests that these domesticates were better suited to this area than the wheat-barley combination on which Harappan civilization was built.[27] Thus, after the Harappan period, settlements may have shifted to areas more suitable for millet cultivation.

A more romantic suggestion is that the Harappan civilization was destroyed by repeated invasions of seminomadic peoples coming out of Central Asia and Iran. The *Rig Veda,* the oldest surviving Vedic Sanskrit literature, describes the conquest of the dark-skinned natives of the Indus Plain by lighter-skinned Aryan invaders, and the Harappans have traditionally been associated with the former. The translation of Sanskrit lit-

10.7 These skeletons found lying in a public area of Mohenjo-daro were interpreted by some as evidence of an invasion, but there is little proof of this.

erature, first accomplished in the sixteenth century, revealed major similarities between Sanskrit, Greek, and European and central Asian language families. These similarities were eventually traced to origins in the Caucasus Mountains of southern Russia and adjacent areas and associated with tall, long-headed, fierce peoples collectively referred to as Aryans or Indo-Europeans, who shortly after 1900 B.C. apparently invaded and influenced the cultures of India, Central Asia, Western Asia, and Europe. How they were able to do this is one of the great unresolved problems of history.

Bronze weapons and other artifacts traditionally associated with the Indo-Europeans have been found in the upper levels of some Harappan sites, and some scholars have identified these with the invaders referred to in the *Rig Veda*. It is difficult to substantiate these invasions, and many now believe that such invasions would have been at most only a minor part of the Harappan collapse. Kennedy specifically rejects the notion that the bodies Sir Mortimer Wheeler found sprawled in the streets of Mohenjo-daro were massacred.[28]

The "invasion" of largely pastoral peoples into agricultural areas in the Harappan case and in similar circumstances elsewhere probably was more a gradual infiltration than a brutal massive invasion of the sort so often led by Yul Brenner and his colleagues in cinematographic histories. Overall, there may have been a gradual shift of power away from the Harappan heartland to peripheral groups in the south and east, where the evolutionary potential may have been higher because of the emergence there of rice or millet agriculture. Political and cultural influence may well have gravitated to those areas outside the primarily wheat-growing regions of the Harappan sphere of influence.[29] Colin Renfrew has challenged the old notion of Indo-European invaders who sacked the Indus civilizations, and suspects that the people of the Indus Plain were already Indo-European speakers.[30] Rather than a violent spread of Indo-European storm-troopers, there may well have been a spread of the Indo-European *languages*, along with elements of crops and agriculture.

The end of Harappan civilization was probably a result of many factors. Continued pressure from peripheral groups, the altering courses of the Indus River, droughts, floods, earthquakes—all may have contributed to the gradual abandonment of Harappan centers.

After the Harappan downfall, many diverse cultures appeared throughout India and Pakistan, ranging from hunters and gatherers to highly sophisticated urban-based civilizations. The centers of power and influence gradually shifted from the Indus Valley to the great Ganges River Valley where, after about 1100 B.C., large cities were built and state-level political systems were formed. Many Harappan elements appeared in these later societies, including aspects of metallurgy, architecture, pottery styles, and agriculture.

The Indus Valley Civilization: Conclusions

Scholars dispute general interpretations of the Indus culture history, some stressing the ecological, demographic, and technological bases of life on the Indus Plain,[31] others sociopolitical factors.[32]

It is difficult to see an obvious equation between the simple ecological circumstances of life on the ancient Indus Plain and the differences that distinguish Harappan settlements from those in Mesopotamia or Egypt. The rigid planning and execution of Harappan settlements bespeaks a powerful centralized authority, perhaps rivaling that of ancient Egypt, but there is no evidence of the great tombs, palaces, and pyramids that accompanied theocratic states in Egypt and elsewhere. Some have suggested that the familiar Indian caste system was already in effect during the Harappan period and that this would have conferred the social control evident in the architecture.[33]

Because of its brevity, lack of monumental architecture, and absence of obvious wealth differential and militarism, some people have argued that the Harappan civilization was not a true state, merely a chiefdom.[34] But as Jacobson notes, recent discoveries cast doubt on this interpretation.[35] Settlement pattern studies show at least a four-level site size hierarchy, consistent with Gregory Johnson's definition of the Mesopotamian state as a society in which there are at least three levels in the administrative hierarchy.[36] There was a form of writing which, even if not developed, conveyed a great deal of administrative and economic information, and recent excavations show a degree of military fortification and economic stratification not previously suspected.

Shaffer concludes that in some ways the analytical models applied to Mesopotamia and other early states are not really applicable to Harappan culture.[37] He notes, for example, that compared to Mesopotamia, the Harappan culture did not seem to exhibit the same degree of cyclical rise and fall of state governments, and that

> in the Indus Valley, a technologically advanced, urban, literate culture was achieved without the usually associated social organization based on hereditary elites, centralized political government (states, empires) and warfare.[38]

Long-distance trade, particularly the flow of goods from the Indus Valley to Mesopotamia, has frequently been suggested as a key factor in the development—and decline—of Harappan civilization. Harappan seals and seal impressions have been found in limited quantities in Mesopotamia and along the Persian Gulf, and there clearly was some commerce between these areas, perhaps by way of ships sailing along the coast and caravans transversing the Iranian Plateau. Exactly which commodities would have been shipped from the Indus Valley westward is unknown, although steatite and a few other minerals and semiprecious stones would have been likely trade items. Commerce appears to have been very one-sided, however, with little going from Southwest Asia to the Indus area, and this has suggested to some that Harappan civilization may have been established and maintained mainly by Mesopotamian or Iranian states. Indeed, some Mesopotamian elements exist in the Indus cultures, such as carved stone boxes, dice, faience, wheeled vehicles, shaft-hole axes, religious art motifs, and the "ram-style" sculptural motif. But, taken as a whole, long-distance trade seems to have had little importance in the evolution of cultural complexity in the Indus Valley. The volume of product exchange was very low and mainly in luxury items, and the movement of goods appears to have been accomplished through intermediaries in the Iranian Plateau, rather than through deliberate and directly administered trade between Harappans and Mesopotamians.

As in the case of Egypt and Mesopotamia, rates of population growth in the Indus Valley appear to have been slow, and there is no evidence of pressure on the agricultural systems until long after the Harappan civilization had collapsed.

In summarizing Harappan civilization, Daniel Miller suggests that the city plans and other architecture are designed as an order "that opposes both the natural environment and the human"; and that in religious practices, burials, and other artifacts, the Harappan culture "represents a standardization of and around the mundane." Miller looks for the explanation of the Harappan culture within its own ideology, an ideology he links to later, historical cultures in this area, and in which personal differences are suppressed. The end of the Harappan civilization he links to "a contradiction between its refusal to acknowledge or represent change and its actual history which could only finally manifest itself with the revolutionary overthrow of the entire state."[39]

So, in the case of the Indus civilizations, we again see the divergence in modern archaeological interpretation. For some, the ecological, technological, and demographic factors are primary and interpretations are limited quite closely to aspects of administration, class differences, trade, and other aspects of the society that can be related directly to artifacts. But Miller and others consider such analyses to be limited, and they stress, instead, the ideological elements that determine a society. In turn, some archaeologists consider Miller's attempts to use inferences about the importance of the "mundane" in the minds of the Harappan peoples to be ultimately speculative and untestable.

Notes

1. Strabo, *Geography*, xv.I.19; quoted in Saggs, *Civilization Before Greece and Rome*, p. 303.
2. Allchin, "India from the Late Stone Age to the Decline of Indus Civilization," p. 337.
3. Vishnu-Mittre, "Discussion on Local and Introduced Crops."
4. Miller, "Ideology and the Harappan Civilization," p. 38.
5. Ibid., p. 39; Jarrige and Lechevaleier, "Excavations at Mehrgarh, Baluchistan: Their Significance in the Prehistorical Context of the Indo-Pakistan Borderlands"; Gupta, *Archaeology of Soviet Central Asia and the Indian Borderlands;* Sankalia, *The Prehistory and Protohistory of India and Pakistan;* Durante, "Marine Shells from Balakot"; Fairservis, *The Roots of Ancient India.*
6. Wright, "The Evolution of Civilizations," p. 325.
7. Ibid., p. 337.
8. Miller, "Ideology and the Harappan Civilization," p. 40.
9. Mughal, "The Early Harappan Period in the Greater Indus Valley and Northern Baluchistan."
10. Jansen, "Settlement Patterns in the Harappa Culture," p. 252; Agrawal, *The Archaeology of India*, p. 135; Miller, "Ideology and the Harappan Civilization," p. 40.
11. Shaffer, "Harappan Culture: A Reconsideration."
12. Fentress, *Resource Access, Exchange Systems and Regional Interaction in the Indus Valley*, pp. 136–37.
13. Miller, "Ideology and the Harappan Civilization."
14. Sarcina, "The Private House at Mohenjodaro"; Sarcina, "A Statistical Assessment of House Patterns at Mohenjodaro."
15. Miller, "Ideology and the Harappan Civilization," p. 48; Sarcina, "The Private House at Mohenjodaro"; Sarcina, "A Statistical Assessment of House Patterns at Mohenjodaro."
16. Fairservis, *The Roots of Ancient India*, pp. 246–47.
17. Possehl, "The End of a State and the Continuity of a Tradition in Proto-Historic India."
18. Rao, *Lothal and the Indus Civilization.*
19. Possehl, "The End of a State and the Continuity of a Tradition in Proto-Historic India."
20. Ibid.
21. Miller,"Ideology and the Harappan Civilization," pp. 56–57; but see also Fentress, "Indus Charms and Urns, a Look at the Religious Diversity at Harappa and Mohenjodaro."
22. Wright, "The Evolution of Civilizations," p. 340.
23. Allchin, "India from the Late Stone Age to the Decline of Indus Civilization," p. 343.
24. Frankfort and Pottier, "Sondage Preliminaire sur l'Establissement Protohistorique Harapeen et Post-Harapeen de Shortugai"; Miller, "Ideology and the Harappan Civilization," p. 54.
25. Miller, "Ideology and the Harappan Civilization," p. 55.
26. Raikes, "The Mohenjo-Daro Floods."
27. Possehl, "African Millets in South Asian Prehistory."
28. Kennedy, "Skulls, Aryans and Flowing Drains: The Interface of Archaeology and Skeletal Biology in the Study of the Harappan Civilisation."
29. Fairservis, *The Roots of Ancient India*, p. 311.

30. Renfrew, "Archaeology and Language."
31. Shaffer, *Prehistoric Baluchistan;* Possehl, "African Millets in South Asian Prehistory."
32. Miller, "Ideology and the Harappan Civilization."
33. Service, *Origins of the State and Civilization,* p. 246.
34. Ibid.
35. Jacobson, "Recent Developments in South Asian Prehistory and Protohistory."
36. Johnson, *Local Exchange and Early State Development in Southwestern Iran.*
37. Shaffer, "Harappan Culture: A Reconsideration."
38. Ibid., p. 49; Miller, "Ideology and the Harappan Civilization."
39. Miller, "Ideology and the Harappan Civilization," p. 64.

Bibliography

Agrawal, D. P. 1971. *The Copper Bronze Age in India.* New Delhi: Manoharlal.
———. 1982. *The Archaeology of India.* Guildford: Curzon Press.
Agrawal, D. P. and S. D. Kusumgar. 1974. *Prehistoric Chronology and Radiocarbon Dating in India.* New Delhi: Manoharlal.
Allchin, B. and F. R. Allchin. 1968. *The Birth of Indian Civilization.* Baltimore: Penguin.
Allchin, F. R. 1960. *Piklihal Excavations.* Hyderabad: Government of Andhra Pradesh.
———. 1961. *Utnur Excavations.* Hyderabad: Government of Andhra Pradesh.
———. 1963. *Neolithic Cattle-keepers of South India.* Cambridge, England: Cambridge University Press.
———. 1968. "Early Domestic Animals in India and Pakistan." In *Man, Settlement and Urbanism,* eds. P. J. Ucko, R. Tringham, and G. W. Dimbleby. London: Duckworth.
———. 1974. "India from the Late Stone Age to the Decline of Indus Civilization." *Encyclopaedia Britannica* 9:336–48.
Allchin, F. R. and D. K. Chakrabarti, eds. 1979. *A Source Book of Indian Archaeology,* Vol 1. New Delhi: Munshirman Maroliarlal.
Dales, G. F. 1966. "Recent Trends in the Pre- and Protohistoric Archaeology of South Asia." *Proceedings of the American Philosophical Society* 110:130–39.
———. 1979. "The Balakot Project: Summary of Four Years Excavations in Pakistan." In *South Asian Archaeology 1977,* ed., M. Taddei. Naples: Istituto Universitario Orientale.
Durante, S. 1979. "Marine Shells from Balakot, Shahr-i Sokhta and Tepe Yahya: Their Significance for Trade and Technology in Ancient Indo-Iran." In *South Asian Archaeology 1977,* ed. M. Taddei. Naples: Instituto Universitario Orientale.
Fairservis, W. A., Jr. 1956. "Excavations in the Quetta Valley, West Pakistan." *Anthropological Papers of the American Museum of Natural History* 45(2). New York.
———. 1975. *The Roots of Ancient India.* 2nd Ed. Rev. Chicago: University of Chicago Press.
Fentress, M. 1976. *Resource Access, Exchange Systems and Regional Interaction in the Indus Valley: An Investigation of Archaeological Variability at Harappa and Mohenjodaro.* Ann Arbor: University Microfilms.
———. 1979. "Indus Charms and Urns, a Look at the Religious Diversity at Harappa and Mohenjodaro." *Man and Environment* 3:9–104.
Frankfort, H.-P. and M.-H. Pottier. 1978. "Sondage Preliminaire sur l'Establissement Protohistorique Harapeen et Post-Harapeen de Shortugai." *Ars Asiatique* 34:29–86.
Gupta, S. P. 1978. *Archaeology of Soviet Central Asia and the Indian Borderlands,* Vol. 1. New Delhi: B. R. Publishing Co.
Ikawa-Smith, F., ed. 1978. *Early Paleolithic in South and East Asia.* The Hague: Mouton.
Jacobson, J. 1979. "Recent Developments in South Asian Prehistory and Protohistory." *Annual Review of Anthropology* 8:467–502.

Jansen, M. 1979. "Architectural Problems of the Harappa Culture." In *South Asian Archaeology 1977*, ed. M. Taddei. Naples: Instituto Universitario Orientale.

———. 1981. "Settlement Patterns in the Harappa Culture." In *South Asian Archaeology 1979*, ed. H. Hartel. Berlin: Dietrich Reimer.

Jarrige, J.-F. 1981. "Economy and Society in the Early Chalcolithic/Bronze Age of Baluchistan." In *South Asian Archaeology 1979*. Berlin: Dietrich Reiner.

———. 1984. "Chronology of the Early Period of the Greater Indus as Seen from Mehrgarh, Pakistan." In *South Asian Archaeology 1981*, ed. B. Allchin. Cambridge, England: Cambridge University Press.

Jarrige, J.-F. and M. Lechevaleier, 1979. "Excavations at Mehrgarh, Baluchistan: Their Significance in the Prehistorical Context of the Indo-Pakistan Borderlands." In *South Asian Archaeology 1977*, ed. M. Taddei. Naples: Instituto Universitario Orientale.

Johnson, G. A. 1973. *Local Exchange and Early State Development in Southwestern Iran*. Museum of Anthropology, Anthropological Papers, University of Michigan, no. 51. Ann Arbor.

Kennedy, K. 1982. "Skulls, Aryans and Flowing Drains: The Interface of Archaeology and Skeletal Biology in the Study of the Harappan Civilisation." In *Harappan Civilization*, ed. G. Possehl. Warminster: Aris & Phillips.

Kennedy, K. and G. L. Possehl, eds. 1976. *Ecological Backgrounds of South Asian Prehistory*. South Asia Occasional Papers and Theses, no. 4.

Lamberg-Karlovsky, C. C. 1967. "Archaeology and Metallurgical Technology in Prehistoric Afghanistan, India and Pakistan." *American Anthropologist* 69:145–62.

Leshnik, L. S. 1968. "The Harappan 'Port' at Lothal. Another View." *American Anthropologist* 70:911–22.

McEvedy, C. 1967. *The Penguin Atlas of Ancient History*. Harmondsworth: Penguin.

Miller, D. 1985. "Ideology and the Harappan Civilization." *Journal of Anthropological Archaeology* 4(1):34–71.

Mughal, M. R. 1970. "The Early Harappan Period in the Greater Indus Valley and Northern Baluchistan." Unpublished doctoral dissertation, University of Pennsylvania.

———. 1972. "Present State of Research on the Indus Valley Civilization." In *Proceedings of the International Symposium on Mohenjo Daro*. Karachi: National Book Foundation.

Perrtula, T. 1977. "Between the Indus and Euphrates: The Comparison of the Evolution of Complex Societies." Seattle: Mimeographed.

Pfeiffer, J. E. 1977. *The Emergence of Society*. New York: McGraw-Hill.

Piggott, S. 1950. *Prehistoric India*. London: Pelican.

Possehl, G. L. 1974. "Variation and Change in the Indus Civilization." Unpublished doctoral dissertation, University of Chicago.

———. 1976. "Lothal: A Gateway Settlement of the Harappan Civilization." In *Ecological Backgrounds of South Asian Prehistory*, eds. A. P. Kennedy and G. L. Possehl. South Asia Occasional Papers and Theses 4:118–31. Ithaca.

———. 1980. "African Millets in South Asian Prehistory." Mimeographed.

———. n. d. "The End of a State and the Continuity of a Tradition in Proto-Historic India." In *Realm and Region in India*, ed. R. Fox. Durham: Duke University.

———. ed. 1982. *Harappan Civilization*. Warminster: Aris & Phillips.

Raikes, R. 1965. "The Mohenjo-Daro Floods." *Antiquity* 39:196–203.

Rao, S. R. 1973. *Lothal and the Indus Civilization*. Bombay: Asia House.

Renfrew, C. 1988. "Archaeology and Language." *Current Anthropology* 29(3)437–468.

Saggs, H. W. F. 1989. *Civilization Before Greece and Rome*. New Haven: Yale University Press.

Sankalia, H. 1974. *The Prehistory and Protohistory of India and Pakistan*. Pune: Deccan College.

Sarcina, A. 1979a. "The Private House at Mohenjodaro." In *South Asian Archaeology 1977*, ed. M. Taddei. Naples: Instituto Universitario Orientale.

———. 1979b. "A Statistical Assessment of House Patterns at Mohenjodaro." *Mesopotamia* 13/14:155–99.

Service, E. 1975. *Origins of the State and Civilization*. New York: Norton.

Shaffer, J. 1978. *Prehistoric Baluchistan*. Delhi: B. R. Publishers.

———. 1972. "Harappan Culture: A Reconsideration." In *Harappan Civilization*, ed. G. L. Possehl. New Delhi: Oxford and Ibh.

Tosi, M. 1984. "The Notion of Craft Specialisation and Its Representation in the Archaeological Record of Early States in the Turanian Basin." In *Marxist Perspectives in Archaeology*, ed. M. Spriggs. Cambridge, England: Cambridge University Press.

Van Lohnizen-de Leeuw, J. E. and J. M. M. Ubagns, eds. 1974. *South Asian Archaeology*. Leiden: Brill.

Vishnu-Mittre. 1977. "Discussion on Local and Introduced Crops." In *The Early History of Agriculture: A Joint Symposium of the Royal Society and the British Academy*, organized by J. Hutchinson. London: Oxford University Press.

Wheeler, Sir. M. 1966. *Civilizations of the Indus Valley and Beyond*. New York: McGraw-Hill.

———. 1968. *The Indus Civilization*. 3rd Ed. Supplementary volume to *The Cambridge History of India*. Cambridge, England: Cambridge University Press.

Wheeler, R. E. M. 1950. *Five Thousand Years of Pakistan*. London: Royal India and Pakistan Society.

Wright, H. T. 1986. "The Evolution of Civilizations." In *American Archaeology Past and Future*, eds. D. J. Meltzer, D. D. Fowler, and J. A. Sabloff. Washington: Smithsonian Institution Press.

Wright, H. T. and S. Pollock. 1985. "Regional Socio-economic Organization in Southern Mesopotamia: The Middle and Late Fifth Millennium B.C." In *La Mesopotamie: Pre- et Protohistoire*, ed. J.-L. Huot. Paris: C.N.R.S.

Yule, P. 1982. *Lothal*. Munich: C. H. Beck.

11

Early Chinese Civilization

Let the past serve the present.

Mao Zedong

At about 1800 B.C., just a century or so after Hammurabi had established a great empire in Southwest Asia, the people of North China began a period of development that was to take them from simple agricultural tribes to one of the most brilliant and complex civilizations of antiquity. Because scientific archaeology in China is only a few decades old and has been interrupted by wars and revolution, we still see only the broad outlines of these developments, but it is clear that China offers interesting similarities and contrasts to other early states.

Compared to other early civilizations, ancient China was perhaps most similar to ancient Egypt, particularly in its florescence of art and in its self-absorption. But China has a far more varied physical landscape than Egypt—or Mesopotamia, for that matter—and this is reflected in the complexity of its cultural evolutionary patterns, where from an early date different regions interacted and competed for dominance.

Mesopotamia gave the world many great technological inventions, but China is at least as rich in this regard, having contributed the magnetic compass, gunpowder, printing, paper, paddle wheel propulsion, and many other inventions.[1]

The Ecological Setting

Modern China incorporates an area slightly larger than the contemporary United States. Within this territory are vastly different environments, ranging from the Himalayas to the Pacific shore, and people from many

CHINA

AD 618	Tang dynasty unifies China; empire based on Chang'an, world's largest city (1 million inhabitants)
AD 220	Collapse of Han dynasty; China divided into three independent states
AD 150	Buddhism reaches China
AD 105	Paper comes into use in China
50 BC	Chinese silk traded to Romans
108 BC	Chinese take control of Korea, establishing military outposts
119 BC	Chinese iron industry becomes state monopoly
206 BC	Accession of the Han dynasty
210 BC	Emperor Shih Huang-ti interred in vast mausoleum with terracotta army
	Shih Huang-ti, first Ch'in emperor, builds the Great Wall
221 BC	Unification of China by the Ch'in dynasty.
350 BC	Crossbow invented
403 BC	Warring States Period; finds of bronze and iron weapons and city defenses
500 BC	Beginning of coinage
550 BC	First significant iron production
770 BC	Beginning of Eastern Chou
1027 BC	Chou dynasty replaces the Shang
1400 BC	First written inscriptions, on oracle bones and bronze ritual vessels
	Urban settlements
1800 BC	Emergence of Shang civilization in northeast China
	Production of wheel-thrown pottery
2500 BC	Emergence of walled settlements in Longshan
	First bronze artifacts
2700 BC	Evidence of silk weaving
3000 BC	Finds of possible ploughshares in Hangzhou area

11.1   A cultural chronology of ancient China.

of these areas came to play some role in shaping Chinese civilization. K. C. Chang—whose work and writings provide much of what is known about China in the West—defined three zones as crucial areas in the evolution of ancient Chinese civilization: the Huang Ho valley, the southern deciduous zone, and the northern forests and steps.[2] He notes that for some thousands of years after the end of the Pleistocene at about 12,000 years ago, China's climate was somewhat warmer and moister than it is today, and much of the country was probably heavily forested, from the temperate forests of the north to the jungles in the south. Chang observes that because of the mountainous nature of China and the heavy post-Pleistocene vegetative cover, people were probably restricted to the river valleys at first, since these would have been the richest sources of food and would offer the most open land for farming. Developments in these areas eventually overlapped, and in later periods of antiquity the vast populations of nomads on the north and western frontier of China greatly influenced the character of Chinese civilization.

Wheat and barley supplied much of the energy that ran Mesopotamian and Egyptian societies, but in China rice and millet were in many instances as important as wheat or barley. North China's developmental leadership was closely tied to its agricultural potential. Pleistocene winds blowing off the Gobi Desert covered parts of North China with a layer of loess (a fine grain sediment) that reached a depth of several hundred meters. The Hwang Ho (or Yellow River—from the color given it by the loess it carries) cuts through these loess plains, frequently changing its course, and through flooding and draining it has created a rich agricultural zone of lakes, marshes, and alluvial fields. Loess is the agricultural soil par excellence: it is organically rich, requires little plowing, and retains near the surface much of the sparse rain that falls on North China. Moreover, it can yield large crops with little fertilization, even under intensive cultivation. The southern alluvial plains, with their hot, humid climate, eventually became the great rice heartlands.

Early Farmers (8000–5000 b.c.)

K. C. Chang notes that "one of the most significant discoveries in the prehistoric archaeology of North China in recent years" is the discovery in eastern Shensi Province and other areas of microlithic stone tools dating perhaps (the dating is uncertain) to about 13,000 years ago. The distinctive artifact styles of these and other finds indicate that a large population of pre-agricultural peoples were widely distributed over much of China about 10,000 years ago. Most of these people were hunter-gatherers, with some groups specializing in fishing near the coasts and large lakes.

There are gaps in the archaeological record in the centuries just after these microlithic industries were in use, but during the past decade many sites have been found that appear to have been farming communities. These villages date to between about 6500 and 5000 b.c. and have been grouped into regional clusters (Figure 11.2). Known as the P'ei-li-kang, Tz'u-shan, and related cultures, these settlements were located along the lower terraces of the western highlands, in the valleys of the Wei, Huang Ho, and a few other rivers. Their populaces lived mainly in small (2–3 meter diameter) subterranean houses, and made most of their living by farming millet and a few other crops, collecting walnuts, hazelnuts, and other wild plant foods, and by raising pigs and fowl and hunting deer. They already knew how to make pottery at this time, and used an impressive diversity of ground stone grinders, bone needles, and many other implements of stone, bone, and wood.

The earliest agricultural sites in south China are only a bit later than those in the North, but our knowledge of them is much less advanced. Several caves in China and Thailand have deeply buried remains from this period, and some of these early levels contain cord-marked pottery and what may be the remains of domesticated plants.

Regional Neolithic Developments in North China (5000–3000 B.C.)

The largest and best known early agricultural Chinese communities are those of the Yang-shao culture (Figure 11.2) The Yang-shao covered the same approximate area as did the earlier P'ei-li-kang culture, and the sim-

11.2 The earliest Neolithic cultures of China appeared in several distinct regions, shown here in darker shading, with the approximate boundaries of their artifact style distributions shown in lighter shading.

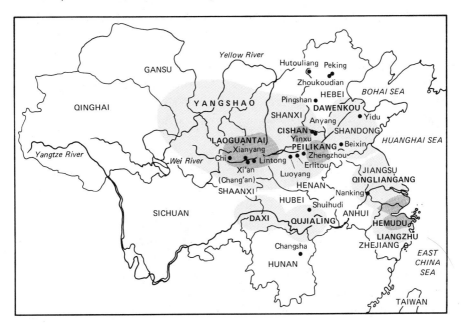

11.3 This reconstruction of a Neolithic house at Pan-p'o-ts'un illustrates how early Chinese farmers protected themselves against the harsh north China winters by living in pit-houses.

ilarities between these two indicate cultural continuity. Most of the Yang-shao settlements were distributed along the banks of the Huang Ho and other river systems in central China. Bones of most of the larger animals native to this part of China have been found in Yang-shao settlements, from rabbits to rhinos. But it was probably maize that provided most of the calories. Nutritionally as rich as wheat, millet is also quick-maturing and drought-resistant—making it admirably suited to the cold, arid plains of North China—and its stems can be used for food, fuel, and fodder. Sorghum and a few other crops were also cultivated. Like their counterparts in the forests of temperate Europe, early Chinese farmers probably cleared land for farming by slashing down vegetation with stone axes and then burning it off when it had dried.[3]

And like other early cultivators they took full advantage of game animals, wild plants, and fishing—as evidenced by the hundreds of hooks, fish gorges, and net weights found. Pigs, domestic fowl, and later, cattle, sheep, and goats were favored sources of meat.

In a real sense, the first ancestors of the civilization we know imprecisely as "China" were the sturdy peasants of Yang-shao period (5000–3000 B.C.) villages like Pan-p'o. Wearing rough clothes made of hemp, bunkered down in houses set several feet below ground (Figure 11.3) to shield themselves from winter snows and summer heat, the people of Pan-p'o followed the familiar agricultural cycle of China. Millet was the staple, pigs recycled wastes into what was even then perhaps a world-class cuisine, and cattle, sheep, and goats contributed their all in the form of milk, draft power, leather, and wool. The few hundred people at Pan-p'o, like most of their neighbors, were probably self-sufficient in everything that really mattered except marriageable women.

Yang-shao villages seem to have been abandoned and reoccupied periodically—probably because these people did slash-and-burn farming, wearing out the soil in the area around their community, then relocating the village, and eventually returning to the abandoned areas when the fertility was renewed by regrowth of vegetation.

Already by the time of Pan-p'o the Chinese aesthetic sense is evident in the lines and color of pottery, jade-carvings, and other artifacts. Stylized fish and animals grace some of the burnished pottery, and some pots were marked with symbols that may be early expressions of written Chinese. Around the houses were many deep pits, presumably for storing millet and other commodities. No doubt most villagers were full-time agriculturalists, but some engaged in silkworm cultivation, pottery manufacture, jade carving, and leather and textile production.

Pan-p'o has been converted into an impressive museum, and more than two hundred of the original inhabitants of this community are still there, their skeletons protruding from house floors and pits. Most of the site has been enclosed in a building, with some of the most important houses and artifacts left in place. At some Yang-shao sites the bodies of people seem to have been reburied in groups, K. C. Chang notes that at Yuan-chun-miao, fifty-seven graves were arranged "in six north-south rows, the earliest being the first and fourth rows, the latest the third and sixth rows. . . . This layout and chronological order have led to the reasonable speculation that the cemetery was divided into two moities, reflecting the two-clan composition of the village."[4]

In the middle of Pan-p'o was a distinctive building, much larger than the typical house and of different construction. To the Chinese excavators, this building seems most likely to have been a communal meeting hall; but one might also suspect that a bit of capitalistic inequality had infected the Yang-shao culture, and that this was the house of a "big man" or chief.[5] However, there are no evident signs of varying status in other artifacts at the site.

There are some differences between North and South China in the Neolithic, but because climates were in general warmer and wetter than they are today, these differences were not as pronounced as they were in later Chinese history.

More than a hundred Neolithic sites (c. 5000–3000 B.C.) have been found in the Lake T'ai-hu area, southeast of Nanking (Figure 11.2). Rice cultivation provided the main staple, but the range of animals hunted was impressive, including boars, elephants, alligators, deer, and many other species. One example of an early southern Neolithic community is at He-mudu, in northern Zhejiang province, where excavations in 1973–1978 revealed a 7000-year-old village of wooden houses built on piles along the shore of a small lake. The preservation of objects at this site is spectacular. Excavators found bone hoes, wooden shuttles for weaving, ivory carvings, and many plant and animal remains, including rice, bottle gourds, water chestnuts, domesticated water buffalo, pigs, dogs, and wild animals such as tigers, elephants, and rhinoceroses.

The extension of pottery styles over much of North China during the period between 5000 and 3000 B.C. is probably the clearest early sign that by this time people were developing the corporate economic and political institutions on which the later Chinese states were constructed.

Early Complex Chinese Societies
(3000–2000 B.C.)

In the early third millennium B.C., the people of North and central China—at least on the basis of our limited archaeological evidence—seem the very picture of egalitarian, peaceful villagers; but by the beginning of the second millennium B.C., signs of social rank and violence are everywhere.

The transformation of the simple farming communities of third millennium B.C. China into more complex social and political forms began shortly after 2400 B.C., with the emergence of the Lungshan culture and related cultures. Lungshan cultures, like the Yang-shao, are defined on the basis of similar styles of artifacts—in the Lungshan case by highly burnished, wheel-made, thin-walled black pottery in many different vessel forms; these pots are found, with minor stylistic variations, from the southeastern coast of China to the northern provinces. Very early Lungshan pottery and other diagnostic artifacts are found over most of the old Yang-shao heart-land.

The Yang-shao communities in some places seem to have given rise almost directly to Lungshan communities, but the linkages between these periods and peoples are still quite obscure.[6]

As with the Yang-shao, the Lungshan peoples lived mainly in villages made up of pit houses arranged around a central "long house," and virtually every Lungshan adult male was probably still a millet farmer who supplemented the family fortunes with hunting, collecting, and part-time craft production of pottery, jade, or another commodity. But Lungshan villages were, on the average, significantly larger than those of the Yang-shao period, and were likely occupied for longer periods of time.

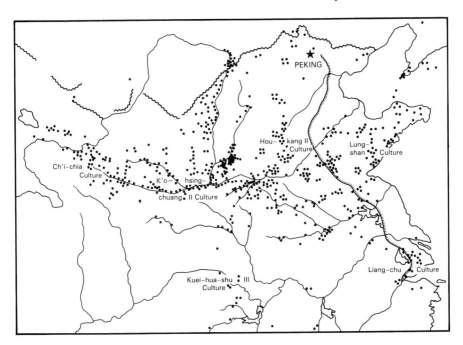

11.4 Archaeological sites of the Lungshan and related cultures.

Like their counterparts in other early civilizations, ancient Chinese farmers eventually made a momentous transition: instead of slash-and-burn agriculture, shifting from year to year, they began to cultivate the same fields each year. The costs of doing this are high: one has to use animal manures, clovers, and discarded plants to renew fertility, or else irrigation water must be led to the fields through canals to replenish soil fertility. In some places, of course, in China as in Egypt, the annual floods renew soil fertility, but in all ancient civilizations a point was eventually reached where intensification of agriculture was required to sustain life. Lungshan agriculture, although still based on millet and a few domestic animals, seems to have been more intensive than that of the Yang-shao. The slash-and-burn, shifting agricultural system of Yang-shao times probably gave way to a permanent field system in the Lungshan period. Domestic poultry, sheep, and cattle became more important, and there is evidence of the increasing significance of rice agriculture in some southern areas.

There are also signs of change in social organization. Compared with Yang-shao graves, Lungshan burials seem to correspond to richer and poorer people, in terms of grave goods. Lungshan pottery and jade ornaments are so technologically sophisticated and beautiful that they suggest at least some semispecialized craftsmen.

Lungshan people sought knowledge through scapulimancy, the art of writing signs on bones, applying heat to the bone to crack it, and then interpreting the pattern of cracking to foretell the future. Archaeologist K. C. Chang interprets the appearance of this art in Lungshan times as a reflection of the rise of at least a semiprofessional class of shamans, an interpretation made credible by the fact that the character meaning "book" already appears on these bones and may signify the existence of specialized scribes.

Chang also notes that the first traces of town walls—which became a standard feature of later communities—are found at the Lungshan site of Ch'eng-tzu-yai. It was built with the "stamped-earth" technique, in which layer after layer of fine loess silts and clays were stamped by workers into a compact wall. The wall at Ch'eng-tzu-yai has largely deteriorated, but from the traces of it the excavators estimate its height at 6 meters and its width at the top to be 9 meters.

Lungshan and related communities were probably still largely self-sufficient and independent. Settlement patterns of this period are poorly known, but there is little evidence of population aggregation or regular spacing of settlements according to economic or administrative principles. Nonetheless, the transition to life in permanent villages and established agricultural fields was made over much of North China in the Lungshan period, from the western highlands into the northern Manchurian highlands and well into southern China.

Chang has summarized the overall developments of China in the third millennium b.c. in terms of eight features that taken together indicate evolving cultural interaction and complexity: (1) the beginnings of copper metallurgy; (2) the widespread use of sophisticated potter's wheels and presumably the emergence of a specialized class of potters; (3) the stamped-earth walls found at some sites, and their implication of the need for civil defense; (4) burials of bodies in wells, mass graves, and other conditions that suggest warfare; (5) the use of animal and other motifs in artworks in ways that indicate emerging status differences; (6) decorated jade masks and other objects that suggest an emerging ritual system practiced over a large part of the country; (7) the widespread use of scapulimancy, also suggesting ritual interaction over a large area; (8) signs of social ranking and wealth differences in burials.

We can easily draw parallels here with Egypt, Mesopotamia, and the Indus Valley: the Chinese use of jade, scapulimancy, and so forth are distinctive in their specifics, but the general pattern is one of emerging and evolving class-consciousness, functional specialization, and administrative complexity.[7]

The First Civilizations (c. 2000–1100 B.C.)

After about 2000 B.C., large towns and cities began to replace or emerge from the tens of thousands of villages that marked earlier, simpler times. During the second millennium B.C. China really became "China," in the sense that this period marked the first widespread use of the distinctively Chinese forms of writing, architecture, art, and ideology. Also during this period all the correlates of cultural complexity, such as monumental architecture, large population concentrations, occupational specialization, written records, gross differences in wealth, power, and prestige, and large public-works projects, appeared in full measure.

One of the earliest sites of this transitional period to complexity is Erh-li-t'ou (Yen-shih) (Figure 11.5). Here lies a vast complex of ruins, scattered over an area about 1.5 by 2.5 kilometers, and extending to a depth of 3 meters in some places.

Among the most impressive features at Erh-li-t'ou are the foundations of two buildings that some have called "palaces"; one is 108 by 100 meters and is associated with a large earthen platform. In one area of the foundation was a group of burials, including one person who seems to have been buried with his hands bound. The other, smaller palace was drained with a system of pottery pipes and associated with a large earthen tomb—long since plundered of most of its wealth. The richness of Erh-li-t'ou's jade, bronze, pottery, and other artifacts, the indications of human sacrifice, and the large buildings all suggest a level of emerging cultural complexity greater than that of previous periods. Other sites of the same age as Erh-li-t'ou have been found, and these together may represent the transitional period between the late Neolithic and the indisputable cultural complexity of the Shang period.

Erh-li-kang Phase (c. 1650–1400 B.C.)

With the Erh-li-kang phase we move into the historical period in ancient China, and from the end of this period on some written documents supplement the archaeological record.

The Erh-li-kang phase is best documented archaeologically at the cluster of settlements near Chengchou. The central area of Erh-li-kang phase settlement was a roughly rectangular arrangement of buildings extending about 3.4 square kilometers, much of which was enclosed by a pounded-earth wall some 36 meters wide at the base and 9.1 meters in height—as estimated from the segments still remaining.[8] The central area of the site is thought to have been the residence and ceremonial center of the ruling elite, and around it were thousands of pit houses, animal pens, shops, storage pits, and other features that make it clear that many of the people

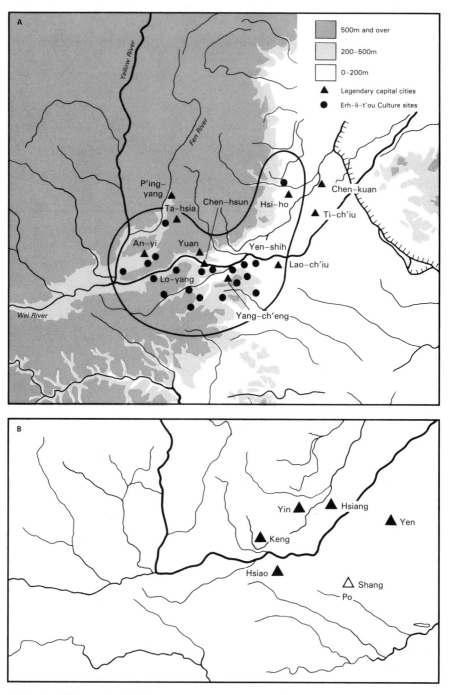

11.5   K. C. Chang's (1986) reconstructions of: (A) Major Erhli-t'ou Culture sites and traditional Hsia Dynasty capitals. (B) Traditional Shang Dynasty capitals.

of Shang China lived lives radically different from those of Neolithic times. Hundreds of skilled, full-time craftsmen probably resided at Chengchou. In one area thousands of pieces of animal and human bone were recovered, much of it already fashioned into fish hooks, awls, axes, and hairpins. In another area were more than a dozen pottery kilns, each surrounded by masses of broken and oven-fired pottery. No jade, leather, or textile workshops have been found, but the circulation of these products at this time is well documented elsewhere.

But the Chengchou craftsmen really displayed their skill in working bronze. Large areas were given over to workshops for casting fish hooks, axes, projectile points, and various ornaments. A kind of mass production was achieved by using multiple molds, made by impressing a clay slab with the forms of six arrowheads, with each impression connected by a thin furrow to a central channel. A second clay mold was placed over the first, the two bound together, and then molten bronze was poured in. After cooling, the individual points could be sawn off from the central stem.

In Western societies especially, there is a tendency to think of cultural changes as often being driven by technological change. The invention of the locomotive, for example, seems to have radically altered nineteenth-century Western society. When we look at ancient China, there is a temptation to see a culture shaped and formed by the invention of bronze-working and iron-working. Bronze-working seems to have been an indigenous development in Shang civilization, probably made possible by the high-temperature kilns first used to fire pottery. The requisite copper and tin could be found within several hundred kilometers of the Shang homeland, and various processes using clay models and the lost-wax process were known. The first great diffusion of Shang artifacts, which probably marked both the rise of national consciousness and ramifying political and economic networks that underlay the emergence of Shang civilization, was also marked by the spread of bronze artifacts.

But the use of bronze seems to have been mainly a stylistic phenomenon rather than a great technological advance. Though lovely to look at, the myriad bronze Shang vessels offer few culinary advantages over ceramics, and there is no evidence that bronze weapons revolutionized Shang-period warfare. And for clearing forests—an important part of the Shang expansion—stone axes and hoes would probably have worked at least as well as bronze tools which were, in any case, a luxury item.

The Shang people probably may have learned the basic processes of bronze metallurgy from others. Already by 4500 b.c. people in southeast Asia, at sites like Ban Chiang, in Thailand, were using cast-bronze spearheads and ornaments.[9] We know relatively little about these early Southeast Asian bronze-using cultures, but it seems likely that their precocious

11.6 Royal Shang tombs were often accompanied with scores of sacrificed people, as evident in this tomb at Hou Chia Chuang, Honan province.

use of bronze was a result of the proximity of copper and tin deposits. Their evolution to cultural complexity followed that of the north in all essential elements and occurred much later.

Both archaeological evidence and ancient documents written after the Shang period indicate that society during the Erh-li-kang phase was headed by a king, who ruled through a hierarchically arranged nobility. Commoners were conscripted for public works and military service; there were highly organized and incessant military campaigns; and many settlements were apparently integrated into an organized intervillage system of commerce.[10] It has not been determined if there were large-scale irrigation systems, but at Chengchou at least a canal system was in use, perhaps to carry water to the settlement or else to remove drainage water or sewage from the complex.[11]

The great mass of Shang people, however, lived much as their ancestors had, in villages of pit houses located along river systems, subsisting on the same kinds of crops and agricultural technology as people of previous millennia.

Yin Phase (c. 1400–1112 b.c.)

The last and most brilliant phase of Shang civilization, the Yin phase, seems to have begun about 1384 b.c., when the Shang king, P'an-keng, is reported to have moved his capital to the city of An-yang, in Hunan province.

Excavations at An-yang and contemporary sites in this area have been conducted intermittently since the 1920s, but the publication of this research is far from complete. Scores of sites within an area of about 24 square kilometers have been tested, and the evidence suggests that the complex at An-yang includes a large ceremonial and administrative center surrounded by smaller dependent hamlets and craft centers.[12] True to tradition, most peasants still lived in small pit houses—not very different from those of 2000 years earlier. Scattered throughout the settlement were granaries, pottery kilns, storage pits, bone and bronze workshops, animal pens, ditches, and familiar features of ancient Chinese life.

No city wall has been found at An-yang, but monumental buildings were constructed. The largest of these was about 60 meters long, rectangular in form, with large stone and bronze column bases, and founded on a large platform of compacted earth. There were at least fifty-three structures of this type (though somewhat smaller) in one group at An-yang, arranged in three main clusters. Although not lavish in construction, these buildings are surrounded by scores of human and animal sacrificial burials, as well as many pits containing royal records written on oracle bones and numerous small structures thought to be for service personnel.[13] Near the cluster of buildings is a cemetery with eleven large graves, replete with lavish, expensive burial goods and many human sacrificial burials—the whole complex surrounded by 1200 smaller, much less lavish burials. Elsewhere, a complex of ceremonial buildings at Hsiao-T'un was dedicated with the sacrifice of 852 people, 15 horses, 10 oxen, 18 sheep, 35 dogs, and 5 fully equipped chariots and charioteers.[14] Generally, many burials here would be certain to arouse the suspicions of a coroner.

The Shang ceremonial and administrative structures are perhaps not as impressive in size or cost as the ziggurats and temples of Mesopotamia, but the level of occupational specialization, the immense wealth of the burials, and the intensity of organization of the agricultural and economic systems remind one of the Mesopotamian city-states of the late fourth millennium B.C.

Little urbanism and only a thin distribution of bronze-working and a few other Shang cultural traits appear in most of non-Shang China at this time, and most of these non-Shang cultures were probably still at a predominantly Neolithic level of development.

Late in the Yin phase of the Shang dynasty (about 1200 B.C.), the written language had evolved to the point that texts from this period give us a detailed portrait of Shang life. Over 3000 phonetic, ideographic, and pictographic characters were in use, of which about 1200 have been identified, and more than 160,000 inscribed shells (of which only some have been translated) and numerous inscriptions in bronze or stone date to this period.[15]

According to the texts, the late Shang rulers held sway over a territory extending from the Pacific shore to Shensi province in the west, and from the Yangtze River in the south to southern Hupeh in the north. At the apex of Shang society was the king, who ruled directly on many affairs of state and was assisted by a complex hierarchy of nobles possessing considerable local autonomy in their respective territories. These lords were charged with defending the homeland, supplying men for armies and public-works projects, and collecting and contributing state taxes. Toward the end of the Shang period, many nobles apparently achieved almost feudal status and were virtually independent in their own domains. But the king was still considered to have superior supernatural powers and to be the pivot of all ritual procedures. The kingdom was ringed with "barbarians," and Shang kings often granted them almost complete autonomy in exchange for peaceful relationships.[16]

On occasion, royal armies of up to about 30,000 men were conscripted and led by the nobles against insurgent "barbarians" and neighboring principalities. The basis of the army was the horse-drawn chariot, supported by infantrymen equipped with bronze-tipped arrows and laminated bows, and royal records indicate that military campaigns often incurred and inflicted frightful casualties. Staggering quantities of plunder were often taken, along with thousands of prisoners, most of whom were apparently sacrificed or enslaved.[17]

The agricultural system seems to have been essentially the same as previously, with millet, wheat, rice, and vegetables the major crops, and cattle, sheep, pigs, and poultry, with the only "new" domestic animal being the water buffalo. The proportions of these crops and animals may have been shifted somewhat, with wheat and rice expanding their range at the expense of millet, but the evidence for this is questionable. There is little record of large irrigation systems anywhere in the Shang domain, and hunting and gathering still supplied a large part of the diet. The mammalian faunal remains from An-yang include massive quantities of boar, deer, bear, and other hunted animals, including a few elephants, rhinoceroses, leopards, and even part of a whale.[18] Apparently intravillage trade in foodstuffs was voluminous.

Local occupational specialization was considerable. Many villages lacked one or more of the more important handicraft workshops, suggesting that products were exchanged among these settlements.[19] The discovery of large caches of agricultural implements (3500 stone sickles, new and used, in a single pit at one site, for example) may indicate a degree of centralized management of both agriculture and craft production.[20] The Shang even had a type of money, in the form of strings of cowrie shells.

Toward the end of the Shang period, there were many walled towns and villages in North and central China, and, compared to earlier periods, a much greater proportion of the populace lived in these semiurban settings. But if we compare the settlement size distribution of late Shang China—or rather what we estimate it to have been—with those of Mesopotamia or Mesoamerica at a comparable level of development, it is clear that Shang China was a much less urbanized society. There were no settlements the size of Ur or Teotihuacan.

We cannot know the mind of the Shang Chinese in any detailed sense, but their artifacts give us clues as to what they thought of the world. Of all ancient peoples, they seem to have been the most concerned with symmetry. All temples and tombs were oblong or square and oriented to the four cardinal directions. Bronzes were always symmetrical, and even the messages inscribed on turtle shells for divinations were repeated on the right and left sides. The world was conceived of as square, the wind as blowing from four quarters, and four groups of foreigners were thought to live on China's borders. Throughout their art, architecture, and literature, elements appear two by two, four by four, and in other intricate but symmetrical arrangements.

The texts and inscriptions tell us that the Shang nobles felt in constant and clear communication with their ancestors, whom they consulted through oracles about the best actions to take in multitudes of situations.

A recent bumper sticker declared "If I can't take it with me, I ain't going"—a sentiment probably very agreeable to Shang rulers. Tomb after tomb is stocked with everything from chariots to rice. And they also employed that most efficient of all incentives to a devoted domestic staff: at the ruler's own death, all his domestic retainers were apparently killed and buried with him. Unlike feudal European nobility, few Shang nobles feared poisoning by members of their households.

Early Imperial China (1100 B.C.–A.D.220)

The early phases of Chinese civilization are often called the "Three Dynasties," the Erh-li-t'ou/Hsia being the first, the Shang the second, and the Chou the third.

Yet these three dynasties ran parallel for part of their courses, and, as K. C. Chang notes, the developmental pattern was one of "Early Shang, interacting with Erh-li-t'ou, and proto-Chou, interacting with Shang."[21]

Chang goes on to quote a poem dating to the Chou Dynasty:

> August is God on high;
> Looking down, he is majestic;
> He inspected and regarded the [states of] the four quarters,
> He sought tranquillity for the people.
>
> These two kingdoms [of Hisa and Shang],
> Their government had failed;
> Throughout those states of the four [quarters] he investigated and
>   estimated;
> God on high brought it to a settlement;
> Hating their extravagance,
> He looked about and turned his gaze to the West,
> And there he gave an abode.[22]

The "West" of this lyric, chosen by God, was the Chou state. Had we the means to see it in all its splendor, ancient imperial China under the Chou and later states may well have appeared to us as the most complex and colorful civilization of all antiquity. After 1100 B.C., the Chou empire (Figure 11.7), and its successors arranged much of China in a feudal system that led to the growth of cities, great cycles of peace followed by warfare, and baroque administrative hierarchies.[23] Families of nobles or commoners would rise to power, make war on their neighbors, extend their kingdoms, and then collapse under the onslaught of competing warlords. Through it all, exquisite bronzes, porcelains, pots, and jewelry were made and lavished on the rich; untold thousands of people were sacri-

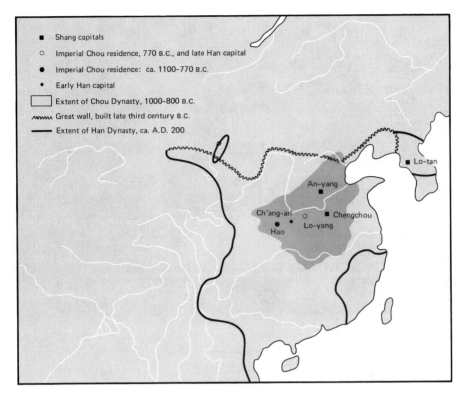

■ Shang capitals

○ Imperial Chou residence, 770 B.C., and late Han capital

● Imperial Chou residence: ca. 1100–770 B.C.

• Early Han capital

☐ Extent of Chou Dynasty, 1000–800 B.C.

ᴧᴧᴧᴧᴧ Great wall, built late third century B.C.

▬▬ Extent of Han Dynasty, ca. A.D. 200

Lo-tan

An-yang

Ch'ang-an   Chengchou

Hao   Lo-yang

11.7   From the original Shang-period state, China evolved into true empires during the Chou and Han Dynasties, whose extents are indicated here.

ficed to be buried with their rulers; and millions lived and died in the eternal agricultural cycle of rural China.

By 500 B.C. iron-working became widespread, and iron agricultural tools were in common use. Iron weapons, mass burials, and military annals tell of a savage form of warfare, not at all like the depersonalized modern combat of tanks, missiles, and automatic weapons. Men in armor fought at close quarters with swords and knives on battlefields swarming with chariots, cavalry, and bowmen.

At about the time iron was introduced, Confucius (551–479 B.C.) lived and taught and, together with the spread of Buddhism in the third century A.D., moved much of Chinese thought forever beyond the reach of the foreigner.

Amid the panoply of arts, crafts, and religion, the prosaic elements of agriculture were also changing. After 500 B.C. great irrigation works were brought into use, allowing the intensive cultivation of wet-rice species as well as many other crops. This in turn supported the tremendous popu-

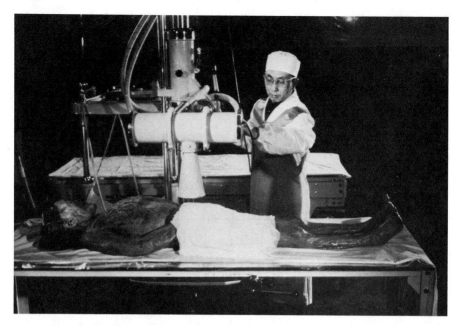

11.8   Some bodies in Chinese tombs were so well preserved that autopsies can reveal causes of death. Here a Chinese radiologist X rays Lady Dai, a Han Dynasty noblewoman who died in 141 B.C. of a heart attack.

lation densities of China. Along with the introduction of the ox-drawn plow and the evolution of precise crop-rotation practices, Chinese agriculture underwrote the impressive cultural elaborations we associate with the Qin (221–207 B.C.) and Han (202 B.C.–A.D. 120) dynasties.

The great richness and opulence of the Qin Dynasty is lavishly displayed in the Museum of Qin Shi Huang, in Shaanxi Province. In a project worthy of Egyptian pharaohs, the emperor Qin Shi Huang—the First Emperor of China—drafted 700,000 people to build a huge mausoleum and palace complex. The work lasted for thirty-nine years and when it was finished, it was full of rare and beautiful objects. Historical records of the time say that the complex was lighted with lamps fueled by the fat of giant salamanders. In 206 B.C. the whole complex was burned in a popular revolt. When archaeologists excavated they found 56.25 square kilometers of buildings and features, including "tombs of immolated slaves, table pits, stone material processing workshops, tombs of criminals, pits for the execution of slaves, pits of life-size terra cotta warriors and horses, and pits of bronze chariots and horses."[24] The site so far has revealed thousands of life-size statues of warriors and horses (Figure 11.9), all rendered in exquisite detail.

One of the world's first imperial censuses was conducted in China in

A.D. 1–2 and tallied 57,700,000 people, at least 10% of whom lived in rectangular wooden towns with populations of up to 250,000. Coins circulated, schools flourished, and a rich store of literature was created.

Through it all, along the northern and western frontiers, the nomadic horsemen and herders of Asia pressed on the periphery of successive empires. The Great Wall, built about 100 B.C., was meant to keep the nomad out and the farmer in, but it was probably not too successful. Throughout Chinese history there was a constant interchange between sedentary and nomadic cultures.

We tend to think of China as a static civilization with ancient and fixed forms. But from the long-term perspective, it has been a varied, ineffably complex cultural pageant.

11.9 Some ancient Chinese burials contained hundreds of life-size figures of people and horses. The figures were of individual people—each one different in details.

The Origins of Cultural Complexity
in North China: Summary and Conclusions

By Mesopotamian standards, at least, early Chinese society was less urban and less given to monumental architecture than the Sumerian peoples, although in later Chinese history the country became heavily urbanized.

Many hypotheses have been suggested for why people aggregate in settlements of different sizes under different conditions. Unfortunately, the archaeological evidence from China during the critical periods is not sufficiently complete to evaluate the several possibilities. Some have thought that warfare was important in determining the settlement size distribution in ancient China, and in fact warfare did have a profound influence on Chinese developments. But warfare, too, seems to be simply another— and somewhat variable—expression of more fundamental changes going on within societies as their complexity and differentiation increase.

One factor of continual importance in all Chinese developments was the great spread of nomadic and seminomadic peoples who lived on the borders of the agricultural heartland. Some of the earliest documents attesting to the rise of the Chinese state, for example, imply that the Chou Empire was initiated by the movement of the warlord Tai Wang to the central Shaanxi province under pressure from his nomadic neighbors.[25] And Owen Lattimore's comprehensive history of relationships between nomad and farmer in China illustrated the crucial role of this relationship in the formation of all later Chinese empires.[26]

Turning to the major similarities between ancient Chinese developments and those elsewhere, in China too one of the first steps was an increasingly intensive and productive agricultural system, coupled with sedentary village communities. As elsewhere, this stage was followed by the rapid extension of pottery styles over a large region and a concurrent increased variability in residential architecture and mortuary complexes. And while monumental architecture may have been somewhat minimized and belated in ancient China, the great wall around Cheng-chou and other features are examples of at least some investment in "wasteful" monumental projects.

K. C. Chang argues that one of the principal ways in which China is somewhat different from other early civilizations is the extraordinary absorption with *shamanism*—the connection to another world, a world of the gods and magical powers, through priests. Chang also offers this condensed hypothesis about Chinese cultural evolution: "the wealth that produced the civilization was itself the product of concentrated political power, and the acquisition of power was accomplished through the accumulation of wealth."[27]

Chang suggests that Chinese civilization operated on a profoundly re-

11.10　Like their contemporaries in Egypt and Mesopotamia, early Chinese complex societies expended enormous amounts of wealth in mortuary cults. This Chinese princess of the late second millennium B.C. was entombed in a suit made of 2000 jade tiles tied together with gold wire.

ligious basis, with kings, nobles, and commoners arranged in a layered universe, under a pantheon of gods, with communication between the physical and spirit worlds channeled through priests. Hence, the abiding concern with divining the will of the gods through oracles—the king himself was credited with oracular powers. And, as Chang notes, technology has a reduced importance in such a fundamentally religious and abstract cosmology, which may explain even some of the contemporary differences between East and West.

China eventually became locked in the same cyclical pattern of expansion and collapse we documented in other ancient civilizations. Indeed, the apparent cyclicity of Chinese history has been the basis for many general models of cultural complexity, ranging from Marx to Marvin Harris.[28] In these ancient empires it is as if some internal limiting factors exist that restrain growth past a certain point, at least until certain evolutionary conditions are present. Perhaps these were primarily technological limits in these early empires. Without modern means of information collection, storage, retrieval, and dissemination, political systems have very definite limits in terms of the numbers of people, projects, and provinces they can successfully integrate.

Notes

1. De, "Classical Chinese Contributions to Shipbuilding."
2. Chang, *The Archaeology of Ancient China*, pp. 95–99.
3. Triestman, *The Prehistory of China.*

4. Chang, "In Search of China's Beginnings: New Light on an Old Civilization," pp. 152–53.
5. Watson, *Ancient China.*
6. The following discussion of early complex Chinese societies is taken mainly from Chang, *The Archaeology of Ancient China.*
7. In addition to Chang, *The Archaeology of Ancient China,* see also Triestman, *The Prehistory of China.*
8. Chang, *The Archaeology of Ancient China,* pp. 286–94.
9. Bayard, "The Chronology of Prehistoric Metallurgy in North-East Thailand: Silabhumi or Samrddhabhumil."
10. Chang, *Early Chinese Civilization: Anthropological Perspectives.*
11. Wheatley, *The Pivot of the Four Quarters.*
12. Ibid.
13. Chang, *Early Chinese Civilization: Anthropological Perspectives,* p. 10.
14. Wheatley, *The Pivot of the Four Quarters,* p. 93.
15. Chang, *Early Chinese Civilization: Anthropological Perspectives,* p. 48.
16. Wheatley, *The Pivot of the Four Quarters.*
17. Ibid., p. 63.
18. Ibid.
19. Chang, *The Archaeology of Ancient China,* pp. 138–39.
20. Wheatley, *The Pivot of the Four Quarters,* p. 76.
21. Chang, "In Search of China's Beginnings: New Light on an Old Civilization," p. 156.
22. Ibid.
23. Zilin, "The Museum of Qin Shi Huang," p. 141.
24. Chang, "In Search of China's Beginnings: New Light on an Old Civilization."
25. Lattimore, *Inner Asian Frontiers of China.*
26. Ibid.
27. Chang, *The Archaeology of Ancient China,* p. 414.
28. See, e.g., Harris, *Cannibals and Kings.*

Bibliography

Anonymous. 1982. *Neolithic Site at Banpo Near Xian.* Trans. Du YouIiang [sic]. [Guide to the Banpo Museum, printing data written in Chinese].

Barnard, N. 1972. "The First Radiocarbon Dates from China." *Monograph on Far Eastern History 8.* Canberra: School of Pacific Studies, A. N. U.

Bayard, D. 1979. "The Chronology of Prehistoric Metallurgy in North-East Thailand: Silabhumi or Samrddhabhumil." In *Early South-East Asia,* eds. R. B. Smith and W. Watson. London: Oxford University Press.

Chang, K. C. 1970. "The Beginning of Agriculture in the Far East." *Antiquity* 44:175–85.

———. 1976. *Early Chinese Civilization: Anthropological Perspectives.* Cambridge, Mass.: Harvard University Press.

———. 1977. *The Archaeology of Ancient China.* 3rd Ed. New Haven: Yale University Press.

———. 1981. "In Search of China's Beginnings: New Light on an Old Civilization." *American Scientist* 69:148–58.

Chang, S. 1963. "The Historical Trend of Chinese Urbanization." *Annals of the Association of American Geographers* 53:109–43.

Chang, T. 1957. *Archaeological Studies in Szechwan.* Cambridge, England: Cambridge University Press.

———. 1963. *Archaeology in China, Vol. 3: Chou China.* Cambridge: Heffer.

De, Zhou Shi. 1987. "Classical Chinese Contributions to Shipbuilding." *Endeavour* NS. 11(1):2–4.

Harris, M. 1977. *Cannibals and Kings.* New York: Random House.

Lattimore, O. 1951. *Inner Asian Frontiers of China.* Boston: Beacon Press.

Li, C. 1957. *The Beginnings of Chinese Civilization.* Seattle: University of Washington Press.

Meacham, W. 1977. "Continuity and Local Evolution in the Neolithic of South China: A Non-Nuclear Approach." *Current Anthropology* 18:419–40.

Nai, H. 1957. "Our Neolithic Ancestors." *Archaeology* 10:181–87.

Service, E. 1975. *Origins of the State and Civilization.* New York: Norton.

Skinner, G. W. 1964. "Marketing and Social Structure in Rural China." *Journal of Asian Studies* 24:3–43.

Stover, L. E. 1974. *The Cultural Ecology of Chinese Civilization.* New York: Pica Press.

Toynbee, A., ed. 1973. *Half the World.* New York: Holt.

Triestman, J. 1972. *The Prehistory of China.* Garden City, N.Y.: Natural History Press.

Watson, W. 1960. *Archaeology in China.* London: Parrish.

———. 1971. *Cultural Frontiers in Ancient East Asia.* Edinburgh; Edinburgh University Press.

———. 1974. *Ancient China.* Greenwich, Conn.: New York Graphic Society.

Wheatley, P. 1971. *The Pivot of the Four Quarters.* Chicago: Aldine.

Wittfogel, K. A. 1957. *Oriental Despotism: A Comparative Study of Total Power.* New Haven: Yale University Press.

Zilin, W. 1985. "The Museum of Qin Shi Huang." *Museum* 147:140–47.

12

Secondary Old World States

> The day may come when, contemplating a
> world given back to the primeval forest, a hu-
> man survivor will have no means of even
> guessing how much intelligence Man once im-
> posed upon the forms of the earth, when he
> set up the stones of Florence in the billowing
> expanse of the Tuscan olive-groves. No trace
> will be left then of the palaces that saw Michel-
> angelo pass by, nursing his grievances against
> Raphael; and nothing of the little Paris cafes
> where Renoir once sat beside Cezanne, Van
> Gogh beside Gauguin. Solitude, viceregent of
> Eternity, vanquishes men's dreams no less than
> armies, and men have known this ever since
> they came into being and realized that they must
> die.
>
> Andre Malraux[1]

To refer to the ancient glories of Greece, Germany, Thailand, and other cultural centers as "secondary" states, as compared to the "primary" states of, for example, Mesopotamia, is to imply, perhaps, more of a difference than existed. In many ways these secondary states were independent developments—creations of their own environments, peoples, and unique histories. But temperate Europe, sub-Saharan Africa, Southeast Asia, and many other areas were in fact heavily influenced by the first civilizations. And in important ways these secondary states were less complex, at least until the modern era.

It is beyond the scope of this book to review these later, less complex political systems in detail, and they are described briefly here to illustrate the extent to which after about 3000 B.C. the Old World became a regionally integrated complex of evolving cultures, none of which developed in isolation. From the last few millennia B.C. onward, Old World history is a dizzingly complex tapestry in which to separate individual strands and even patterns is to do some degree of violence to the whole. The reader may also wish to consider anew the biblical prophecy that "the last shall be first," a sentiment repeatedly demonstrated in Old World culture history, where "peripheral" political systems often rose to imperial power.

The Aegean and Western Anatolia

> The grand object of travelling is to see the shores of the Mediterranean. On those shores were the four great Empires of the world: the Assyrian, the Persian, the Grecian, and the Roman. All our religion, almost all our law, almost all our arts, almost all that sets us above the savages, has come to us from the shores of the Mediterranean. (Samuel Johnson, A.D. 1709–1784)

Old World prehistory after about 4000 B.C. is really an account of the integration of the world into fewer and fewer great empires, and the slow

12.1 A cultural chronology of Aegean and Anatolian civilizations.

YEARS BC		ANATOLIA	AEGEAN AND CYPRUS
IRON AGE	0 / 500	PERSIAN RULE	
		IRON AGE NEO-HITTITE STATES Carchemish URARTU TUSHPA (Van) SEA PEOPLES →	
MIDDLE BRONZE AGE	1000	LATE BRONZE AGE (1500-1200) HITTITE NEW KINGDOM *Agriculture and metallurgy Expansion into Syria* Hattusha Western Anatolia AHHIJAWA (ACHAEANS?) Arzawa	LATE MINOAN NEW PALACE PERIOD *Minoan settlements abroad* Knossos, Phaistos, Malia, Zakro, Khania / LATE HELLADIC *Mycenaean palaces-commerce* Mycenae, Pylos, Tiryns / LATE CYPRIOT II-III *Commercial centera-trade* Enkomi, Kitlon, Hala Sultan Tekke, Aylos Dhimitrios
	1500	MIDDLE BRONZE AGE OLD ASSYRIAN COLONIES *Karum Kanesh* HITTITE OLD KINGDOM INDO-EUROPEANS Hattusha	CRETE: MIDDLE MINOAN *First palaces-commerce* Knossos, Malia, Phaistos / GREECE: MIDDLE HELLADIC *Shaft graves* Mycenae / CYPRUS: MID-LATE CYPRIOT *Urbanism-metallurgy* Enkomi
	2000	EARLY BRONZE AGE Troy	EARLY BRONZE AGE (Aegean) *Development toward palatial centers* Knossos / EARLY BRONZE AGE (Cyprus) Philia Sotira Kaminoudhia
	2500	EARLY BRONZE AGE Troy	EARLY BRONZE AGE (Aegean) *Urban Centers* Lerna Tiryns Crete Cyclades / CHALCOLITHIC (Cyprus) Erimi
	3000		

YEARS BC	ANATOLIA	AEGEAN AND CYPRUS
3000	NEOLITHIC Kum Tepe (Troad)	FINAL NEOLITHIC Kea Lemnos Knossos Phaistos (Crete) / CERAMIC NEOLITHIC Aylos Epiktitos Vrysi Sotira Teppes
4000	NEOLITHIC Beycesultan Can Hassan	MIDDLE/LATE NEOLITHIC *Large Aegean islands settled* Euboea, Chios, Rhodes, Samos / *Occupation gap (Failed colonization?)* Dhali Agridi
5000	NEOLITHIC *Early village sites* Can Hassan III Hacilar	EARLY NEOLITHIC (AEGEAN) *Early village sites* Nea Nikomedeia Franchthi Cave, Lerna Knossos (Crete) / ACERAMIC NEOLITHIC (CYPRUS) Khirokitia Tenta Cape Andreas Kastros
6000	NEOLITHIC *Early village sites* Catal Huyuk Hacilar *Obsidian production/trade Wall paintings and molded relief sculpture* Catal Huyuk	EPIPALEOLITHIC *First island settlement* Crete Cyprus?
7000	NEOLITHIC *Small agricultural settlements Early use of metals* Cayonu	EPIPALEOLITHIC
8000	UPPER PALEOLITHIC	UPPER PALEOLITHIC (AEGEAN) *Sea travel and resource exploitation* Franchthi Cave Melos (Cyclades)
9000		

economic intermeshing of Europe, Asia, and Africa. At 2500 B.C. a citizen of urban Egypt and a hunter-forager working the north coast of France had probably not the slightest impact on each other's lives, not even as the end points of trading networks in exotic commodities. But by about A.D. 200, northern France and Egypt were provinces in the Roman empire, and French veterans of Roman armies were sometimes paid their pensions in the form of farms in Egypt's Fayyum Oasis. Today, of course, Chinese drink Pepsi-Cola, Iraq relies on French armaments, Great Britain's economy is heavily dependent on its citizens of Pakistani and Indian origin—the whole world, in short, is becoming interconnected.

When we left the early complex societies of Southwest Asia in Chapter 8, this process of growing functional interdependence was at a rather restricted regional level, with only the most attenuated links connecting, for example, Egypt and Iran.

Anatolian sites like Çatal Hüyük (p. 326) were part of the first appearances of social complexity in Southwest Asia, but when the evolutionary focus moved to Mesopotamia, Anatolia became somewhat peripheral to the great states and empires of the Tigris-Euphrates Alluvium. As we saw in Chapter 8, the great civilizations of Mesopotamia and Egypt exploited the richness of the Aegean world from their very beginnings, but by about 3000 B.C. indigenous cultures in the Aegean region began to evolve into powerful complex societies in their own right.

The "wine dark sea" of which Homer rhapsodized is the rich expanse of the Aegean between Greece and Anatolia that encloses Lemnos, Lesbos, and hundreds of other islands. The ragged coastlines of these islands and the Greek and Anatolian mainlands lead back to arid but still very productive farm lands and to abundant sources of metals, woods, and other materials. Domesticated wheat, barley, sheep, goats, grapes, olives, and other crops were established all over the Mediterranean by 5000 B.C., and by 3000 B.C. great volumes of commodities were flowing throughout the eastern Mediterranean world. By about 2700 B.C. Troy, Poliochni on the island of Lemnos, and other communities were already cities, with many scores of stone buildings, roads, and a specialized and lucrative industry in gold, tin, and other metals (Figure 12.2).

These metals were not really indispensable to the economies of the Aegean world of this age. Some bronze tools and weapons were no doubt useful, but so much of this trade seems to have been in luxury goods without direct economic utility—gold figurines, for example. Probably the greatest volumes of trade were in commodities long since consumed and therefore archaeologically invisible, such as the millions of gallons of wine no doubt shipped all over the Mediterranean. And it is really not quite accurate to think of golden figurines as economically useless: status, and

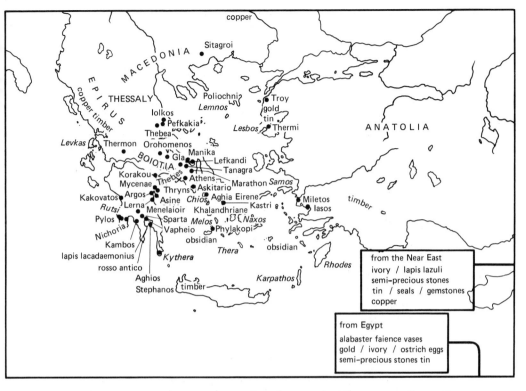

12.2   Early states of the Aegean were based on trade. The commodity exchange patterns included hundreds of kinds of goods traded by ship not only in the central Aegean but also along the Palestinian coast and as far west as Spain.

the pursuit and social reinforcement of status were primary economic stimuli all over the ancient world.

By about 2500 B.C., the Aegean islands, including Crete, and the Anatolian and Greek mainlands were dotted with hundreds of towns and villages. Gold was a central element in trade among these communities, but painted pottery, obsidian bowls, marble statues and figurines, dyed textiles, and hundreds of other products were not only widely traded, they also appear in rich tombs that indicate societies already stratified into the rich and the poor, the powerful and the weak. Hissarlik, the "Troy" of the Classical world, was a city at 2300 B.C., with fortress walls and many larger stone buildings.

No area of the world has contributed more to world civilization than the eastern Mediterranean, especially in the form of Greek science, arts, and letters. But throughout the history of this region it also seems to have been the most fractious, with at least a 5000-year history of warfare and resistance to political unification. Some of the communities of later third millennium B.C. seem to have been redistributive centers for towns and villages around them, but for the most part this was an era of hundreds of small and competitive cities.

At about 2200 B.C. many communities on the Greek mainland and some of the islands appear to have been burned at about the same time, and when they were rebuilt some of the styles of architecture and pottery seem quite different from the earlier periods. Contemporary archaeologists try to avoid explaining major cultural changes in terms of simple, convenient "just so" stories such as invasions, but the Mediterranean world may well have been transformed at this time by the invasion at about 2300 B.C. of Indo-European peoples—perhaps proto-Greek speakers.

Crete seems not to have been affected much by these developments on the mainland, and eventually Crete came to dominate much of the Aegean world. People have lived on Crete in farming villages since at least 6000 B.C., and in succeeding millennia they combined a productive agricultural economy with a thriving maritime trade. By about 1900 B.C. the many small towns and villages on Crete seem to have become linked increasingly into a state-level political, economic, and religious system. The general pattern of change is similar to that of all the other early civilizations: monumental architecture, increasingly interdependent economies, rising population densities, and everywhere the signs of rank, wealth, and power. The settlement at Knossos included a large palace that probably functioned as a production center for many commodities, such as pottery, goldwork, and lovely vessels in serpentine and other hard stones. A writing system, known as Linear A, was developed, probably as a way to manage this complex economy.

The culture of Crete of the second millennium B.C. is generally known as the Minoan Civilization, and throughout this era the wealth and power of Crete seem to have increased. Minoan goods are found all over the Aegean and the Greek and Anatolian mainlands, and as far away as the Levant and Egypt. Knossos probably had a population of 10,000 people just in the central area of the settlement. Great palaces and rich tombs are the most obvious signs of the character of this society, but it was all based on a thriving economy. In the warehouses at Knossos hundreds of huge pottery storage vessels were found with enough capacity to hold tens of thousands of gallons of wine and olive oil.[2]

About 1700 B.C. an earthquake destroyed the palace at Knossos, but the community was rebuilt and became ever more prosperous and powerful. In 1500 B.C. a volcano erupted on the island of Thera, just 70 miles away, and probably destroyed some Minoan towns on Crete.

Toward the end of the second millennium B.C., Minoan culture and power was supplanted over most of the Aegean world by Mycenaean civ-

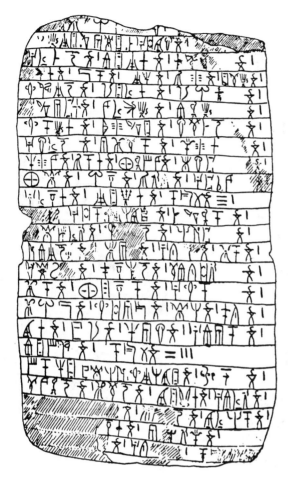

12.3 This clay tablet from Knossos was written in Minoan Linear B script, which was probably an early mixture of hieroglyphic and syllabic characters. Most of the Knossos texts are simple accounts of goods and labor.

ilization. The origins of Mycenaean culture seem to have been on the Argos plain on the Greek mainland. The stone fortress at Mycenae (Figure 12.5) appears to have been the center of a warrior kingdom, based on an extensive trade network. The rich shaft graves found at Mycenae reflect a luxurious yet violent society, caught up in competition for economic and political dominance of not only the mainland but the Aegean islands and western Anatolia as well.

Scholars will never be able to resolve how much of Homer's *Iliad* and *Odyssey* is fiction and how much fact, but these epics probably reflect much of the character of late Mycenaean civilization.

At about the same time the Mycenaean culture was flourishing in the Aegean region, the Hittites, an Indo-European people, had established a large rich state in central and eastern Anatolia. The Hittites eventually

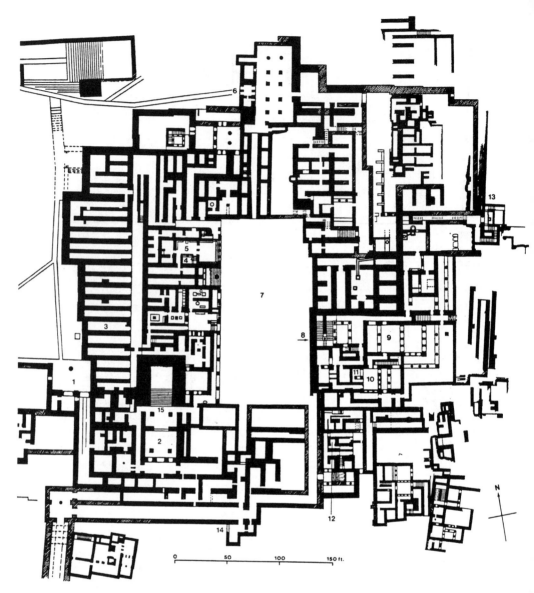

12.4 The Minoan palace at Knossos was a complex of shrines, residential units, storage rooms, and other buildings arranged around a large central courtyard. Such constructions were typical of the powerful trading states of the Aegean area in the second and third millennia B.C. (1) west entrance; (2) south anteroom; (3) west storerooms; (4) lustral chamber; (5) throne room; (6) north entrance; (7) central court; (8) grand staircase; (9) king's megaron; (10) queen's megaron; (11) queen's bath; (12) Shrine of the Double Axes; (13) east entrance; (14) south entrance; (15) stairway to second story.

458

12.5 Heinrich and Sophia Schliemann seated to the right of the legendary Lion Gate at Mycenae in 1876.

became a major player in the complex political and economic competition between Egypt and Assyria.

In the early first millennium B.C., the eastern Mediterranean world entered a period of complex political and economic interaction. Greek city-states continued their millennia-old traditions of combat, the Phoenicians of the Levant established a rich trading empire that reached as far west as Spain, and the Etruscans of Italy began to lay the political and economic foundations of what would become the Roman Empire.

Temperate Europe

Until the seventh millennium B.C., "Barbarian Europe"—the great forests and grasslands beyond the Aegean Sea and extending north through Britain, Scandinavia, and Russia—was a wilderness inhabited only by hunters and foragers whose major cultural achievements included phenomenal skill at deer hunting and a theology based on the worship of oak trees.

One of the great questions of comparative history is how it happened that for thousands of years much of the creativity of the world was centered in Mesopotamia, China, the Indus, and the Mediterranean, and then, after the first few centuries A.D., the cultural center of the world seemed to shift so decisively to western Europe. In his review of this issue, Steven Shenan notes that many early scholars, especially V. Gordon Childe and the Marxists saw history as a case in which "the Olympic torch of development was passed from East to West."[3] During the past two decades, research suggested that many innovations in fact were independently invented in Europe, and in general there has been a critical reevaluation of the notion that to understand European prehistory one should look first at Mediterranean civilizations.[4]

The archaeology of post-glacial hunter-foragers of Mesolithic Europe is just beginning to be treated systematically.[5] The Mesolithic of Europe is defined in terms of a shift in stone tool types, especially the replacement of larger tanged stone points by much smaller geometric ones—which may reflect primarily the introduction and improving efficiency of bows and arrows.[6] As Douglas Price has noted, this transition may have been mainly the result of different raw materials available as the glaciers retreated and the evolution of tool-making skills, rather than profound social or cultural changes.[7] He also points out that we have lost a lot of the Mesolithic archaeological record because sea levels in northern Europe did not reach current levels until about 6000 B.C.

Peter Bogucki's analysis of agricultural origins in the northern European Plain, particularly in the region of Poland, suggests that for centuries foragers and primitive agriculturalists in the Mesolithic were increasingly symbiotic in their economies, and that the first agriculturalists, the Funnel Beaker People, involved the "selective incorporation of cultigens and domestic animals into an essentially Mesolithic way of life."[8] He also suggests that the great stone monuments and tombs that eventually appeared all over Europe were in fact territorial markers, which became important considerations as agriculture spread.

On the Mediterranean fringes of European farming communities, agricultural villages were established on a reliable productive mixture primarily of cereals, olives, and grapes in combination with sheep and goats.

The processes by which domesticates and farming spread from the Near East and Mediterranean areas into the great temperate forests of northern Europe remains one of the most interesting and intensively researched subjects in contemporary archaeology.

There are great environmental differences between the semiarid Mediterranean zone and the wetter, cooler regions of Britain, the Low Countries, northern Germany and southern Scandinavia. A prehistoric farmer in southern Italy needed only a few hand tools to farm successfully, but a peasant in northern Germany required capital-intensive equipment, such as the horse-drawn mouldboard plow. As Graeme Barker has observed,

12.6   The spread of Celtic languages toward the end of the first millennium B.C. and some of the trade routes that connected temperate Europe to Mediterranean civilizations.

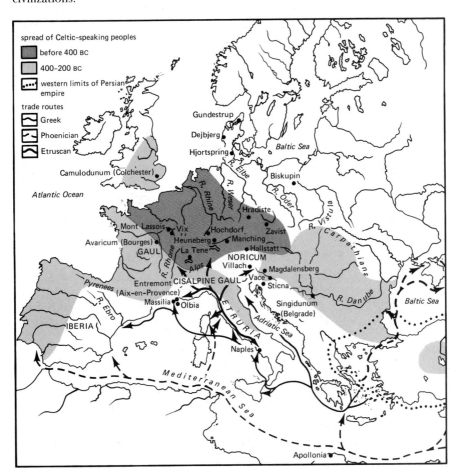

The object of Mediterranean ploughing has always been to conserve the moisture content of the soil by scuffling the surface, whereas in northern Europe heavy equipment is needed to break up the thick mass of roots in order to release the soil nutrients from the lower depths. . . . The analysis of the early history of agriculture in particular, and of the productive economy in general, in the two areas has been characterized by a definite sense of entrepreneurial risk takers in the Mediterranean contrasting with their honest but stolid contemporaries in the north, rather like the familiar contrast between the respective football teams of southern and northern European countries today.[9]

Agriculture's spread across Europe can be seen in Figure 6.14, and this dispersal pattern raises some interesting questions. To a certain extent, this spread can be assumed to be determined by the blunt forces of environment, technology, and demography. The genetic composition of cereals sets a strict limit to the rate at which species adapt to different environments, and without artificial fertilizers the yields of crops—and therefore human population densities—are narrowly constrained. But when we look at the problem of how small-time European farmers eventually produced complex societies in Europe, we can begin to see how complexly interrelated are four dimensions of these ancient societies: climate-environment, farming technology, human population fluctuations, and sociopolitical institutions.[10] By 6000 B.C., a Mediterranean complement of domestic grains, cows, and sheep—and the technology to exploit them—had moved up the Danube into central Europe. People lived in small clusters of wooden huts, often with their animals, and farmed narrow plots near rivers and streams. Some of the spread of this form of agricultural settlement in Europe is marked after 5000 B.C. by pottery decorated with incised, linear patterns—the Bandkeramik style. Unlike that of the semiarid plains of the Middle East and the Mediterranean littoral, however, the agricultural way of life in Europe proper required clearing thick forests and grasslands, at first by cutting down the trees, later by plowing.

location and composition of Neolithic European sites and has shown that the distribution of loess and alluvium and the forces of trade and social stratification produce considerable variations in the density and sizes of settlements across the European Plain.[11]

By 3000 B.C. agricultural villages were to be found from Great Britain far into eastern Russia. There were hundreds of local resource specializations, but most of these settlements subsisted mainly on domesticated cereals and cattle. Copper axes and ornaments were widely distributed, especially in central Europe and the Balkans.

Colin Renfrew has recently proposed that the spread of farming across Europe between 6000 and 3000 B.C. was associated with the replacement of local languages by Indo-European languages.[12] Population densities were heaviest in the central European areas where rivers supplied broad alluvial fields, but after 3000 B.C. villages appeared in light to heavy densities over almost every area where the environment permitted stock raising and cereal agriculture. The areas where this kind of subsistence was possible were greatly expanded by the introduction (perhaps from the Near East) of the animal-drawn plow. In the absence of an arid climate or alluvial soils, it is difficult for the agriculturalist to compete with wild vegetation and renew the fertility of the soil. But the plow gave these early European farmers a way to expose the soil for seeding, and it contributed to weed control and soil fertility by plowing under vegetation (although the deep plowing of modern times was a later invention). Susan Gregg has devised a complex computer model that simulates the kinds of economic interactions that hunter-foragers and Neolithic agriculturalists may have established. She suggests that long-term stable economic relationships may have developed between foragers and immigrant farmers, as these farmers colonized more and more of temperate Europe.[13]

Once most Europeans became agriculturalists, the stage was set for the evolution of complex forms of social organization. Some of the factors that transformed Europe were local, others international.

Some of the earliest signs of social change in temperate Europe are, predictably, in southern Europe, on the fringes of the Mediterranean world. For example, the Varna cemetery near the Black Sea coast of Bulgaria, and dating to about 4000 B.C., contained massive hoards of golden and copper tools, suggesting not only great differences in economic classes in this society, but also the existence of a metallurgical industry of considerable sophistication.[14] Varna may well have flourished on the basis of a Black Sea trade.

Because they are so common and preserve well, metal and ceramic objects have been used as the defining characteristics of European peoples before the Romans. The "Battle Ax People," for example, is the general term for people who lived between about 3500 and 2300 B.C., whose copper and stone axes are found all over temperate Europe, mainly in graves, along with their distinctive cordmarked pottery. These people seem to have originated in central and eastern Europe, and they may have begun using the domestic horse and wheeled carts as early as 3000 B.C. As usual, it is difficult to determine if people in eastern Europe physically migrated westward *en masse,* or if their art styles and economic impact spread across Europe much in advance of any actual population movements. If Indo-European languages originated in the Caucasus Mountains, then the spread of the Battle Ax peoples may well have been one in which people moved rapidly westward, displacing or intermarrying with the natives.

Another spread of artifact styles involved the "Beaker Culture," named after the distinctive pottery they produced. The Beaker Culture was evident in much of western and central Europe just after 3000 B.C., and eventually the Battle Ax and Beaker cultures seem to have coalesced.

In talking about these cultures we must remember, of course, that nearly all the people of this era were simple agriculturalists, with a few families living together in tiny hamlets. Most would have lived out their lives in small houses and farms. Houses were simple structures of logs, clay, and thatch.

After about 2000 B.C. temperate Europe began to evolve slowly into somewhat more complex social forms. An unmistakable sign of this is Stonehenge (Figure 12.7). Many other great stone monuments were built in Europe during the second millennium B.C., most apparently the center of rituals and religion, although their use in astronomical calculation is at least possible. The organization and planning required to build monuments like Stonehenge probably reflect a society with at least incipient rank, class, and occupational specialization, and all through the last few millennia B.C. expanding trade between Europe and the Mediterranean and Aegean regions would have provided some stimulus to cultural change.

Lewthwaite used the term "Mafia model" to describe views of the emergence of complex societies in Europe in which "the motor for social change

12.7 Stonehenge began as a series of ritual excavations about 2700 B.C. and was greatly elaborated between 2000 and 1000 B.C. The arranged stones probably had astronomical significance. Stonehenge is just one example of thousands of stone monuments built in this period in western Europe.

12.8 In the first millennium A.D. wooden roads connected towns in the wet low-lands of northern Germany and Scandinavia. Carts drawn by oxen and horses moved large volumes of food and other commodities over these routes.

is the ambition of certain individuals to dominate and control their fellow human beings, an ambition that can be realized in certain ecological . . . circumstances."[15]

From burials not only of bodies but also of hoards of precious metals and artifacts, we can speculate that warrior tribes contended all across Europe in the late second millennium B.C., with their fortunes rising and falling to some extent with their success in maintaining trading links with the Mediterranean and Aegean societies. Baltic amber, salt, metal tools, and hundreds of other products flowed along these ancient trade routes.

Until about 1000 B.C. the secret of iron-making was a monopoly of the Hittite state in Anatolia, but after this time knowledge of the technique spread north and west into Europe. Iron is a much more utilitarian metal than bronze or gold: iron tools could make a major difference in efficiencies of plows, hoes, awls, knives, swords, arrowheads, and a thousand other products.

One early European iron-working culture was the Hallstatt, which after about 700 B.C. is reflected in the distinctive artifacts of this tradition, from the Low Countries.

The Hallstatt and La Tene cultures of first millennium B.C. Europe flourished in a period of rapid population expansion, migrations of various ethnic groups, reformation of trade routes along lines dictated by Mediterranean cultures, and other major social changes.

With the expansion of Roman power between 200 B.C. and well into the first millennium A.D., the character of Europe was forever changed. It became a hybrid mix of Greek, Middle Eastern, and native elements.

Africa

Africa is so rich in gold, silver, iron, jewels, ivory, palm oil, and other resources, that one might expect it to have been a center of initial origins of cultural complexity, as it was for the first three million years of our genus. But Africa has formidable barriers that afflict the agriculturalist more than the hunter-gatherer.[16] The enormous Sahara Desert, expanding and contracting with Pleistocene and post-Pleistocene climatic changes, has almost always isolated most of Africa from the critical mass of Near Eastern and Mediterranean cultures. And even in the more humid sub-Saharan regions the dense vegetation, poor soils, and unpredictable rainfall make large-scale intensive agriculture unproductive for the primitive cultivator. Nor do the great rivers, like the Niger, have the regular regimes, large semiarid alluvial plains, or latitude that make the Nile, Tigris and Euphrates, and Yellow river valleys so productive. To make matters worse, Africa has a veritable horror show of diseases, including a tsetse fly-borne cattle illness that barred pastoralists from many areas, and malaria, which recurrently wiped out human populations in the more tropical areas.

Given these and other ecological problems, early African societies are an object lesson in human inventiveness and adaptability in the face of an extreme environment.

Archaeologists used to think that sub-Saharan Africa was a cultural backwater—a veritable "cultural museum"—in the Pleistocene, but recent research indicates that an advanced stone-tool technology, pottery, and animal domestication were indigenous developments in sub-Saharan Africa.[17]

Sub-Saharan Africa of the last few millennia B.C. was probably a rich blend of hunting, foraging, fishing, and agricultural societies, with extensive trade networks in gold, salt, and foodstuffs. We know comparatively little about the earliest phases of these because of the problems of preservation and inadequate archaeological research. It seems that everywhere that competent surveys are undertaken, the development of African civilizations is revealed. Surveys in the mid-1980s in Equatorial Guinea

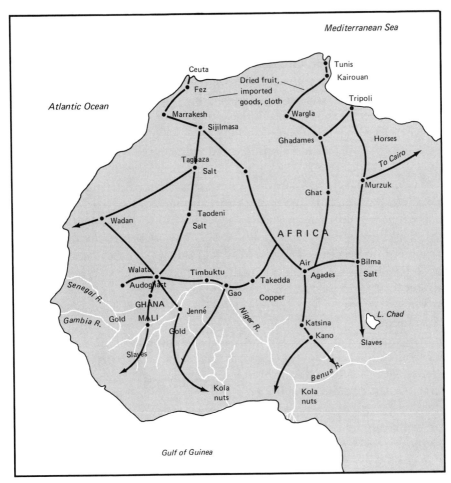

12.9   In the first millennium A.D., trade routes in various commodities connected west African kingdoms with each other and with coastal traders from the Middle East, Europe, and the Mediterranean.

and Gabon, for example, revealed Neolithic Bantu villages already well-acquainted with iron-smelting by the mid-first millennium B.C.[18]

The character of all these cultures was changed radically, however, by the introduction (beginning about 500 B.C.) of an iron-working technology. In an extensive review of the origins of iron-working in west Africa, McIntosh and McIntosh conclude that the evidence for an independent indigenous origin there is "exceedingly slim at present."[19] Iron artifacts were used extensively there from the end of the first millennium B.C.

Complex societies quickly followed the introduction of iron in west Africa, and first millennium B.C. tombs in Nigeria and Senegal reflect great differences in wealth. At Igbo-Ukwu in Nigeria, one burial chamber contained cast-bronze objects and thousands of glass and carnelian beads.[20] Arab accounts of these societies in the eighth through the eleventh centuries A.D. describe complex empires, with armies, kings, and massive craft industries.[21] Many scholars have concluded that it was Arab commerce that stimulated west African cultural complexity, but McIntosh and McIntosh conclude that the spread of Arab goods and culture "reflects the grafting of Arab commerce onto a preexisting infrastructure of Saharan and sub-Saharan networks. Such an explanation implies that earlier phases of development at Igbo-Ukwa were characterized by indigenous processes of trade expansion and increasing social stratification."[22]

The first five centuries A.D. saw the rapid spread over much of Africa of agriculturalists using iron tools and weapons and subsisting in part on indigenous domesticates like sorghum and squash, and on domestic cattle and other animals. Only in a few arid wastes were these agriculturalists unsuccessful in displacing the stone-age hunters who were the heirs of the millions-of-years-old tradition of African hunting-foraging. Formerly, it was widely accepted that there was a great migration of Bantu-speaking peoples who changed sub-Saharan societies, but Hall has effectively challenged the notion of a great southward spread of Bantu farmers.[23]

In contrast to the earlier-held view that the first African states were "contact-states," created by the economic and political "touch" of the established North African and Egyptian states, it now seems more reasonable to see sub-Saharan indigenous states as products of their own cultural matrix. The rapid population growth associated with the introduction of agriculture to these areas, the rising spiral of trade in gold copper, salt, and slaves, and the sporadic influences of distant Mediterranean and Asian empires probably all combined to produce the rich chiefdoms and states known in Africa in the second millennium A.D. Archaeologists are just beginning to unravel the history of these complex cultures, and it is beyond the scope of this book to review them in detail. Most seem to have been marked by some public buildings and were composed of a ruler and an upper class who organized and taxed a complex economic system composed of farmers, herders, miners, ironworkers, traders, and religious leaders. Cemeteries often show great variations of wealth and, presumably, social status.

By the sixteenth century, much of west Africa was divided among several states, all locked into trade networks that ultimately fed hundreds of products to the empires of the Mediterranean world and Asia.

After the sixteenth century, Africa fell increasingly under the control of non-Africans, and only in the last fifty years have most Africans retrieved their autonomy. They were held in thrall for so long by outsiders because of their inferior economic position, but also because European-introduced diseases killed many people and because of slavery: the African slave trade was organized and profited from by Africans well before the sixteenth century; some have estimated that nearly as many African slaves were shipped north into the Near Eastern and Mediterranean areas as were eventually carried to the New World.

The Eurasian Periphery

Nomadic pastoralists ranged from the eastern shores of the Black Sea to the Gobi Desert and eventually came to represent a major force in European and Asian developments. Domestic horses and camels were in use by the mid-second millennium B.C. in much of this area, and with the addition of two-wheeled carts, the development of a diet based on milk products rather than meat, and the emergence of an aggressive chiefdom-like command structure, Eurasian nomads terrorized Europe and sacked China time and time again.

Japan's cultural isolation during many periods of its history offers scholars something like the experimental control that laboratory scientists in other disciplines can exert. That is, Japan's rich history was influenced by China at various critical points, but it has always been so distinctive and self-contained that on some levels it can only be understood in its own terms. But in Japan, too, the familiar developmental stages of ancient cultural complexity were followed. The initial spread of pottery styles were followed by monumental architecture and vast wealth differentials, culminating in the development of the most rigid and complex social hierarchies the world has ever seen.

One of the many remarkable things about ancient Japan was the early date at which hunter-foragers began living in permanent communities. Melvin Aikens and Takeru Akazawa note that already by 11,000 B.C. Japanese of the Jomon culture were using pottery—the earliest known extensive ceramics industry in the world—and living for all or most of the year in communities of pit-houses, subsisting on a rich and varied diet of deer, bear, whale, salmon and many other fish, seabirds, shellfish in abundance, and numerous kinds of berries, nuts, and other plants.[24] The rather minimalist modern Japanese diet centered on rice, kelp, and raw fish was a very late development, a product of massive population growth and an extraordinarily intensive agricultural system. As in other areas of the world, after the end of the Pleistocene Japanese hunter-foragers began exploiting some species of plants with enough intensity to begin changing these plant species genetically. There is some evidence that by 5000 B.C. rice, barley, buckwheat, and millet were being intensively exploited in some areas of Japan. But not until the first millennium B.C. did rice agriculture supplant the ancient foraging-collecting economy in Japan. Rice agriculture seems to have spread in a very intricate pattern, at different rates and in different ways, based on the ecological differences in the various areas of Japan. Rice agriculture spread most quickly into those zones that offered poorer and less stable wild resources—another example, in other words, of the close cost-benefit calculations cultural evolution seems to make.

Analyses of skeletons of early Japanese show that there was continued mixing with Korean and Chinese peoples, but Japan began to develop a very distinctive national culture in the last few millennia B.C.

To those familiar with its culture, Japan—like ancient Athens and many other early civilizations—is an object lesson showing that the blunt analytical tools of archaeology, with their fixation on economy, environment, and architecture, can reveal very little of the most fascinating aspects of these societies.

In Southeast Asia, not only do we not know many of the most crucial links in the culture history (in part because of poor climatic conditions for archaeological preservation), but what we do know flagrantly contradicts our most comforting archaeological expectations. As Bronson notes, recent Southeast Asian culture history contains such disconcerting paradoxes as cannibals and collecting groups with alphabets of their own (the Batak of Sumatra and the Tagbanwa of Palawan), hunter-gatherers with the ability to make excellent steel (some Punan groups in Borneo), and even (as in nineteenth-century northern Vietnam) complex and populous states without settlements that can readily be called cities.[25] At least some of these developments, however, are the direct result of outside contacts and, generally, only northern Vietnam can be said to have the usual hallmarks of cultural complexity before about 100 B.C. The great monuments of Thailand, Cambodia, and Vietnam are all products of the late first and second millennia A.D. and are intertwined with Chinese and Indian history. Charles Higham has argued persuasively, though, that although India was a powerful force in, for example, the origins of complex societies in southeast Asia, many important developments were the result of local factors.[26]

One of the most interesting aspects of the rise of states and cities in Thailand, for example, is the extent to which these developments seem to have been tied to trade. People who lived mainly on wet-rice agriculture and practiced both iron and metal working had settled in the Non Chai area of Northeastern Thailand by the first millennium B.C. Bronze ornaments, salt, salted and fermented fish, marine shells, and many other commodities were probably traded in great volumes in the lower Mekong Valley, in association with chiefdoms or small states.[27]

Founder groups from mainland southeast Asia made the sea-crossing

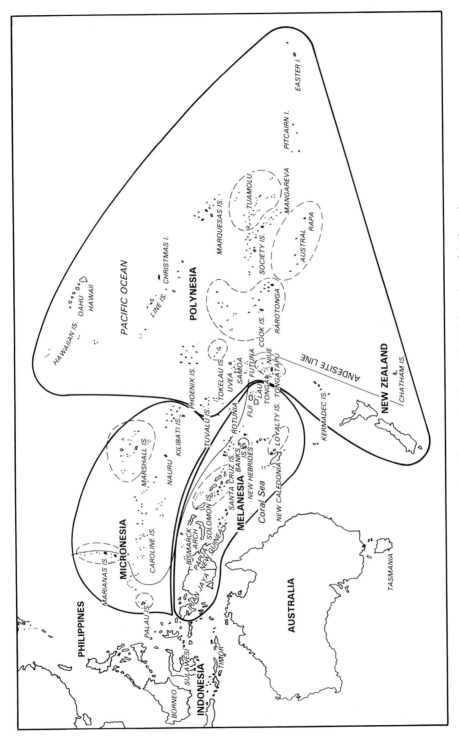

12.10 Cultural areas of Micronesia, Melanesia, and Polynesia.

471

to Australia and New Guinea at least 40,000 years ago, based on recent evidence, and hunter-gatherers across southeast Asia gradually evolved somewhat different physical characteristics.[28]

From about 3000 B.C. onward, there seems to have been a continuous expansion into much of southeast Asia of farmers whose cultural roots are in subtropical Mainland Asia.[29]

> This southern Mongoloid population is evidence in the recent (Neolithic and Metal periods) skeletal record, and the reasons for the expansion can be related to the high demographic densities, and territorial demands of an agricultural life-style, together with superior transport technology (i.e., sailing canoes) . . . this expansion reached its limits in Polynesia during the first millennium A.D.[30]

By definition, archaeology can never be a truly experimental science. But the distribution of societies across the great arch of the southern Pacific are as close as we might hope to get. Island by island, people colonized this part of the world, and the subtle variations of local climates, the different times of colonization have produced a varied cultural landscape that is, in the cliché, a giant cultural laboratory.

But these areas have great significance for a truly evolutionary analysis of the past. In Polynesia, for example, during the last three millennia peoples spread out from island to island, and the evolution of their languages, adaptations, and physical types reflects the workings of basic evolutionary mechanisms.[31] Patrick Kirch has argued that, in some ways, the evolution of complex societies in Hawaii, Tonga, and other areas of Polynesia agrees well with the "circumscription hypothesis" of Robert Carneiro (see Chapter 7).

Scholars have long wondered how ancient peoples sailed across the tropical Pacific from west to east, when the prevailing winds and ocean currents move from east to west. Various simulation models have been considered,[32] and different colonizing rates and processes are possible. Finney reports a voyage during 1986–1988 of the *Hokule'a*, a 19-meter double-hull canoe under the command of Hawaiian navigator, Nainoa Thompson.[33] By exploiting occasional spells in which the prevailing winds shifted, the canoe was sailed from American Samoa east to Tahiti, along the track the ancient settlers of these islands might have traveled.

Conclusions

The cultures on the periphery of the Old World, from northern Scandinavia to Hawaii, repeat the common fact that most aspects of cultural complexity are the partial products of technology and agricultural productivity. Reindeer herders in Lapland and the yam farmers and pig herders of Oceania represent gradients of cultural complexity, but nowhere in these marginal areas do we find the towns, the libraries, the many-leveled hierarchical arrangements of functional interdependence, or the regional political powers that mark states and empires.

Notes

1. Barber, *The Cyclades in the Bronze Age.*
2. Pfeiffer, *The Emergence of Society,* p. 282.
3. Shennan, "Trends in the Study of Later European Prehistory"; see also Childe, *The Prehistory of European Society.*
4. See, e.g., Harding, *The Myceneaeans and Europe.*
5. See, e.g., Rowley-Conwy et al., *Mesolithic Northwest Europe: Recent Trends;* Price, "The Mesolithic of Western Europe"; Nygaard, "The Stone Age of Northern Scandinavia: A Review."
6. Fischer, Hansen, and Rasmussen, "Macro and Micro Wear Traces on Lithic Projectile Points."
7. Price, "The Earlier Stone Age of Northern Europe," p. 3.
8. Bogucki, "The Establishment of Agrarian Communities on the North European Plain," p. 11.
9. Barker and Gamble, *Beyond Domestication in Prehistoric Europe,* p. 9.
10. Grigg, *The Dynamics of Agricultural Change;* Barker and Gamble, *Beyond Domestication in Prehistoric Europe,* p. 22.
11. Whittle, "Neolithic Settlement Patterns in Temperate Europe."
12. Renfrew, "Archaeology and Language."
13. Gregg, *Foragers and Farmers.*
14. Renfrew, "Archaeology and Language."
15. Lewthwaite, "Comment on A. Gilman, 'The Development of Social Stratification in Bronze Age Europe.'"
16. Harlan et al., *Origins of African Plant Domestication.*
17. Reviewed in McIntosh and McIntosh, "Current Directions in West African Prehistory."
18. Clist, "Early Bantu Settlements in West Central Africa: A Review of Recent Research."
19. McIntosh and McIntosh, "From Stone to Metal: New Perspectives on the Later Prehistory of West Africa," p. 106.
20. Ibid., p. 110.
21. Ibid.
22. Ibid., p. 120.
23. Hall, *The Changing Past: Farmers, Kings, and Traders in Southern Africa, 200–1860.*
24. Akazawa and Aikens, *Prehistoric Hunter-Gatherers in Japan;* Aitkens, Ames, and Sanger, "Affluent Collectors at the Edges of Eurasia and North America"; Nishida, "The Emergence of Food Production in Neolithic Japan."
25. Bronson, "South-east Asia: Civilizations of the Tropical Forest," p. 262.
26. Higham, *The Archaeology of Mainland Southeast Asia.*
27. Charoenwongsa and Bayard, "Non Chai: New Dates on Metalworking and Trade from Northeastern Thailand."
28. Bellwood, "The Prehistory of Island Southeast Asia: A Multidisciplinary Review of Recent Research."
29. Ibid.
30. Ibid., p. 191.
31. Kirch and Green, "History, Phylogeny, and Evolution in Polynesia"; Kirch and Hunt, *Archaeology of the Lapita Cultural Complex: A Critical Review.*

32. Reviewed in Terrell, *Prehistory in the Pacific Islands.*
33. Finney, "Voyaging Against the Direction of the Trades: A Report of an Experimental Canoe Voyage from Samoa to Tahiti."

Bibliography

No attempt is made here to provide a bibliography for the many secondary states and empires of the ancient world.

Aitkens, C. M., K. M. Ames, and D. Sanger. 1986. "Affluent Collectors at the Edges of Eurasia and North America: Some Comparisons and Observations on the Evolution of Society among North-Temperate Coastal Hunter-Gatherers." In *Prehistoric Hunter-Gatherers in Japan,* Akazawa, T. and C. M. Aikens, eds., Tokyo: University of Tokyo Press.

Akazawa, T. and C. M. Aikens, eds. 1986. *Prehistoric Hunter-Gatherers in Japan.* Tokyo: University Museum, University of Tokyo, Bulletin No. 27.

Barber, R. L. N. 1988. *The Cyclades in the Bronze Age.* Iowa City: University of Iowa Press.

Barker, G. and C. Gamble. 1985. "Beyond Domestication: A Strategy for Investigating the Process and Consequence of Social Complexity." In *Beyond Domestication in Prehistoric Europe,* eds. G. Barker and C. Gamble. New York: Academic Press.

Barker, G. and C. Gamble, eds. 1985. *Beyond Domestication in Prehistoric Europe.* New York: Academic Press.

Bellwood, P. 1987. "The Prehistory of Island Southeast Asia: A Multidisciplinary Review of Recent Research." *Journal of World Prehistory* 1(2):171–224.

Bintliff, J. ed. 1984. *European Social Evolution.* Bradford: Bradford University Press.

Bogucki, P. 1987. "The Establishment of Agrarian Communities on the North European Plain." *Current Anthropology* 28(1):1–24.

———. 1988. *Forest Farmers and Stockherders.* Cambridge, England: Cambridge University Press.

Bronson, B. 1980. "South-east Asia: Civilizations of the Tropical Forest." *Cambridge Encyclopedia of Archaeology,* pp. 262–66. New York: Crown Publishers/Cambridge University Press.

Charoenwongsa, P. and D. Bayard. 1983. "Non Chai: New Dates on Metalworking and Trade from Northeastern Thailand." *Current Anthropology* 24(4)521–23.

Childe, V. G. 1958. *The Prehistory of European Society.* Baltimore: Penguin.

Clist, B. 1987. "Early Bantu Settlements in West Central Africa: A Review of Recent Research." *Current Anthropology* 28(3):380–82.

Connah, G. 1987. *African Civilizations.* Cambridge, England: Cambridge University Press.

Ekholm, K. 1981. "On the Structure and Dynamics of Global Systems." In *The Anthropology of Pre-Capitalist,* eds. J. Kahn and J. Lobera. London: Macmillan.

Finney, B. 1988. "Voyaging Against the Direction of the Trades: A Report of an Experimental Canoe Voyage from Samoa to Tahiti." *American Anthropologist* 90(2):401–5.

Fischer, A., P. V. Hansen, and P. Rasmussen. 1984. "Macro and Micro Wear Traces on Lithic Projectile Points." *Journal of Danish Archaeology* 3:19–46.

Fitzhugh, W. 1975. *Prehistoric Maritime Adaptations of the Circumpolar Lone.* The Hague: Mouton.

Gilman, A. 1981. "The Development of Social Stratification in Bronze Age Europe." *Current Anthropology* 22:1–8.

Gregg, S. A. 1988. *Foragers and Farmers.* Chicago: University of Chicago Press.

Grigg, D. B. 1982. *The Dynamics of Agricultural Change.* London: Hutchinson.

Hall, M. 1987. *The Changing Past: Farmers, Kings, and Traders in Southern Africa, 200–1860.* Cape Town: David Phillip.

Harding, A. F. 1984. *The Myceneaeans and Europe*. New York: Academic Press.

Harlan, J. J. DeWet, and A. Stemler, eds. 1976. *Origins of African Plant Domestication*. The Hague: Mouton.

Higham, C. 1988. *The Archaeology of Mainland Southeast Asia*. Cambridge, England: Cambridge University Press.

Kirch, P. V. 1984. *The Evolution of Polynesian Chiefdoms*. Cambridge, England: Cambridge University Press.

———. 1988. "Circumscription Theory and Sociopolitical Evolution in Polynesia." *American Behavioral Scientist* 31(4):416–27.

Kirch, P. V., ed. 1986. *Island Societies*. Cambridge, England: Cambridge University Press.

Kirch, P. V. and R. C. Green. 1987. "History, Phylogeny, and Evolution in Polynesia." *Current Anthropology* 28(4):431–56.

Kirch, P. V. and T. L. Hunt. 1988. "Archaeology of the Lapita Cultural Complex: A Critical Review." Seattle: Burke Museum.

Lewthwaite, J. 1981. "Comment on A. Gilman, 'The Development of Social Stratification in Bronze Age Europe.' " *Current Anthropology* 22:14.

McIntosh, S. K. and R. J. McIntosh. 1983. "Current Directions in West African Prehistory." *Annual Review of Anthropology* 12:215–58.

McIntosh, S. K. and R. J. McIntosh. 1988. "From Stone to Metal: New Perspectives on the Later Prehistory of West Africa." *Journal of World Prehistory* 2(1):89–133.

Milisauskas, S. 1978. *European Prehistory*. New York: Academic Press.

Nishida, M. 1983. "The Emergence of Food Production in Neolithic Japan." *Journal of Anthropological Archaeology* 2(4):305–22.

Nygaard, S. E. 1989. "The Stone Age of Northern Scandinavia: A Review." *Journal of World Prehistory* 3(1):71–116.

Pfeiffer, J. E. 1977. *The Emergence of Society*. New York: McGraw-Hill.

Price, T. D. 1986. "The Earlier Stone Age of Northern Europe." In *The End of the Paleolithic in the Old World*, ed. L. G. Straus. Oxford: B.A.R. International Series 284.

———. 1987. "The Mesolithic of Western Europe." *Journal of World Prehistory* 1(3):225–305.

Renfrew, C. 1988. "Archaeology and Language." *Current Anthropology* 29(3):437–68.

Rowlands, M., Larsen, M. and K. Kristiansen, eds. 1987. *Center and Periphery in the Ancient World*. Cambridge, England: Cambridge University Press.

Rowley-Conwy, P., M. Zvelebil, and H. P. Blankholm. 1987. *Mesolithic Northwest Europe: Recent Trends*. Sheffield: John R. Collis.

Shennan, S. J. 1987. "Trends in the Study of Later European Prehistory." *Annual Review of Anthropology* 16:365–82.

Straus, L. G. 1986. "The End of the Paleolithic in Cantabrian Spain and Gascony." In *The End of the Paleolithic in the Old World*, ed. L. G. Straus. Oxford: B.A.R. International Series 284.

Terrell, J. 1986. *Prehistory in the Pacific Islands*. New York: Cambridge University Press.

Weissleder, W. 1978. *The Nomadic Alternative: Models and Models of Interaction in the African-Asian Deserts and Steppes*. The Hague: Mouton.

Wemick, R. and the Editors of Time-Life. 1973. *The Monument Builders*. Alexandria, Va.: Time-Life.

Whittle, A. 1987. "Neolithic Settlement Patterns in Temperate Europe." *Journal of World Prehistory* 1(1):5–52.

Zvelebil, M. and P. Rowley-Conwy. 1984. "Transition to Farming in Northern Europe: A Hunter-gatherer Perspective." *Norwegian Archaeological Review* 17:104–28.

13

The Evolution of
Mesoamerican Civilization

And after they had slain the captives, only [then] Uixtociuatl
['s impersonator] followed; she came only at the last. They
came to the end and finished only with her.

And when this was done, thereupon they laid her down
upon the offering stone. They stretched her out upon her
back. . . . They laid hold of her; they pulled and stretched
out her arms and legs, bending [up] her breast greatly, bending down her head taut, toward the earth. And they bore
down upon her neck with the tightly pressed snout of a sword
fish, barbed, spiny, spined on either side.

And the slayer stood there; he stood up. Thereupon he cut
open her breast.

And when he opened her breast, the blood gushed up high;
it welled up far as it poured forth, as it boiled up.

And when this was done, then he raised her heart as an
offering [to the god] and placed it in the green jar, which was
called the green stone jar.

And as this was done, loudly were the trumpets blown. And
when it was over, then they lowered the body and the heart
of [the likeness of] Uixtociuatl, covered by a precious mantle.

> From the sixteenth century Aztec account
> of the ritual execution of a woman who played
> the role of the goddess Uixtociuatl
> (from the Dresden Codex)[1]

In Easter week of 1519, Hernan Cortez landed on the coast of Veracruz,
Mexico, and began a military campaign that would end in the crushing
defeat of the Aztec civilization. For perhaps 20,000 years before Cortez's
arrival, the peoples of the Old and New Worlds had had so little contact
that they were physically different and spoke entirely different languages.
But here is the curious thing: when Cortez traveled the road from Veracruz to the Aztec capital near Mexico City, he passed through cities, towns,
villages, markets, and irrigated fields; he saw slavery, poverty, potentates,

farmers, judges, churches, massive pyramids, roads, boats, pottery, and textiles; in short, he encountered a world whose almost every aspect he could understand in terms of his own experience as an urban Spaniard of the sixteenth century.

There were of course many dissimilarities between the Spanish and Aztec peoples. The Aztecs had no ocean-going ships, or advanced metallurgy, for example. And the psychological differences between the Aztecs and their European conquerors were particularly profound. The Spanish, despite their imperialism and murderous ferocity in warfare, viewed the Aztecs' preoccupations with death and human sacrifice with abhorrence, and the Aztecs found many aspects of Christianity both evil and incomprehensible.

Yet despite profound differences in their respective morals and ideas, the Spanish and the Aztecs were fundamentally culturally alike: they lived in hierarchically organized, class-structured, complex, expansionistic empires, with state churches, intensive agricultural and industrial systems, and many other features in common.

Spanish military superiority effectively ended the indigenous developments of Mesoamerican civilizations, and it is interesting to speculate about what the Europeans would have found in the Americas if they had come a few centuries later; would there have been a great Pan-American empire to rival that of Spain's? Or is it possible that in, say about A.D. 1850, people living on the coast of England, or France, or Spain might have awakened to the sight of Aztec warships in *their* harbors?

In any case, the rise of agricultural and complex societies in the New World offers scholars a "Second Earth," a comparative case: by examining these cultural developments in two somewhat different environments and times, the causes of these forms of cultural evolution are more clearly revealed.

The Ecological Setting

As in other areas of early civilization, the first complex societies did not appear in just one area of Mexico and then gradually expand to occupy Mesoamerica; instead, there were several developmental centers, including the South Gulf Coast, the Valley of Mexico, the Valley of Oaxaca, and the Mayan lowlands. Eventually some of these regions coalesced into larger political polities, but only after centuries of competition, expansion, and collapse.

In each area, cultural evolution was much influenced by three general ecological conditions: (1) the millions of years of mountain-building volcanic activity that left Mesoamerica a still trembling land of towering mountains and circumscribed valleys, and which in many areas com-

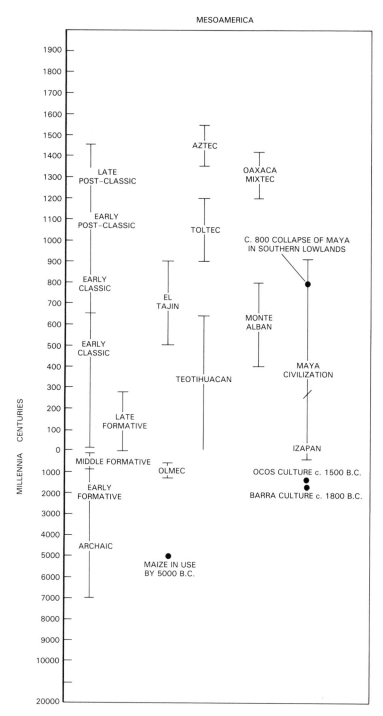

13.1 A cultural chronology of Mesoamerica. People were probably in Mesoamerica before 10,000 B.C. but the evidence is controversial.

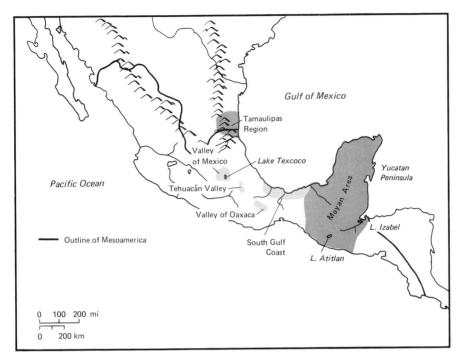

13.2 The geography of Mesoamerica. The darker shading identifies developmental centers of early cultural complexity.

pressed extremely different flora, fauna, and climates into close proximity, thereby rewarding interregional exchange, even though it made transport and communication difficult; (2) the absence of any domesticable animal suitable for providing milk, transport, or draft power; and (3) the virtual absence of large rivers and alluvial plains in warm latitudes, where year-round irrigation agriculture was productive.

All civilizations can be understood to a limited extent purely in thermodynamic terms. From the food that keeps our bodies at proper metabolic temperature to the draft animals, engines, or nuclear reactors that propel our vehicles, the connection between energy and culture is close and causal. And, as in all early civilizations, in ancient Mesoamerica a few plant and animal species were, in effect, the power base of cultural evolution. And not just any domesticated plants and animals would do: there had to be a reliable, voluminous carbohydrate source and nutritionally complementary plants and animals. In Mesoamerica these foods were, principally, maize, beans, and squash, augmented by protein from rabbits, deer, dogs, and in some places, fish and shellfish.[2] These rather unpromising foodstuffs were enlivened in antiquity by the use of cacao (from which chocolate is derived), incindiary peppers, numerous herbs and spices, and several natural hallucinogens. And like all ancient farmers, they used the magic of fermentation to effect vast improvements in various high-sugar plants—cactus being the principal source of alcohol in Mesoamerica.

The Archaeological Record of Early Complex Mesoamerican Societies (c. 1600 B.C.–A.D. 1524)

As we saw in Chapter 6, between 9000 and 4000 years ago most of Mexico was inhabited mainly by hunter-foragers whose seasonal dispersal into microbands and ways of making a living kept population densities low, mobile, and nonagricultural. A few groups along the margins of the lake in the Valley of Mexico may have been sedentary villagers, as were some groups along the coasts, and their role in the domestication of plants and animals and the eventual spread of agriculture is unclear.[3] Kirkby sug-

13.3 Mexico's first agriculturalists lived in wattle-and-daub houses much like these contemporary homes in Morelos, Mexico. The people in this village speak Nahuatl, which is derived from the language of the Aztecs.

gests that selection for larger maize cob size was such that about 4000 years ago it became feasible over large areas of the Mexican highlands for people to subsist mainly on maize.[4] Grinding corn kernels into a powder and making tortillas out of it may not seem to be a profoundly creative development, but maize requires a fairly sophisticated storage and processing technology if it is going to provide most of a community's calories every year. In any case, shortly after about 1800 B.C. sedentary agricultural communities appeared at about the same time in many different areas. From the hot, wet Guatemalan lowlands to the arid Tehuacán Valley, the earliest villages were quite similar in size and contents.[5] Almost all houses (Figure 13.3) were built using the wattle-and-daub method—walls were constructed of woven reed sheets, plastered with mud, and dried by the hot Mexican sun. Roofs were of thatched materials. Houses were seldom larger than 4 by 6 meters, with a tamped clay floor on which fine sand was scattered.

Most of the earliest farming communities were tiny hamlets, just ten to twelve houses, which were home to about fifty to sixty people, but some were larger. Most houses that have been excavated have yielded the same remains, mainly grinding stones, storage pits, pieces of large ceramic storage jars, bones of cottontail rabbits, carbonized maize fragments, and broken pieces of ceramic charcoal braziers.[6] In addition, ovens, middens, and graves are very common. While the proportion of plant and animal foods varied somewhat, all villages probably grew maize, beans, squash, peppers, and some other crops and hunted deer and rabbits. Each village, or each extended family, may have had a specialist who did pressure flaking of stone, leatherworking, or a similar craft, and individual villages may have concentrated on specialties like salt production, feather weaving, shellworking, grinding stone manufacture, and the like.

As in Mesopotamia, China, and elsewhere, the background to the origins of complex society in Mesoamerica was a great scatter of relatively simple agricultural villages, where the mechanics of producing a reliable, expandable food supply had been mastered. Many areas of Mesoamerica contributed to the overall rise of the first Mesoamerican states, but four areas appear to have been particularly important: the South Gulf Coast, the Valley of Mexico, the Valley of Oaxaca, and the Mayan Lowlands.

Early Complex Societies on the South Gulf Coast (c. 1500–400 B.C.)

The earliest and most radical break with the simple village farming tradition of Mesoamerica occurred in the sweltering lowlands of the South Gulf Coast. Here, shortly after 1350 B.C., people built massive pyramids and platforms, lived in groups of hundreds or even thousands in small

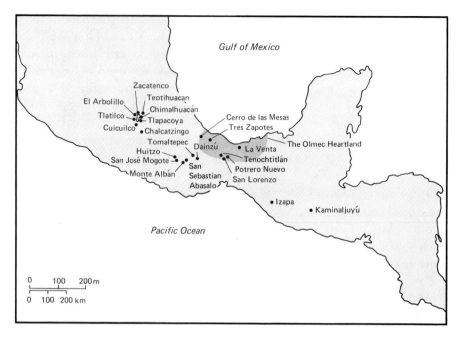

13.4   Centers of initial cultural complexity in first millennium B.C. Mesoamerica.

towns, intensively farmed a variety of ecological zones, and turned out what many regard as some of the most arresting art of all time.

These people are known to us as the Olmec, a name derived from an ancient American word for rubber—doubtless a reference to the rubber trees that grow in this area—but a name these people themselves probably did not use.

Some scholars have argued that the Olmec culture was the *cultura madre* (mother culture) of all later complex societies in Mesoamerica, and that they were directly responsible for transforming their neighbors by military, political, religious, or economic means into complex societies. Others have argued that the Olmec represent only one of several largely independent cases of the evolution of social complexity in Mesoamerica.

The Olmec Heartland is a coastal strip approximately 350 kilometers in length, extending inland about 100 kilometers. It was created by the alluviation of several rivers that run to the sea from the highlands. Except for a few areas, the region is thickly forested. Torrential rains fall during the summer, but the area is dry in the winter, which permits swidden, or slash-and-burn agriculture, as it is sometimes called. Swidden agriculture involves cutting down all the vegetation in a particular area and then waiting for the dry season so that the cut vegetation can be burned, which returns nutrients to the soil—an important contribution when manure and artificial fertilizers were not available, and, unlike the peasants of Mesopotamia, the Olmec had no cattle to graze on fields and replenish them with manure. After burning, the land is sown, and the crops germinate and come to maturity in the rainy season. After one or two years of exploitation, however, the land must be left fallow, sometimes for twenty years or more. If the cycle is accelerated, productivity falls rapidly. In the flat lowlands of the Olmec Heartland as much as 70 to 90% the land is fallow at any one time.[7] Maize, beans, and squash were agricultural staples in Early Formative times, supplemented by hunting and fishing and collecting wild plant foods. In coastal areas mussels and other rich resources of the marine interface could also be collected. Some river levees near the coast are annually inundated with water-borne silt of such fertility that two crops a year are possible using swidden techniques. Indeed, the precocity of the Olmec in developing the first complex Mesoamerican culture was probably tied directly to the great agricultural potential and rich floral and faunal resources of these riverine environments.

After many years of research, we finally have substantial data about the Early Formative cultures of the South Gulf Coast, although we will probably never know with great precision the population densities in the countryside in many periods because the tiny hamlets they apparently lived in are covered by vegetation and easily missed in archaeological surveys.[8]

Thus, we know the Olmec primarily from their larger ceremonial centers, and on this basis they are an impressive culture. At San Lorenzo, for example, in southern Veracruz, people who made a somewhat non-Olmec looking pottery had been living for a substantial time. Then, shortly after 1300 B.C., people at San Lorenzo carried thousands of tons of clay, sand, and rock in baskets to make a large platform some 600 meters by 100 meters and at least 7 meters thick, on which they made scores of basalt sculptures, numerous artificial ponds, and clay platforms.[9]

San Lorenzo actually is a group of sites within a diameter of about 5 kilometers, a complex that reached its peak between about 1150 and 900 B.C. How different the Olmec were from their hunter-forager ancestors is illustrated both by the huge platform and by the Olmec "heads" (Figure 13.5) and other monumental sculptures, such as free-standing figures of kneeling men and carved stelae and "altars," all carefully executed from massive basalt blocks. On them are engraved fantastic mythical creatures representing hybrids of snakes, jaguars, and humans.

The basalt for these carvings was imported from about 80 kilometers away, probably by being floated down rivers on rafts. Some of the heads weigh 40 metric tons, so scores of people must have been involved.

Plant remains do not preserve well in the South Gulf Coast, but we can assume most of the people were maize farmers. They ate some peccary and deer, but fish and domestic dogs provided most of their protein. Hu-

13.5 The Olmec produced scores of these stone sculptures and transported them many miles to ceremonial centers.

man bones found at San Lorenzo with obvious burning and butchering marks suggest cannibalism, either ritual or for more secular motives. So many bones from a species of marine toad *(Bufo marinus)* have been found at San Lorenzo that archaeologists have long suspected that the Olmec used these animals for the hallucinogen they are known to produce. These toads are restricted to the very moist environments of coastal areas, and the discovery of their remains in Oaxaca—a very dry area several hundred kilometers distant—suggests that they may have been a trade item in the Early Formative. Obsidian, imported in large volumes from the Mexican highlands to the west, was used for arrowheads, knives, and many other tools.

Much of Olmec agriculture may have been swidden-based, but very productive farming was possible on the river levees. The twenty or more artificial ponds and lagunas built at San Lorenzo before 900 B.C. were apparently drained by an elaborate system of deeply buried basalt troughs covered with slabs. Several possible uses have been suggested for these ponds, ranging from water storage to intensive hydraulic agriculture.[10]

At another Olmec site, La Venta, located on a small island in a coastal swamp near the Tonal River, the Olmec constructed a series of mounds, platforms, courts, and pyramids covering more than 5 square kilometers. Much of this has been destroyed by looters and an oil well/processing installation, but excavations in the 1940s revealed a large portion of this site's plan. Dominating the area is a pyramid of clay, 128 by 73 meters at the base and 33.5 meters high. Two long, low mounds extend out to the north from the pyramid, with a circular mound between them. All these mounds are oriented eight degrees west of true north. Parts of the mounds were painted in red, yellow, and purple, and the use of clays of different colors must have made them a gaudy sight.

The most impressive artworks at La Venta are four "Olmec heads." Like those found at San Lorenzo, these heads are as large as 3 meters high and depict a human head with a serious, not to say sneering, facial expression, and they usually are shown wearing a "football helmet" (Figure 13.5). Since the Olmec had no metal tools, we assume they worked with grinding and pecking stone implements, and it is difficult to believe that these sculptures were made by anyone other than skilled specialists. These mounds and sculptures, with their unearthly figures, all set in the riot of jungle life of the steamy Olmec area, seldom leave the visitor unaffected. These heads may, in fact, be connected to a ball game of sorts.

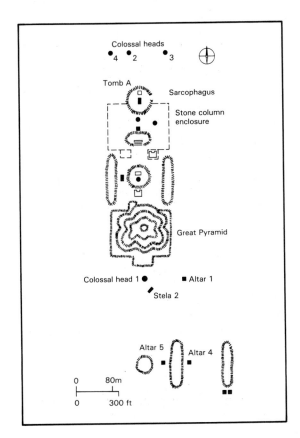

13.6 The Olmec complex at La Venta included a 30 meter high pyramid and several other lower mounds.

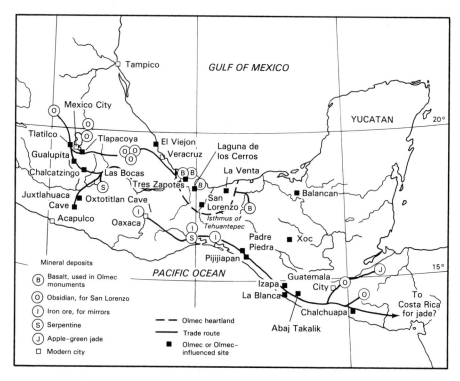

13.7   Major sites and mineral deposits in the Olmec area.

Ritual ball games were a major part of Mesoamerican societies until the time of the European Conquest, although these "games" were more like fights, apparently, and involved human sacrifice and slaughter.

Among the spectacular discoveries at La Venta were three superimposed "pavements," each composed of about 485 serpentine blocks (about the size of small construction bricks), laid out in a traditional Olmec design, a jaguar mask, and then carefully buried.

Unfortunately, the acidic, damp soils here do not preserve bones well, and the only burial information we have from La Venta comes from a tomb in a large mound near the central pyramid. The tomb was elaborately constructed of basalt slabs, and on its limestone floor two juveniles were laid out in fabric bundles heavily coated with red paint. Buried with the bodies were jade figurines, beads, a shell ornament, a stingray spine, and a few other items, and these burials may reflect inherited wealth and prestige. Other types of evidence relating to social complexity, such as residential architecture and settlement patterns, are not well represented at La Venta. There is little residential debris there except for pottery and a few clay figurines. Apparently, most of the people who built La Venta did not live there permanently.

Other Olmec ceremonial centers were built at Laguna de los Cerros and elsewhere (Figure 13.7), and by 900 B.C. Olmec culture had reached its peak on the South Gulf Coast. It is possible to have large buildings and other trappings of social complexity on the basis of a very simple economic system, but by 900 B.C. the Olmec had an intensive and productive agricultural system and they were trading jade, iron ore, obsidian, bitumen, magnetite mirrors, shark teeth, stingray spines, and perhaps cocoa and pottery, and many other goods in complex patterns between the Olmec Heartland and highland Mexico and as far south as Guatemala.

Other signs indicate evolving complexity. Tres Zapotes, which had stone heads and pottery like that of La Venta, also had the famous "Stela C"—widely regarded as perhaps the oldest writing in the New World. One side of this basalt slab was decorated with a design similar to Olmec motifs and the other had a date engraved on it in the Long-Count—the ancient Mex-

13.8 These Olmec ceramic "baby" figurines were distributed over a large region of central Mesoamerica.

ican calendar, discussed in detail below. When reconstructed (one piece was found long after most of the rest of the stela was discovered), the date on the stela can be read as the equivalent to September 3, 32 B.C. Many scholars believe that it was the Maya of later Mexican prehistory who invented calendars and writing, but Michael Coe thinks it is possible that the Olmec invented these.[11]

Judging from their art style, the Olmec seem to have believed that at some distant time in the past a woman mated with a jaguar and gave issue to a line of half-human, half-feline monsters, or "were-jaguars." These are portrayed in pottery, stone, and other media in a highly stylized way, usually as fat infants of no discernible sexuality. Their snarling mouths, toothless gums, and cleft heads give them a strikingly bizarre quality that some have explained as an imitation of a birth defect of the neural tubes, or as the deformities one would expect of a mating between a human and a jaguar.[12]

"Olmec" ceramic and sculptural designs have been found far outside the borders of the South Gulf Coast. Some bas-relief rock carvings at Las Victorias in highland El Salvador strongly resemble those at La Venta, and similar sculptures have been discovered in the highlands of Guerrero and Morelos in western Mexico. At Chalcatzingo, in Morelos, cliff sculptures include a standard Olmec motif of a woman seated in what may have been meant as the mouth of cave or a steaming monster, apparently to get out of a "rain" of penises (Figure 13.9). Even the most devout of the post-processual archaeologists (Chapter 1) might have difficulty reading a convincing social logic into this sculpture. Also in Morelos, at the burial sites of San Pablo Pantheon and La Juana, David Grove has found hollow ceramic "baby" figurines and other ceramics that closely resemble those from San Lorenzo. The rather sinister quality that many people see in these styles of Olmec baby figurines must reflect something about our own culture. Olmec styles of pottery, worked stone, jade, and other artifacts have also been found at several sites in the Valley of Oaxaca, at Tlatilco and Tlapacoya near Mexico City, in Guatemala, and elsewhere in Mesoamerica.

By any standards, the Olmec were far different from their hunter-forager ancestors, and if we match them against the checklist of monumental architecture, burials of juveniles with wealth, voluminous craft production and trade, writing, etc., the Olmec were clearly a complex society. Their relatively low population densities and small population numbers in centers is probably a direct result of the constraints of their agricultural methods, although we may have an inaccurate sense of their agriculture which may have been more intensive than many think. Between 900 and 400 B.C. the Olmec civilization seems to have come to an end as a distinctive entity in Mesoamerica. The appearance at San Lorenzo at about 700 B.C. of new pottery wares, figurines, and art styles, and the apparent intentional destruction of the Olmec monuments may mean that the settlement was overrun by non-Olmec invaders. So much effort was expended in destroying this community and burning the monuments in long ditches that we might suspect that it was done as the culmination of a cataclysmic battle, perhaps after many years of simmering dispute with outsiders. Or, the Olmec may have foreseen military defeat and destroyed and buried these sculptures themselves.[13]

Olmec culture continued for hundreds of years after 700 B.C. at Tres Zapotes, La Venta, and other sites, suggesting that all the ceremonial centers were not part of a unified system. But by 400 B.C. the Olmec Heart-

13.9   Relief 1, Chalcatzingo, Morelos. One interpretation of this Olmec motif is that it is a woman ruler within a cave or a stylized monster's mouth, which gives off steam or smoke, while raindrops fall from above.

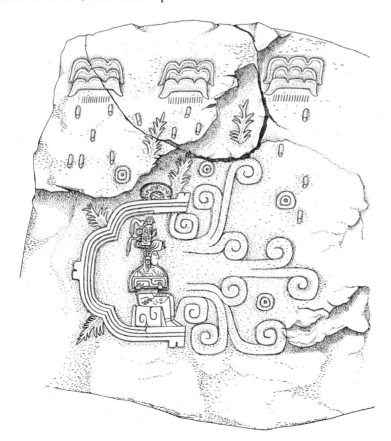

land was a cultural backwater, and only in recent years has this area been brought into more or less under the control of the central Mexican government.

Between about 900 and 200 B.C., other areas of southern Mesoamerica evolved small political polities that we might consider chiefdoms, and in some cases economic exchange and specialization seem quite complex.[14]

Most people's lives involve sufficient continuity that it is difficult to appreciate the transitory nature of social systems. We live in an age where the very complexity of our political system and its accumulated diversity make it unlikely that—barring cosmic calamities—our essential social and cultural fabric will disintegrate. But if the archaeological record is any guide, early complex societies were rather more evanescent. These early Olmec chiefdoms may have matured and then died out more from an unlucky conjunction of disease, drought, and internal revolt than from interregional battles. Some have argued (e.g., Earle[15]) that for at least part of the time, these different Olmec centers were competing, with first one then another becoming dominant; but, alternatively, they might have developed sequentially, with successive centers dying out without much conflict among them.

Early Complex Cultures in the Valley of Mexico

The Valley of Mexico is a large basin with no external drainage and rimmed on three sides by high mountain walls cut by only a few passes; even in the north where there are no mountains, the valley is delimited by a series of low hills. The valley has often been considered a "natural" analytical unit, bounded as it is by such impressive natural barriers, but archaeological research has revealed that almost from their arrival here the people of this area interacted with cultures far beyond the valley itself.[16]

Much of the Valley of Mexico lies beyond the temperature limits of maize agriculture, but until the last four hundred years, a large lake covered the low central portion of the valley, providing rich resources in the form of fish, fowl, turtles, algae, and reeds.

There is not a single navigable stream or river in the whole Valley of Mexico today, and maize agriculture in most places in the Valley would have depended on rainfall and small streams.[17] Rainfall is sharply seasonal and varies considerably from north to south. The upper slopes of the Valley of Mexico provide today, as they did in the past, many wood products, and in earlier times they supported large deer herds that were an important part of the prehistoric and early historic diet.

Between 1100 and 800 B.C., when the Olmec were rapidly developing relatively high population densities, monumental architecture, and spectacular art, there were still relatively few settlements in the Valley of Mexico, and most of these were small villages and hamlets. Only a few sites, such as Tlatilco and Cuicuilco, were larger; the latter was estimated at approximately 25 hectares (with a population of perhaps 500).[18] Cuicuilco was covered with lava by a volcanic eruption around 150 B.C. and therefore little is known about it. Tlatilco has been largely destroyed by looters and the activities connected with a modern brick factory built on the site. How much of the occupation there dates to the period between 1100 and 800 B.C. will probably never be known.

The apparent two-tiered site size hierarchy, with a few large towns like Tlatilco and Cuicuilco, and many small villages, suggests a simple, perhaps tribal, organization, and the distribution of settlements does not point to any political or social spacing. Settlements seemed to be located principally around the edge of the great lake, although there were a few small villages in the highlands in areas where the soil is particularly rich and deep. Differences in settlement size seem to be a result of local variations in agricultural potential.

Nor is there much evidence of complex architecture at these settlements. A few small mounds and platforms may date to before 800 B.C., but none is on the scale of the pyramids, platforms, and other structures found on the South Gulf Coast. No evidence of elaborate residential structures or monumental sculptures has been found.

The cemetery at Tlatilco provides little evidence that the occupants of the Valley of Mexico were living in complex cultures. Burial goods include pottery, shell ornaments, obsidian tools, figurines, bone tools, and jade and serpentine objects, and some women seem to have been buried with more numerous and more expensive objects than other people in the cemetery, perhaps even with sacrificed men and children. But there are no lavish mortuary cults.

From about 800 B.C. to about 500 B.C., the population density of the Valley of Mexico increased considerably. At least ten sites were larger than 50 hectares (each inhabited by about 1000 people), and one, Cuicuilco, probably had a population of about 2,500. All the larger sites are located along the lake margin, while scattered small hamlets are in the highlands.

The larger settlements along the lake are fairly evenly spaced at 8 to 10 kilometer intervals, and they all used similar styles of pottery, suggesting some degree of social or political integration, but probably at a low level.

For the period between about 500 and 200 B.C., however, there is persuasive evidence of changing cultural complexity in the Valley of Mexico. Population density rose considerably, and people were now living in larger settlements. Cuicuilco may have had as many as 7500 people at this time—

an unmanageable size without considerable social organization and control. Many other settlements of 80 to 100 hectares existed, and intermediate and small settlements also increased in number. Small "temple" platforms of stone and clay, some 3 to 4 meters high, appeared in several areas, and there were large stone structures at Cuicuilco and other sites.

So much of the early complex societies in the Valley of Mexico have been destroyed by the expansion of Mexico City that we may never know to what extent they paralleled and interacted with the contemporary Olmec peoples.

Early Complex Societies in Oaxaca
(c. 1600–200 B.C.)

The basis of cultural evolution in the Olmec area was the productivity of the coastal zone and river alluvium, and in the Valley of Mexico it was the lake shore, but in Oaxaca it seems to have been the diversity of ecological zones and some small river valleys. On the valley floor are large fertile alluvial areas of flat land with a water table sufficiently high for irrigation to be easily accomplished. Grading up into the mountains, the piedmont areas are less fertile than the alluvium, but they can be productively farmed by diverting water from the perennial streams that run toward the valley floor. The higher mountains are cooler and wetter than the other zones and are still covered with pine and oak trees.

Frost is not nearly the limiting factor on agriculture it is in the Valley of Mexico, although it can be a significant determinant of productivity. Since Formative times irrigation by means of canals and wells has been an important aspect of agriculture in Oaxaca.

The bedrock underlying parts of the valley is travertine, a smooth, hard stone that can be fashioned into ornaments, and this material was traded at least as far away as the Valley of Mexico and the South Gulf Coast. Also, some of the streams in the valley have extremely high salt content and were used for salt production until about 1910; presumably this production (and trade) would have been possible in the Formative period as well.

Some of the most important resources of Oaxaca were the native iron ores, including magnetite, ilmenite, and hematite. Small pieces of these materials were polished and used as mirrors and ornaments, which were then traded widely over Mesoamerica and used as marks of status.

Shortly after 1400 B.C., the most productive areas of the piedmont and the alluvium in the Valley of Oaxaca were occupied by small villages composed of perhaps fifty people living in tiny wattle-and-daub structures. The first significant deviation from this pattern of egalitarian farmers occurred sometime between 1350 and 1150 B.C., when the inhabitants of at least one site (San José Mogote) built several "public buildings" that together covered about 300 square meters. Although these structures average only 5.4 by 4.4 meters each, they are interpreted as public buildings because the floors were carefully covered with a distinctive white lime plaster and swept clean, in contrast to the average house of this period, whose floors were usually stamped clay and sand and covered with household debris.

Other evidence suggests that these buildings at San José Mogote may have been intended for special functions: they are oriented eight degrees west of true north, about the same as the major monumental constructions at La Venta, in the Olmec Heartland; they were repaired and reused over longer periods of time than the obviously residential structures; and at least one of them had an "altar" or step against one wall.

Most of the other Formative villages in Oaxaca lack such public structures, although one, Tomaltepec, was found to have a large prepared mudbrick platform.[19] In the floor of this structure was a storage pit, considerably larger than any of the others at the site, containing relatively large quantities of obsidian, ornamental seashell, and deer and rabbit bones.[20]

Between 1400 and 1000 B.C., overlapping in time with the construction of this platform, a large cemetery was created at Tomaltepec. Eighty burials containing a total of about 100 individuals were found, and most of these burials had almost the same goods, mainly ceramics and a few other small items. In four of the burials, however, small quantities of obsidian, magnetite, and jade were found, but these differences in grave goods seem fairly small in view of the overall similarity. And, interestingly, no juveniles or infants were buried here; all were adults, suggesting that this society had not yet achieved significant social stratification.

Analysis of trade items in the valley between 1400 to 1000 B.C. also reinforces this impression of low-level community organization. A few items, such as obsidian, were traded, but in small amounts, and the trade was probably organized through individual households.[21] But archaeological evidence after 1000 B.C. suggests major changes in the cultural organization of Oaxacan society. The largest site of this period, San José Mogote, had several successive public buildings of earth and adobe construction.[22]

More crafts were apparently performed at San José Mogote than at other settlements in the valley at this time. Debris from working obsidian, jade, magnetite, shell, and other substances is found here in concentrations proportionately greater than at other sites. There was a major increase in the volume of "exotic" traded materials in Oaxaca at this time also, perhaps in response to increasing social stratification: these materials seem to have been used most frequently to make ornaments that reflect differences in social rank. There are clear resemblances between Olmec figurines and ones of comparable age from Oaxaca, particularly the baby-face style. In addition, some pottery with Olmec motifs reminiscent of the South Gulf Coast is found in Oaxaca. Certainly, magnetite and obsidian were moved in substantial volumes between Oaxaca and the South Gulf Coast between 1100 and 850 B.C.

Burial evidence during the period from 1100 B.C. to 850 B.C. implies some low-level ranking.[23] Marc Winter reports that of fifty Formative burials in the valley, decorated "Olmec" pottery definitely occurs with adult male burials, but rarely or never with female burials.[24]

After about 850 B.C., variation in settlement size was greater than it had been earlier. By 550 B.C., San José Mogote, for example, grew to fifteen times the size of the next largest community. Many settlements excavated have public architecture, and their distribution seems to mirror the growing importance of social and political factors in determining site location. Between about 600 and 200 B.C., Oaxaca produced what some would classify as Mesoamerica's first state—a development discussed in more detail below.

In summary, the Valley of Mexico, the South Gulf Coast, and the Valley of Oaxaca all show similar patterns during these early developmental periods. In areas of naturally good agricultural productivity, or where production could be easily intensified, settlements grew large and the first expressions of monumental architecture, differential consumption of goods, and a more intricate flow of commodities and aesthetic styles can be detected.

Despite some evidence to the contrary,[25] it seems unlikely that cultural evolution in either Oaxaca or the Valley of Mexico was directly instigated by the Olmec through military imposition, economic exploitation, or slavish imitation. The wide distribution of Olmec styles of ceramics, figurines, and sculpture and the construction in Oaxaca and elsewhere of public buildings with astronomical orientations similar to those of the Olmec buildings seem to reflect interregional trade networks and perhaps the circulation of important people, but the archaeological evidence does not support the idea of strong political ties between these three regions.[26]

The Rise of Mesoamerican States (c. 600 B.C.–A.D. 900)

The Valley of Oaxaca

Whereas the South Gulf Coast seems to have been the leader in the initial stages of Mesoamerican cultural evolution, it is perhaps the Valley of Oaxaca that produced the first "state"—at least as defined by multilevel settlement size hierarchies, the complex intertwinings of economic specializations to

13.10 These bodies at Fabrica San José were buried (c. 100 B.C.–A.D. 100) with many ceramic vessels and other goods. Note that two skulls were separated from their bodies and placed on a stone ledge at the time of burial.

the extent that many settlements were no longer self-sustaining entities, and the florescence of public art and architecture in the context of religious and political elites.

Shortly after about 500 B.C., Oaxaca was transformed. Richard Blanton notes that at least the following changes occurred: the population density increased at a more rapid rate than in any previous period; agricultural intensification in the form of canal irrigation became important; pottery manufacture became specialized and perhaps was part of the Valley's first market system; the settlement at Monte Albán, on a high plateau in the center of the Valley, grew to become a major regional center (Figure

13.11 Like their contemporaries in Teotihuacan and the Mayan areas, the people of Oaxaca built huge stone plazas and platforms as ceremonial centers of their political systems.

13.11).[27] Blanton has argued that the capital at Monte Albán was founded in an unpopulated area between various important settlements as an expression of a new confederation between people who previously may have been linked by rather low-level social and economic ties.[28]

Blanton and his colleagues estimate that Monte Albán grew from an unoccupied, waterless wasteland to a great religious and political complex of 5000 people in fewer than two hundred years.[29] To sustain this rate of increase, they suggest, people would have had to move in from other areas and have intensified agriculture in the valley and piedmont to support the town.

By 200 B.C. the population of Monte Albán reached about 17,000—a great city, by ancient standards. Agriculture was intensified around Monte Albán as well, and rural population densities soared. Craft specialization in pottery and other commodities was sharpened, and the location and products of kilns indicate some degree of regional administrative control, although Blanton concludes that Monte Albán itself remained mainly a ceremonial center with few economic functions.[30]

In one of the Main Plaza buildings at Monte Albán is a gallery of carved stone reliefs whose main theme seems to be the commemoration of the torture and killing of enemies. Scores of bodies are depicted with open mouths, closed eyes, and blood streaming from them in flowery patterns.[31] The depiction of genital organs on these figures seems to have been an added insult, as nudity was regarded as scandalous in most of Mesoamerica at this time.[32]

In succeeding centuries population densities and settlement patterns varied considerably, but the investments in monumental architecture, the hierarchical arrangements of towns in terms of the distribution of goods and services, and other markers of cultural complexity persisted and became more elaborate. Thus, by 200 B.C. or shortly thereafter, the first Mexican state, by almost any definition, may have been operating in Oaxaca.

It may be overstating the case to say that by 200 B.C. we have also moved from prehistory to history in Mesoamerica, but the engraved signs and symbols at Oaxaca and elsewhere in Mexico seem to be well on the way to the development of a true writing system.

Although the rough outlines of Oaxacan culture history have been established, scholars argue about what forces determined this culture history. Sanders and Nichols have presented an ecological deterministic interpretation that stresses the importance of irrigation agriculture and population growth as the main causes of the Valley's florescence.[33] In contrast, Kowalewski and Feinstein, in a lengthy assessment of change in economic systems in Oaxaca, conclude that much of this "economic variation was largely determined by the changing functions and degrees of chiefly or state power."[34] They reconstruct a situation in which the power and predelictions of elites changed, and with them the permeability of the borders of Oaxacan political systems.

These arguments turn on the accuracy of population estimates and other economic data we do not yet have, but the great diversity of opinion on this problem mirrors the controversy in anthropology in general about the sufficiency of cultural ecological explanations.[35]

The Valley of Mexico

Oaxaca may have had the first "state" in Mesoamerica, but the slightly later and parallel developments in the Valley of Mexico are in some ways much more impressive.

Thanks to William Sanders, Jeffrey Parsons, Richard Blanton, and their students, who—despite rattlesnakes and the epidemiological horrors of the Playa Azul North Bar—have done systematic archaeological surveys of the Valley of Mexico, we can now reconstruct over four thousand years of settlement history in this area.[36]

In most of these archaeological surveys the procedure was to collect a sample of pottery from each site and then estimate the site size, periods of occupation, and distinctive architectural features. The limitations of such data have been intensively reviewed by Tolstoy and others, but this information is by far the most useful and systematic archaeological evidence in Mesoamerica.[37] As such, it has been used for the application of several kinds of quantitative analyses.

Before 200 B.C. the Teotihuacan area of the Valley of Mexico had been relatively unimportant culturally, but it has large areas suitable for irrigated agriculture and possesses large springs capable of supplying irrigation systems. Obsidian is also available nearby, and the area is thought to have supported large stands of edible maguey cactus and nopal, a plant species that is home to an insect that can be rendered into a red dye highly prized in prehispanic Mexico. Moreover, Teotihuacan stands along a natural trade route to eastern Mesoamerica—an important advantage given the difficult terrain of this region.

Between about 500 and 150 B.C., Teotihuacan supported a few small villages with a combined population of at most 3000, but between 150 and 1 B.C. the population growth rate exceeded that of any other period, and a city extending some 6 to 8 square kilometers formed and reached about one-third its eventual maximum size.[38] Between about A.D. 1 and 150 the growth rate was still high but had slowed; the average population during this period was probably between 60,000 and 80,000[39] and rose to between 100,000 and 200,000 by A.D. 500.

During this time work was completed on the massive Pyramids of the Sun and the Moon (Figure 13.12) and on at least twenty other important temple complexes.[40] The Pyramid of the Sun is over 200 meters long on each side of its base—as large as the great pyramid of Khufu in Egypt— and rises 60 meters (half the height of the Khufu pyramid). The interior is filled with approximately 1 million cubic meters of sun-dried bricks and rubble. In volume, this probably equaled 2 million cubic meters of uncompressed fill, which would have required the excavation, transport, and shaping of the soil in an area 1.4 kilometers square to a depth of 1 meter —a considerable effort by any standards. The Pyramid of the Moon is somewhat smaller (150 meters at the base, 45 meters high), but of greater architectural sophistication, with a series of inset trapezoidal platforms. Pottery fragments in the fill of these pyramids indicate that the pyramids were constructed by using material from earlier occupations near the city. Considering the size of these structures, it is little wonder that the Aztecs believed that the pyramids had been constructed by giants and that some of the gods were buried beneath them.

Already by about A.D. 100 Teotihuacan had hundreds of workshops with perhaps as much as 25% of its population employed as craft specialists, making products in obsidian, ceramics, precious stones, slate, basalt,

13.12 The Pyramid of the Sun (upper left) and the Pyramid of the Moon (from which this photograph was taken) were the ceremonial center of Teotihuacan. At times the city probably had a population of more than 100,000, and there were hundreds of workshops and houses.

seashells, feathers, basketry, leather, and other materials. Massive public constructions were underway; considerable variability existed in mortuary complexes and residential architecture; and the settlement patterns in the surrounding areas were heavily influenced by the city.

The rest of the city is perhaps even more significant in terms of its evidence of cultural complexity. It was laid out in quadrants, formed by the Street of the Dead intersected by streets running east to west. Some of the quadrants were more densely occupied than others, and very different architectural styles and artifacts are found in various zones of the city. Along the main north–south street are elaborate residences, presumably for societal elites, as well as large and small temple complexes. Many of the more impressive buildings are built on platforms and often face inward on patios and courtyards. Most buildings are one story high. In some temple complexes the walls are decorated with beautiful murals depicting religious themes, warfare, imaginary animals, and scenes from daily life.

The basic residential unit of Teotihuacan appears to have been the large, walled, windowless compound, made of adobe bricks and broken-up volcanic rock and faced with a fine plastered clay. Many such compounds measured 50 meters or more on a side and internally were divided into many rooms, porticoes, patios, and passageways. In some, open patios let in sun and air and drained the compounds through underground stone troughs. Many walls were decorated with frescoes of jaguars, coyotes, trees, gods, and people in naturalistic settings.

Some residential complexes at Teotihuacan were found to have concentrations of artifacts characteristic of distant areas of Mesoamerica. The Oaxaca Barrio, for example, included ceramics, funeral urns, burials, and other elements indistinguishable from the artifacts used in Oaxaca—over 400 kilometers to the south—and very much in contrast to the distinctive artifacts of the Teotihuacan natives. These foreign "barrios" appear to have remained culturally distinct and intact for at least several centuries and may have been trade entrepôts or ethnic "ghettos," but no persuasive explanation of these features has been made.

The city's people apparently ate large quantities of nopal and other kinds of cactus, as well as maize, beans, squash, and a variety of other domesticated and nondomesticated plants and animals. Even at Teotihuacan's peak, however, there was considerable hunting, as evidenced by about 80% of the animal remains found being deer bones. The discovery of many burned and cracked human bones in settlements near Teotihuacan and the depiction of human sacrifice in Teotihuacan murals suggest that the diet may have been augmented occasionally with human flesh.

Rebecca Storey analyzed 67 bodies from burials in one of the poorer compounds at Teotihuacan and found that, as it did for their fellow early city dwellers in Rome and elsewhere, living in early preindustrial cities shortened life: half the people died before about 15 years of age, and only a very few lived past 30.[41]

By the time Teotihuacan reached its maximum size, it had apparently depopulated much of the rest of the Valley of Mexico: only one other major settlement appears in the valley at about A.D. 500, and it is but a small fraction of the size of Teotihuacan. In fact, the abandonment of rural sites correlates so closely with Teotihuacan's growth that it appears likely that populations were either drawn or coerced directly into the city; overall population growth in the Valley of Mexico at this time (A.D. 100 to 600) was probably very minor and absorbed by Teotihuacan. Also, sometime in the first or second century B.C., the city of Cuicuilco was buried under 5 meters of lava and the surrounding agricultural areas rendered worthless by layers of volcanic ash, removing Teotihuacan's only large competitor.[42] By A.D. 500 Teotihuacan-style ceramic vases were placed in the richly furnished burials of apparently high-status individuals on the Gulf Coast, in Oaxaca, and elsewhere.

Teotihuacan so dominated the Valley of Mexico that some have wondered if its power extended far beyond the Valley of Mexico. It seems unlikely that the few hundred thousand people at Teotihuacan were able to extend military control over the millions of people living in the rest of Mesoamerica: fighting a military campaign in the rough terrain of these distant areas would have been suicidal. More likely, the Teotihuacanos were tied to the many other areas through trade networks. The city has no major defensive fortifications, but it does have what appear to be large market areas, and the ecological diversity of Mesoamerica would have put a high premium on large-volume trade in basic agricultural and technological commodities. By circulating these many products, people would have had a much higher standard of living and much greater protection against food shortages.

Sometime before A.D. 600, Teotihuacan's size and influence began to decline. As the city shrank in population, new centers and settlements appeared throughout the Valley of Mexico, particularly on its edges. Significantly, after A.D. 600 Teotihuacan styles in pottery, architecture, and other artifacts disappeared from the rest of Mesoamerica. It is as if a complicated exchange network had been beheaded, and local cultures began developing their own distinct traditions.

Archaeologist George Cowgill has argued that Teotihuacan stopped growing in size and power after A.D. 600 because it had reached the limits of productivity of its sustaining area: there are only a few thousand hectares of land near Teotihuacan suitable for canal irrigation, and once populations at the city reached about 100,000, much of the food would have had to have come from riskier, less productive forms of agriculture and from collecting and hunting wild plants and animals.[43]

There have been various challenges to explanations of Teotihuacan on the basis of its ecological circumstances.[44] Teotihuacan may have been outcompeted by other political systems based on more productive agricultural and economic resources: there is some evidence that at least part of its sphere of influence was encroached on by emerging states in the Mayan areas. Even closer to home, political systems centered at Tula, Xochicalco, and elsewhere in the Valley of Mexico may have begun to block Teotihuacan's access to needed raw materials and foodstuffs.

Elizabeth Brumfiel attempted to test the hypothesis that "population pressure" was an important factor in cultural evolution in the Valley of Mexico between 500 B.C. and A.D. 1—the important interval just before the florescence of Teotihuacan.[45] She did this by estimating the agricultural productivity and the potential for intensification of each settlement known for this period. Her argument is complex and mathematical, but one of her conclusions is quite simple: an important factor in the growth of smaller towns and villages was the imposition of tribute by elites in the largest communities. She found little convincing evidence of "population pressure" at the critical period—although, as usual in archaeology, the data are not conclusive.[46]

Several other analyses of these Valley of Mexico settlement patterns have been set in mathematical-locational geographical terms,[47] and the results have stimulated a great deal of discussion about Mesoamerican culture history and applicability of these kinds of analytical techniques.

The Maya

In Mexico the gods ruled, the priests interpreted and interposed, and the people obeyed. In Spain, the priests ruled, the king interpreted and interposed, and the gods obeyed. A nuance in an ideological difference is a wide chasm.
(Richard Condon)

At about the same time the civilization of Teotihuacan was developing in the Valley of Mexico (200 B.C.–A.D. 600), Mayan civilization was emerging in southern Mexico, Guatemala, Belize, and Honduras.

Early Mayan civilization may have been influenced to some minor degree by Teotihuacan, but it was largely its own creation. The Maya devised a complex writing system, their temples and palaces are spectacularly beautiful monumental constructions, and they are thought to have organized vast areas and many peoples under a centralized government. The Maya have traditionally been considered a civilization without cities, but Mayan centers such as Tikal and Dzibilchaltun had thousands, perhaps tens of thousands of people in permanent residence—although these settlements never attained the size of Teotihuacan, and they had less diversity in residential architecture and perhaps less occupational specialization than existed in Teotihuacan or Oaxaca at a comparable time.

It is difficult to fault archaeologists from spending most of their time excavating the spectacular Mayan ceremonial centers instead of surveying the snake-infested jungles for the rural settlements that supported these centers, but the result is that we still know comparatively little about regional economic and demographic aspects of the Maya. But a lot of valuable work has recently been done,[48] and the general picture is much clearer than previously.

Much of the Mayan homeland is hot, semitropical forest, but large areas are highlands created by a string of snow-capped volcanoes that extend from southeastern Chiapas toward lower Central America. In the highlands volcanic ash and millennia of wind and water erosion have created a rich thick layer of soil spread over a convoluted landscape of ravines, ridges, and valleys. Hard seasonal rains make this a fairly productive agricultural zone.

The tropical lowlands at the steamy heart of Mayan civilization cover the Petén and the Yucatán Peninsula, a massive limestone shelf built up out of the seas over millions of years. The land is rugged toward the southern part of the Petén, but most of the peninsula is flat. There are few rivers or lakes because the porous limestone quickly drains away surface water.

The lowland climate is hot and humid for most of the year, but drought can be a severe problem because the rainfall is seasonal and localized. Recent research has revealed large irrigation systems in some areas and, during at least some periods, permanent field agriculture with annual cropping was probably very important economically in the lowlands. In much of the Mayan area, the fields used for maize, beans, squash, tomato, and pepper cultivation must be fallowed for four to eight years after about three years of production. Even then, the Mayan homeland is sufficiently productive that some estimates of the land's carrying capacity on the basis of traditional swidden agriculture are up to sixty to eighty people per square kilometer.[49] Also, exploitation of the nuts of the ramón or breadnut tree *(Brosimum alicastrum)* may have been a major part of the Mayan agricultural strategy. Marvin Harris claims that when "archaeologists speak dramatically of having to hack away the jungle in order to expose the wonders of Maya architecture and sculpture, they generally neglect to say that they were hacking away at an overgrown [breadnut] orchard."[50] Dennis Puleston estimated that 80% of the calories consumed by the people of Tikal in some periods came from breadnuts.[51]

Studies of contemporary Mayan cultivators have been summarized by Matheny and Gurr, and when all the evidence of canals, raised fields, botanical remains from sites, and other data are considered, it's clear that permanent field, intensified agriculture, not swidden, must be seen as the basis for Mayan civilization.[52] Most peasant cultivators use animals, whether wild or domestic, to convert brush and hedgerow vegetation into usable form, and the small Yucatec deer—which was intensively hunted—apparently filled this role in the Mayan lowlands, along with domestic dogs, rabbits, and wildfowl.

Early archaeological remains suggestive of Mayan culture have been found at Izapa, near the Mexico–Guatemala border on the Pacific Coast, where there may have been some mounds and plazas as early as 1500 b.c. But it is many centuries later that the main elements associated with the

Maya are evident. From about 200 B.C. on at Izapa, over eighty temple mounds were built, many of them faced with cut stone, and some of the core elements of Mayan art appeared.[53]

Another highland center was Kaminaljuyú, now ravaged by the expansion of Guatemala City. Kaminaljuyú was apparently a large settlement by about 800 B.C., and between 100 B.C. and A.D. 200 it reached its zenith. Beautiful sculptures in the style of that at Izapa were executed, some bearing glyphs, or signs, that are assumed to convey historical and calendrical information. Kaminaljuyú also contained many temple mounds and rich burials, suggesting that it was a ceremonial center for many small dependent hamlets in the surrounding countryside.

One of the most important early Mayan sites was Chiapa de Corzo, in Chiapas. Already occupied at 1400 B.C., by 550 B.C. small pyramids and other civic buildings existed at this site, and by 150 B.C. there is clear evidence of gross wealth and status differences in residential architecture and mortuary complexes. Pottery and other artifacts indicate that very early on the people at Chiapa de Corzo were in contact with their neighbors at Kaminaljuyú and in the Valley of Oaxaca.

Settlements in the Petén, the Yucatán Peninsula, and other lowland areas may have been inhibited by the thick vegetative cover of this area and by other environmental factors. Xe (pronounced shay) on the Pasion River and a few other sites have evidence of occupation as early as 1000 B.C., but they have no monumental buildings or advanced artistic traditions.

Between 550 and 300 B.C., population densities increased markedly in the lowlands, and villages with some ceremonial architecture were established at Dzibilchaltun, Becan, Tikal, and elsewhere. The ceramics found in the lowlands at this time are quite similar, suggesting that the people were participating in at least some generalized exchange systems, but there is little evidence of political federation or voluminous economic exchange. As we have seen in the case of the Harappan (Indus Valley) and Halafian (Mesopotamian) ceramics, however, the spread of a distinctive, uniform ceramic style over large areas often precedes rapid and fundamental cultural change. By about A.D. 1, a distinctively styled pottery was in use over the entire 250,000 square kilometers of the Mayan lowlands, and pyramids, platforms, and other large public buildings were being constructed at Yaxun, Dzibilchaltun, Uaxactun, and elsewhere.

Between A.D. 300 and 900, Mayan civilization reached its climax as hundreds of beautiful pyramids, temples, and other buildings were completed, and painting and sculpture flourished.

The first part of this period corresponds to the florescence of Teotihuacan, and some see the Maya as developing principally under Teotihuacan's stimulus.[54] Trade between the Mayan area and Teotihuacan may have been considerable, but much of Mayan development was autonomous and distinctive. After A.D. 600, when Teotihuacan rapidly began to lose influence and population, the Maya began a three-hundred year period of intense development. Hundreds of temple complexes were constructed and beautiful stone sculptures executed—many dated and inscribed.

But despite these material indications of their brilliant achievements, the Mayan peoples themselves seem to have lived much as their ancestors had. They still had not aggregated in cities, and although thousands of people must have cooperated to construct these great complexes, most still lived in small undifferentiated rural hamlets. Stephen Houston has used archaeological evidence and interpretations of Mayan writing to estimate that most Mayan polities were organized within areas of 70 kilometers diameter or less—with most of the settlements within a day's walk of the major center.[55] These dispersed agricultural hamlets were grouped around small ceremonial centers that included a small temple pyramid and a few other stone constructions. Several districts of small ceremonial centers were congregated around the major ceremonial centers of Tikal, Uaxactun, Palenque, Uxmal, and other sites. Figure 13.13 shows the complexity of construction at one of the major sites, and all over southern Mesoamerica at this time there were beautiful, gleaming white limestone pyramids and temples surrounded by marvelously executed stone sculptures and decorated with wall paintings. Most centers also had ball courts made of stucco-faced rock where, apparently, people played a game in which a rubber ball was meant to be thrown or batted through a stone ring protruding from one of the inclined walls of the court.

Tikal contains a large (70 meters tall) pyramid-temple complex, many inscribed stelae, and several rich tombs. Some of the greatest of all known Mayan frescoes are at Bonampak, dating to the end of the eighth century A.D. Here, in several rooms of murals, the paintings tell a story of warfare, the torturing of prisoners of war, and celebration. The carefully drawn mutilated bodies, marching bands, richly dressed figures, and men with weapons convey an extraordinarily vivid sense of militarism, royalty, and religion. Mary Ellen Miller has argued that these scenes are in part a depiction of raids to take prisoners. Unfortunately, these murals have been obscured by seepage through the composite limestone, so that they are only clearly visible when they are splashed with water or kerosene.[56]

Themes of military triumph, the torture of captives, and the power of the ruling classes were also commonly depicted in bas-relief sculpture throughout the Classic period—even in the Valley of Oaxaca and the peripheral areas of the Mayan sphere of influence. Individuals of presumably higher status were juxtaposed in stone carvings with persons of lower status, and differences of dress, bearing, and position sharpened the contrast. In some cases, representations of prisoners and commoners were carved into the facings of stone steps, so that they were trod on by the nobility—a not too subtle visual pun.[57] Computer simulations have been used to suggest that as many as 77,000 people may have lived in Tikal's immediate environs at its peak.[58] In Quintana Roo, the Classic Mayan metropolis of Coba may have been one of the largest, most populous settlements in the Classic Maya period. Folan et al. have interpreted variation in architecture, such as walled and unwalled compounds, as indicative of elaborate social stratification.[59]

13.13   The Mayan ceremonial structure at Tikal.

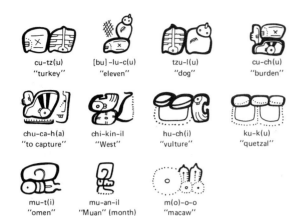

cu-tz(u) "turkey"	[bu]-lu-c(u) "eleven"	tzu-l(u) "dog"	cu-ch(u) "burden"
chu-ca-h(a) "to capture"	chi-kin-il "West"	hu-ch(i) "vulture"	ku-k(u) "quetzal"
mu-t(i) "omen"	mu-an-il "Muan" (month)	m(o)-o-o "macaw"	

13.14 Some Mayan glyphs and the phonetic values assigned to them by Y. V. Knorosov. Most surviving inscriptions were royal records of reigns and wars.

In recent years we have also seen greatly increased research on the smaller Mayan areas, such as Cerros, in Belize.[60]

Given the social and economic benefits that accrue from public-works projects, it is not at all surprising that the Maya should have built temples, pyramids, and platforms. They apparently had no metal, winches, hoists, or wheeled carts, and never developed the barrel-vault or arch constructions that allowed Old World civilizations to build multistoried temples and palaces. Thus, as noted of other early civilizations, the only things they could build to any height were the basic geometric forms: the coincidence of pyramids in many ancient civilizations seems largely accountable in terms of these basic facts of construction capabilities—especially for the Maya, whose homeland rested on a gigantic layer of limestone, which when wet could easily be cut with flint tools.

Mayan writing is in many ways a more impressive achievement than their pyramids and temples. Unfortunately, only three Mayan "books" survive; the Dresden, Madrid, and Paris codices. These are long strips of bark paper covered with a layer of plaster and folded like screens. There are also, fortunately, many stone inscriptions in the same writing system.

The sixteenth-century Spanish cleric Brother Landa thought Mayan writing (Figure 13.14) at first to be entirely alphabetic. Scholars quickly realized that Mayan writing was to be read in double columns, from left to right and from top to bottom, and by the turn of the last century Mayan glyphs had been identified for the "zero" and "twenty" signs, the cardinal points of the compass, the basic colors, Venus, the months of the year, and the "Long Count," the system of reckoning by which the Maya figured how many years had elapsed since the beginning of their time.

Unlike the case with Egyptian hieroglyphs and Mesopotamian cunei-form, Mayan writing had to be deciphered without the aid of "parallel scripts"—that is, expression of the same text in two different languages, one of which is known.[61] As noted in Chapter 8, *phoneticization* is a key step in the evolution of a written language, for once the elemental sounds of the spoken language are represented by abstract characters, the complete language can be written. The Russian linguist Yuri Knorosov was the first to demonstrate that Mayan writing was indeed a phonetic system, and since his initial translations, many inscriptions have been almost entirely translated.[62]

The Mayans apparently began to write substantial texts between 600 and 400 B.C., and the first scripts evolved into four different types: the Zapotec and Mayan versions, which began as hieroglyphic stone inscriptions, and the Mixtec and Aztec systems, which were written mainly on paper or hides.[63]

In the Middle East the first writings are almost unrelievedly economic, but in Mesoamerica the surviving documents are primarily calendrical and historical, recording, for example, when a temple was begun, when a king defeated a rival, and what lands were under the control of the state. About the dull details of maize and men, they seemed much less concerned.

The Maya were sophisticated mathematicians. They used a base-20 system, expressing the quantity 39, for example, as 19 numbers after 20, and the value 60 as 3 twenties.[64] They discovered the concept of zero and used a place-value notation system that allowed them to express numbers beyond 100,000,000. They had no way to express fractions in mathematical notation, but they computed the length of the solar year to 365.242000 days, compared to our own Gregorian calendar figure of 365.242500 days (the true value is approximately 365.242198 days). The Maya used two calendars. One was the familiar solar calendar in which a year equaled 365 days, but whereas we intercalate an extra day every four years to compensate for the year being actually 364.25 days long, the Maya blithely ignored this and let the seasons creep around the calendar. And in contrast with our system of twelve months of from 28 to 31 days, the Maya had eighteen named months of 20 days each, with 5 days, which were considered highly unlucky, added to the end.

The second calendar involved a 260-day year, composed of the intermeshing of the sequence of numbers from 1 to 13 with 20 named days (Figure 13.15). These two calendars ran parallel, and thus every particular day in the 260-day calendar also had a position in the solar calendar. The calendars' permutations are such that each named day would not reappear in the same position for 18,980 days, or fifty-two of our solar years. Every day on the Mayan calendar had its omens, and activities were rigorously scheduled by their astrological significance.

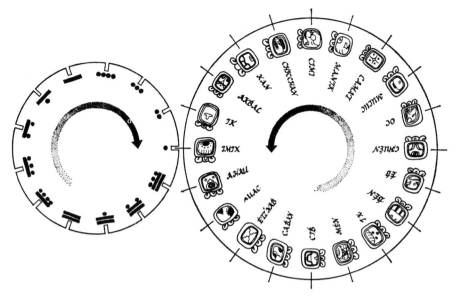

13.15   One of the Mayan calendars was calculated from the correlation of the numbers from one to thirteen and twenty named days.

In some of the later adaptations of the Mayan calendar, it appears that people were named after the day on which they were born and could not marry if they shared a numerical coefficient (e.g., Mr. "8 deer" could not marry Ms. "8 flower").[65] Crops were planted and harvested only on calendrically favorable days, and the whole operation of society was astrologically scheduled (several computer programs are now available to convert anyone's birthdate into its corresponding Mayan calendrical date, if the reader wishes to determine the validity of Mayan astrology).[66]

For reasons not entirely clear but having to do with the accomplishment of a certain number of calendrical cycles, the Maya believed that the world was created on August 13, 3114 B.C. and would end on December 24, A.D. 2011.

Although we do not yet fully understand why and how Mayan civilization declined, we do know when it happened. Each new monumental building of the Maya was usually accompanied by a stone stele engraved with the date of its construction, and thus we know that while many buildings were completed during the eighth and ninth centuries A.D., by A.D. 889 only three sites were under construction, and by about A.D. 900, construction seems to have ended for good. On the basis of ceramics and other information, we know that depopulation of the countryside and centers apparently followed quickly.

The effects of the collapse were felt far south of the main areas of the Lowland Maya, particularly in areas where local economies and traditions were connected to those in the main Mayan areas.[67]

Shortly after A.D. 900 the Toltecs, a people whose culture was centered at Tula in the Valley of Mexico, apparently established feudal control over some areas of the lowlands—an event setting off many years of internal dissension and local revolt. During this time various centers were built and some major population concentrations developed; but in population density, construction projects, art styles, inscribed stelae, settlement patterns, and mortuary complexes, the power and glory of Mayan culture began to fade.

Many factors have been suggested as causes of the development, and also the collapse, of Mayan civilization, and new interpretations sporadically appear. Not surprisingly, many involve the ecology of the Mayan homeland, which seems at first glance to be a major barrier to cultural evolution.

William Rathje argued that a major stimulus to Mayan developments was trade in obsidian, stone, and salt.[68] Others have argued that there probably was no need to trade for these commodities in most of the Mayan area.[69]

The discovery that the Maya had intensive permanent field agriculture with canals, reservoirs, and moats has led some to conclude that irrigation agriculture was a powerful stimulus of Mayan developments;[70] some Mayan canals are an impressive 1.6 kilometers in length, 30 meters in width, and 3 meters deep. We do not know how many such irrigation systems remain undiscovered in the tropical forests of the Mayan lowlands, and thus cannot estimate how important they may have been in the evolution of Mayan cultural complexity, or even what the water carried by the canals was used for. Matheny proposed drinking water, irrigation water for small garden plots, and a source of mud for renewing soil fertility as possible uses, and Marvin Harris suggests that irrigation to grow two crops a year may have also been part of the motivation for these projects.[71]

A traditional archaeological approach to explaining the development of something is to imagine a set of problems to which the development is a solution. We might ask, given the Mayan environment, under what conditions would it have been advantageous to organize into larger political and social units? Perhaps the answer lies in the necessity of local exchange to meet the threat of drought, disease, or disturbance. Rainfall is quite variable within the Mayan area, and many other things can adversely affect each community's agricultural system. Because the communities were all so similar in the crops they grew and their techniques for growing them, a major drought, such as happens in this area every eight to ten years, could result in the starvation of many people in hundreds of hamlets. But this could be avoided if many villages established exchange networks that spread the risks. Each year earthquakes, droughts, disease, floods, warfare, or some other combination of calamities might wipe out some sectors of the subsistence system, but if a village belonged to an organization that included many hamlets, it could get help or give help depending on its fortunes. Because of the inability to drastically intensify agricultural production in any one area, cities were, of course, out of the question. Similarly, population-control regulators were very important, hence the monumental construction projects in this most unlikely of places.

Whatever is at the root of the Mayan cultural evolution, the collapse of this culture poses equally interesting questions. Warfare seems to have increased toward the end of the Mayan period, and we might ask why it became more prevalent then and why the Maya were unable to fight off its effects at this time, after so many centuries of successful dealings among themselves and with their neighbors. No appreciable land shortages or overpopulation seem to have occurred, nor is there much evidence of foreign military pressures on these people.

We have argued that the evolution of Mayan society could probably be tied to the necessity of spreading the "risk" of life in this area by integrating many different settlements under a centralized authority. But, by the same token, such an integrated system might eventually have encountered a series of catastrophes and internal problems spaced so closely together and in such a sequence that their effects could not be successfully fought off. One of "Murphy's Laws" is that if anything can go wrong, it will. This is especially true for cultural systems spanning millennia. Earthquakes, disease, warfare, drought, crop disease—all have certain periodicities, and unfavorable conjunctions must necessarily arise if the system is sufficiently long-lived.

It is also possible that Mayan civilization collapsed in bloody civil wars and class conflicts. As Lowe points out, coups d'etat are not exactly rare in Central America, and there have been some interesting mathematical analyses of the rate and direction of the Mayan collapse that are not at variance with the ideas of a proletarian revolution.[72]

But the rather weak data we have from the Mayan areas can be made to fit many different hypotheses about the source and kind of cultural changes involved. Many more years of patient research, particularly continued slogging through the jungle lowlands on archaeological surveys, will be required for any comprehensive understanding of the determinants of Mayan culture history.[73]

Post-Classic Mesoamerica (A.D. *900–1524*)

The captain Alonso Lopez de Avila, brother-in-law of the
adelantado Montejo, captured during the war . . . a young
Indian woman of lovely and gracious appearance. She had
promised her husband, fearful lest they should kill him in
the war, not to have relations with any other man but him,
and so no persuasion was sufficient to prevent her from
taking her own life to avoid being defiled by another man;
and because of this they had her thrown to the dogs.[74]

At about A.D. 900, when Mayan political power in the lowlands was beginning to wane, much of highland Puebla, Mexico, and Hildago was apportioned among several competing power centers. One of these, Cholula, in
Puebla, stands along a route connecting the Valley of Mexico with the
lowlands to the east, and the city's importance may have derived from
defensive and commercial functions attendant on this route. After A.D.
900 a massive pyramid, covering 16 hectares and rising to a height of 55
meters, was constructed at Cholula, along with many other buildings.

At about A.D. 968 the Toltecs established a military empire centered at
Tula and embracing towns north and west of the Valley of Mexico, and
they may soon have come into conflict with the power centers at Cholula
and elsewhere in the Valleys of Mexico and Puebla. Eventually, through
military and other means the Toltecs were able to dominate most of these
rivals.

The Toltecs established trade and military outposts in many areas of
northern and western Mexico and exported metal, gemstones, and other
commodities as far north as Arizona and New Mexico. To the south, the
Toltecs established administrative control over Chichen Itza and perhaps
other towns in the Mayan lowlands where the collapse of the Mayan civilization was delayed for a century or two.

Eventually, however, Toltec power weakened, and, under the onslaught
of invading Chichimec, a group from the north, the Toltecs broke up into
many smaller, competitive groups. Tula itself was almost entirely destroyed by invaders at about A.D. 1156. Succeeding centuries saw the rise
of various other cultural traditions in central Mexico, such as the Tarascan state.[75] And on the periphery of Mexico, indigenous cultural traditions continued to be strong in some areas. The island of Cozumel, on
the east coast of Yucatán, for example, was a major trading center between A.D. 1250 and the arrival of the Spanish in 1519.[76]

One of the last tribes to invade central Mexico from the north and west
was the Aztecs. Aztec histories and legends, as recorded by the Spanish,
tell of their arrival in the Valley of Mexico as rag-tag foragers and primitive agriculturalists who at first were forced by the established residents
of the valley to live in the swamps around the lake, subsisting on flies,

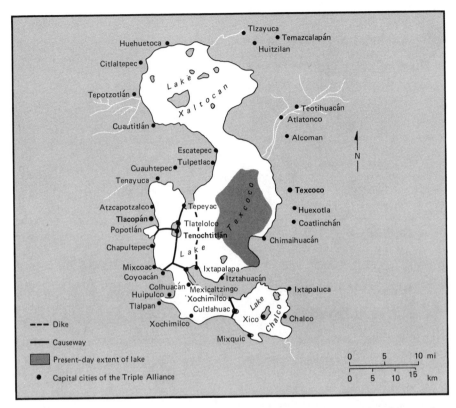

13.16 The most important Aztec cities were built around the lake that used to cover part of the Valley of Mexico. The lake margins were intensively farmed, and boats conveyed great quantities of food and craft products among these cities.

snakes, and vermin. According to legend, rival political groups in the valley enlisted the Aztecs in their campaigns but avoided other contacts with them because of the Aztecs' predilections for human sacrifice and other barbarisms. At war with various groups, the Aztecs were forced to take refuge on islands in the lake where, according to legend, they built their first city, Tenochtitlan. In time Tenochtitlan grew to become a massive complex of pyramids, courts, and other buildings, now largely buried beneath the streets of Mexico City.

As allies of the Tepanec kingdom of Atzcapotzalco, the Aztecs conquered many of the surrounding cities, and at about A.D. 1427 they turned on their erstwhile allies and through savage warfare brought most of central Mexico under their control. Military expeditions conquered peoples all the way to the Guatemalan border, and garrison towns were established from the Pacific Coast to the Gulf of Mexico.

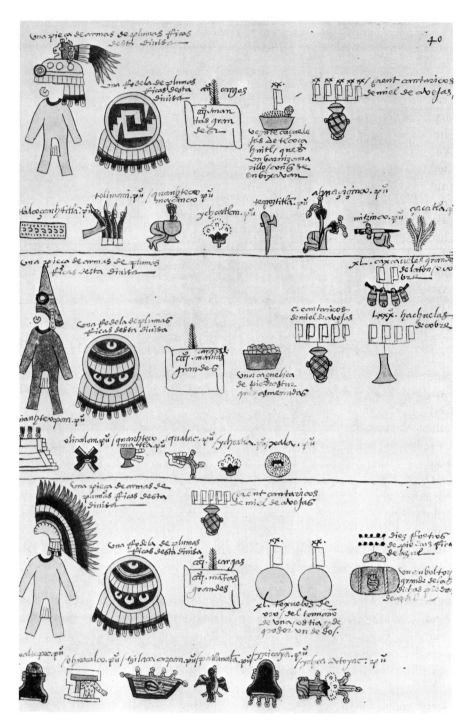

13.17 A Mexican tribute list (from the Codex Mendoza) describing the tribute paid to the government by towns in Guerrero Province. Among the commodities described are uniforms, shields, jade, gourds, sage, and amaranth.

Although the Aztecs are usually associated with militarism, they also created an impressive civil and commercial administration. Between about A.D. 1300 and 1520 they drained large areas of the Valley of Mexico, transforming them into productive agricultural plots. Michael Smith has argued that the settlement pattern of the Valley of Mexico during Aztec times was one of a hierarchically arranged marketing system of products with intense local specialization in goods and services.[77] Many commodities, including salt, reeds, fish, stone, cloth, various crops, ceramics, gold, and wood were exchanged among hundreds of communities. In fact, the improbable location of Tenochtitlan—on an island in the middle of the lake—is probably best understood in terms of its central role in these redistributive networks.[78] In 1519 Tenochtitlan is estimated to have had about 200,000 to 300,000 inhabitants, five times the population of London at that period, and there were many other large cities within the Aztec domain.[79] Many cities had broad avenues, causeways, temples, pyramids, and other large buildings, often interspersed with gardens, courtyards, and large markets.

It is estimated that between one and two million people lived in the Valley of Mexico in late Aztec times.[80] In the southern areas of the valley's lake system, maize, beans, squash, tomatoes, and other crops were grown on *chinampas*, long rectangular plots of ground created out of the lake bed by piling up layers of aquatic weeds, mud, human feces, garbage, and other materials. If crops were transplanted and carefully tended, as many as seven per year could be grown. The lake also provided great and reliable quantities of food in the form of fish, waterfowl, and salamanders.

The Aztecs were organized into a highly stratified class system headed by a god-king. Beneath the king were the nobles, the *pilli*, all of whom belonged to the royal house, while the great mass of the populace were commoners and were organized in large clans, called *calpulli*. The *calpulli* were the basic units of Aztec society. Each was composed of several lineages, totaling several hundred people, one of whom was designated the *calpule*, or leader. Members of a *calpulli* usually lived in the same village or ward, fought together as a unit if drafted for war, held and worked land in common, paid taxes as a unit, and worshiped at the shrine maintained by the *calpulli*. The leaders of the *calpulli* were the direct link between the imperial government and the people.

The *calpulli* differed from one another in social rank. There was some social mobility for individuals—usually by virtue of extraordinary service to the state in warfare, trade, or religion. At the bottom of the social scale were slaves, who worked the fields, performed other menial tasks, and were sacrificed in enormous numbers to various gods.

The Aztecs believed that the present world was just one in a succession of creations by the gods and that constant effort was required to forestall

A

B

C

D

13.18 These early drawings (from the Codex Florentino) describe the sequence of ritual human sacrifice. (A) The priest is dressed in the skin of a sacrificed person; another person (B) was given a shield, mirror, and other ritual items and played the part of the god Tezcatlipoca; the individual was then sacrificed (C) and eaten (D).

the extinction of the sun and the utter disappearance of humanity. Human blood was an essential part of the ritual (Figure 13.18) whereby the end of the world was postponed, and each time a human heart was ripped from a sacrificed person, another small step was taken toward prolonging the daily rebirth of the sun.[81] At times long lines of sacrificial victims snaked down the steps of the major pyramid mounds, on the top of which priests, hour after hour, cut the heart from each person. After the heart and blood had been offered to the gods, the body was thrown down the steps of the pyramid and subsequently flayed and then, perhaps, eaten. Other victims were pitted in gladiatorial contests, or beheaded, drowned, or cast into fires. The Spanish conquistadores may have exaggerated the numbers of people sacrificed, but it seems inescapable that the Aztecs annually killed many tens of thousands and perhaps hundreds of thousands of people. This slaughter was not only accepted by the common people, it seems to have been widely supported. All war captives knew their fate, and it was an act of honor to accept a sacrificial death. Young men were selected each year to lead a life of luxury surrounded by complaisant young women and feasting on the best of food, realizing full well that at the end of the year they would be sacrificed. And throughout the land parents turned over infants and children to government officials for use in annual sacrificial rites.

Many of the sacrificial victims, as well as soldiers who died in battle, people struck by lightning, and mothers who died in childbirth, were thought to spend eternity in various paradises, cosseted with the pleasures of this world.

With its emphasis on death, blood, and cosmic cataclysm, it is little wonder that Aztec theology struck the Spanish as somewhat heterodox. Even anthropologists, renowned for their cultural relativism, are impressed with the violence of Aztec religion. But human sacrifice is an old and recurrent theme in the evolution of complex cultures; in Mesopotamia, China, North America, and most other places, warfare and slaughter can be found which equaled that of the Aztecs in form, if not in intensity.

Why did so many ancient people consider it necessary to kill each other, and why did the Aztecs outstrip most previous cultures in this regard? Michael Harner has argued that the key to Aztec sacrifice is the contribution the cannibalism of sacrificial victims made to the Aztec diet.[82] Mesoamerica lacks any large domesticated animals that could have been effectively integrated with Aztec agricultural strategies, and this animal protein and fat deficiency may have been compensated for by cannibalism. There is little doubt that the Aztecs engaged in cannibalism, since several sixteenth-century Europeans described it as it happened, but we have insufficient evidence with which to evaluate Harner's thesis.

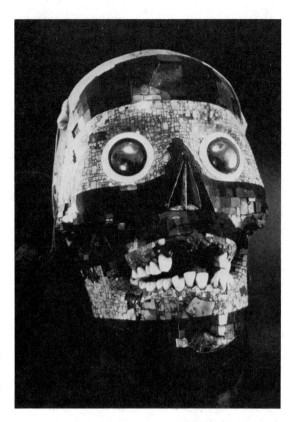

13.19 The supreme Aztec divinity was Tezcatlipoca, whose skull is modelled here in jade and crystal.

Despite their death cults, the Aztecs in everyday life were a colorful and in some ways engaging people. The Spanish remarked on their love of flowers and natural beauty, and their poetry contains many references to the joys of the natural world. The Spanish were amazed to find that Aztecs bathed their entire bodies most days—level of personal cleanliness that would have struck even most eighteenth- and nineteenth-century Europeans as bizarre and unhealthy.

Dress for men and women was often a loincloth and a woven cloak, and brightly colored cotton fabrics were used for ornamentation. In the countryside women often went about naked to the waist, but middle-and upper-class urban women wore decorated blouses.[83]

The diet of the Aztecs centered on maize, beans, squash, and tomatoes, although the wealthier people could eat various fruits, nuts, meats, and other exotic foods. The relatively unvaried diet was enlivened with peyote and other natural hallucinogens, and by tobacco and pulque, a cactus-derived alcoholic drink possessing near-miraculous powers to revive and nourish the weary peasant (or archaeologist).

The Spanish Conquest

The melancholy history of the conquest of Mesoamerica by Spanish adventurers in the early sixteenth century was recorded in detail by the Spanish themselves. Accounts on both sides of this meeting of worlds is fascinating in terms of a clash of cultures.[84]

In 1519 Cortez left Cuba with a sizable force of ships, men, armaments, and horses, and sailed to the coast of Veracruz. With the advantage of horses, cannons, war dogs, and an extraordinary esprit de corps, Cortez and his men were able to march directly into the Aztec capital at Tenochtitlan, where they were at first welcomed by the Aztec king, Moctezuma, who was under the delusion that the Spanish were gods returning to their ancestral homeland. He could hardly have been more wrong. Within a short time, the Spanish had kidnapped and jailed him and were forming alliances with local non-Aztec peoples, who were only too happy to help the Spanish displace the Aztecs. Moctezuma and many of his people were eventually killed in a fierce battle at Tenochtitlan, after which Aztec resistance stiffened; but within a few years the Spanish had captured most of the Aztec heartland. In 1524 they hanged the last Aztec king, and thereafter Spanish domination of Mexico was rapid.

Summary and Conclusions

It is better to know some of the questions than all of the answers. (James Thurber, 1894–1961)

As in all the other cases of early cultural complexity we have considered, it is clear that simple aspects of economy, ecology, and demography explain a lot of what happened in the New World. The Olmec, Teotihuacanos, Mayans, Aztecs, and other Mesoamerican societies appeared where they did and not in Newfoundland or Nebraska because of the exceptional productivity of the South Gulf Coast, Oaxaca, and other areas of Mesoamerica, given a simple farming technology and the maize-beans-squash complex.

We see in Mesoamerica, too, the stimulus to development provided by irrigation agriculture, environmental circumscription, interregional exchange, and other factors.

But once we get beyond this simple ecological level of analysis, we encounter a welter of variability in sociopolitical forms, economic histories, settlement patterns, and the other elaborations in these complex societies. In the absence of written documents it is difficult to apply Marxian or other sociopolitical models to Mesoamerica. So we are left with a case of cultural evolution that is so striking a parallel to what happened in, for example, Mesopotamia, that we must consider both areas examples of a single developmental pattern. But efforts to discern the intrinsic combinations of causes in any of these early complex cultural sequences remain a largely unrealized goal of archaeology.

And, from a nonscientific point of view, in some ways the most interesting thing about early Mesoamerican societies is the psychological contrast they make compared to ourselves. We may be sure that the average Olmec town of 500 B.C. had the full range of human personality types and an agricultural existence we can all imagine. But in their cosmology and world and life views, ancient Mesoamericans were profoundly different from the Spanish who first met them, and probably from any of the other Western European traditions.

Notes

1. Quoted in Harris, *Cannibals and Kings,* p. 101.
2. Harner, "The Ecological Basis for Aztec Sacrifice."
3. Niederberger, "Early Sedentary Economy in the Basin of Mexico."
4. Kirkby, *The Use of Land and Water Resources in the Past and Present Valley of Oaxaca, Mexico.*
5. Flannery, "The Early Mesoamerican House," p. 13–15.
6. Flannery and Winter, "Analyzing Household Activities," p. 36.
7. Pelzer, *Pioneer Settlement in the Asiatic Tropics.*
8. Sisson, "Settlement Patterns and Land Use in the Northwestern Chontalpa, Tobasco, Mexico: A Progress Report."
9. Coe and Diehl, *The Land of the Olmec: The People of the River.*
10. Sanders and Price, *Mesoamerica: The Evolution of a Civilization,* p. 57.
11. Coe, *Mexico,* p. 77.
12. Murdy, "Congenital Deformities and the Olmec Were-Jaguar Motif."
13. Drennan, "Religion and Social Evolution in Formative Mesoamerica," pp. 362–63; Weaver, *The Aztecs, Maya, and Their Predecessors,* p. 52.
14. Stark et al., "El Balsamo Residential Investigations: A Pilot Project and Research Issues."
15. Earle, "A Nearest-Neighbor Analysis of Two Formative Settlement Systems."
16. Sanders, Parsons, and Santley, *The Basin of Mexico: Ecological Processes in the Evolution of a Civilization.*
17. Nichols, "A Middle Formative Irrigation System Near Santa Clara Coatitlan in the Basin of Mexico."
18. Parsons, "The Development of a Prehistoric Complex Society: A Regional Perspective from the Valley of Mexico," p. 91.
19. Whalen, "Zoning Within an Early Formative Community in the Valley of Oaxaca."
20. Whalen, *Excavations at Santo Domingo Tomaltepec: Evolution of a Formative Community in the Valley of Oaxaca, Mexico.*
21. Winter, "The Archaeological Household Cluster in the Valley of Oaxaca."
22. Flannery, "Contextual Analysis of Ritual Paraphernalia from Formative Oaxaca," p. 335.
23. Pyne, "The Fire-Serpent and Were-Jaguar in Formative Oaxaca: A Contingency Table Analysis."
24. Winter, "Tierras Largas: A Formative Community in the Valley of Oaxaca, Mexico."
25. Coe, "The Olmec Style and Its Distribution."
26. Blanton et al., *Ancient Mesoamerica, A Comparison of Change in Three Regions,* pp.

180–83; Flannery, "The Olmec and the Valley of Oaxaca: A Model for Inter-Regional Interaction in Formative Times"; Grennes-Ravits and Coleman, "The Quintessential Role of Olmec in the Central Highlands of Mexico"; Flannery, Marcus, and Kowalewski, "The Preceramic and Formative of the Valley of Oaxaca."

27. Blanton, "Advances in the Study of Cultural Evolution in Prehispanic Highland Mesoamerica," pp. 261–62; Blanton et al., *Ancient Mesoamerica, A Comparison of Change in Three Regions*, p. 67.

28. Blanton, "Cultural Ecology Reconsidered."

29. Blanton et al., *Ancient Mesoamerica, A Comparison of Change in Three Regions.*

30. Ibid. See also Blanton, "Advances in the Study of Cultural Evolution in Prehispanic Highland Mesoamerica," p. 263.

31. Marcus, "The Iconography of Militarism at Monte Albán and Neighboring Sites in the Valley of Oaxaca."

32. Coe, *Mexico*, p. 82.

33. Sanders and Nichols, "Ecological Theory and Cultural Evolution in the Valley of Oaxaca."

34. Kowalewski and Feinstein, "The Economic Systems of Ancient Oaxaca: A Regional Perspective," p. 425.

35. See *Current Anthropology* 29(1):52–80 for a discussion of this point.

36. See, e.g., the works of William Sanders (Sanders, Parsons, and Santley, *The Basin of Mexico: Ecological Processes in the Evolution of a Civilization*), Jeffrey Parsons (Parsons, *Prehistoric Settlement Patterns in the Texcoco Region, Mexico;* idem., "The Development of a Prehistoric Complex Society: A Regional Perspective from the Valley of Mexico"), and Richard Blanton (Blanton, "Prehistoric Adaptation in the Ixtapalapa Region, Mexico"); O'Brien et al., "On Interpretive Competition in the Absence of Appropiate Data."

37. Tolstoy, "Advances in the Basin of Mexico, pt. 1."

38. Cowgill, "Quantitative Studies of Urbanization at Teotihuacan."

39. Ibid.

40. Millon, "The Study of Urbanism at Teotihuacan, Mexico," p. 42.

41. Storey, "An Estimate of Mortality in a Pre-Columbian Urban Population."

42. Parsons, "The Development of a Prehistoric Complex Society: A Regional Perspective from the Valley of Mexico."

43. Cowgill, "Quantitative Studies of Urbanization at Teotihuacan."

44. Kurtz, "The Economics of Urbanization and State Formation at Teotihuacan"; Blanton, "Advances in the Study of Cultural Evolution in Prehispanic Highland Mesoamerica."

45. Brumfiel, "Regional Growth in the Eastern Valley of Mexico: A Test of the 'Population Pressure' Hypothesis."

46. See Tolstoy, "Advances in the Basin of Mexico, pt. 1."

47. Alden, "A Reconstruction of the Toltec Period Political Units in the Valley of Mexico."

48. Siemens and Puleston, "Ridged Fields and Associated Features in Southern Campeche: New Perspectives on the Lowland Maya"; Coe and Diehl, *The Land of the Olmec: The People of the River;* Turner and Harrison, *Prehispanic Maya Agriculture.* Reviewed in Matheny and Gurr, "Variation in Prehistoric Agricultural Systems of the New World"; Cowgill, "An Agricultural Study of the Southern Maya Lowlands"; Flannery, *Maya Subsistence. Studies in Memory of Dennis E. Puleston;* Ashmore, *Lowland Maya Settlement Patterns.*

49. Cowgill, "An Agricultural Study of the Southern Maya Lowlands."

50. Harris, *Cannibals and Kings*, p. 91.

51. Puleston and Puleston, "An Ecological Approach to the Origins of Maya Civilization"; see also Dickson, "Further Simulations of Ancient Agriculture and Population at Tikal, Guatemala."

52. Matheny and Gurr, "Variation in Prehistoric Agricultural Systems of the New

World"; Turner and Harrison, *Pulltrouser Swamp: Ancient Maya Habitat, Agriculture, and Settlement in Northern Belize.*

53. Coe, *The Maya*, p. 61.
54. Sanders and Price, *Mesoamerica: The Evolution of a Civilization.*
55. "Archaeology and Maya Writing," p. 26.
56. Miller, *The Murals of Bonampak.*
57. Marcus, "The Iconography of Power Among the Classic Maya," p. 92.
58. Dickson, "Further Simulations of Ancient Agriculture and Population at Tikal, Guatemala."
59. Folan, Kintz, and Fletcher, *Coba. A Classic Maya Metropolis.*
60. Robertson and Freidel, *Archaeology at Cerros, Belize, Central America. Volume I: An Interim Report.*
61. Chippindale, Hammond, and Sabloff, "The Archaeology of Maya Decipherement."
62. Knorosov, *Maya Hieroglyphic Codices.*
63. Marcus, "The Origins of Mesoamerican Writing," p. 36.
64. Ifrah, *From One to Zero*, p. 404.
65. Marcus, "The Origins of Mesoamerican Writing."
66. See, e.g., Doty, "A New Mayan Long Count-Gregorian Conversion Computer Program."
67. Joyce, "Terminal Classic Interaction on the Southeastern Maya Periphery."
68. Rathje, "Classic Maya Development and Denouement: A Research Design."
69. Fry, *Models and Methods in Regional Exchange.*
70. Matheny, "Maya Lowland Hydraulic Systems."
71. Harris, *Cannibals and Kings: The Origins of Cultures*, p. 91.
72. Lowe, "On Mathematical Models of the Classic Maya Collapse: The Class Conflict Hypothesis Reexamined"; see also Bove, "Trend Surface Analysis and the Lowland Classic Maya Collapse."
73. Webster and Gonlin, "Household Remains of the Humblest Maya."
74. Diego de Landa, *Relacion de las Cosas de Yucatán*, 32 (quoted in Todorov, *The Conquest of America*).
75. Pollard, "Central Places and Cities: A Consideration of the Protohistoric Tarascan State."
76. Freidel and Sabloff, *Cozumel. Late Maya Settlement Patterns.*
77. Smith, "The Aztec Marketing System and Settlement Pattern in the Valley of Mexico: A Central Place Analysis."
78. Parsons, "The Development of a Prehistoric Complex Society: A Regional Perspective from the Valley of Mexico," p. 107.
79. Coe, *Mexico*, p. 151.
80. Parsons, "The Development of a Prehistoric Complex Society: A Regional Perspective from the Valley of Mexico."
81. Soustelle, *Daily Life of the Aztecs*, p. 97.
82. Harner, "The Ecological Basis for Aztec Sacrifice."
83. Soustelle, *Daily Life of the Aztecs*, p. 135.
84. Todorov, *The Conquest of America;* Collier, Rosaldo, and Wirth, *The Inca and Aztec State.*

Bibliography

Adams, R. E. W. 1977. *Prehistoric Mesoamerica.* Boston: Little, Brown.
Adams, R. E., W. E. Brown, Jr., and T. P. Culbert. 1981. "Radar Mapping, Archaeology, and Ancient Mayan Land Use." *Science* 213:1457–63.
Alden, J. R. 1979. "A Reconstruction of the Toltec Period Political Units in the Valley

of Mexico." In *Transformations: Mathematical Approaches to Culture Change,* eds. C. Renfrew and K. L. Cooke. New York: Academic Press.

Armillas, P. 1971. "Gardens on Swamps." *Science* 174:653–61.

Arnold, J. E. and A. Ford. 1980. "A Statistical Examination of Settlement Patterns at Tikal, Guatemala." *American Antiquity* 45:713–26.

Ashmore, W., ed. 1981. *Lowland Maya Settlement Patterns.* Albuquerque: University of New Mexico Press.

Blanton, R. E. 1972. "Prehistoric Adaptation in the Ixtapalapa Region, Mexico." *Science* 175:1317–26.

———. 1976. "The Origins of Monte Albán." In *Cultural Change and Continuity,* ed. C. E. Cleland. New York: Academic Press.

———. 1978. *Monte Albán. Settlement Patterns at the Ancient Zapotec Capital.* New York: Academic Press.

———. 1980. "Cultural Ecology Reconsidered." *American Antiquity* 45:145–51.

———. 1983. "Advances in the Study of Cultural Evolution in Prehispanic Highland Mesoamerica." In *Advances in World Archaeology, Vol. 2,* eds. F. Wendorf and A. E. Close. New York: Academic Press.

Blanton, R. E., S. A. Kowalewski, G. Feinman, and J. Appel. 1981. *Ancient Mesoamerica, A Comparison of Change in Three Regions.* Cambridge, England: Cambridge University Press.

Bove, F. J. 1981. "Trend Surface Analysis and the Lowland Classic Maya Collapse." *American Antiquity* 46:93–112.

Brumfiel, E. 1976. "Regional Growth in the Eastern Valley of Mexico: A Test of the 'Population Pressures' Hypothesis." In *The Early Mesoamerican Village,* ed. K. V. Flannery. New York: Academic Press.

Brush, C. F. 1969. "A Contribution to the Archaeology of Coastal Guerrero, Mexico." Unpublished doctoral dissertation. Columbia University, New York.

Chippendale, C., N. Hammond, and J. Sabloff. 1988. "The Archaeology of Maya Decipherement." *Antiquity* 62(234):119–22.

Coe, M. D. 1965. *The Jaguar's Children: Pre-Classic Central Mexico.* New York: Museum of Primitive Art.

———. 1965. "The Olmec Style and Its Distribution." *Handbook of Middle American Indians* 3:739–75.

———. 1966. *The Maya.* New York: Praeger.

———. 1968. "San Lorenzo and the Olmec Civilization." In *Dumbarton Oaks Conference on the Olmec,* ed. E. P. Benson. Dumbarton Oaks, Washington, D. C.

———. 1970. "The Archaeological Sequence at San Lorenzo Tenochtitlan, Veracruz, Mexico." *Contributions of the University of California Archaeological Research Facility,* n. 8, pp. 21–34.

———. 1984. *Mexico.* New York: Thames and Hudson.

Coe, M. and R. Diehl. 1980. *The Land of the Olmec: The People of the River,* Vol. 2. Austin: University of Texas Press.

Collier, G. A., R. I. Rosaldo, J. D. Wirth. eds. 1982. *The Inca and Aztec State.* New York: Academic Press.

Conrad, G. W. and A. A. Demarest. 1984. *Religion and Empire. The Dynamics of Aztec and Inca Expansionism.* Cambridge, England: Cambridge University Press.

Cook, S. 1946. "Human Sacrifice and Warfare as Factors in the Demography of Precolonial Mexico." *Human Biology* 18:81–102.

Cowgill, G. 1974. "Quantitative Studies of Urbanization at Teotihuacan." In *Mesoamerican Archaeology: New Approaches,* ed. N. Hammond, pp. 363–96. Austin: University of Texas Press.

———. 1976. Public Lecture. Seattle.

Cowgill, U. 1962. "An Agricultural Study of the Southern Maya Lowlands." *American Anthropologist* 64:273–86.

———. 1971. "Some Comments on Manihot Subsistence and the Ancient Maya." *Southwest Journal of Anthropology* 27:51–63.

Culbert, T. P. 1973. *The Classic Maya Collapse*. Albuquerque: University of New Mexico Press.

———. 1988. "Political History and the Decipherment of Maya Glyphs." *Antiquity* 62(234):135–52.

Dickson, D. B. 1981. "Further Simulations of Ancient Agriculture and Population at Tikal, Guatemala." *American Antiquity* 46:922–26.

Doty, D. C. 1979. "A New Mayan Long Count-Gregorian Conversion Computer Program." *American Antiquity* 44:780–83.

Drennan, R. D. 1976. "Religion and Social Evolution in Formative Mesoamerica." In *The Early Mesoamerican Village*, ed. K. V. Flannery. New York: Academic Press.

Earle, T. 1976. "A Nearest-Neighbor Analysis of Two Formative Settlement Systems." In *The Early Mesoamerica Village*, ed. K. V. Flannery. New York: Academic Press.

Ferdon, E. M. 1959. "Agriculture Potential and the Development of Cultures." *Southwestern Journal of Anthropology* 15:1–19.

Flannery, K. V. 1968. "The Olmec and the Valley of Oaxaca: A Model for Inter-Regional Interaction in Formative Times." In *Dumbarton Oaks Conference on the Olmec*, ed. E. P. Benson. Dumbarton Oaks, Washington, D. C.

———. 1976a. "The Early Mesoamerican House." In *The Early Mesoamerican Village*, ed. K. V. Flannery. New York: Academic Press.

———. 1976b. "Evolution of Complex Settlement Systems." In *The Early Mesoamerican Village*, ed. K. V. Flannery. New York: Academic Press.

———. 1976c. "Linear Stream Patterns and Riverside Settlement Rules." In *The Early Mesoamerican Village*, ed. K. V. Flannery. New York: Academic Press.

———. 1976d. "Contextual Analysis of Ritual Paraphernalia from Formative Oaxaca." In *The Early Mesoamerican Village*, ed. K. V. Flannery. New York: Academic Press.

Flannery, K. V., ed. 1982. *Maya Subsistence. Studies in Memory of Dennis E. Puleston*. New York: Academic Press.

Flannery, K. V., J. Marcus, and S. A. Kowalewski. 1981. "The Preceramic and Formative of the Valley of Oaxaca." *Supplement to the Handbook of Middle American Indians*, pp. 48–93. Austin: University of Texas Press.

Flannery, K. V. and M. C. Winter. 1976. "Analyzing Household Activities." In *The Early Mesoamerican Village*, ed. K. V. Flannery. New York: Academic Press.

Flash, W. L., Jr. 1988. "A New Look at Maya Statecraft From Copan, Honduras." *Antiquity* 62(234):157–69.

Folan, W. J., E. R. Kintz, and L. A. Fletcher. 1983. *Coba. A Classic Maya Metropolis*. New York: Academic Press.

Fox, J. W. 1987. *Maya Postclassic State Formation*. Cambridge, England: Cambridge University Press.

Freidel, D. A. and J. A. Sabloff. 1984. *Cozumel. Late Maya Settlement Patterns*. New York: Academic Press.

Fry, R., ed. 1980. *Models and Methods in Regional Exchange*. SAA Papers no. 1. Washington.

Fry, R. and S. Cox. 1974. "The Structure of Ceramic Exchange at Tikal, Guatemala." *World Archaeology* 6:209–25.

Graham, I. 1988. "Homeless Hieroglyphs." *Antiquity* 62(234):122–26.

Grennes-Ravits, R. and G. Coleman. 1976. "The Quintessential Role of Olmec in the Central Highlands of Mexico." *American Antiquity* 41:196–205.

Grove, D. C. 1968. "The Pre-Classic Olmec in Central Mexico: Site Distribution and Inferences." In *Dumbarton Oaks Conference on the Olmec*, ed. E. P. Benson. Dumbarton Oaks, Washington, D. C.

Grove, D. C., ed. 1987. *Ancient Chalcatzingo*. Austin: University of Texas Press.

Harner, M. 1977. "The Ecological Basis for Aztec Sacrifice." *American Ethnologist* 4:117–35.

Harris, M. 1977. *Cannibals and Kings: The Origins of Cultures.* New York: Random House.

Houston, S. D. 1988. "The Phonetic Decipherment of Mayan Glyphs." *Antiquity* 62(234):126–35.

———. 1989. "Archaeology and Maya Writing." *Journal of World Prehistory* 3(1):1–32.

Hunt, R. C. and E. Hunt. 1976. "Canal Irrigation and Local Social Organization." *Current Anthropology* 17:389–411.

Ifrah, G. 1985. *From One to Zero.* Trans. L. Bair. New York: Viking Penguin.

Joyce, R. A. 1986. "Terminal Classic Interaction on the Southeastern Maya Periphery." *American Antiquity* 1986: 313–29.

Kelley, D. H. 1981. *Mayan Culture History as Process.* Ithaca: Cornell University Press.

Kirkby, A. 1973. *The Use of Land and Water Resources in the Past and Present Valley of Oaxaca, Mexico.* Ann Arbor: Memoirs of the Museum of Anthropology, University of Michigan, no. 5.

Kowalewski, S. A. and L. Feinstein. 1983. "The Economic Systems of Ancient Oaxaca: A Regional Perspective." *Current Anthropology* 24(4):413–41.

Knorozov, Y. 1982. *Maya Hieroglyphic Codices.* Trans. by S. D. Cole. Austin: University of Texas Press.

Kurtz, D. V. 1987. "The Economics of Urbanization and State Formation at Teotihuacan." *Current Anthropology* 28(3):329–53.

Landa, D. de. *The Maya Account of Affairs of Yucatan.* Chicago: J. O'Hara.

Lees, S. H. 1973. *Sociopolitical Aspects of Canal Irrigation in the Valley of Oaxaca.* Ann Arbor: Memoir of the Museum of Anthropology, University of Michigan, no. 6.

Lowe, J. W. G. 1982. "On Mathematical Models of the Classic Maya Collapse: The Class Conflict Hypothesis Reexamined." *American Antiquity* 47:643–52.

MacNeish, R. S. 1962. *Second Annual Report of the Tehuacan Archaeological-Botanical Project.* Andover, Mass.: R. S. Peabody Foundation for Archaeology.

———. 1964. "Ancient Mesoamerican Civilization." *Science* 143:531–37.

Marcus, J. 1973. "Territorial Organization of the Lowland Classic Maya." *Science* 180:911–16.

———. 1974. "The Iconography of Power Among the Classic Maya." *World Archaeology* 6:83–94.

———. 1976. "The Size of the Early Mesoamerican Village." In *The Early Mesoamerican Village,* ed. K. V. Flannery. New York: Academic Press.

———. 1976. "The Origins of Mesoamerican Writing." *Annual Review of Anthropology* 5:35–68.

———. 1976. "The Iconography of Militarism at Monte Alban and Neighboring Sites in the Valley of Oaxaca." In *The Origins of Religious Art and Iconography in Preclassic Mesoamerica,* ed. H. B. Nicholson. Latin American Center, U.C.L.A.

Matheny, R. T. 1976. "Maya Lowland Hydraulic Systems." *Science* 193:639–46.

Matheny, R. T. and D. L. Gurr. 1983. "Variation in Prehistoric Agricultural Systems of the New World." *Annual Review of Anthropology* 12:79–103.

Meggers, B. J. 1954. "Environmental Limitation in the Development of Culture." *American Anthropologist* 56.

Miller, M. E. 1986. *The Murals of Bonampak.* Princeton: Princeton University Press.

Millon, R. 1974. "The Study of Urbanism at Teotihuacan, Mexico." In *Mesoamerican Archaeology: New Approaches,* ed. N. Hammond. Austin: University of Texas Press.

Millon, R., R. B. Drewitt, and G. L. Cowgill. 1973. *Urbanization at Teotihuacan, Mexico,* Vol. 1, Parts 1 and 2. Austin: University of Texas Press.

Murdy, C. N. 1981. "Congenital Deformities and the Olmec Were-Jaguar Motif." *American Antiquity* 46:861–69.

Nations, J. D. 1980. "The Evolutionary Potential of Lacondon Maya Sustained-Yield Tropical Forest Agriculture." *Journal of Anthropological Research* 36:1–30.

Nichols, D. L. 1982. "A Middle Formative Irrigation System Near Santa Clara Coatitlan in the Basin of Mexico." *American Antiquity* 47:133–44.

Niederberger, C. 1979. "Early Sedentary Economy in the Basin of Mexico." *Science* 203:131–46.

O'Brien, M. J., J. A. Ferguson, T. D. Holland, and D. E. Lewarch. 1989. "On Intrepretive Competition in the Absence of Appropiate Data: Monte Alban Revisited." *Current Anthropology* 30(2):191–99.

O'Brien, M. J., R. D. Mason, D. E. Lewarch, and J. A. Neely. 1982. *A Late Formative Irrigation Settlement Below Monte Alban*. Austin: University of Texas Press.

Offner, J. A. 1981. "On the Inapplicability of 'Oriental Despotism' and the 'Asiatic Mode of Production' to the Aztecs of Texcoco." *American Antiquity* 46:43–61.

Palerm, A. and E. Wolf. 1957. "Ecological Potential and Cultural Development in Mesoamerica." In *Studies in Human Ecology*. Social Science Monographs 3:1–38.

Parsons, J. R. 1968. "Teotihuacan, Mexico, and Its Impact on Regional Demography." *Science* 162:872–77.

———. 1971. *Prehistoric Settlement Patterns in the Texcoco Region, Mexico*. Ann Arbor: Memoir of the Museum of Anthropology, University of Michigan, no. 3.

———. 1974. "The Development of a Prehistoric Complex Society: A Regional Perspective from the Valley of Mexico." *Journal of Field Archaeology* 1:81–108.

Peebles, C. S. 1988. "From Art to Algebra in Maya Studies." *Antiquity* 62(234):170–72.

Pelzer, R. J. 1945. *Pioneer Settlement in the Asiatic Tropics*. New York: American Geographical Society, Special Publications, no. 29.

Pires-Ferreira, J. W. 1975. *Formative Mesoamerican Exchange Networks with Special Reference to the Valley of Oaxaca*. Ann Arbor: Memoirs of the Museum of Anthropology, University of Michigan, no. 7.

Pollard, H. P. 1980. "Central Places and Cities: A Consideration of the Protohistoric Tarascan State." *American Antiquity* 45:677–97.

Porter, M. N. 1953. *Tlatilco and the Pre-Classic Cultures of the New World*. New York: Viking Fund Publications in Anthropology, no. 19.

Puleston, D. E. and O. S. Puleston. 1971. "An Ecological Approach to the Origins of Maya Civilization." *Archaeology* 24:330–36.

Pyne, N. M. 1976. "The Fire-Serpent and Were-Jaguar in Formative Oaxaca: A Contingency Table Analysis." In *the Early Mesoamerican Village*, ed. K. V. Flannery. New York: Academic Press.

Rathje, W. L. 1971. "The Origin and Development of Lowland Classic Maya Civilization." *American Antiquity* 36:275–85.

———. 1973. "Classic Maya Development and Denouement: A Research Design." In *The Classic Maya Collapse*, ed. T. P. Culbert. Albuquerque: University of New Mexico Press.

Robertson, R. A. and D. A. Freidel, eds. 1988. *Archaeology at Cerros, Belize, Central America. Volume I: An Interim Report*. Dallas: Southern Methodist University Press.

Sahagun, F. B. de. 1976. *A History of Ancient Mexico*. Trans. F. R. Bandelier from the Spanish version of Carlos Maria de Bustamante. Glorieta, New Mexico: Rio Grande Press.

Sahlins, M. 1968. "Notes on the Original Affluent Society." In *Man the Hunter*, eds. R. B. Lee and I. DeVore. Chicago: Aldine.

Sanders, W. T. 1973. "The Cultural Ecology of the Lowland Maya: A Re-Evaluation." In *The Classic Maya Collapse*, ed. T. P. Culbert. Albuquerque: University of New Mexico Press.

Sanders, W. T. and D. L. Nichols. 1988. "Ecological Theory and Cultural Evolution in the Valley of Oaxaca." *Current Anthropology* 29(1):33–80.

Sanders, W. T., J. R. Parsons, and M. H. Logan. 1976. "Summary and Conclusions." In *The Valley of Mexico*, ed. E. Wolf. Albuquerque: University of New Mexico Press.

Sanders, W. T., J. R. Parsons, and R. S. Santley. 1979. *The Basin of Mexico: Ecological Processes in the Evolution of a Civilization.* New York: Academic Press.

Sanders, W. T. and B. Price. 1968. *Mesoamerica: the Evolution of a Civilization.* New York: Random House.

Sheets, P. D., ed. 1983. *Archeology and Volcanism in Central America.* Austin: University of Texas Press.

Siemens, A. H. and D. E. Puleston. 1972. "Ridged Fields and Associated Features in Southern Campeche: New Perspectives on the Lowland Maya." *American Antiquity* 37:228–39.

Sisson, E. B. 1970. "Settlement Patterns and Land Use in the Northwestern Chontalpa, Tabasco, Mexico: A Progress Report." *Ceramica de Cultura Maya* 6:41–54.

Sjoberg, G. 1965. "The Origin and Evolution of Cities." *Scientific American* 213(3):54–63.

Smith, M. 1979. "The Aztec Marketing System and Settlement Pattern in the Valley of Mexico: A Central Place Analysis." *American Antiquity* 44:10–24.

Soustelle, J. 1961. *Daily of the Aztecs.* Trans. P. O'Brian. Stanford: Stanford University Press.

Stark, B., L. Heller, F. W. Nelson, R. Bishop, D. M. Pearsall, D. S. Whitley, and H. Wells. 1985. "El Balsamo Residential Investigations: A Pilot Project and Research Issues." *American Anthropologist* 87(1):100–11.

Steponaitis, V. P. 1981. "Settlement Hierarchies and Political Complexity in Non-market Societies: The Formative Period in the Valley of Mexico." *American Anthropologist* 83:320–63.

Storey, R. 1985. "An Estimate of Mortality in a Pre-Columbian Urban Population." *American Anthropologist* 87(3):519–35.

Thomas, P. M. 1981. *Prehistoric Maya Settlement Patterns at Becan, Campeche, Mexico.* Publication 45, Middle American Research Institute, Tulane University, New Orleans.

Todorov, T. 1984. *the Conquest of America.* Trans. R. Howard. New York: Harper & Row.

Tolstoy, P. 1981. "Advances in the Basin of Mexico, pt. 1." *The Quarterly Review of Archaeology* 2:33–34,36.

Tolstoy, P. and A. Guinette. 1965. "Le Placement de Tlatilco dans le Cadre du Pre-Classique du Basin de Mexico." *Journal de la Societe des Americanistes.* (Paris) 54:47–91.

Tolstoy, P. and L. Paradis. 1970. "Early and Middle Preclassic Culture in the Basin of Mexico." *Science* 167:344–51.

Turner, B. L. 1974. "Prehistoric Intensive Agriculture in the Maya Lowlands." *Science* 185:118–24.

Turner, B. L. and P. D. Harrison, eds. 1978. *Prehispanic Maya Agriculture.*

———. 1984. *Pulltrouser Swamp: Ancient Maya Habitat, Agriculture, and Settlement in Northern Belize.* Austin: University of Texas Press.

Urban, P. A. and E. M. Schortman, eds. 1986. *The Southeast Maya Periphery.* Austin: University of Texas Press.

Weaver, M. P. 1972. *The Aztecs, Maya, and Their Predecessors.* New York: Seminar Press.

Webster, D. and N. Gonlin. 1988. "Household Remains of the Humblest Maya." *Journal of Field Archaeology* 15(2):169–90.

Whalen, M. E. 1974. "Community Development and Integration During the Formative Period in the Valley of Oaxaca, Mexico." Paper read at the Annual Meeting of the American Anthropological Association, Mexico City.

———. 1976. "Zoning Within an Early Formative Community in the Valley of Oaxaca." In *The Early Mesoamerican Valley,* ed. K. V. Flannery. New York: Academic Press.

———. 1981. *Excavations at Santo Domingo Tomaltepec: Evolution of a Formative Community in the Valley of Oaxaca, Mexico.* Ann Arbor: Museum of Anthropology, University of Michigan, no. 12.

Winter, M. 1972. "Tierras Largas: A Formative Community in the Valley of Oaxaca, Mexico." Unpublished doctoral thesis, University of Arizona.

————. 1976. "The Archaeological Household Cluster in the Valley of Oaxaca." In *The Early Mesoamerican Village,* ed. K. V. Flannery. New York: Academic Press.

Wolf, E. 1959. *Sons of the Shaking Earth.* Chicago: University of Chicago Press.

Wolf, E., ed. 1976. *The Valley of Mexico.* Albuquerque: University of New Mexico Press.

14

Andean Civilizations

In the lands assigned to Religion and to the Crown, the Inca kept overseers and administrators who took care in supervising their cultivation, harvesting the products and putting them in storehouses. The labor of sowing and cultivating these lands and harvesting their products formed a large part of the tribute which the taxpayer paid to the king. . . . The people assembled to cultivate them in the following way. If the Inca himself . . . or some other high official happened to be present he started the work with a golden [spade] . . . and following his example, all did the same. However the Inca soon stopped working, and after him the other officials and nobles stopped also and sat down with the king to their banquets and festivals which were especially notable on such days.

The common people remained at work . . . each man put into his section his children and wives and all the people of his house to help him. In this way, the man who had the most workers finished his *suyu* first, and he was considered a rich man; the poor man was he who had no one to help him in his work and had to work that much longer. . . .

Father Bernabe Cobo (c. A.D. 1639)

When European explorers reached Peru and the other countries of Andean South America, they found a civilization of great wealth and complexity. As Father Cobo described in the above passage, the Inca Empire was rigidly hierarchical, constituted in the same socioeconomic class structure that we have seen in other early political systems. The Spanish *Conquistadores*, however, found the Inca Empire on the verge of revolution, with long-subjugated ethnic groups and provinces plotting ways to overthrow the central government.

As with other New World political systems, one has to wonder how it would all have turned out had the Spanish not "beheaded" this civilization in the sixteenth century through murder, warfare, and the spread of European diseases. The Spanish themselves did not know about the origins of smallpox and other diseases that so terribly afflicted aboriginal Ameri-

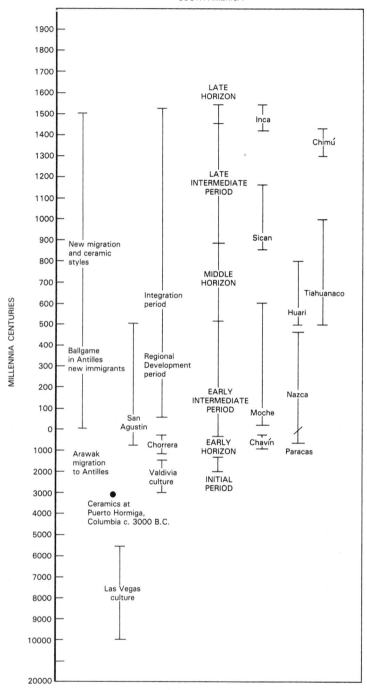

SOUTH AMERICA

14.1 A cultural chronology of South America. People probably were in South America before 10,000 B.C. but the evidence is controversial.

cans, but their slavery and intentional destruction of South American civilizations makes it difficult to maintain a dispassionate anthropological perspective.

Andean South American civilizations paralleled other early complex cultures in their intensive farming systems, massive pyramids and temples, large cities, powerful armies, and hierarchies of wealth, power, and prestige. Yet they were interestingly different from those of Europe and Asia in, for example, that they were the only empire of this scale ever to function without a written language. Yet the civilizations of Andean South America were the largest political systems ever to evolve in the Precolumbian New World, and they left a rich legacy of pottery, gold-smithing, textiles, and stone architecture.

The Ecological Setting

[E]cological complementarity was a major human achievement, forged by Andean civilizations to handle a multiple environment, vast populations, and hence high productivity. It helps us understand the unique place of the Andean achievement in the repertory of human histories; it may even point to future possibilities.(John Murra)[1]

Like all the other areas where complex societies developed independently, Andean South American civilization was possible because of—indeed, to an extent can be defined as—the economic integration of the resources from highly varied physical environments.[2] The Andes rise so sharply from the Pacific that only a thin strip of land, less than about 60 kilometers at its widest point, separates the mountains from the sea. And because the Andes shield the coast from the rain-bearing air currents crossing the continent from the Atlantic, most of this coastal strip is one of the world's driest deserts, where rain falls only once or twice every five years. In a few places winter fogs along the coast keep skies overcast and, in most years, provide enough moisture through condensation to support vegetation zones (called *lomas*). But most of the coastal strip is utterly dry, and when the wind blows, dunes can quickly cover houses and choke irrigation canals.

This desert is habitable only because of the fifty or so small rivers that flow down from the mountains, across the plain, and into the sea. Most contain water during only part of the year, but the larger, permanent ones support forests and shrubs and their attendant wildlife, and there are areas where the rivers keep the water table sufficiently high so that cultivation is possible without irrigation. In some valleys, rivers have created broad alluvial plains of potentially rich farmlands. Near the mouths of the rivers are fish, freshwater shrimp, and other resources. Human life along the coast is tied directly to these rivers and streams because they provide the only drinking water. The coast has extraordinarily rich concentrations of fish, birds, birds' eggs, sea mammals, mollusks, crustaceans, kelp, and other plant foods.

These rich marine resources are produced by a fascinating interplay of wind and ocean currents, where winds drive water north along the coast while the earth's rotation from east to west pushes the water westward, creating an upwelling of water from the ocean floor. Carried with these deep waters are tremendous concentrations of phosphates and other nutrients that support countless billions of microscopic plants, and these form the basis of a complex food chain comprising anchovies and other small fish that eat the plants; larger fish, birds, and sea mammals that eat the anchovies; and, ultimately, people, who exploit many links in the chain.

Occasionally, shifts in wind and water change water temperatures and the plants die, cutting off the base of the food chain, and when this happens rotting plant and animal life fills the air with clouds of hydrogen sulfide that can blacken ships and houses.[3] Several years may pass before the fertility of the sea is restored. The frequency in prehistory of *el Niño* (a reference to the Christ child), as this disturbance is called, is unknown, but it has occurred several times in the last half-century, and, assuming approximately the same conditions in ancient times, *el Niño* may have been a limiting factor on human population densities in coastal Andean South America.

In the mountains are lush valleys, large basins, and high grassy plateaus (called *punas*). Hunters and gatherers here were succeeded after 1800 B.C. by farmers of potatoes, maize, quinoa, and other crops. For much of Andean South American prehistory, these *punas* were inhabited to 4500 meters and higher by hunters and gatherers, who were later replaced by llama and alpaca herders.

The eastern slopes of the Andes, the *montaña,* are wet and heavily forested, and the combination of steep slopes and intense rain apparently limited exploitation by prehistoric peoples. East of the Andes is the Amazon Basin, a tropical rainforest whose rubber, feathers, and other products were brought into early Andean South American economic systems, but which was never controlled by Andean peoples.

Early Hunters and Gatherers

A few South American sites have been dated (controversially) to 20,000 to 15,000 years ago, but not until about 10,000 years ago is there substantial evidence of people in the mountains and coasts of Andean South America. John Rick has surveyed large areas of these uplands, and in caves and rock shelters he has found projectile points, scrapers, knife blades, and other traces of these early Peruvians; they ate a lot of deer, guanaco, and vicuna (animals related to the camel), in some cases perhaps were even able to live year round in small areas.[4] In the beginning they also hunted giant ground sloths and a few other animals that became extinct at about 10,000 years ago.

Some of them were probably "transhumant," meaning that they moved up and down the mountains to exploit various resources as they came in season. Many people made these seasonal moves once alpacas and llamas were domesticated, because these animals require constant tending and frequent moves to new pasturages. The "thin" air, intense cold, blizzards, and thick fogs of the highlands make movement difficult, and over millennia of adapting to these conditions, natural selection has produced Andean peoples with extraordinary cardiovascular systems. Genetics and lifelong exposure to the strains of life at high altitudes produce people who can work hard in air so low in oxygen that people unadapted to this environment can hardly function.

As indicated in Chapter 6, already by 7500 B.C., tubers, beans, fruits, peppers, guanacos, and llamas were staples of the highland Andean diet.[5]

Maize—perhaps introduced as a crop from Central America—was being eaten by at least 3200 B.C., and in the Andes as in Mesoamerica there seems to have been a period of thousands of years in which maize was just one of several plant foods. The maize from levels of Guitarrero Cave dating to between 5000 and 3000 B.C., as well as maize remains from the Ayacucho region are of a size and uniformity suggesting that maize and people were already somewhat dependent on each other.

Several highland sites suggest that already by about 4000 B.C. guinea pigs had been domesticated—if they were not domesticated by this time, then these Andean peoples were formidable hunters of these rodents, for guinea pig bones are thickly spread through many layers of occupational refuse. Guinea pigs do so well in captivity, are so prolific and easy to feed, and are so suitable in terms of size for the modern diet, that one wonders why international food conglomerates have not renamed and successfully marketed them.

But potatoes were the primary staple in many Andean areas. Potatoes do not preserve well in modern storage, and much less so in the archaeological record, but some remains have been found in Pikimachay Cave in levels dating to between 4550 to 3100 B.C.

We may never know if coastal Andean South America was occupied before or at the same early date as the mountains, because long stretches of what were beaches and inland areas before 3000 B.C. have been flooded by rising sea levels.[6] In areas that escaped flooding there is evidence of communities as early as 10,000 years ago, so some scholars suspect that coastal populations may have been substantial at an early date.[7]

One reason archaeologists are so fixated on analyses of diets of ancient peoples is that usually they are assessing the possibility that changes in diet were associated with major cultural changes. What is particularly intriguing about Andean South America in this regard is the possibility that the earliest phases of its coastal civilization were based not primarily on agriculture, but on intensive fishing and foraging along the rich Pacific beaches.

As usual in such analyses, we have two contending schools of thought and not enough data to evaluate either argument. We know that by at least 6000 B.C. quite a few people lived on the Pacific coast, probably moving between the river valleys, the *lomas* (fog-oases), and the coast.[8] For several millennia thereafter, these societies seemed to have changed little, as they adapted to this rich complex of environments. Anyone who has had the good fortune to live at the interface between sea, plains, and mountains where oysters, clams, and fish can be combined with the wealth of terrestrial foods can appreciate the stability and diversity of such economies, and the Andean coast is among the richest such zones in the world.

Recent excavations at Paloma,[9] on the coast of the Chilca Valley, 65 kilometers south of Lima, illustrate some aspects of life in such regions. Paloma was located about 3.5 kilometers from the ocean and 7.5 kilometers from the Chilca River, on the edge of the *lomas* zone, where fogs supported some vegetation. Beginning about 8000 years ago, people were living there in reed huts, probably already putting together a varied economy of plant collecting, hunting, and foraging.

But shortly before 3000 B.C., people apparently began to abandon the *lomas* to concentrate on the interface of coast and river valley, and small, dispersed settlements began to appear in these zones.[10] By about 2500 B.C. many small sedentary communities had appeared along the Andean coast. In some cases, the diet seems to have been based mainly on marine resources, with wild or domestic plants of only secondary importance,[11] but there was great variability from site to site.

At Alto Salaverry,[12] which was occupied from about 2500 to 1800 B.C., people cultivated and subsisted mainly on plants, but a lot of the calories in their diet came from sharks, bonito, mussels, and many other types of marine foods. In several communities of this period skeletons were found that show a pattern of bone growth in the inner ear that is common among people who spend a lot of time diving in cold water.[13]

At Chilca, the primary meat source appears to have been sea lions, but mussels, other invertebrates, and a variety of collected plants were also important. At other sites the remains of sharks, rays, cormorants, gulls, pelicans, and other animals attest to the importance of the resources of the coastal shallows, as does the presence of fish hooks, nets, and lines. No boats have been found at any of these sites, but the kinds of fish and invertebrates usually eaten along the coast are easily taken with simple nets.

By about 2000 B.C., at least a hundred communities dotted the Andean coastline, many of them on river deltas and bays. Most of these seem to have been quite simple foraging-based communities, and few had more than several hundred inhabitants. At Chilca the burials of thirty adults and twenty-two children and adolescents had only minor differences in grave goods or positioning of the corpses. Some people were interned with spindles and spindlewhorls, others with fish hooks and lines, still others with cotton and weaving tools or a pointed stick and spatula kit that may have been used in shellfish gathering.

Once the Andean coast was fairly thickly settled, people here yielded to the same impulses as did peoples in Egypt, Mesopotamia, China, and other early developmental centers: they began building more and more monumental architecture. It is absurd to think that people generally have some innate desire to build big stone buildings, but in Andean South America, too, no sooner had people devised economies of some reasonable reliability and richness than they began "wasting" massive amounts of their wealth in temples, tombs, and pyramids. A principal example is El Paraiso (also known as Chuquitanta), on the banks of the Rio Chillon, about 2 kilometers inland from the sea. This site is dominated by a large "temple" complex consisting of a central large structure flanked by two protruding wings, with the whole complex enclosing a large patio. There are eight or nine distinct structural units at the site, and the two largest complexes were built on artificial mounds more than 250 meters in length and 50 meters in width, rising to a height of over 5 meters. It has been estimated that there are over 50 hectares of such constructions at this site.[14] One complex of rooms, over 450 meters long, is among the largest buildings ever constructed in ancient Andean South America.[15]

The significance of El Paraiso is difficult to assess. Most of the occupational refuse at the site has been destroyed by centuries of plowing. The paucity of residential remains at the site suggests it may have been mainly a ceremonial center (Moseley[16] estimates its population at 1500 to 3000 people), but it is difficult to determine how much of the site was occupied at a single time or how long it took to build the whole complex.[17]

There are other early monumental buildings along the coast in the Preceramic period, and most seem to have had a clear ritual purpose. Their alignment with mountain peaks, small inner rooms, internal burials, and other characteristics suggest ritual purposes, although like other early ceremonial structures around the world, they probably were centers for economic exchange and administration. Religion and economy, as we have noted, are inextricably intertwined throughout early human history.

Michael Moseley has argued at some length that early Andean South

American complex societies, as represented by El Paraiso and similar sites, were initially based on an essentially nonagricultural economy.[18] If so, this would be interesting but would not alter our basic understanding of cultural complexity, for it is clear that what is important in cultural evolution at this stage is overall productivity of an environment rather than the specific ways in which it is productive.

David Wilson challenged the idea that most of the calories that powered Andean South America's first complex societies were from maritime foraging linked with inland hunting and gathering.[19] On the basis of archaeological data and a complicated statistical estimate of resources availability, Wilson argues that the largest Formative period Andean South American sites, like El Paraiso and Huaca Prieta, are neither large enough nor internally complex enough to be considered representative of state-level societies. He concludes that even if we consider these large settlements to be the center of chiefdoms, the evidence suggests they were based on maize agriculture.[20]

Quilter and Stocker show that at most times marine resources potentially could have supported large sedentary communities along the Andean South American coast in the Preceramic period, but they note that a key question here is the *reliability* of marine resources.[21] With *el Niño* as a periodic threat, perhaps maize agriculture would have represented a more stable (if somewhat more boring) resource than the riches of the sea. Maize, especially if grown with irrigation canals carrying reliable spring runoff waters from the melting snows of the Andes, would have the advantages of great reliability, productivity, and storability—at least compared to oysters and anchovies. But as Quilter and Stocker point out, fish can be dried or converted to a paste that has at least some shelf life. People who know anchovies only from the oversalted horrors on pizzas may underestimate the appeal and nutritive potential of this staple of Andean South American life, which was served fresh, dried, and converted into the kind of paste that, as every cook knows, is a marvelous addition to many tomato-based sauces. Sea mammals, too, can be rendered into storable oil.

Although the evidence from Paloma (see above) indicates almost total reliance on marine animal products rather than hunted or domesticated animals,[22] with little evidence of an agricultural economy, we still have far too little data to resolve this debate.

Sheila and Thomas Pozorski's history of the Casma Valley, on the north central coast, between 2500 and 200 B.C. suggests that the coastal areas developed complex societies earlier than the highlands, but both highlands and coastal areas were to some extent linked culturally and changing in the same direction.[23]

Quilter and Stocker argue that the highlands and lowlands were complexly interrelated culturally during the Preceramic Period.[24] Potatoes, cereals, fish, and many other remains were probably traded in substantial quantities, but in addition, there is a similarity in the style of shell ornaments and other artifacts that suggests "a common ideological system that was shared by populations from the Pacific shore to the tropical forest, and that accompanied economic interactions."[25]

As noted above, between about 5500 and 4000 B.C., people at Ayacucho and elsewhere in the highlands were probably subsisting on several species of domestic plants and animals, including llamas, guinea pigs, gourds, squash, quinoa, amaranths, and chili peppers.[26] Domestic maize is found on the coast by about 2000 B.C. and, if this plant were brought into Andean South America from Mesoamerica by way of the Andes highlands, we would expect future research to locate maize remains in the these areas to date to long before 2000 B.C.

People in the Andean highlands also built monumental structures. At Kotosh, a site at about the 1800-meter level on the eastern slopes of the Andes, a large temple complex may have been begun before 1800 B.C. The earliest structure was built on a stone-faced platform some 8 meters high. At least ten superimposed building levels were found at Kotosh, and the intensity of occupation and large scale of these constructions probably reflect the emergence of a chiefdom.

The earliest of these "temples" may even have been erected by non-agriculturalists.[27] Most of the later constructions appear to have been the work of maize farmers who relied heavily on domesticated animals, but the paucity of habitation refuse near these massive stone constructions may indicate that they were largely ceremonial, perhaps acting as focal points of trade and social exchange between widely spaced agricultural groups.[28]

The Initial Period (1800–900 B.C.)

As with other early civilizations, few aspects of the rise of the first Andean states seem clear or simple. Scholars dispute every aspect of this developmental pattern, from its chronology to its economic basis. But the outlines of Andean evolutionary developments have been somewhat clarified in the past decade.[29] Between 1800 and 900 B.C., the way of life in Andean South America changed drastically. Maize—possibly indigenous, but more likely an import from the north via the Andean South American highlands—was brought into intensive cultivation along the coast. In the highlands, where maize does not do well, tubers, quinoa, and other crops were the primary foods. Excavations at sites dating to shortly after 1800 B.C. show that along the coast people were probably eating fewer fish, shellfish, and other littoral, foraged foods, and increasing their consumption of maize, manioc, sweet potatoes, beans, peanuts, and other crops. The domestication of the llama around 1800 B.C. provided meat and wool as well as important transport power—something totally absent in Mesoamerica. The increased importance of containers in agricultural societies is reflected in the first wide-scale distribution of pottery in Andean South America shortly after 1800 B.C., and there seems little need to explain early Andean South American ceramics—as some have—in terms of contacts with Japanese fishermen or other sources. Archaeology has no record of people who, having made the transition to agriculture with a concurrent need of containers, failed to learn how to make them.

These various subsistence and technological changes were paralleled by changes in the distribution of settlements, as many of the fishing communities along the coast were slowly abandoned, and people moved inland to take up agriculture. The distribution of settlements in the inland valleys and the botanical remains found at these sites suggest that simple irrigation canals were being constructed from about 1800 to 1200 B.C. to grow maize, squash, legumes, beans, sweet and white potatoes, and peanuts.[30]

The transition to inland settlement and an agricultural economy required major social and technological changes, in the form of a technology for irrigation, ground preparation, harvesting, and storage, as well as organized, coordinated labor groups.[31]

Pyramids were still being built close to the shore, but the largest constructions were now inland. At Cerro Sechin, for example, sometime before about 100 B.C. a platform mound was built that stands over 30 meters high and is 550 by 400 meters at the base, making it one of the largest buildings of its type anywhere in Andean South America.[32] Carvings and sculpture at Cerro Sechin seem to be scenes of soldiers being killed, trophy heads, and other designs[33] that remind us that war and conquest seem to be as much a part of early cultural complexity as pottery making.

The Early Horizon Period (900–200 B.C.)

As we saw in the cases of Egypt, Mesoamerica, and elsewhere, "art" is not just a minor peripheral part of the rise of ancient civilizations: art was an integral part of this process. In all early states the cities, rank-weaith hierarchies, and functionally interdependent economies were presaged by the spread of an art style, usually expressed mostly directly in pottery.

Peru was no exception to this pattern. After about 900 B.C. people living at Chavín de Huantar (Figure 14.2) and other sites in the highlands of northern Andean South America began to use the same styles of decoration in their pottery, architecture, and other artifacts, and over suc-

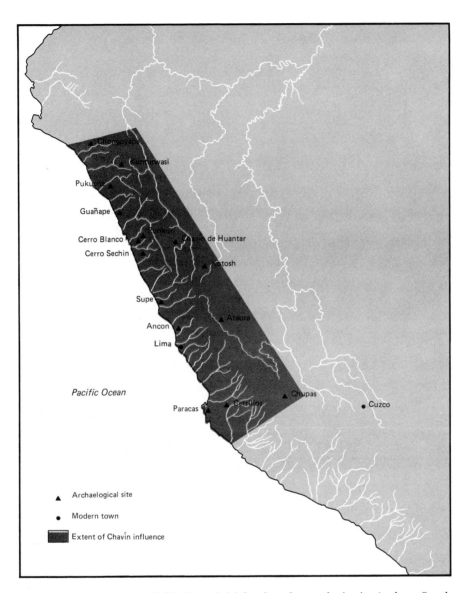

Chongoyape

Kunturwasi

Pukuch

Guañape

Cerro Blanco

Pankorú

Chavín de Huantar

Cerro Sechin

Kotosh

Supe

Ataura

Ancon

Lima

Pacific Ocean

Chupas

Cuzco

Paracas

Curpilos

▲ Archaelogical site

● Modern town

Extent of Chavín influence

14.2 As in other early civilizations, initial cultural complexity in Andean South America was prefaced by the rapid spread of particular styles of pottery and other artifacts. Chavín-style artifacts have been found throughout the area indicated here.

14.3   One of the traditional Chavín motifs was this feline god.

ceeding centuries tens of thousands of people over a great area were participating in the *Chavín Horizon,* as this complex of stylistic elements was called. The main motifs of Chavín art are decidedly nonrepresentational, bizarre depictions of hybrid combinations of people and jaguars, as well as snakes, bats, fish, crabs, and crocodile-like figures. These motifs spread about the same time in some areas as apparent increases in product exchange, activity specialization, population densities, and investments in monumental buildings, but the general tenor of the Chavín diffusion is reminiscent of the initial spread of Olmec art in Mesoamerica—a relatively simple extension of aesthetic and perhaps religious traditions, in the absence of elaborate political hiearchies or economic elites.

Chavín de Huantar, the site after which the art style is named (though it is probably not be the earliest or even most important Chavín settlement), boasts a ceremonial complex composed of two low platform mounds, a massive terraced platform and a sunken court, 48 meters on a side and paved with stone. Interior support for the Castillo is provided by skillfully combined adobe walls and cut stone beams, and many interior rooms appear to have been decorated with painted designs. Some rooms apparently served as repositories for offerings, since hundreds of finely crafted ceramic pots were found there, as well as masses of llama, guinea pig, and fish bones and seashells imported from distant coasts.[34]

Gold was the medium of the finest art in the Chavín era, as craftsmen cut, embossed, annealed, cast, and welded it into ear spools, nose ornaments, plaques, crowns, and face coverings for corpses. Copper and silver were also extensively used for making ornaments, and weaving became a fine art as well.

The central religious symbols of Chavín were widely distributed over the northern highlands and the northern and central coasts, and Chavín-style ceramics and architecture are found even in small villages.

The Chavín cult, like other such early religions, was probably an effective means of stimulating people to act in coordinated ways, because the expenses of large buildings and the "furs and feathers" of office are cheap, compared to their power in directing the population toward specific economic and political goals. The Amazonian animals typical of Chavín designs and the distribution of these designs suggest that the center at Chavín de Huantar may have been an intermediate point on routes connecting the coasts with the exotic, rich world of the interior Amazon. Llama herders from the mountains, maize farmers from the valleys, fishermen, artisans, and other specialists may have derived from the Chavín cult the kinds of administration and political support and integration that set the stage for the evolution of state societies.

There is some evidence in the form of fortresses, however, that the Chavín period may have been more violent than originally thought: redistribution centers and a unified religious iconography do not necessarily mean an era of ecumenism and peace.[35]

The Early Intermediate Period (200 B.C.–A.D. 1000)

In the first millennium A.D., Andean South American societies were transformed from relatively simple, small political units that we might call chiefdoms, into much much larger and more populous militaristic cultures that we can legitimately term states.[36] Within this period the population of Andean South America rose from a few hundred thousand to approximately four or five million, large cities appeared in scores of places, armies conquered thousands of square kilometers, irrigation systems brought rich harvests to desert and mountains, and the ceramic, architec-

14.4   Early Andean South Americans had elaborate textile arts and crafts, as il-lustrated by this cloth from the Paracas Culture of the first millennium B.C.

tural, metallurgical, and textile arts (Figure 14.4) reached such heights that archaeologists have traditionally referred to this period as the Classic.[37]

This transformation seems to have arisen out of the disintegration of the Chavín cult at about 200 B.C., which was followed by the emergence of as many as fifteen different centers of regional development. Ceremonial centers can be found in many places in the southern Andean South American highlands at this time, as well as in the Nazca Valley and other coastal regions. Less work has been done on the residential areas of these centers than on the monumental architecture, and while it appears that some of these were sizable towns, such as Tambo Viejo in the Acari Valley (Nazca area), which contains hundreds of rectangular rooms, most of which seem to have been residences, Helaine Silverman has shown that the extent to which the Nazca state was urbanized has probably been exaggerated.[38] Also numerous along the northern coast were great fortresses of terraced adobe platforms with room complexes and defensive peripheral walls.

The complexity of Classic period agriculture is well illustrated by the irrigation system of the Mochica state, where mud canals were built high in the hills, diverting water through kilometers of canals that snaked along the mountainside, down to the valleys. Because the Mochica worked only with mud, the construction of these canal systems had to be done with great precision; for if the water flowed too slowly, silt would accumulate so rapidly as to make the canal a vast waste of effort, or, if it flowed too fast, the whole system could be eroded. Cleaning the wind-blown sand from these systems probably required the annual orchestration of thousands of laborers.

We know much about life in Andean South America during this period because the people recorded their activities in great detail in ceramics, sculpture, paintings, and tapestries. Pottery vessels depict people hunting deer with spears and clubs, fishermen putting to sea in small canoes, blowgun hunters taking aim at birds, weavers working under the direction of a foreman, and many people engaging in war, human sacrifice, and violence. People are also shown being carried on sedan chairs, seated on thrones, receiving tribute, and presiding at executions.

But most famous of all the aesthetic expressions of this period are the frank depictions—usually in pottery—of sexual practices. While every conceivable sexual variation is amply illustrated, the vast majority involve acts still considered illegal in the state of Georgia. Pots representing sexual themes in the most explicit terms may have been used in ordinary daily life, and to drink from many of them is to perform, symbolically at least, somewhat nontraditional sexual acts. If the sexual practices depicted in pottery are in any way a reflection of the proclivities of the people—and reports of the Spanish and the Inca suggest this was the case—then the Mochica may have devised a very efficient system of birth control.[39] For although there are few depictions of homosexuality (most involving lesbian relationships), procreative acts of sexual intercourse are much less celebrated than nonprocreative acts in this pottery.

Another central element of life in these early Andean South American states was warfare. Every well-surveyed coastal valley has been found to have fortresses and fortified settlements dating to this period, and weapons are common in these sites, particularly along the southern coast. The art of this period is also full of bellicose themes, with depictions of warriors, battle scenes, and mutilations. Trophy heads and mummified corpses showing signs of violence are frequently found in cemeteries.

In the Nazca Valley people created huge figures on the desert floor (Figure 14.5) by turning over patinated stones. Interpretations of these have ranged from the absurd (Erich von Daniken suggests they were beacons for ancient astronauts) to the economic (make-work projects to keep idle peasants employed in the off season).

The capital of Moche seems to have been flooded following an earthquake that changed the course of the Moche River between A.D. 400 to 500.[40] At about A.D. 600, the many rival "states" of Classic Andean South America began to give way to several larger competing political systems, one centered at Huari in the Manteco Basin, another at Tiahuanaco, at the southern end of Lake Titicaca, and a third in the Moche-Chimú area. In these and perhaps other areas, wars of conquest brought large territories under centralized, hierarchically organized governments and lessened regional isolation.[41]

Tiahuanaco is one of the first and largest "states" to be based in large part on potatoes, which were intensively cultivated with other crops on fields reclaimed from the lake marshes. The people of Tiahuanaco included master stonecarvers, whose monumental gates, statues, and other buildings are some of the most impressive in all Andean South America.

Huari existed as a political system for only a century or two, but at its high point it carried out political and economic activities over most of the coast and highlands between Cajamarca in the north and Sicuani in the south. The evidence for the Huari "empire" comes mainly from the distribution of specific art styles and religious symbols over a wide area of

14.5 Elaborate huge designs were created in Peruvian deserts by turning over patinated rocks and arranging them. These designs would not have been visible from ground level. Scholars have suggested they were calendrical or religious symbols.

the central highlands. Significantly, these motifs show up most frequently in the burials of individuals whose associated mortuary goods appeared to reflect particularly high status. The city of Huari expanded to an impressive 10 square kilometers, making it one of the largest residential sites in the ancient New World.

It is probably significant that some of ancient Peru's major roadways may have been constructed during this period, for such roads would have been very important in facilitating the exchange of goods and services over an area as large as that apparently administered from Huari.

Huari and Tiahuanaco may have been fierce enemies or dual capitals of a single political polity—the evidence is not clear.

Regional States (A.D. 1000–1476)

With the collapse of the Huari and Tiahuanaco political systems between A.D. 800 and 1000, at least seven different areas became power centers in Andean South America, the best known and most developed of which was the Chimú state centered in the Moche Valley on the northern coast. A capital of the Chimú political system was the beautiful city of Chan-Chan, a planned settlement covering nearly 8 square kilometers—one of the largest Pre-Columbian cities in the New World. It was divided into ten rectangular sectors, each containing houses, terraces, reservoirs, parks, roads, and public buildings. By the time Chan-Chan was built, Andean societies were rigidly stratified. Fiedel notes that in one area of Chan-Chan the skeletons of between two hundred and three hundred young women were found, "probably members of the royal harem, [who] were sacrificed either at the time of the deceased ruler's funeral or at later commemorative ceremonies."[42]

Goldworking and silverworking, ceramics, weaving, and sculpture were all highly developed crafts. Chimú society seems to have been rigidly stratified according to wealth and prestige, and the extension of political and economic control appears to have been based on a highly efficient army.[43]

Perhaps the most significant development in Andean South America during this period (A.D. 1000–1476) was the multiplication of urban centers. Much of southern Peru remained largely rural, but in the northern half of the country some of the greatest cities of the preconquest period were built.[44]

The Imperial Transformation (A.D. 1476–1525)

The largest and most highly integrated ancient political system ever to appear in the New World evolved in Andean South America within the space of only eighty-seven years. Centered in the Cuzco Valley, the Inca Empire (now often spelled "Inka," and more properly known as the Empire of Tahuantinsuyu ["world of the four quarters"]) eventually stretched from Colombia to central Chile and from the Pacific to the eastern jungles, tying together under the administration of a single royal lineage many diverse regional economic and political systems. At its height, as many as six million people may have been living under Inca rule in one of the most intricately ordered societies of all time.[45]

Native and Spanish accounts say that the Inca began their rise to power out of the dissolution of the many small competing Andean South American states of the thirteenth and fourteenth centuries A.D. The people of Cuzco were attacked by a rival state at about A.D. 1435 and managed to prevail. Succeeding monarchs at Cuzco added new provinces to the empire by conquest, treaty, and simple annexation. The oral histories of the Inca—recorded by the Spanish—speak of military campaigns in which Inca kings smashed the rival power of Chan-Chan in the 1460s, put down large-scale revolts in the 1470s, and greatly expanded the empire in the 1480s.[46]

The economic basis of the Inca Empire (Figure 14.6) was a highly integrated system of fishing, herding, and farming. Rivers were channeled through stone-lined canals, and lowland irrigation systems, which had existed for thousands of years, were extended and brought under a centralized authority. Llamas and alpacas were raised for wool, while dogs, muscovy ducks, and guinea pigs provided most of the meat. But the staple foods were maize, beans, potatoes, quinoa, oca, and peppers.

The food storage methods used by the Inca were very important in establishing imperial food reserves. Potatoes were alternately dried and frozen to produce a black, pulpy product called *chuño*, meat was turned into jerky, and grain was brewed into *chica*, a nutritious beer. Archaeologists familiar with this combination say it is not as bad as it sounds.

The people of the empire were complexly organized according to a decimal system in which there were administrators for every unit of taxpayer from 10 to 10,000. Most people were members of large kin groups, called *ayllu;* marriages were between members of the same *ayllu*. The *ayllu* were usually economically self-sufficient units, and were bound together by complex patterns of reciprocal obligations, such as requiring members to work each other's lands when one was absent and to support widows and the infirm. Farmers worked a certain amount of time on state-owned plots, while craftsmen and specialists, such as runners, weavers, and goldsmiths, contributed according to their particular talents.

Records of taxes, transactions, and census figures were kept with the aid of the *quipu*, a set of strings tied into knots at different levels (Figure 14.7) according to a decimal notation system that could be used by a special hereditary class of accountants to memorize the information.[47] A writ-

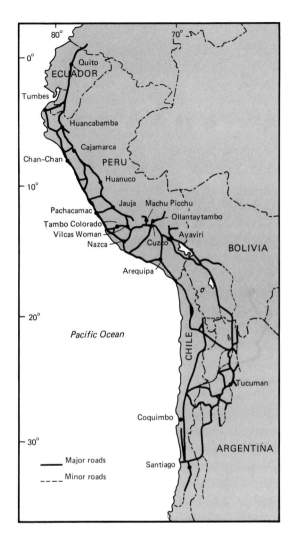

14.6 The Inca road system connected almost every settlement to two main north-south routes and facilitated voluminous transports of goods and effective administration.

ing system of the type used in early Mesopotamia would no doubt have conveyed more religious and philosophical information, but for simple information storage and retrieval, the *quipu* appears to have been an adequate substitute for writing, when complemented by the enormous Inca bureaucracy.[48]

Gold, fabrics, and other luxury goods were collected from over the empire for distribution among the elites. Women, too, were treated as commodities. Government agents visited each village periodically and took selected girls of about age ten back to provincial capitals where they were taught spinning, weaving, and cooking, and then were apportioned out as wives for the emperor and the nobles.

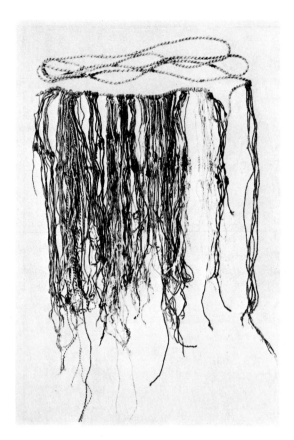

14.7 The Inca never developed a written language, but administrators used the *quipu,* a set of knotted strings, to keep records.

The Inca Empire both created and was created by its system of roads.[49] Most villages were largely self-sufficient, but the flow of goods and information and, most important, the armies required to create the empire were dependent on the road system, comprising an overall network of about 40,000 kilometers of paved roads. Road beds were excavated through hillsides, swamps were crossed by drained causeways, walls were built along roadways to protect the traveller from the fierce gales of the uplands, and wide rivers and ravines were crossed by suspension bridges made of woven vines hung from stone towers. All along the roads were storehouses and administrative outposts, and runners stationed about a kilometer apart were reputed to carry messages over distances as great as 2400 kilometers in just five days.[50]

Although they were master builders, the Inca stressed the rural, village way of life. Typical Inca residential units were rectangular walled houses of stone or adobe, subdivided into smaller units. Most public constructions were not for houses in the cities, but in the form of palaces, temples, granaries, fortresses, barracks, and highway stations. The skill used in these constructions is amazing, considering the simple tools employed. The Inca cut stones into huge blocks simply by chipping and abrading them with harder stones, and they then fitted them together (without the use of mortar) so precisely that, as the cliché goes, a knifeblade cannot be inserted between them.

The cultural order and social structure of the Inca were expressed not only in its public monuments, but even in its domestic architecture. Susan Niles has catalogued how in one particular case, the buildings constituting a fifteenth-century estate of a noble family are arranged to reflect the rigid class hiearchy of the Inca polity.[51]

The capital city of Cuzco was an orderly arrangement of houses, monumental buildings, and streets, well provided with a municipal water and drainage system. The great temple of Qori Kancha here had exterior walls measuring 68 by 59 meters, and a semicircular annex that rose to a height of more than 34 meters. A gold frieze about a meter wide ran along the exterior wall, and the entranceway was heavily sheathed in gold plate. Many other structures at the capital were lavishly decorated with gold and silver.

The comparatively great internal security of the empire made it unnecessary to defend most settlements, except with occasional hilltop forts. The heart of the Inca army was the common foot soldier armed with club, mace, battle axe, or lance. Slings, bolas, and spear throwers were used prior to the main attack, but it was brutal hand-to-hand combat that usually decided the issue. One successful tactical innovation of the Inca was the practice of holding back a large body of troops who were thrown in at a critical juncture—a simple tactic similar in a way to Napoleon's successful use of reserves.

The European Conquest

Archaeological research in some areas of Andean South America has recently been suspended because revolutionary groups have taken archaeologists captive for short times, and to understand the origins of such revolutionary fervor one must look at the colonial history of Andean South America.

After sporadic, occasionally hostile contacts in the A.D. 1520s, the Spanish under Francisco Pizzaro set out toward the provincial capital at Cajamarca, the residence of the Atahualpa, the Inca king. Why they were never intercepted and massacred remains a mystery, but warfare between rival claimants to the Inca throne at this time was probably a factor. In any case, the Andean South Americans soon had cause to regret their diffidence. Pizzaro and his men entered the city on 15 November 1532, and found it to be a massive, fortified center, but, surprisingly, nearly deserted. After establishing himself in a fortress with a couple of cannons and his few score soldiers, Pizarro elected to wait until the Inca king made an effort to visit him. Eventually the emperor came, borne on a litter and preceded by thousands of soldiers, attendants, and subjects. The first Spaniard to approach the king was the chaplain who, as part of Pizarro's contract with the king of Spain and the Pope, was charged with spreading the Christian faith. The chaplain immediately began to harangue the king, through an interpreter, about the creation of the world, the fall of Adam and Eve, the Virgin Birth, the establishment of the papacy, and other dogma, culminating with the announcement that the Pope had given the Inca Empire to King Charles of Spain.

Not surprisingly, the Inca king took exception to parts of the chaplain's speech, and he wanted to know how the Pope could give away something that was not his, and how it had happened that the god of the Christians had died, since the Inca deity, the Sun, was immortal.[52] When the Atahualpa asked how the chaplain knew all these things, he was handed a breviary. The king looked briefly and no doubt uncomprehendingly inside, and then threw it away. At this point the Spanish attacked, and then the inexplicable happened: instead of killing the Spanish, the Inca fled, dropping their weapons and killing themselves in their panicked flight, and the Spanish were able to dispatch hundreds and capture the king with little trouble. They remained fortressed in Cajamarca for some months, detaining the king, who tried to win his release by offering to fill a room (supposed to have been 6.5 by 4.5 meters) with gold. The Spanish meanwhile took masses of gold and silver in raids, most of it in the form of exquisitely wrought figures, which they melted into ingots.

Rumors of insurrections in the countryside convinced the Spanish to execute the Inca king, and they did so, considering themselves enlightened for giving him the option of being garroted rather than burned at the stake—a reward to the king for allowing himself to be baptized. With the Atahualpa's death and the ensuing factionalism among rival claimants to the throne, as well as the devastation brought on by introduced diseases and the horror wrought on the populace through warfare and the destruction of the irrigation system, the population of Andean South America is thought to have dropped from over six million to fewer than two million within a few decades of the conquest.

The Origins of Complex Cultures in Andean South America: Conclusions

The reader by now will not be surprised to learn that most scholars interpret the rise of civilization in Andean South America to be the result of a multiplicity of factors.

Robert Carneiro's hypotheses linking warfare, population growth, and environmental circumscription (Chapter 7) to the rise of states seems particularly applicable to Andean South America. There are some suggestions that warfare was in fact an important "stress" that stimulated some kinds of complexity (though perhaps not in the manner suggested by Carneiro).[53] But sustained warfare seems to have been more of an important factor well after the appearance of such things as monumental buildings, the coordination of regional economies, craft specialization, and the rise of great religious traditions.

How are we to account for Andean South American complex societies? Obviously, the rich maritime and agricultural resources were essential ingredients in this development. In only a few areas of the world is it possible to produce and gather enough food to run complex cultures on the basis of primitive technologies, and Andean South America is one of these.

An important "negative" element in the evolution of Andean South American appears to have been the fact that, unlike ancient China, Mesopotamia, or the Indus Valley, Andean South America was geographically isolated from other highly complex political systems. Evolving Old World civilizations soon came into contact with one another, and their political, economic, and social interchanges appear to have transformed each of them to some degree. But, except for Mesoamerica—which was very distant and cut off by ocean and jungle—Andean South America evolved alone.

The absence of a domesticable draught animal also was a limit on Andean South American development. Llamas compensated for this to a degree, but they cannot compare with the transport abilities of horses, mules, or oxen. It is difficult to judge the effects the presence of a domesticable draught animal in Andean South America would have had, but it may be significant that almost all agricultural areas of Andean South America today are plowed.

Nor is it easy to weigh the effect on Andean South American developments of the scarcity of accessible iron ore. In the Old World iron-working seems to have been intimately associated with the expansion of great empires, and, had it been available, it might also have changed the character of later Andean South American developments.

Thus, in summary of Andean South American prehistory, we see that cultural developments there paralleled those in other centers of independent complex society formation in most important details, including the initial spread of a religious cult, the importance of a highly productive economy, the widespread occurrence of monumental architecture, and the gradual emergence of highly stratified, economically integrated state and imperial political systems.

Notes

1. Murra, *Andean Ecology and Civilization,* p. 11.
2. Masuda et al., eds., *Andean Ecology and Civilization.*
3. Idyll, "The Anchovy Crisis."
4. Rick, *Prehistoric Hunters of the High Andes.*
5. Lynch, *Guitarrero Cave: Early Man in the Andes.*
6. Richardson, "Maritime Adaptations on the Peruvian Coast: A Critique and Future Directions," pp. 140–45.
7. Quilter and Stocker, "Subsistence Economies and the Origins of Andean Complex Societies," p. 547.
8. Patterson, "Central Peru: Its Population and Economy."
9. Benfer, "The Challenges and Rewards of Sedentism: The Preceramic Village of Paloma, Peru"; Reitz, "Faunal Remains from Paloma, an Archaic Site in Peru."
10. Quilter and Stocker, "Subsistence Economies and the Origins of Andean Complex Societies," p. 545.
11. Parsons, "Preceramic Subsistence on the Peruvian Coast," p. 297.
12. Pozorski and Pozorski, "Alto Salaverry: A Peruvian Coastal Preceramic Site."
13. Benfer, "The Challenges and Rewards of Sedentism."
14. Engl and Engl, *Twilight of Ancient Peru.*
15. Lanning, *Peru Before the Incas,* p. 71.
16. Moseley, *The Maritime Foundations of Andean Civilization.*
17. Quilter and Stocker, "Subsistence Economies and the Origins of Andean Complex Societies," p. 554.
18. Moseley, *The Maritime Foundations of Andean Civilization.*
19. Wilson, "Of Maize and Men: A Critique of the Maritime Hypothesis of State Origins on the Coast of Peru."
20. See also Scott, "The Maritime Foundations of Andean Civilization: A Reconsideration."
21. Quilter and Stocker, "Subsistence Economies and the Origins of Andean Complex Societies."
22. Reitz, "Faunal Remains from Paloma, an Archaic Site in Peru."
23. Pozorski and Pozorski, *Early Settlement and Subsistence in the Casma Valley, Peru.*
24. Quilter and Stocker, "Subsistence Economies and the Origins of Andean Complex Societies."
25. Ibid., p. 555, citing Lathrap, "The Antiquity and Importance of Long-Distance Trade Relationships in the Moist Tropics of Pre-Columbian Sough America."
26. Kaplan, Lynch, and Smith, "Early Cultivated Beans *(Phaseolus vulgaris)* from an Intermontane Peruvian Valley."
27. Lumbreras, *The Peoples and Cultures of Ancient Peru,* p. 47.
28. Moseley, "The Evolution of Andean Civilization."
29. Haas, et al., eds., *The Origins and Development of the Andean State.*
30. Moseley, *The Maritime Foundations of Andean Civilization,* p. 105.
31. Ibid., p. 106.
32. Ibid., p. 107.
33. Fiedel, *Prehistory of the Americas,* p. 319.
34. Lumbreras, *The Peoples and Cultures of Ancient Peru,* p. 62.
35. Greider, *The Art and Archaeology of Pashash.*
36. Reviewed in Keatinge, *Peruvian Prehistory;* also see Haas, Pozorski, and Pozorski, *The Origins and Development of The Andean State;* Fiedel, *Prehistory of the Americas;* Donnan, ed. *Early Ceremonial Architecture in the Andes.*
37. Lanning, *Peru Before the Incas,* pp. 114–15.
38. Silverman, "Cahuachi: Non-Urban Cultural Complexity on the South Coast of Peru."

39. Donnan, *Moche Art and Iconography.*
40. Moseley, "The Evolution of Andean Civilization."
41. Lanning, *Peru Before the Incas,* p. 127.
42. Fiedel, *Prehistory of the Americas,* p. 333.
43. Keatinge, "Chimú Rural Administration Centers in the Moche Valley, Peru," p. 79.
44. Moseley and Day, *Chan Chan: Andean Desert City.*
45. Collier, Rosaldo, and Wirth, *The Inca and Aztec States, 1400–1800.*
46. Lanning, *Peru Before the Incas,* pp. 159–60.
47. Ibid., pp. 166–67.
48. Ascher and Ascher, *Code of the Quipu: A Study in Media, Mathematics, and Culture.*
49. Hyslop, *The Inka Road System.*
50. Von Hagen, "America's Oldest Roads."
51. Niles, *Callachaca.*
52. Engl and Engl, *Twilight of Ancient Peru,* p. 119.
53. Wilson, "The Origins and Development of Complex Prehispanic Society in the Lower Santa Valley, Peru."

Bibliography

Ascher, M. and R. Ascher. 1982. *Code of the Quipu: A Study in Media, Mathematics, and Culture.* Ann Arbor: University of Michigan Press.

Benfer, R. A. 1984. "The Challenges and Rewards of Sedentism: The Preceramic Village of Paloma, Peru." In *Paleopathology and the Origin of Agriculture,* eds. M. N. Cohen and G. J. Armelagos. New York: Academic Press.

Benson, E. P., ed. 1971. *Dumbarton Oaks Conference on Chavín.* Washington, D.C.: Dumbarton Oaks Research Library and Collection.

Browman, D. L. 1974. "Pastoral Nomadism in the Andes." *Current Anthropology* 15:188–96.

———. 1975. "Trade Patterns in the Central Highlands of Peru in the First Millennium B.C." *World Archaeology* 6:322–30.

Bushnell, G. H. S. 1963. *Peru.* Rev. ed. New York: Praeger.

Carneiro, R. 1970. "A Theory of the Origin of the State." *Science* 169:733–38.

Collier, G. A., R. I. Rosaldo, and J. D. Wirth, eds. 1982. *The Inca and Aztec States, 1400–1800.* New York: Academic Press.

Conrad, G. W. 1981. "Cultural Materialism, Split Inheritance, and the Expansion of Ancient Peruvian Empires." *American Antiquity* 46:3–26.

Donnan, C. B. 1973. *Moche Occupation of the Santa Valley, Peru.* Berkeley: University of California Publications in Anthropology, no. 8.

———. 1976. *Moche Art and Iconography.* Los Angeles: U.C.L.A. Latin American Center Publications.

Donnan, C. B., ed. 1985. *Early Ceremonial Architecture in the Andes.* Washington, D. C.: Dumbarton Oaks.

Engel, F. 1957. "Early Sites on the Peruvian Coast." *Southwestern Journal of Anthropology* 13:54–68.

Engl, L. and T. Engl. 1969. *Twilight of Ancient Peru.* Trans. A. Jaffe. New York: McGraw-Hill.

Feldman, R. A. 1980. *Aspero, Peru: Architecture, Subsistence Economy, and other Artifacts of a Preceramic Maritime Chiefdom.* Ph.D Dissertation, Department of Anthropology, Harvard University.

Fiedel, S. J. 1987. *Prehistory of the Americas.* Cambridge, England: Cambridge University Press.

Flannery, K. V. 1973. "The Origins of Agriculture." *Annual Review of Anthropology* 2:271–310.

Greider, T. 1978. *The Art and Archaeology of Pashash.* Austin: University of Texas Press.

Gross, D. R. 1975. "Protein Capture and Cultural Development in the Amazon Basin." *American Anthropologist* 77:526–49.

Haas, J., S. Pozorski, and T. Pozorski, eds. 1987. *The Origins and Development of the Andean State.* New York: Cambridge University Press.

Hyslop, J. 1984. *The Inka Road System.* New York: Academic Press.

Idyll, C. P. 1973. "The Anchovy Crisis." *Scientific American* 228:22–29.

Japanese Scientific Expedition to Nuclear America. 1979. *Excavations at La Pampa in the North Highlands of Peru, 1975.* Tokyo: University of Tokyo Press.

Jones, G. D. and R. R. Kautz, eds. 1981. *The Transition to Statehood in the New World.* Cambridge, England: Cambridge University Press.

Kaplan, L., T. Lynch, and E. E. Smith, Jr. 1973. "Early Cultivated Beans *(Phaseolus vulgaris)* from an Intermontane Peruvian Valley." *Science* 179:76–77.

Keatinge, R. W. 1974. "Chimú Rural Administration Centers in the Moche Valley, Peru." *World Archaeology* 6:66–82.

———. 1988. *Peruvian Prehistory: An Overview of Pre-Inca and Inca Society.* New York: Cambridge U. Press.

Lanning, E. P. 1967. *Peru Before the Incas.* Englewood Cliffs, N. J.: Prentice-Hall.

Lathrap, D. W. 1968. "Relationships Between Mesoamerica and the Andean Areas." In *Handbook of Middle American Indians,* Vol. 4. Austin: University of Texas Press.

———. 1973. "The Antiquity and Importance of Long-Distance Trade Relationships in the Moist Tropics of Pre-Columbian South America." *World Archaeology* 5:170–86.

Lumbreras, L. G. 1974. *The Peoples and Cultures of Ancient Peru.* Trans. B. J. Meggers. Washington, D. C.: Smithsonian Institution Press.

Lynch, T., ed. 1980. *Guitarrero Cave: Early Man in the Andes.* New York: Academic Press.

MacNeish, R. S., R. K. Vierra, A. Nelken-Terner, R. Lurie, and A. G. Cook. 1983. *Prehistory of the Ayacucho Basin, Peru, Volume IV: The Preceramic Way of Life.* Ann Arbor: University of Michigan Press.

Masuda, S., I. Shimada, and C. Morris, eds. 1985. *Andean Ecology and Civilization.* Tokyo: University of Tokyo Press.

Morris, C. 1985. "From Principles of Ecological Complementarity to the Organization and Administration of Tawantinsuyu." In *Andean Ecology and Civilization.* Masuda, S., I. Shimada, and C. Morris, eds, Tokyo: University of Tokyo Press.

Moseley, M. E. 1972. "Subsistence and Demography: An Example of Interaction from Prehistoric Peru." *Southwestern Journal of Anthropology* 28:25–49.

———. 1975. *The Maritime Foundations of Andean Civilization.* Menlo Park, Calif.: Cummings.

———. 1983. "The Evolution of Andean Civilization." In *Ancient Native Americans,* ed. J. D. Jennings. San Francisco: Freeman.

Moseley, M. E. and R. C. Day, eds. 1982. *Chan Chan: Andean Desert City.* Albuquerque: University of New Mexico Press.

Murra, J. 1958. "On Inca Political Structure." In *Systems of Political Control and Bureaucracy in Human Society,* ed. V. F. Ray. Seattle: University of Washington Press.

———. 1965. "Herds and Herders in the Inca State." In *Man, Culture and Animals.* Washington, D. C.: American Association for the Advancement of Science.

———. 1985. "The Limits and Limitations of the 'Vertical Archipelago' in the Andes." In *Andean Ecology and Civilization.* Masuda, S., I. Shimada, and C. Morris, eds, Tokyo: University of Tokyo Press.

Murra, J. and C. Morris. 1976. "Dynastic Oral Tradition, Administrative Records, and Archaeology in the Andes." *World Archaeology* 7:269–79.

Niles, S. A. 1988. *Callachaca.* Iowa City: University of Iowa Press.

Ortloff, C. R., M. E. Moseley, and R. A. Feldman. 1982. "Hydraulic Engineering Aspects of the Chimú/Chicama-Moche Intervalley Canal." *American Antiquity* 47:572–95.

Parsons, J. 1968. "An Estimate of Size and Population for Middle Horizon Tiahuanaco, Bolivia." *American Antiquity* 33:243–45.

———. 1977. Personal communication.

Parsons, J. and N. Psuty. 1975. "Sunken Fields and Prehispanic Subsistence on the Peruvian Coast." *American Antiquity* 40:259–82.

Parsons, M. 1970. "Preceramic Subsistence on the Peruvian Coast." *American Antiquity* 35:292–303.

Patterson, T. C. 1966. "Early Cultural Remains on the Central Coast of Peru." *Nawpa Pacha* 4:145–55.

———. 1971. "Chavín: An Interpretation of Its Spread and Influence." In *Dumbarton Oaks Conference on Chavín,* ed. E. Benson. Washington, D.C.: Dumbarton Oaks.

———. 1971. "Central Peru: Its Population and Economy." *Archaeology* 24:316–21.

Patterson, T. C. and E. P. Lanning. 1964. "Changing Settlement Patterns on the Central Peruvian Coast." *Nawpa Pacha* 2:113–23.

Patterson, T. C. and M. E. Moseley. 1968. "Late Preceramic and Early Ceramic Cultures of the Central Coast of Peru." *Nawpa Pacha* 6:115–33.

Pickersgill, B. 1969. "The Archaeological Record of Chile Peppers *(Capsicum spp.)* and the Sequence of Plant Domestication in Peru." *American Antiquity* 34:54–61.

Pickersgill, B. and A. Bunting. 1969. "Cultivated Plants and the Kon-Tiki Theory." *Nature* 222:225–27.

Pozorski, S. and T. Pozorski. 1979. "Alto Salaverry: A Peruvian Coastal Preceramic Site." *Annals of the Carnegie Museum of Natural History* 49:337–375.

———. 1988. *Early Settlement and Subsistence in the Casma Valley, Peru.* Iowa City: University of Iowa Press.

Prescott, W. H. 1908. *History of the Conquest of Peru.* London and New York: Everyman's Library.

Quilter, J. and T. Stocker. 1983. "Subsistence Economies and the Origins of Andean Complex Societies." *American Anthropologist* 85(3):545–62.

Reitz, E. J. 1988. "Faunal Remains from Paloma, an Archaic Site in Peru." *American Anthropologist* 90(2):310–22.

Richardson, J. B., III. 1981. "Maritime Adaptations on the Peruvian Coast: A Critique and Future Directions." Paper Presented at the 47th Annual Meeting of the Society for American Archaeology, San Diego, California.

Rick, J. W. 1980. *Prehistoric Hunters of the High Andes.* New York: Academic Press.

Rowe, J. H. 1946. "Inca Culture at the Time of the Spanish Conquest." *Bureau of American Ethnology Bulletin,* n. 143:183–331. Washington, D. C.: Smithsonian Institution.

———. 1967. "Form and Meaning in Chavín Art." In *Peruvian Archaeology: Selected Readings.* Palo Alto: Peek Publishers.

Scott, R. J. 1981. "The Maritime Foundations of Andean Civilization: A Reconsideration." *American Antiquity* 46:806–21.

Shady, R. and A. Ruiz. 1979. "Evidence for Interregional Relationships During the Middle Horizon on the North-Central Coast of Peru." *American Antiquity* 44:670–84.

Silverman, H. 1988. "Cahuachi: Non-Urban Cultural Complexity on the South Coast of Peru." *Journal of Field Archaeology* 15(4): 403–30.

Vescelius, G. 1981. "Early and/or Not-so-Early Man in Peru: Guitarrero Cave Revisited." *The Quarterly Review of Archaeology* 2:8–13.

Von Hagen, V. W. 1952. "America's Oldest Roads." *Scientific American* 187:17–21.

———. 1965. *The Desert Kingdoms of Peru.* London: Weidenfeld and Nicolson.

Willey, G. R. 1962. "The Early Great Art Styles and the Rise of Pre-Columbian Civilizations." *American Anthropologist* 64:1–14.

Wilson, D. J. 1981. "Of Maize and Men: A Critique of the Maritime Hypothesis of State Origins on the Coast of Peru." *American Anthropologist* 83:93–114.

———. 1983. "The Origins and Development of Complex Prehispanic Society in the Lower Santa Valley, Peru: Implications for Theories of State Origins." *Journal of Anthropological Archaeology* 2:209–76.

Wing, E. S. 1973. "Utilization of Animal Resources in the Andes." Report NSF GS-3021. Gainesville, Fla.: Florida State Museum.

———. 1973. "Animal Domestication in the Andes." Paper presented at the Twelfth International Congress of Prehistoric and Protohistoric Sciences, Chicago.

15

Early Cultural Complexity in North America

You must explain to the natives of [North America] that there is only one God in heaven, and the emperor on earth to rule and govern it, whose subjects they must all become and they must serve.

Instructions of the Viceroy
to Fray Marcos De Ninza (1538)

As Europeans invaded North America in the sixteenth and seventeenth centuries, they encountered what looked to them like the relics of ancient civilizations: thousands of large earthen mounds, some nearly as large as the Egyptian pyramids, dominated the Ohio and Mississippi River valleys, and in Arizona, Colorado, and New Mexico large, neatly planned but abandoned towns that had obviously once been inhabited by thousands were found along the major rivers and even in the arid highlands.

The Europeans of this age were among the greatest egotists and ethnocentrists the world has known, so it is not surprising that they assumed that these impressive works had been constructed by their own ancestors—the Celts, Romans, or perhaps the Vikings; they simply could not believe that the ancestors of the poor and "degenerate" Native Americans they saw about them might have had something to do with these great monuments. Some Europeans even blamed the Indians for massacring what the Europeans believed to be an ancient and extinct "superior" American race—the ultimate in adding insult to injury, given the Indians' eventual fate.

The truth, of course, is that these mounds and abandoned settlements had indeed been built by Native Americans many centuries before the Europeans arrived, and we now know that these ancient Americans had begun to travel much the same road to cultural complexity as had the people in Mesoamerica, Egypt, and elsewhere.

The people of "PreColumbian" North America were influenced by

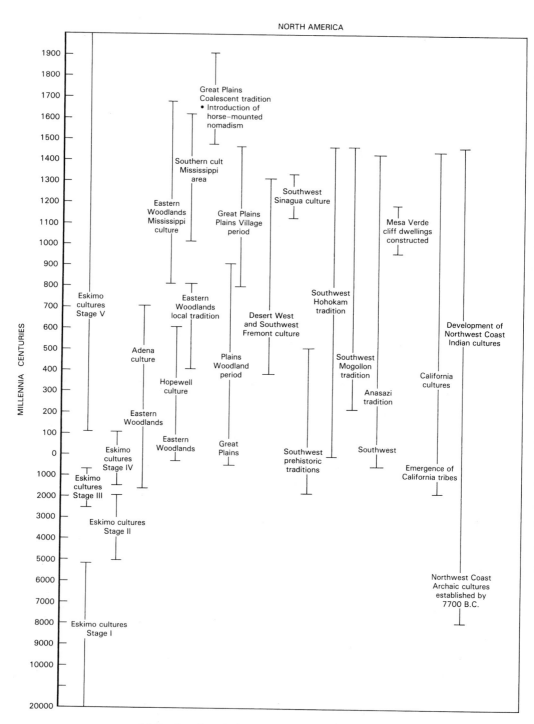

15.1 A cultural chronology of North America.

contact with the complex cultures of ancient Mesoamerica, and it is also true that, according to some criteria, aboriginal American cultures never reached the level of complexity attained in "independent" ancient centers of cultural evolution, such as Mesoamerica and Mesopotamia. But the native North Americans did develop some aspects of cultural complexity, and they did so largely independently. Moreover, the ecological and cultural reasons North American cultures did not exactly parallel those of Mexico or Peru provide some insights into the general evolution of culture.

For more than a century archaeologists have been systematically digging North American sites, from Alaska to Tierra del Fuego, and we should know more about developments there than virtually anywhere else in the world, given the high ratio of archaeologists per square kilometer. But, unfortunately, early American archaeologists also yielded to the temptations of excavating mainly rich tombs and large towns, resulting in a very biased view of the American past. And because of past decades of turning ancient Native American cultural resources into parking lots and hamburger stands, and because of the looting that continues today, so much of the archaeological record of North America has been destroyed that much of the American past will forever remain a mystery.

The old, sometimes perjorative myths about Indians are now giving way in some circles to new ones stressing the social harmony, ecological purity, and superior metaphysics of ancient Native Americans. As with all cultures, there is much to admire about ancient North Americans, but the evidence suggests that for much of their antiquity, aboriginal Americans were like prehistoric peoples in every other part of the world—harried small-time farmers, worrying about this year's drought and next year's deluge, shifting their crops from year to year as they exhausted the soil and timber of one place after another, and as enmeshed in slavery, exploitation, and warfare as any ancient Chinese, Egyptian, or European.

The North American East

The most complex ancient cultures in North America evolved in the great river valleys of the North American east, and the more than four hundred years of industrialization in this area have blotted out most traces of pre-European life here. Thus we know only in rough outlines the ancient cultures of this region.

The Ecological Setting

We saw in Chapter 5 that about 12,000 years ago the huge glacial ice sheets that once covered much of eastern North America retreated into

Canada, and the distribution of plant and animal species in this region became similar to that of the recent past. Between about 8000 and 6500 B.C., the annual temperature of much of the North American east was probably 2.5 degrees (Centigrade) cooler than at present, and after that time climates probably fluctuated in ways that directly affected aboriginal Americans, such as the Hypsithermal (4000 B.C.), when climates seem to have been cooler and moister.[1] Subtle shifts of climate and effective rainfall may have had important effects on the distribution of nut-bearing trees, which were a very important food source, and also on the size and stability of rivers and lakes—rich environments around which early Americans often built their hunting-foraging economies.

Temperate eastern North America is a rich niche for hunter-foragers, but its abundant agricultural potential can only be realized with an industrial technology. The American southwest and west include some very rich areas, but here, too, there are no large areas with the same agricultural potential—given primitive technologies—as that, for example, of the Nile Valley or Mesopotamia. The modern United State's agricultural economy is so productive that it is is somewhat surprising just how poor a place it is for a preindustrial subsistence farmer. First, most of it is far from the equator and therefore has less solar radiation and fewer frost-free days, and thus lower absolute agricultural potential, than tropical environments—if one does not have modern plows and tractors to rip open the rich grasslands of the Plains and Midwest. Also, the Mississippi and the Ohio rivers are not very similar to the Nile or the Indus in that the land near these American rivers is thickly forested and relatively cool—unlike the great warm arid plains that so richly rewarded Old World cultivators.

The Background to Agriculture:
The Late Archaic Period (c. 5000–800 B.C.)

As we saw in Chapter 5, by 5000 B.C., people had colonized most of North America and had begun to differ considerably on the kinds of resources and technologies on which they based their economies. Until late in the first millennium B.C., hunting and foraging continued to be the way of life for most people, but through these many millennia after the Ice Age, the overall trend was toward increasing numbers of sedentary communities that specialized first in hunting and intensive plant collecting and later in maize and bean agriculture.

Richard Yerkes calls the last three millennia B.C. the "Transitional Period" because of the spectrum of cultural changes that slowly but fundamentally set the stage for the elaboration of American cultures in the last centuries before European colonization. He notes that it was in this period of three thousand years in the American East that people increased exploitation of the river valleys in which their descendants would eventually establish agricultural villages, and that they also were the first to domesticate native plants, such as squash, gourds, goosefoot, sunflowers, and tobacco. People also increased the kinds and volumes of goods they exchanged. Bruce Smith has shown how, beginning about 3000 B.C., people in some parts of the East developed a way of life in which they migrated between summer and winter base camps, to take advantage of different resources as they changed with the seasons.[2]

The productive patches of maize, beans, and squash that eventually allowed Native Americans to live in great numbers all over North America were only the last, most productive stages of long slow shift, away from simple plant collecting and hunting, and toward an ever-growing reliance on fewer and fewer kinds of plants. Long before maize reached temperate North America, people there had specialized in exploiting several kinds of plants in early forms of agriculture. So, the questions about North American agriculture are the same as those that can be applied to other areas. Why did agricultural economies appear when and where they did, through what processes, and with what results, and what was the long-term relationship between local domesticates and those introduced from elsewhere?

Perhaps the earliest domesticates found in the North American woodlands are gourds and squash. Charred squash rinds from west central Illinois have been dated to 5000 and 4000 B.C.,[3] and some from Missouri, Tennessee, and Kentucky have been dated to at least 2500 B.C. These plants may have been introduced from Mexico or were locally domesticated. They were probably used as containers and for seeds, and occasionally eaten directly, and we may suppose that their use in this capacity goes back many millennia before we have archaeological evidence of them.

Domestication is just the name for the "mutualism" that occurs when people and some plant or animal species become intimately dependent on one another, and this is usually the byproduct of centuries of systematic collection: thus many native plants were probably in the process of domestication once sizable populations and post-Pleistocene climates were established. Undomesticated varieties of some North American plants were smaller and had smaller seeds than modern varieties, and during the domestication process average seed size was increasing in some species— presumably as part of the domestication process. Remains of goosefoot, sunflowers, sumpweed, and other plants with starchy or oily seeds seem to have been grown in small gardens beginning at least by 2000 B.C. and they remained important food sources until most of them were displaced by European crops.

Michael O'Brien has developed a "Coevolutionary Model" to account

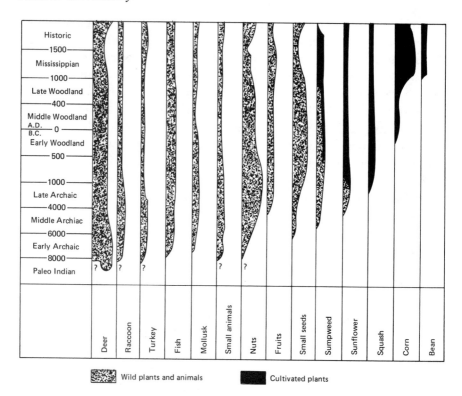

15.2 In the prehistoric American Middle West the transition to the agricultural way of life was a process in which native crops and wild plant foods were replaced by maize, beans, and squash (the thickness of the shaded and black areas reflect relative importance of the different resources).

for the transition to agriculture in one area, near the Illinois/Missouri border.[4] Based on David Rindos' work on plant domestication,[5] O'Brien suggests that the consistent but low-intensity use of sunflower seeds, goosefoot, marshelder, and other native annuals during the period up to about 300 B.C. created a relationship between these plants and people that resulted in a form of domestication and agriculture. He argues that initial exploitation of these plants would not have been of great significance in the overall economy, but the costs of exploiting them were very low, since they did not conflict with other activities and one could carry these seeds as a minor food without too much effort. But then he says, "by 200 B.C. the interaction between human groups and their resource base had reached the point that sedentism became a viable option."[6] He notes that the risks of concentrating on these plants was considerable, in the way of droughts and floods. But they also had potential for supporting increased population densities.

David Braun has shown that during the Middle and Late Woodland periods people began to make more and more pottery in ways that seem particularly designed to cook starchy seeds, such as those of goosefoot. Such pottery has to withstand high heat and sudden temperature changes, and all over the world people making these kinds of pots invented the same globular shapes and kinds of tempering and wall thicknesses that facilitate the long slow simmering of seeds and other foods in liquids. Stews and soups are not the most exciting foods in most cuisines (Szechuan and French Mediterranean excepted), but soups and stews are extraordinarily nutritious—and, more than that, they make excellent foods out of things that normally people cannot much profit from, such as dry goosefoot seeds and a picked-over rabbit skeleton: the best way to increase the nutrition of these annuals is to boil them, which "may increase palatability as well"[7] (it couldn't hurt it, in this case).

Simply by devising ways to slow-cook the abundant seeds in their environments, Native Americans seem to have enlarged their food resources to the point that a less mobile, less risky way of life was possible for more and more people. The pervasive effects of the shift to an economy based in part on starchy seeds have been documented over much of the North American East.[8]

Some have speculated that the techniques for making ceramic vessels were introduced into North America from Mexico, but it is more likely that pottery was independently invented in eastern North America at least once, and probably several times, as containers became increasingly important in plant collection and preparation.

The collection and domestication of these various plant species seem always and everywhere to have been combined with relentless hunting, fishing, and nut gathering, and there is little evidence that agriculture was being practiced at this time. That is, little energy was invested in modifying the environments of sunflowers and these other plants through hoeing, watering, or weeding. Melvin Fowler noted that most of the native plants apparently domesticated in the East are species that prefer open, disturbed areas, such as would have been provided by the refuse piles that must have become increasingly available as hunters and gathers cluttered the landscape with snail and clam shells, fish bones, and other debris.[9] When and to what extent the kind of natural association between these plants and human disturbances was converted into farming, in the sense of planting and caring for these crops, will probably always be difficult to measure.

One of the earliest and most dramatic indications that American cultures were becoming more complex than those of the previous millennia is the Poverty Point site, in Louisiana (Figure 15.3), where vast earthworks were constructed between 1300 and 700 B.C.

The six nested ranks of earthen mounds at Poverty Point, and the huge associated earthen mound, were probably mortuary cult centers, and some of the deceased of this society were buried under elaborate mounds. Some of the Poverty Point people were skilled artisans who made vessels out of steatite and sandstone, pipes out of clay and stone, and axes, adzes, saws, and weights from hard kinds of stone.

Elsewhere in the ancient world, such as Jericho, we have seen that agriculture is not a prerequisite for substantial architecture—all that is required is sufficient productivity. The Poverty Point people may have done some limited agriculture of native plants, but it was probably a rich blend of hunting and foraging that supported these mound-builders.

After about 800 B.C. North America east of the Mississippi Valley became a mosaic of varying adaptations, with the use of ceramics and an economy based, perhaps, somewhat on local and imported domesticates, but primarily on an Archaic mix of hunted and collected resources.[10]

Like their contemporaries in other parts of the world, ancient Ameri-

15.3 One of the most impressive Archaic sites in North America is the complex of earthworks at Poverty Point, Louisiana. Between about 1500 and 700 B.C. hunter-gathering peoples in this area built six nested octagons of piled earth. The outer octagon is about 1300 meters in diameter. Part of this complex was later eroded away by a river. A small mound near these earthworks and other occupations produced many beautiful artifacts, as well as thousands of fired clay balls that were apparently heated and then dropped into containers to boil foods.

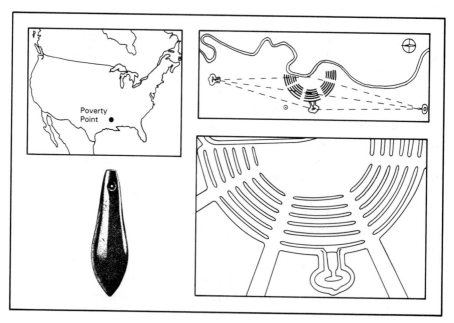

Poverty
Point ●

cans used the occasion of death to express social attitudes and practices. Beginning at about 2000 B.C., people all over the east began burying their dead with exotic items, such as copper, marine shells, and ground stone jewelry—often covering the corpse with red ocher, a mineral pigment.[11] Yerkes suggests that these "individuals would go to their graves with the exotic materials that they had accumulated, but it is unlikely that they were able to pass their achieved status on to their descendants."[12]

Evolving Cultural Complexity: The Early and Middle Woodland Periods (c. 800 B.C.–A.D. 800)

Between about 800 B.C. and A.D. 800, population densities in many parts of eastern North America increased sharply, thousands of gigantic earthworks were constructed, interregional trade expanded, and large villages were built. This era of change, usually referred to as the Woodland Period, is associated with two major cultural traditions: the Adena, centered in southern Ohio, and extending into Indiana, Kentucky, Pennsylvania, and West Virginia, and the Hopewell, centered in southern Ohio but extending widely over much of eastern North America. These cultures overlapped in time, with the Adena somewhat earlier than Hopewell in some areas. Both are defined on the basis of styles of pottery, engraved stone tablets, textiles, and worked bone and copper.

The mound-building and artifact styles of the Adena seem concentrated in the interval between about 500 B.C. and A.D. 700, and the richness of some of these sites, in the burial goods found in the mounds, as well as the scale of the mounds themselves (Figure 15.4), originally led some scholars to suppose that these people were already chiefdoms, built on the sweat and labor of exploited farmers.

But the Adena seem to have been still mainly hunter-forager peoples who moved seasonally. Maize may have been cultivated on a small scale, but it was probably not very important.[13] And impressive as the mounds are, they could easily have been built by small groups adding a layer or two each year, without much supervision.[14]

Most of the mounds were mortuary centers in which one or more corpses were placed in log tombs or clay pits, sometimes with beautiful stone tools and other goods, and then covered with earth. Often the bodies were cremated before burial, others seem to have been exposed before the mound was heaped up over them.

These mounds may have been centers for goods exchange or ceremonial sites, but they also could have served as territorial markers.[15]

Stone tobacco pipes are found in some Adena sites, and smoking probably played an important role in rituals. The native strains of tobacco were much more powerful than modern strains and could have produced

15.4   Great Serpent Mound, an Adena construction near Cincinnati, Ohio.

narcotic effects much more exhilarating than a Pall Mall and a Scotch and soda.[16]

It is unfortunate that so many Adena sites were destroyed by modern constructions, and also that archaeologists excavated scores of mounds but almost no ordinary Adena settlements. Those few that have been excavated suggest simple, somewhat temporary structures, probably made by setting series of posts in the ground and covering them with hides or wood—the "wigwam," in fact.[17]

The Hopewell culture dominated the Middle West from about 0 A.D./ B.C. to A.D. 750. The Ohio Hopewell was centered in southern Ohio, and the Havana Hopewell dominated the central and lower Illinois River Valley.

Grave goods are not usually profuse in Adena mounds, but Hopewell tombs often contain finely worked copper, pipestone, mica, obsidian, me-

15.5   Adena-Hopewell and related cultures. The Adena culture was centered in Ohio, but Adena trade networks reached as far away as New York. Adena culture appears to have peaked at about 100 B.C. and was waning by A.D. 400. By about 100 B.C. Adena cultures in Ohio were being replaced by Hopewell cultures and for next nine centuries the Hopewell cultures spread over much of the East.

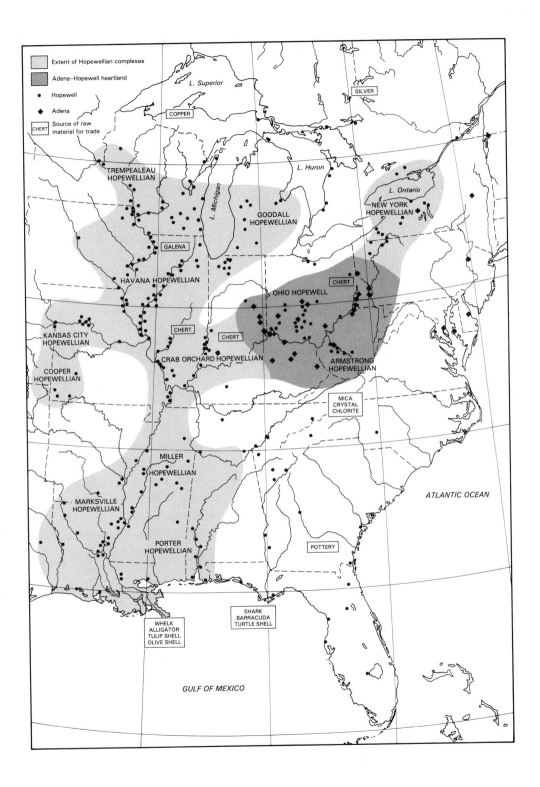

Extent of Hopewellian complexes

Adena–Hopewell heartland

• Hopewell

♦ Adena

CHERT Source of raw material for trade

L. Superior

SILVER

COPPER

L. Huron

L. Ontario

NEW YORK HOPEWELLIAN

TREMPEALEAU HOPEWELLIAN

L. Michigan

GOODALL HOPEWELLIAN

GALENA

HAVANA HOPEWELLIAN

CHERT

OHIO HOPEWELL

CHERT

KANSAS CITY HOPEWELLIAN

CHERT

CRAB ORCHARD HOPEWELLIAN

ARMSTRONG HOPEWELLIAN

COOPER HOPEWELLIAN

MICA CRYSTAL CHLORITE

MILLER HOPEWELLIAN

MARKSVILLE HOPEWELLIAN

ATLANTIC OCEAN

PORTER HOPEWELLIAN

POTTERY

WHELK ALLIGATOR TULIP SHELL OLIVE SHELL

SHARK BARRACUDA TURTLE SHELL

GULF OF MEXICO

teoric iron, shell, tortoise shell, shark and alligator teeth, bear teeth, ceramics, and other commodities; there is also evidence of the ultimate grave good—sacrificed humans. Hopewell settlements and mounds increased in size and number in the centuries before about A.D. 400, and major earthworks were often built near the burial mounds.

These burials may reflect something about the dead person's rank and status within the community—though burials are an uncertain guide to the actual prestige and power relationships of any community. Joseph Tainter carried out a statistical analysis of more than 500 Hopewell burials and concluded that there were six discrete levels of status within the Hopewell community, corresponding to six different forms of burial, ranging from simple holes in the ground to massive mounds in which the corpse was accompanied by the finest goods.[18] In some Adena and Hopewell mounds, infants and juveniles were found buried with great ceremony and rich goods, suggesting the inheritance of rank and prestige and the control of the society's resources by a limited number of its people.

To understand these mound complexes one would have to place them in the context of a local economy—and about this we know comparatively little. Because many later Hopewell buildings and earthworks were precisely planned and relatively expensive to construct, we must suspect that at least some people had become specialized labor-organizers and engineers of sorts, perhaps even constituting a privileged social group. But one of the many puzzling things about these mound complexes is that in many areas they do not seem to have any associated domestic houses. Some archaeologists assumed that any society capable of building such impressive works must certainly have been composed of sedentary agriculturalists, but Hopewell peoples for the most part probably hunted and gathered essentially the same plant and animal species as did their ancestors, with deer, ducks, small mammals, fish, snails, mussels, pigweed, lambsquarter, sunflowers, and nuts providing much of their nutrition. Outside the Adena and Hopewell spheres of influence, the Archaic subsistence and settlement systems appear to have remained essentially unchanged,[19] but this may have been the era in which the bow and arrow replaced the spear and atlatl over much of North America. George Odell studied thousands of stone tools from the Illinois Valley, looking for the kinds of damage to stone edges that projectile points typically sustain. He concluded that archaeologists have often looked at the wrong stone tools in trying to reconstruct ancient hunting: numerous stone tools besides obvious projectile points show evidence of high-energy impact typical of use as arrowheads; in fact, many of the lithics that show high-energy impact damage are crude-looking pieces archaeologists have often called "retouched flakes." The use of such projectile points seems to have increased dramatically after about four thousand years ago, and by the first few centuries A.D. the bow and arrow may have been adding enough extra production to some economies that significantly higher population densities were possible.

In any case, the Hopewell heartland was in no sense so poor that hunters and gatherers were desperately trying to become full-time maize farmers. Estimates of resources around the Hopewell settlement at Scoville, in the lower Illinois Valley, indicate that "within a half-hour's walk . . . there would annually be from 182,000 to 426,000 bushels of nuts and 48,000 to 484,000 bushels of acorns, 100 to 840 deer, 10,000 to 20,000 squirrels, and 200 turkeys. Not computed were seeds, fruits, smaller animals, fish, mussels, and migratory birds (6 million mallards were estimated to be in the Illinois River Valley in 1955)."[20] Given these abundant resources, it is not surprising that maize agriculture apparently appeared first in poorer, more marginal environments and displaced hunting and gathering in the lushest environments only much later.

Hopewell peoples traded extensively in exotic commodities, such as obsidian and Grizzly bear teeth from Montana, silver from Ontario, copper from northern Michigan, and shark teeth and many marine shells and products from the Gulf Coast. By the first few centuries A.D., people all over eastern North America were participants in a great trading network, and by "consuming" these goods by lavishing them on the dead and burying them deeply in mound-tombs, the demand for these treasures was kept high and the trading system in constant operation.

The conspicuous consumption of the Hopewell era, with the great mounds, rich burials, and rich arts and crafts, indicates that Archaic egalitarian social structures were changing, but they seem to have been fundamentally less complexly organized than, for example, Mesoamerican cities like Teotihuacan.[21] Most of the Hopewell and Adena communities were small sedentary groups of hunter-collectors who built the burial mounds and ceremonial complexes for reasons that may fundamentally have had to do with distributing resources. Whereas Archaic hunters and gatherers met their needs by following seasonal rounds and exploiting a diversity of resources, Hopewell communities probably accomplished the same thing by exchanging products among themselves. Like the Adena mounds, however, these constructions could have also been territorial markers.

But we know comparatively little about daily life for the mass of the Hopewell or Adena peoples. The few houses excavated seem to be rectangular or ovoid constructions supported on posts and covered with bark or mats. Most people apparently lived in small hamlets composed of a few such houses and a few storage pits and garbage dumps. Some people were obviously specialists in flint working and a few other arts and crafts, and some of these communities may have been organized around a "big-man" and his favored lineage. But at the Murphy Site, Licking County,

Ohio—one of the few Hopewell sites that has been excavated and analyzed for evidence of functional specialization—William Dancey and Paul Hooge found little evidence of complex economic or social organization. This site was occupied in the first few centuries A.D., probably intermittently or annually during the summer and fall.[22] Dancey and Hooge carefully examined hundreds of stone tools for patterns of wear, and found that most of them were simply chipped from a core, used for a short time, and then pitched in the garbage. They found no evidence of craft-specialization or farming, although these people did use pottery.

The Late Woodland Period

After about A.D. 400 there is some evidence of a decline in mound building and other changes in settlement patterns that used to be considered evidence of a Hopewellian "collapse." But it is not at all certain that there was, in fact, a collapse, at least in the sense of disintegrating economic relationships and declining productivity and population densities.

Richard Yerkes has reviewed studies by David Braun and others that suggest that although the people who lived toward the end of the Hopewell era, indeed, seem to have invested comparatively little effort in rich burials, pottery decoration, and trade of exotic goods, these cultural changes may be reflections of evolving—not disintegrating—cultural complexity.[23] At this time populations seem to have increased and people began producing more of their food through cultivation of native plants. People seem to have moved into most of the major river valleys and established small settlements, perhaps in small nuclear family groups. As we have seen elsewhere, monumental architecture may be a response to stress and a device to control societal growth, rather than just an expression of societal wealth and extravagance.

Mississippian Cultures (A.D. 800–1650)

From about A.D. 600 to about 1650, the agricultural way of life spread over much of eastern North America, and in the North American East the Mississippian culture appeared—the high point of cultural evolution in aboriginal North America, particularly in terms of geographical extent of influence, ceremonialism, public works, technology, population density, and social stratification.

Stoltman has listed some characteristics that define Mississippian culture, including intensive maize-beans-squash agriculture; apparent complex social organization—perhaps with chiefs and elites who could expropriate great amounts of societal wealth and power; a theocratic social organization, with elites having both religious and political power; towns in which one hundred, in some cases, one thousand or more people lived year round, often behind fortification; monumental architecture, including mounds and tombs; some occupational specialization in farming, trading, ritual, and administration; mortuary cults involving certain patterns of burying the dead with "status goods" under earthen mounds; distinctive pottery in shell-tempered wares, sometimes in the form of animals, and often beautifully incised, polished, or in other ways decorated; use of the bow and arrow, with distinctive small triangular stone arrowheads; and certain kinds of houses, built by digging wall trenches and then using clay and thatch to form rooms.[24]

The floresence of Mississippian culture may have been linked with the continuing domestication and improvement of maize. The sensitivity of maize to daylight and temperature is vividly demonstrated each summer along the northern fringe of its present range, such as in Seattle, where except in the warmest, driest summers few varieties can manage more than cigar-size ears. Maize was probably first an agricultural staple in North America only in those southern zones where there are at least two hundred frost-free days a year, but during the Mississippian era newer strains appeared that could be farmed productively with just one hundred twenty frost-free days. And so adaptable is the plant that it was a staple as far north as Ontario, Canada, by the time the Europeans arrived in the sixteenth century. Primitive strains of maize had such small cobs when first introduced from Mexico that considerable selective breeding was required before it was worthwhile to do all the work of clearing land, weeding, and harvesting necessary for successful maize agriculture here. But, as noted in Chapter 6, once domesticated beans were combined with maize in North American fields, thereby supplying critical amino acids lacking in maize, people could live in dense concentrations all over the North American East. The spread of maize farming has been detected by analyzing carbon isotope concentrations in the skeletons of ancient Native Americans. Maize contains relatively high amounts of ^{13}C, so people who eat maize also have high concentrations of this isotope. Several studies have shown that in Missouri, Arkansas, and other areas, skeletons of people who lived before about A.D. 1000 tend to have little ^{13}C, but beginning at about A.D. 900 to 1000, this isotope is found in radically higher concentrations.[25]

Artifacts and mounds of the Mississippian type first appeared in the lower Mississippi Valley, but they soon spread into the Tennessee River drainage and by A.D. 800 to 900 occurred over much of the Ohio and Missouri river valleys. Between A.D. 900 and A.D. 1600, large towns with impressive ceremonial centers were built from Florida to northern Illinois, and from Ohio to eastern Oklahoma, but the heartland of this culture was in the central Mississippi Valley.

The largest prehistoric settlement north of Mexico was Cahokia, in East

St. Louis, Illinois. Beginning at about A.D. 600 the people of Cahokia began building mounds and other features, and by about A.D. 1250 there were over 100 mounds within the 13 square kilometers of the site. Monk's Mound, an earthen pyramid in the center of Cahokia, is over 30 meters high, 241 by 316 meters at the base, and covers an area of more than 6.5 hectares (Figure 15.6). Thirty to forty thousand people are estimated to have lived in the environs of Cahokia at about A.D.1200, in several large towns, a few smaller towns, and more than forty villages; and no doubt people living within a large area around the settlement had some contact with Cahokia.[26]

The beginnings of a class-based society in which elites could control community wealth is evident at Cahokia. One adult male was buried with 20,000 shell beads, 800 arrowheads, and sheets of mica and copper. There is evidence of six separate burial episodes, involving at least 261 people, including 4 mutilated men and 118 women, many or all of whom were probably ritually strangled on the premise that they could thus accompany the "chief" into the next world.

Some elite groups were permanent residents of Cahokia, but most of the people who built and sustained this center were probably maize farmers who lived in surrounding hamlets and who supplemented their farm incomes with hunting and gathering.

Cahokia is an impressive site, but it was probably not functionally similar to Teotihuacan or similar cities in prehistoric complex societies, at least in degree of occupational specialization or the volumes and diversity of products produced, the class-stratification of the society, and the overall productivity of the economy.

Another great Mississippian site was at Moundville, in Alabama, where a central earthen mound is the focus of a settlement in which elaborate burials, fortifications, and other structures were found. Christopher Peebles' statistical analysis of the grave goods unearthed with these burials suggests a society with considerable internal ranking.[27] Moundville had a complex history, and one gets the impression that here was a society with the potential to become a true urban metropolis, the center of a state.[28] Here too, however, the straight ecological arguments are quite convincing as to why this never happened. In a temperate forested environment, with an economy based on extensive cultivation of maize and a few other plants and considerable hunting and foraging, there was not the stimuli of irrigation agriculture, interactions with nomads, and the other features of life in early agricultural Mesopotamia, Egypt, or China. The environment simply does not permit or allow agricultural intensification on the scale possible in the tropical Olmec river basins, the Nile Valley, and other areas of advanced complexity.

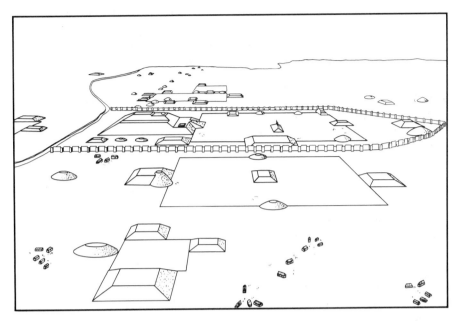

15.6 This reconstruction of Cahokia shows the palisade that enclosed the center of the site and some of the seventeen other major structures that existed at about A.D. 1200. The base of Monks Mound, the largest pyramid, is larger than the Great Pyramid at Giza, Egypt.

Ouside the great ceremonial centers, Mississippian villages were dispersed settlements of a few score wattle-and-daub structures, supported by internal wooden beams, with floors of packed earth.[29] The highest Mississippian population densities were in the rich river bottoms where it was possible to combine maize and bean farming with waterfowl exploitation.[30] In some areas the river annually renewed the fertility of the soil through alluviation, and permanent field agriculture was possible. Without plows or an advanced technology, the Native Americans in many areas relied on a form of swidden cultivation, in which they burned off the vegetation on rich, well-drained alluvial plains, then planted maize and a few other crops, invested some effort in weeding and cultivation during the growing season, and then harvested and stored as much maize as they could. After one or two seasons a given plot of swiden land would lose its fertility, and the areas of cultivation—and perhaps the whole village— would have to be shifted elsewhere.

Where maize was most productive the people probably planted two crops a year in the richest lands, with beans planted next to the maize stalks and encouraged to grow up the stalks. Beans, squash, sunflowers, and other crops were interspersed among the maize fields. The Mississippians were more gardeners than farmers, compared to, for example, the wheat farmers of Southwest Asia. There is little evidence that Mississippian peoples ever did much in the way of flood control or irrigation, but they certainly used hoes and intensive weeding to boost crop yields. The Pilgrims' report of Native Americans putting a fish in each hill of maize seeds has been disputed, but this too would have increased yields.

Our view of Mississippian life comes from both archaeological and ethnological sources, as Mississippian communities were still extant when the Europeans arrived in the sixteenth century. Initial accounts of the aboriginal Americans by Europeans must be viewed with suspicion, because there is every reason to believe that European diseases sped quickly through aboriginal popoulations—far faster in fact than the spread of the Europeans. Some major diseases were introduced by the Europeans, but it now appears that mycobacterial diseases (possibly tuberculosis) were present in prehistoric Peru, and there is some evidence that apparently high incidences of TB in native Americans "may have been the result of enforced changes in ecological and environmental factors rather than of exposure to a new infectious disease."[31] These colonists described intensely stratified societies where the elites were able to draw almost without limit on the resources of the communities.

Warfare among competing Mississippian chiefdoms may have been frequent and brutal, particularly as the best lands were overexploited and fertility began to fall.[32]

Ethnographic accounts of Mississippian communities as they existed in the sixteenth and seventeenth centuries—long after the culmination of Mississippian culture—describe an intensely class-conscious society in which nobles and warriors alternately exploited and abused the "stinkards," or commoners and slaves who made up most of the societies. The upper classes were slavishly obeyed and respected. They frequently married the lower classes, but the aristocrat could divorce or kill the lower-ranking spouse, given even minor cause.

Many late Mississippian and later period mounds and burials contain ornaments, pottery, and other artifacts decorated with motifs almost identical to some Mesoamerican motifs, including plumed serpents, eagles, jaguars, and warriors carrying trophy heads, as well as the fifty-two-year calendar round. Collectively, they are taken as evidence of a southern (Mesoamerican) religious cult. Opinions differ on their significance in terms of contacts with Mesoamerica; some consider them evidence of quite direct and sustained contacts, others see them as minor borrowings with little more than an accidental connection to Mesoamerican cultures. The southern cult, as well as the Adena and Hopewell mortuary ceremonialism, likely achieved importance in North America only after cultural complexity in the North had reached a stage where these elements "made sense" in terms of northern societies.

After A.D. 1000 Mississippian "colonists" began to emigrate from the cultural heartland, and some groups may have founded quasimilitary enclaves in Alabama, Missouri, and elsewhere. Conceivably, these daughter communities were sent with the express purpose of extending Mississippian influence over other cultural groups, but the evidence for this is scanty.[33]

The spectacular Mississippian sites dominate the North American archaeological record, but all over the North American East there were divergent cultural traditions, many of them interacting economically, but also remaining distinctive in essential ways, from language to social organization. To the northeast, the Iroquoian peoples adapted maize-beans-squash farming to the colder uplands of New York State, New England, and southeastern Canada. Their shifting pattern of swidden agriculture seems to have created land shortages and recurrent warfare, forcing people to live in well-fortified villages. To the west, marginal cultivators lived in the richer river valleys of Nebraska, South Dakota, and the other north central states, with bison-hunting groups on the grasslands where maize agriculture could not be done until the introduction by Europeans of deep-bottom plows. The horse-mounted nomadic hunters who range through thousands of Hollywood epics have some basis in reality, but only after the Europeans reintroduced the horse in America. Most of the Desert West was thinly settled by hunter-foragers until European colonization, but rich and sedentary communities were established along the Northwest Coast and areas of California where marine and terrestial resources were so naturally productive that low-level complex societies could be sustained by simple fishing-foraging-hunting.

In contrast to most other ancient cultures, there is little question about the immediate cause of the decline of Mississippian culture. The Indians had no natural immunities to measles, smallpox, and cholera, and the densely settled Mississippian areas provided an ideal medium for the rapid spread of these highly contagious diseases.[34] Le Page du Pratz, who lived with the Natchez from 1718 to 1734, found that even "minor" diseases were devastating:

> Two distempers, that are not very fatal in other parts of the world, make dreadful ravages among them; I mean small-pox and a cold, which baffle all the arts of their physicians, who in other respects are very skillful. When a nation is attacked by the small-pox, it quickly makes great havock; for as a whole family is crowded into a small hut, which has no communications with the external air, but a door about two feet wide and four feet high, the distemper, if it seizes one, is quickly communicated to all. The aged die in consequence of their advanced years, and the bad quality of their food; and the young, if they are not strictly watched,

15.7 A sixteenth-century illustration by Jacques le Moyne of the burial of a Florida Indian chief. Arrows have been driven into the ground around the grave, and the chief's houses (upper left) are being ceremonially burned.

destroy themselves, from an abhorrence of the blotches on their skin. . . . Colds, which are very common in winter, likewise destroy great numbers of the natives. In that season they keep fires in their huts day and night; and as there is no other opening but the door, the air within the hut is kept excessively warm without any free circulation; so that when they have occasion to go out, the cold seizes them, and the consequences of it are almost always fatal.

Ann Ramenofsky has analyzed the archaeological evidence from several regions of the North American East and found that in each area there are significantly fewer settlements that date to the decades just prior to European invasion of these areas, which suggests that epidemic European diseases spread rapidly inland from the first physical encounters between most Native Americans and Europeans.[35] Native Americans died by the hundreds of thousands from diseases introduced by people they never saw.

Within a few decades of European contact in the sixteenth century, the once highly integrated and proud Mississippian people, and other cultures as well, were a much reduced and poverty-stricken group, living amid thousands of abandoned settlements and eroding mounds attesting to their former greatness.

In summary of cultural developments in the North American East, we see the familiar script of cultural evolution. From a hunting and gathering base, specialized hunting and minor plant exploitation gradually gave way to an intricate hunting-foraging economies, in which fishing, nut collecting, shellfishing, and other activities were added to the subsistence repertoire according to what was probably a largely unrecognized but very precise "cost-benefit" analysis; then, after centuries of manipulation and selection, maize-based agriculture displaced less-productive economies in many areas, with consequent increasing population densities and the establishment of large sedentary communities. Once food production reached certain levels, the familiar harbingers of increasing cultural complexity, such as the spread of religious and stylistic traditions, monumental architecture, mortuary cults, and increasingly diverse and interdependent arts and crafts, also appeared. Social and religious hierarchies emerged as "efficient" ways to make the decisions necessary for the perpetuation of these increasingly complex economies, and the institutionalization of prestige and privilege may have arisen as an effective way of reducing competition between these populations and of maximizing administrative efficiency.

In these essentials, little differentiates the sequence of cultural evolution in North America from that in Mesoamerica, except that, given the available domesticates and technology, most of eastern North America had less agricultural potential than Mesoamerica, causing these northern cultures to stabilize at a much lower level of complexity. Had the Europeans not invaded and introduced their diseases, true state-level societies might well have evolved from the remnants of the Mississippian climax, but only if food production could have been increased drastically through technological, agricultural, or administrative innovations.

Prehistoric Agriculturalists in Southwestern North America

The dramatic mountains, crystal skies, and primary colors of the Southwest make it an attractive place to live—especially if one has access to a municipal water system and a modern market and transport system. But from the perspective of a hunter and gatherer or subsistence farmer, the majority of the Southwest is not a lush environment; there are few large rivers or streams and the combination of low rainfall and high elevation renders much of the Southwest a land of searing summers and bitterly cold winters.

In some prehistoric periods the Southwest was wetter than it is today, but for most of the last ten thousand years the Southwest has usually been at least as hot and dry as it is today, and there were short periods of extreme drought.

As a consequence, although people have lived here for at least ten thousands years, and although they eventually adopted maize-based agriculture, aggregated into large towns, and evolved some occupational and administrative specialization, they never produced class-structured, hierarchically organized, economically differentiated societies. Periodic food shortages stimulated the development of small-scale cooperation among groups, but there were not the economic inducements or basic productivity to make such elaborated cultural forms necessary or possible.[36]

Yet the archaeological record of the American Southwest is an extraordinary cultural resource. Its proximity to major universities and excellent aridity-related preservation has meant that the archaeology of this region is among the best known in the world (although here, as in most areas, research has concentrated on ceremonial centers and important segments of the archaeological record are largely unstudied).[37] Many important methodological advances in archaeology have developed in part out of research in the Southwest, including tree-ring dating, regional statistical sampling designs, analyses of site formation processes, and paleobotanical and paleoenvironmental research. Also, the very fact that early Southwest cultures formed low level political and economic systems offers an opportunity to study the initial stages of cultural changes that in more productive environments are obscured archaeologically by the great states and empires that followed.

As we saw in Chapter 6, people have been in the Southwest for at least twelve thousand years, and up to about A.D. 1, people here worked out a marvelous array of arid-land adaptations based on hunting and foraging. Archaeologists used to think that some maize was cultivated here as early as 3000 B.C., but the timing and rate of spread of maize agriculture in the Southwest and its association with the appearance of sedentary villages is a complex problem.[38] Traces of domesticated maize dating to about a 1000 B.C. and even earlier have been found at several sites in the Chaco area of New Mexico, but maize seems to have been a very minor food in these economies. As Alan Simmons notes, these early indications of the knowledge of domestic maize lead one to wonder why it took so long for these Southwesterners to shift to a full-time reliance on maize-farming. It is likely that domesticated maize was available for adoption all along the southern margins of the Southwest for several millennia before it became important in the Southwest. Strains had to be adapted to local Southwest environments and certain thresholds of technology and demography would have had to be crossed before maize-farming could be a practical option for most Southwesterners.

15.8 A 1930s-era photograph of a then-typical Indian village in the American Southwest. This form of adobe construction goes back many centuries in the Southwest.

Foragers of the Cochise culture, heirs to a tradition thousands of years old of desert plant collecting, may have been exploiting maize for a long time, but it was only about 500 to 300 B.C. that they began using a productive, drought-resistant strain of maize—called Chapalote maize—that eventually formed the basis for widespread reliable maize cultivation in the Southwest. Chapalote maize combined with domesticated beans and squash became the dietary staples, but even today Southwesterners still hunt deer, rabbits, and other animals, and augment their farm produce with high-calorie nuts, berries, and seeds. Like all marginal cultivators, when drought and crop failure struck, the people of the Southwest intensified exploitation of hardy wild and partially domesticated plants, such as cactus and mesquite.

The Hohokam

The Hohokam peoples, who flourished in the Salt and Gila river valleys (Arizona) between about A.D. 500 and A.D. 1200 (arguments about these dates remain)[39] seem to represent the confluence of Archaic hunting and gathering traditions and Mesoamerican cultures. Some archaeologists (e.g., Haury[40]) have concluded that the Hohokam were more than local peoples whose culture was transformed by contacts with the high civilizations of Mexico; on the basis of late Hohokam styles of architecture, construction, ceramics, turquoise ornaments, and other artifacts, they conclude that Hohokam styles are so similar to Mesoamerican examples that they can only be the result of an immigration of Mexican people who transplanted their way of life directly to the Hohokam area, where previously there had been only hunters and gatherers.

One of the most interesting aspects of the Hohokam is their agricultural system. Their homeland is set in some of the driest deserts of North America, where summer temperatures have impressed even archaeological veterans of the Near East. Beginning at about 300 B.C., the Hohokam channelled water from the Salt and Gila rivers to irrigate their gardens of maize and other crops. Modern buildings have erased much of the Hohokam irrigation system, but two canals near Phoenix were over 16 kilometers long, several meters wide, and about 60 centimeters deep when first constructed. Tightly woven grass mats were probably used as gates to open and close canal segments, and earthen dams on the rivers in some places diverted water through canals for more than 50 kilometers across the desert floor, with many small branches serving individual fields.

In the recent past, southwestern Indian irrigation systems, as well as comparable agricultural systems elsewhere in the world, have been operated by a few thousand people in relatively simple tribal organizations in which no coercion, permanent authorities, or police agencies were necessary, and this might well have been the case among the Hohokam.

Although the largest irrigation systems were apparently built sometime after A.D. 800, by about A.D. 300 there were already indications of minor increases in the complexity of Hohokam cultures. Some low platform mounds, about 29 meters long, 23 meters wide, and 3 meters high, were built at the Snaketown site, and here and elsewhere people built large sunken ball courts like those found in Mesoamerica. The Hohokam ball courts were east-west oriented oval depressions about 60 meters long, with 4.5- to 6-meter-high sloping earth embankments on a side. Early Spanish observers of the ball game in Mesoamerica report that the objective was to try to knock a rubber ball (several of which have been found in the Southwest) through a goal using knees, elbows, or torso, and that losing players were sometimes executed. The southwestern ball courts were probably not stops on the northern road trips of the Mexican major leagues, but they do indicate very close southern affinities, as do the platform mounds. And between A.D. 900 and 1200, many other Mesoamerican elements were imported, including cotton textiles, certain ceramic motifs, pyrite mirrors, effigy vessels, cast copper bells, ear plugs, etched shell ornaments, and even parrots and macaws—probably imported from the South and kept and prized for their feathers.

Generally, however, there is little evidence in either subsistence practices, settlements, or mortuary ceremonies of evolving rank and wealth differences. Most of the Hohokam lived in small square pit houses roofed

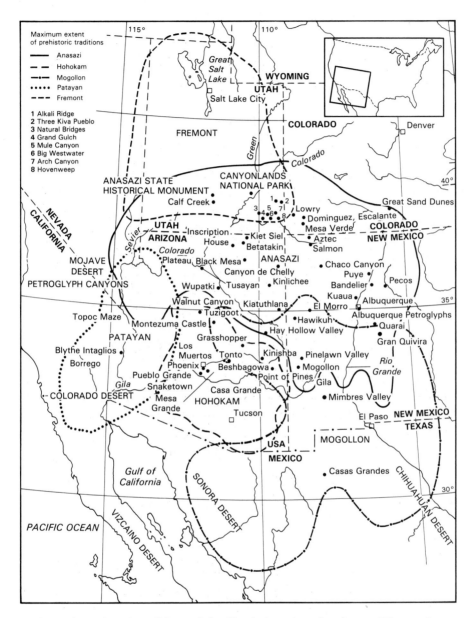

15.9 Major cultural traditions of the North American Southwest. These cultures are shown at their maximum extent, which they did not reach simultaneously.

581

with clay and grass domes supported by a wooden pole framework. Early dwellings appear large enough for several families, but single-family residences became more popular in later periods.

Hohokam settlement patterns are not well known. There are few indications of a master plan in village layouts or of economic specialization of villages. Most villages probably repeated all the economic activities of other villages, except that some favored locations allowed a greater reliance on maize, beans, squash, and other crops. Nonetheless, many atlatl dart points and arrowheads have been found at each Hohokam site, and botanical remains indicate that almost every community supplemented its diet with wild mustard, amaranths, chenopods, cactus fruits, mesquite, screwbeans, and other wild products.[41]

Artifacts in the Hohokam style have been found over a wide area, and various scholars have tried to determine whether these artifact distributions represent migrations, or trading expeditions, or the spread of regular trade networks, perhaps with associated political and social ties.[42]

The Mogollon

The Mogollon peoples (pronounced something like "Mug-ee-yone"), who also were heirs to the Archaic desert foraging cultures of the last several millennia B.C., lived mainly in the mountains of east central Arizona and west central New Mexico. Sedentary villages and ceramics of the Mogollon type may have already been established by the last few centuries B.C., but the distinctive Mogollon red-and-brown pottery is securely dated to about the third century A.D., at which time villages of about fifteen pit houses each were scattered along ridges, bluffs, and terraces. Like their contemporaries in much of the Americas, they relied on the maize-beans-squash complex, supplemented by many wild plants and game. Mogollon burials of this period were often simple inhumations in house floor accompanied by a few pottery vessels, turquoise ornaments, and stone tools typical of unstratified societies.

An important Mogollon site is the Swarts ruins, Mimbres Valley, New Mexico. There river boulders set in adobe were used to construct a large complex of conjoined rectangular rooms, access to which was apparently only through the roof, since there were no exterior doors. Inside walls were plastered with mud, and doorways led from one room to another. Storage bins, shelves, fireplaces, and benches constituted the essential furnishings. Corpses were frequently interred in the floors of abandoned rooms, in occupied rooms, or within the village compound; most burials were very simple.

The Mogollon built no pyramids, ball courts, or major irrigation systems, but they did produce an extraordinarily beautiful array of ceramics, particularly the "Mimbres" ceramic forms, decorated with vivid figures of frogs, insects, fish, deer, and other animal life painted against black backgrounds. Stephen Jett and Peter Moyle have shown that many of the fish species illustrated on Mimbres pottery are marine species that could be found no closer to the Mimbres area than 1500 kilometers southwest, in the Gulf of California. The beauty of their pottery has made the Mogollon victims of widespread looting, a sickening destruction of yet another Native American culture.

Mogollon settlement patterns show some evidence of minor emerging social ranking. Mogollon villages seem to be of two types: a small cluster of fewer than five pit houses, without any large ceremonial kivas, and larger collections of one hundred or more pit houses associated with "great kivas." The largest pit houses in these communities tend to have the greatest amount of storage space, more exotic trade items, and evidence of greater consumption of agricultural products. Lightfoot and Feinman interpret this evidence and other data to mean that a "suprahousehold" decision-making organization had developed among the Mogollon, perhaps as a result of competition among rival political leaders.[43] But it is hard to link changing distributions of artifacts like these to the fairly subtle social changes that would have been involved in the rise of suprahousehold administrative positions. In any case, there is little about the Mogollon to suggest profound social changes in the direction of complexity. Natural limitations on agricultural intensification seem to have restricted the complexity of social and political organization.[44]

One of the major settlement pattern transitions in Southwestern culture history occurred during the first millennium A.D., when people gradually, generation by generation, stopped living in relatively small villages of semisubterranean pit houses and began building their communities in the "pueblo" style of blocks of contiguous above-ground rooms.

For the archaeologist, of course, such a change is immediately seen in adaptive terms: what conditions would induce this sort of long-term, regional change? In a consideration of this question from the perspective of Mogollon settlements in western Texas, Michael Whalen notes that the change is associated with an increase in regional population densities, the appearance of larger community size, a growing reliance on plant cultivation as the main subsistence source, and an increasingly elaborate ceremonialism, as expressed in the design and contents of structures.[45]

Whalen suggests that as populations grew, climates fluctuated, and settlements spread into different microenvironments, people tried to maintain their traditional way of life by extending agriculture into the margins of basins where water was more available. The greater productivity of these basin margins fostered certain kinds of economic agricultural specialization in which the community rather than the family was the basic social and economic unit. McGuire and Schiffer show that the transition to pit houses was the outcome of a long period of experimentation.[46]

Fred Plog analyzed the history of the Colorado Plateau for the period between A.D. 400 and about 1540 and concluded that at various times people in this area formed "alliances" in which there was fairly intensive trade, some emergent social ranking and stratification, and the formation of administrative and trade centers.[47] He distinguishes these Colorado Plateau systems from the more complex societies of Mesoamerica and elsewhere, noting that the Colorado Plateau societies never intensified agricultural production to the level that typically supported more complex societies, nor did they ever seem to include full-time craftsmen working in administered centralized workshops. The small steps the southwesterners made in this direction, Plog concluded, arose out of the needs of addressing imbalances between population densities and resources, but these changes were always strictly limited by the relatively great environmental diversity and generally low agricultural potential of the region.

Some Mogollon sites, such as the Casas Grandes sites in northern Mexico, have what appear to be Mogollon style artifacts along with Mesoamerican items, such as ball courts, platform mounds, and jewelry in turquoise and shell. Here too scholars disagree as to whether such sites represent the direct intervention of Mesoamerican traders in Southwestern economies or a more peripheral, indirect kind of relationship, and just how complex these societies would have been.

Mogollon cultural traditions and population densities appear to have declined rapidly by A.D. 1300, perhaps as droughts contracted the areas in which they could survive, or as a result of invasion or absorption by neighboring Anasazi groups.

The Anasazi

Like the Mogollon and Hohokam, the Anasazi developed out of desert foraging cultures, and their earliest prepottery representatives are widely known as the Basketmaker cultures. The Anasazi of Black Mesa, Arizona, have been particularly intensively studied. It is not entirely certain how mobile early Anasazi groups were, but by the first few centuries A.D. they were apparently living in many sedentary settlements located on old river terraces and mesa tops or in river valleys in the high plateau country of the central Southwest.[48] The earliest houses were circular structures of wattle and daub set on log bases, or semisubterranean houses. These early Anasazi did not use pottery and, although they ate maize, beans, and squash, they invested little labor in cultivating these crops, relying instead on wild foods, such as roots, bulbs, grass seeds, nuts, acorns, berries, cactus fruits, sunflowers, deer, rabbits, antelopes, and wild sheep.

After about A.D. 400, the Anasazi began to use pottery and to build large pit houses, most of which were circular or rectangular, from 3 to 7.5 meters in diameter, and were covered by log and mud roofs supported on center posts. Interior walls were plastered with mud or faced with stone, access was through a descending passageway, and fireplaces and benches were standard furnishings. At some sites large ceremonial pit houses, or kivas, were built. After about A.D. 700, above-ground masonry houses were erected in some Anasazi communities, but the pit house and kiva combination continued to be the basic village type until the end of the thirteenth century A.D.

Defensive considerations, greater exploitation of more productive strains of maize, and climatic changes beginning about A.D. 700 may have spurred the Anasazi into constructing the "cliff cities" for which they are famous.[49] Pollen and geological studies suggest that summer rainfall, in the form of torrential storms, increased after A.D. 700, while winter rainfall decreased, and the resulting changes in water tables and stream flows may have forced the Anasazi to congregate around larger, permanently flowing rivers.[50] Hillsides were terraced to control erosion, and diversion canals and dams were constructed to control and store as much of the vital summer rain-

15.10   Reconstruction of Pueblo Bonito, as it may have appeared at about A.D. 1050.

15.11 Cliff Palace, Mesa Verde, Colorado, is one of the most impressive settlements built under massive cliffs in the American Southwest.

water as possible. On Chapin Mesa, in the Mesa Verde area, for example, the older agricultural fields were extended by an elaborate checkdam system that added 8 or 12 hectares of cultivable land.

In some areas erosion forced frequent settlement relocations, and some communities met the changing agricultural conditions by scattering into small family groups; but the prevailing response was to aggregate into large towns along the major rivers. By about 1100 A.D., prosperous enclaves were established at Mesa Verde, Chaco Canyon, Canyon de Chelly, and elsewhere, and in some ways these communities represent the high point of southwestern culture. There were twelve large towns along Chaco Canyon, of which Pueblo Bonito was the largest, with over eight hundred rooms and perhaps twelve hundred people (Figure 15.10).

By about A.D. 1150, most of the known cliff cities in the Four Corners area had been established, all of very similar construction. Settlements grew by simple accretion; more rooms were added as needed. The remote location of the cliff towns may have had to do with defensive considerations, although defense against what enemy we do not know.

As beautiful as many late Anasazi settlements were, they do not appear to have been the work of a highly complex society. The buildings, while superbly adapted to their environment, are quite crudely constructed. There may have been minor occupational specialization in ceramic manufacture, weaving, and turquoise carving, but most if not all the people were subsistence farmers. Nor is any evidence of differential rank expressed in domestic architecture or in grave goods. Even the irrigation system, while intricate and efficient, was probably administered through simple kinship systems.

Shortly before A.D. 1300, many once prosperous Anasazi communities began to be abandoned, and when the Spanish arrived in the sixteenth century, they found the descendants of the Anasazi living along the Rio Grande in small villages, each a largely autonomous political and economic unit.

In what is now Utah, on the periphery of the Anasazi world, the Fremont culture appeared in the first and early second millennia A.D.—defined by distinctive artifact styles. Another peripheral group were the Patayan of the Colorado River area.

Summary and Conclusions

> Illegal aliens have always been a problem in the United
> States. Ask any Indian. (Robert Orben)

The New World was truly a "second earth," where long millennia of indigenous cultural development produced a rich spectrum of peoples and cultures unaffected by contact with European and Asian peoples. In some ways it is less than satisfying to conclude that the temperate latitudes of North America never produced states and empires because of the simple absence of draft animals, developed metallurgy, extensive warm arid alluvial plains, and a few other techno-environmental factors—but this conclusion seems inescapable. From the perspective of the present, the violent collision between Old and New World cultures also seems an inevitable outcome of the competition dynamic of evolutionary processes and the demographic, techno-environmental, and other conditions of both areas. But to rank Old and New World cultures in this sense of an evolutionary scale is only a useful research device; it should not obscure the great richness of the ancient cultures of both areas.

Notes

1. Yerkes, "The Woodland and Mississippian Traditions in the Prehistory of Midwestern North America," pp. 311–12.
2. Smith, "Middle Mississippi Exploitation of Animal Populations"; Smith, "The Role of *Chenopodium* as a Domesticate in the Pre-Maize Garden System of the Eastern United States," p. 32.
3. Asch and Asch, "Prehistoric Plant Cultivation in West-Central Illinois."
4. O'Brien, "Sedentism, Population Growth, and Resource Selection in the Woodland Midwest: A Review of Coevolutionary Developments."
5. Rindos, *The Origins of Agriculture.*
6. O'Brien, "Sedentism, Population Growth, and Resource Selection in the Woodland Midwest: A Review of Coevolutionary Developments," p. 186; also see Keegan, ed, *Emergent Horticultural Economies of the Eastern Woodlands.*
7. Braun, "Comment on M. J. O'Brien, 'Sedentism, Population Growth, and Resource Selection in the Woodland Midwest: A Review of Coevolutionary Developments,'" p. 189.
8. Braun and Plog, "Evolution of 'Tribal' Social Networks: Theory and Prehistoric North American Evidence."
9. Fowler, "Agriculture and Village Settlement in the North American East: The Central Mississippi Valley Area, a Case History."
10. Farnsworth and Emerson, *Early Woodland Archaeology;* Morse and Morse, *Archaeology of the Central Mississippi Valley.*
11. Stoltman, *Prehistoric Mound Builders of the Mississippi Valley;* Marquardt, "Complexity and Scale in the Study of Fisher-Gatherer-Hunters"; Rothschild, "Mortuary Behavior and Social Organization at Indian Knoll and Dickson Mounds."
12. Yerkes, "The Woodland and Mississippian Traditions in the Prehistory of Midwestern North America," p. 315.
13. Murphy, *An Archaeological History of the Hocking Valley,* pp. 161–63; Yerkes, "The Woodland and Mississippian Traditions in the Prehistory of Midwestern North America," p. 317.
14. Yerkes, "The Woodland and Mississippian Traditions in the Prehistory of Midwestern North America," p. 317.
15. Brown, "Long-term Trends to Sedentism and the Emergence of Complexity in the American Midwest," p. 219.
16. Coe, Snow, and Benson, *Atlas of Ancient America,* p. 50.
17. Bush, "A Ceramic Analysis of the Late Adena Buckmeyer Site"; Baby and Langlois, "Prehistoric Agriculture," p. 3.

18. Tainter, "Social Inference and Mortuary Practices: An Experiment in Numerical Classification."
19. Dragoo, "Some Aspects of Eastern North American Prehistory: A Review," pp. 18–19.
20. Jennings, *Prehistory of North America*, p. 232.
21. Brose and Greber, *Hopewell Archaeology: The Chillicothe Conference.*
22. Yerkes, "The Woodland and Mississippian Traditions in the Prehistory of Midwestern North America," p. 323.
23. Braun, "The Social and Technological Roots of 'Late Woodland.' "
24. Stoltman, "Ancient Peoples of the Upper Mississippi River Valley," pp. 232–33.
25. Lynott et al., "Stable Carbon Isotopic Evidence for Maize Agriculture in Southeast Missouri and Northeast Arkansas."
26. Porter, *The Mitchell Site and Prehistoric Exchange Systems at Cahokia: A.D. 1000–300.*
27. Peebles, "Moundville and Surrounding Sites."
28. Steponaitis, *Ceramics, Chronology, and Community Patterns. An Archaeological Study at Moundville.*
29. Jennings, *Prehistory of North America*, p. 256.
30. Smith, "Middle Mississippi Exploitation of Animal Populations—A Predictive Model."
31. Clark et al., "The Evolution of Mycobacterial Disease in Human Populations."
32. Dickson, "The Yanomamo of the Mississippi Valley? Some Reflections on Larson . . . Gibson . . . and Mississippi Period Warfare in the Southeastern United States."
33. Yerkes, "The Woodland and Mississippian Traditions in the Prehistory of Midwestern North America," p. 346.
34. Stewart, *The People of America.*
35. Ramenofsky, *Vectors of Death.*
36. Minnis, *Social Adaptation to Food Stress.*
37. Upham, "Archaeological Visibility and the Underclass of Southwestern Prehistory."
38. Cordell, *Prehistory of the Southwest;* Simmons, "New Evidence for the Early Use of Cultigens in the American Southwest."
39. See, e.g., Schiffer, "Hohokam Chronology."
40. Haury, *The Hohokam Desert Farmers and Craftsmen.*
41. Washburn, "The American Southwest," p. 124.
42. Reviewed in Cordell, *Prehistory of the Southwest.*
43. Lightfoot and Feinman, "Social Differentiation and Leadership Development in Early Pithouse Villages in the Mogollon Region of the American Southwest," p. 81.
44. Minnis, *Social Adaptation to Food Stress.*
45. Whalen, "Cultural-Ecological Aspects of the Pithouse-to-Pueblo Transition in a Portion of the Southwest."
46. McGuire and Schiffer, "A Theory of Architectural Design."
47. Plog, "Political and Economic Alliances on the Colorado Plateaus, A.D. 400 to 1450."
48. Powell, *Mobility and Adaptation: The Anasazi of Black Mesa, Arizona.*
49. Dickson, "Settlement Pattern Stability and Change in the Middle Northern Rio Grande Region, New Mexico: A Test of Some Hypotheses."
50. Washburn, "The American Southwest," p. 114; but see Rice, "A Systematic Explanation of a Change in Mogollon Settlement Patterns."

Bibliography

Asch, D. L. and N. B. Asch. 1985. "Prehistoric Plant Cultivation in West-Central Illinois." In *Prehistoric Food Production in North America*, ed., R. I. Ford, University of Michigan Museum of Anthropology, Anthropological Papers 75.

Baby, R. and S. Langlois. 1977. "Prehistoric Agriculture: A Study of House Types in the Ohio Valley." *Ohio Historical Society Echoes* 16(2):1–5.

Braun, D. P. 1983. "Pots as Tools." In *Archaeological Hammers and Theories*, eds., A. Keene and J. Moore. New York: Academic Press.

———. 1987. "Comment on M. J. O'Brien, 'Sedentism, Population Growth, and Resource Selection in the Woodland Midwest: A Review of Coevolutionary Developments.'" *Current Anthropology* 28(2):189–90.

———. 1988. "The Social and Technological Roots of 'Late Woodland.'" In *Interpretations of Culture Change in Eastern Woodlands During the Late Woodland Period*, ed. R. W. Yerkes. Dept. of Anthropology, Ohio State University, Occasional Papers in Anthropology 3:17–38.

Braun, D. P. and S. Plog. 1982. "Evolution of 'Tribal' Social Networks: Theory and Prehistoric North American Evidence." *American Antiquity* 47:504–25.

Brose, D. S. and N. Greber, eds. 1979. *Hopewell Archaeology: The Chillicothe Conference.* Kent: The Kent State University Press.

Brown, J. A. 1977. "Current Directions in Midwestern Archaeology." *Annual Review of Anthropology* 6:161–79.

———. 1985. "Long-term Trends to Sedentism and the Emergence of Complexity in the American Midwest." In *Prehistoric Hunter-Gatherers: The Emergence of Cultural Complexity*, eds. T. D. Price and J. A. Brown. Orlando, Florida: Academic Press.

Broyles, B. J. 1971. *Second Preliminary Report: The St. Albans Site, Kanawha County, West Virginia.* West Virginia Geological and Economic Survey Report, Archaeological Investigation 3.

Bullen, R. P. 1971. "The Beginnings of Pottery in Eastern United States as Seen from Florida." *Eastern States Archaeological Federation Bulletin*, no. 30:10–11.

Bush, D. E. 1975. "A Ceramic Analysis of the Late Adena Buckmeyer Site, Perry County, Ohio." *Michigan Archaeologist* 21:9–23.

Caldwell, J. 1958. *Trend and Tradition in the Prehistory of the Eastern United States.* Menasha: Mem. Ser. Am. Anthr. Assoc. 88.

———. 1971. "Eastern North America." In *Prehistoric Agriculture*, ed. S. Struever. Garden City, N.Y.: Natural History Press.

Clark, G. A., M. A. Kelley, J. M. Grange, and M. C. Hill. 1987. "The Evolution of Mycobacterial Disease in Human Populations." *Current Anthropology* 28(1):45–62.

Cleland, C. E. 1966. *The Prehistoric Animal Ecology and Ethnozoology of the Upper Great Lakes Region.* Ann Arbor: Museum of Anthropology, Anthropological Papers, University of Michigan, no. 29.

Coe, M., D. Snow, and E. Benson. 1986. *Atlas of Ancient America.* New York: Facts on File.

Cordell, L. S. 1984. *Prehistory of the Southwest.* New York: Academic Press.

———. 1984. "Southwestern Archaeology." *Annual Review of Anthropology* 13:301–32.

Cordell, L. S. and F. Plog. 1979. "Escaping the Confines of Normative Thought—A Reevaluation of Puebloan Prehistory." *American Antiquity* 44:405–29.

Cressman, L. S. 1977. *Prehistory of the Far West.* Salt Lake City: University of Utah Press.

DeJarnette, D. L. 1967. "Albama Pebble Tools: The Lively Complex." *Eastern States Archaeological Federation Bulletin*, no. 26.

Dickson, D. B. 1975. "Settlement Pattern Stability and Change in the Middle Northern Rio Grande Region, New Mexico: A Test of Some Hypotheses." *American Antiquity* 40:159–71.

———. 1981. "The Yanomamo of the Mississippi Valley? Some Reflections on Larson . . . Gibson . . . and Mississippi Period Warfare in the Southeastern United States." *American Antiquity* 46:909–16.

Doyel, D. E. 1986. "A Short History of Hohokam Research." In Reid, J. J. and D. E. Doyel, eds. *Emil W. Haury's Prehistory of the American Southwest.* Tucson: University of Arizona Press.

Dragoo, D. W. 1963. "Mounds for the Dead." *Annals of Carnegie Museum* 37:1–315.

———. 1976. "Some Aspects of Eastern North American Prehistory: A Review." *American Antiquity* 41:3–27.

Dunnell, R. C. 1967. "The Prehistory of Fishtrap, Kentucky: Archaeological Interpre- -tation in Marginal Areas." Unpublished doctoral dissertation, Yale University.

Farnsworth, K. B. 1973. *An Archaeological Survey of the Macoupin Valley.* Springfield: Illinois State Museum, Reports of Investigations, no 26.

Farnsworth, K. B. and T. E. Emerson, eds. 1986. *Early Woodland Archaeology.* Kampsville Seminars in Archaeology, No. 2. Kampsville, Ill.: Kampsville Archeological Center, Center for American Archeology.

Ford, R. I. 1974. "Northeastern Archeology Past and Future Directions." *Annual Review of Anthropology* 3:385–414.

———. 1981. "Gardening and Farming before A. D. 1000: Patterns of Prehistoric Cultivation North of Mexico." *Journal of Ethnobiology* 1(1):6–27.

Fowler, M. L. 1969. *Explorations into Cahokia Archaeology.* Illinois Archaeological Survey Bulletin 7.

———. 1971. "Agriculture and Village Settlement in the North American East: The Central Mississippi Valley Area, a Case History." In *Prehistoric Agriculture,* ed. S. Struever. Garden City, N.Y.: Natural History Press.

———. 1975. "A Pre-Columbian Urban Center on the Mississippi." *Scientific American* 233(2):93–101.

Glassow, M. A. 1980. *Prehistoric Agricultural Development in the Northern Southwest: A Study in Changing Patterns of Land Use.* Socorro, N.M.: Ballena Press.

Griffin, J. B. 1967. "Eastern North American Archaeology: A Summary." *Science* 156:175–91.

Gumerman, G. J., ed. 1988. *The Anasazi in a Changing Environment.* Cambridge: Cambridge University Press.

Haury, E. W. 1976. *The Hohokam Desert Farmers and Craftsmen.* Tucson: University of Arizona Press.

Hill, J. N. 1966. "A Prehistoric Community in Eastern Arizona." *Southwestern Journal of Anthropology* 22:9–30.

———. 1970. *Broken K Pueblo: Prehistoric Social Organization in the American Southwest.* Tucson: University of Arizona Press.

Hines, P. 1977. "On Social Organization in the Middle Mississippian—States or Chiefdoms?" *Current Anthropology* 18:337–38.

Houart, G. L. 1971. *Koster—A Stratified Archaic Site in the Illinois Valley.* Springfield: Illinois State Museum Reports of Investigations, no. 22.

Hunter-Anderson, R. L. 1981. "Comments on Cordell and Plog's 'Escaping the Confines of Normative Thought.'" *American Antiquity* 46:194–97.

———. 1986. *Prehistoric Adaptation in the American South-West.* Cambridge, England: Cambridge University Press.

Irwin-Williams, C. 1968. "The Reconstruction of Archaic Culture History in the Southwestern United States." In *Archaic Prehistory in the Western United States,* ed. C. Irwin-Williams. Portales: Eastern New Mexico University Contributions in Anthropology.

Jennings, J. D. 1974. *Prehistory of North America.* 2nd Ed. New York: McGraw-Hill.

Jett, S. C. and P. B. Moyle. 1986. "The Exotic Origins of Fishes Depicted on Prehistoric Mimbres Pottery from New Mexico." *American Antiquity* 51(4):688–720.

Keegan, W. F., ed. 1987. *Emergent Horticultural Economies of the Eastern Woodlands.* Occasional Paper No. 7. Carbondale: Center for Archaeological Investigations, Southern Methodist University.

Lightfoot, R. G. and G. M. Feinman. 1982. "Social Differentiation and Leadership Development in Early Pithouse Villages in the Mogollon Region of the American Southwest." *American Antiquity* 47:64–81.

Longacre, W. A. 1975. "Population Dynamics at the Grasshopper Pueblo, Arizona." In *Population Studies in Archaeology and Biological Anthropology: A Symposium,* ed. A. C. Swedlund. *American Antiquity* 40:71–74, memoir 30.

Lynott, M. J., T. W. Boutton, J. E. Price, D. E. Nelson. 1986. "Stable Carbon Isotopic Evidence for Maize Agriculture in Southeast Missouri and Northeast Arkansas." *American Antiquity* 51(1):51–65.

McGuire, R. H. and M. B. Schiffer. 1983. "A Theory of Architectural Design." *Journal of Anthropological Archaeology* 2:277–303.

Madsen, D. B. 1979. "The Fremont and the Sevier: Defining Prehistoric Agriculturalists North of the Anasazi." *American Antiquity* 44:711–22.

Marquardt, W. H. 1985. "Complexity and Scale in the Study of Fisher-Gatherer-Hunters: An Example from the Eastern United States." In *Prehistoric Hunter-Gatherers: The Emergence of Cultural Complexity,* eds. T. D. Price and J. A. Brown. Orlando, Fla.: Academic Press.

Milanich, J. T. and C. H. Fairbanks. 1987. *Florida Archaeology.* New York: Academic Press.

Minnis, P. E. 1985. *Social Adaptation to Food Stress.* Chicago: University of Chicago Press.

Moratto, M. J. 1984. *California Archaeology.* New York: Academic Press. Foreword by F. A. Riddell. Contributions by D. A. Frederickson, C. Raven, and C. N. Warren.

Morse, D. F. and P. A. Morse. 1983. *Archaeology of the Central Mississippi Valley.* New York: Academic Press.

Muller, J. 1986. *Archaeology of the Lower Ohio River Valley.* New York: Academic Press.

Murphy, J. L. 1975. *An Archaeological History of the Hocking Valley.* Athens: Ohio University Press.

O'Brien, M. J. 1987. "Sedentism, Population Growth, and Resource Selection in the Woodland Midwest: A Review of Coevolutionary Developments." *Current Anthropology* 28(2):177–97.

Ortiz, A. 1979. "Southwest." *Handbook of North American Indians, Vol. 9.* Washington, D.C.: Smithsonian Institution.

Peebles, C. S. 1977. "Moundville and Surroundings Sites: Some Structural Considerations of Mortuary Practices, II." In *Approaches to the Social Dimensions of Mortuary Practices,* ed. J. A. Brown. Society for American Archaeology Memoir, no. 25.

Pfeiffer, J. E. 1977. *The Emergence of Society.* New York: McGraw-Hill.

Plog, F. 1983. "Political and Economic Alliances on the Colorado Plateaus, A.D. 400 to 1450." In *Advances in World Archaeology, Vol. 2,* eds. F. Wendorf and A. E. Close. New York: Academic Press.

Powell, S. 1983. *Mobility and Adaptation: The Anasazi of Black Mesa, Arizona.* Carbondale: Southern Illinois University Press.

Porter, J. W. 1969. *The Mitchell Site and Prehistoric Exchange Systems at Cahokia: A.D. 1000–300.* Illinois Archaeological Survey Bulletin, no. 7.

Prufer, O. H. 1964. "The Hopewell Cult." *Scientific American* 211:90–102.

———. 1965. "The McGraw Site: A Study in Hopewellian Dynamics." Cleveland Museum of Natural History Scientific Publications, n.s. 4(1).

Ramenofsky, A. F. 1987. *Vectors of Death. The Archaeology of European Contact.* Albuquerque: University of New Mexico Press.

Reid, J. J. and D. E. Doyel, eds. 1986. *Emil W. Haury's Prehistory of the American Southwest.* Tucson: University of Arizona Press.

Rice, G. E. 1975. "A Systemic Explanation of a Change in Mogollon Settlement Patterns." Unpublished doctoral dissertation, University of Washington.

Rindos, D. 1984. *The Origins of Agriculture: An Evolutionary Perspective.* New York: Academic Press.

Ritchie, W. A. 1969. *The Archaeology of New York State.* 2nd ed. Garden City, N.Y.: Natural History Press.

Rothschild, N. A. 1979. "Mortuary Behavior and Social Organization at Indian Knoll and Dickson Mounds." *American Antiquity* 44:658–75.

Schiffer, M. B. 1982. "Hohokam Chronology: An Essay on History and Method." In *Hohokam and Patayan Prehistory of Southwestern Arizona*, R. H. McGuire and M. B. Schiffer, eds. New York: Academic Press.

Schiffer, M. and J. Skibo. 1987. "Theory and Experiment in the Study of Technological Change." *Current Anthropology* 28(5):595–623.

Schoenwetter, J. 1962. "Pollen Analysis of Eighteen Archaeological Sites in Arizona and New Mexico." In *Prehistory of Eastern Arizona*, eds. P. S. Martin et al. Fieldiana: Anthropology 53:168–209.

Simmons, A. H. 1986. "New Evidence for the Early Use of Cultigens in the American Southwest." *American Antiquity* 51(1):73–89.

Smith, B. 1974. "Middle Mississippi Exploitation of Animal Populations—A Predictive Model." *American Antiquity* 39:274–91.

———. 1985. "The Role of *Chenopodium* as a Domesticate in the Pre-Maize Garden System of the Eastern United States." *Southeastern Archaeology* 4:51–72.

Steponaitis, V. P. 1983. *Ceramics, Chronology, and Community Patterns. An Archaeological Study at Moundville.* New York: Academic Press.

Stewart, T. D. 1973. *The People of America.* New York: Scribner's.

Stoltman, J. B. 1983. "Ancient Peoples of the Upper Mississippi River Valley." In *Historic Lifestyles of the Upper Mississippi River Valley*, ed. J. Wozniak. New York: University Press of America.

———. 1986. *Prehistoric Mound Builders of the Mississippi Valley.* Davenport, Iowa: Putnam Museum.

Struever, S. and G. L. Houart. 1972. "Analysis of the Hopewell Interaction Sphere." In *Social Exchange and Interaction*. Ann Arbor: University of Michigan, Museum of Anthropology, Anthropological Papers, no. 46.

Struever, S. and K. D. Vickery. 1973. "The Beginnings of Cultivation in the Midwest-Riverine Area of the United States." *American Anthropologist* 75:1197–1220.

Tainter, J. 1975. "Social Inference and Mortuary Practices: An Experiment in Numerical Classification." *World Archaeology* 7:1–15.

Tuck, J. A. 1971. "An Archaic Cemetery at Port au Choix, Newfoundland." *American Antiquity* 36:343–58.

Upham, S. 1988. "Archaeological Visibility and the Underclass of Southwestern Prehistory." *American Antiquity* 53(2):245–61.

Upham, S., K. G. Lightfoot, and G. M. Feinman. 1981. "Explaining Socially Determined Ceramic Distributions in the Prehistoric Southwest." *American Antiquity* 46:822–36.

Vogel, J. C. and N. J. Van Der Merwe. 1977. "Isotopic Evidence for Early Maize Cultivation in New York State." *American Antiquity* 42:238–42.

Washburn, D. R. 1975. "The American Southwest." In *North America*, S. Gorenstein et al., eds. New York: St. Martin's Press.

Watson, P. J. 1974. *Archeology of the Mammoth Cave Area.* New York: Academic Press.

Whalen, M. 1981. "Cultural-Ecological Aspects of the Pithouse-to-Pueblo Transition in a Portion of the Southwest." *American Antiquity* 46:75–91.

Willey, G. R. 1966. *An Introduction to American Archaeology, Vol. 1: North and Middle America.* Englewood Cliffs, N.J.: Prentice-Hall.

Yarnell, R. A. 1965. "Early Woodland Plant Remains and the Question of Cultivation." *The Florida Archaeologist* 18:78–81.

Yerkes, R. W. 1988. "The Woodland and Mississippian Traditions in the Prehistory of Midwestern North America." *Journal of World Prehistory* 2(3):307–58.

16

Prehistory in Perspective

What's past is prologue.

Shakespeare

Having been led from the three million-year-old hominid footprints of Laetoli through the European conquest of the Americas, the reader is now invited to return to the question posed in the first chapter: What does the past mean? Or, more precisely, how are we to understand the past? This is, of course, the question of the ages, and by now the reader will not expect a clear, complete answer.

Although it is perhaps a bad pedagogical tactic to raise a question, review the evidence, and then conclude with "we're not really sure," most contemporary archaeologists find themselves in this position. We have a general sense of some of the potentially important causes of various ancient cultural changes, but there is no single, powerful, comprehensive explanation of the past. Thus, rather than demonstrating the power and accuracy of any specific models of culture and history, we have simply explored the limitations and potentials of some of the more obvious explanations of the past.

One of the earliest casualties revealed by our archaeological review is the nineteenth-century view of mechanical cultural evolutionism, with its belief in the gradual improvement of people and their societies, generation after generation, as we "progressed" from rude, murderous savagery to the better, brighter world of industry, liberalism, and democracy. The Victorian proponents of cultural evolutionism—who number among their ranks some of the greatest and most humane intellects of all time—can be forgiven, perhaps, for this belief in progress, for they lived in an age of marvelous changes. But, as Simone Weil said, "The great mistake of the nineteenth century was to assume that by walking straight on one mounted into the air."

Yet even today many people, including anthropologists, tend to see the world and history in these same mechanical, evolutionist terms, with the implication that prehistory and history describe a "progressive" trajectory, leading to ourselves. Our review of the archaeological records reveals, however, that instead of a smooth rising curve of social, physical, techno-logical, and moral evolution, our past has been a series of fitful cycles, where social forms and technologies have reached their limits of growth and then failed, to be replaced by new social forms and technologies, more complex in some ways than their predecessors, but neither permanent nor "better" in any evident moral or philosophical sense. Clearly, there is nothing in the archaeological record to indicate that we are on a straight path to an earthly paradise. Today, two-thirds of the world's population are "involuntary vegetarians,"[1] whose diet, morbidity and mortality rates, and general standard of living compare poorly with those of most Pleis-tocene hunting and gathering bands: and even in the most industrialized countries, the vast majority labors longer for sustenance than did many of the people of prehistory. Moreover, the misery, poverty, wars, and as-sorted atrocities of the past several decades make the idea of humanity's moral evolution so laughable as to require no discussion.

But this lack of "progress" in the Victorian sense should not be taken as an indictment of the general cultural evolutionary model, for the ar-chaeological record also demonstrates that despite repeated reversals and long plateaus, the past three million years have witnessed a gradual in-crease in the complexity of cultures, as measured in terms of ability to divert energy to human use and in the size, differentiation, and interde-pendence of social, political, and economic systems. Moreover, cultural evolution has been patterned: domestication, agriculture, cultural com-plexity, and many specific cultural developments appeared independently but in very similar ways in various times and places. Yet there is nothing inevitable or even predictable about the specifics of ancient cultural change. It is the nature of evolutionary patterns that they only appear to have a direction when one looks backward. Ivy Compton-Burnett observed that "real life seems to have no plots," and in the same sense, evolutionary history has no plots, except in retrospect. For example, the people who first started domesticating plants and animals can be assumed to have had no idea of the eventual outcome of the immense ecological drama in which they were involved.

To discern an evolutionary process in history is not the same as *explain-ing* history, of course. No general theory of explaining our long evolution-ary history has won universal acceptance. Most archaeologists employ a vaguely evolutionary perspective: they assume that cultural innovations are constantly arising in human societies, and that some of these innovations confer an adaptive advantage and are "fixed" within cultural systems and perpetuated until they too are outmoded and replaced. Most archaeologists also adopt a rather "vulgar" materialist determinism in that they try to understand cultural forms and dynamics to some extent as expressions of technological, environmental, and economic variables. Thus, such things as the social egalitarianism of hunters and gatherers in the Pleistocene and the rigid social stratification of emerging states are generally understood as reactions to their respective economic bases. Nonetheless, even some of the most ardent materialists stress the importance of free will, religion, and ideology in "determining" the operations of societies, and no one argues that evolutionary cultural materialism is a complete, explanatory theory of culture.

Among those archaeologists who retain the hope of making archaeology a formal, scientific, mathematical, and fundamentally *explanatory* discipline, there is a growing trend away from the somewhat naive imitation of physics that archaeology aspired to in the late 1960s, early 1970s. Many of the younger archaeologists of the 1960s and 1970s rejected traditional archaeology and traditional archaeologists as "stamp collectors" who had no notions of science or explanation. But more and more archaeologists are in a position like that described by Mark Twain, who said: "When I was a boy of fourteen, my father was so ignorant I could hardly stand to have the old man around. But when I got to be twenty-one, I was astonished at how much the old man had learned in seven years." There is a growing sense that the archaeology of the 1970s and 1980s was not all that much more powerful and scientific than the "prerevolutionary" archaeology of the 1940s and 1950s.

Most archaeologists are no longer searching for mechanical cause-and-effect relationships like those in physics, where changes in some variables (e.g., temperature of a gas) result in predictable uniform changes in other variables (e.g., gas pressure and volume). Instead, and in concert with contemporary perspectives in the natural sciences, those archaeologists who seek a mathematical, analytical explanation of the past are seeking probabalistic explanations—statements that, under specified conditions, estimates can be made about the likelihood that similar variables will give rise to similar consequences. Merilee Salmon[2] has demonstrated how statistical analyses can be applied to problems of reconstructing ancient societies. Salmon considers, for example, the problem of determining whether or not people in coastal Ecuador at about 3000 B.C. ate maize—a question of some significance in the debates about how maize was first domesticated (see Chapter 6). One part of the evidence on this question is a sample of seventy-six teeth recovered from six individuals in a Valdivian cemetery. Salmon notes that dental caries (cavities) are more common in agricultural groups than in hunter-gatherer groups, that eating maize tends to cause a lot of cavities, that in one study 15.5% of all teeth in a group of ancient Central American maize farmers had at least one cavity, but that not a single tooth in the sample of seventy-six from Valididvia had a cavity. So in trying to determine whether or not the people at Valididvia were agriculturalists, she formulates two hypotheses, one that the Valdivian people were agriculturalists, the other, that they were not. To choose between these hypotheses, she uses mathematical probability theory and some estimates to assess the likelihood of finding seventy-six cavity-free teeth among maize agriculturalists and finds it to be very low—that is, it is statistically improbable.

Other archaeologists tried to produce more general principles, not related to any particular site or period. Michael Schiffer, for example, offered a formula to describe and explain changes over time in the rate at which tools were used and discarded in ancient societies.[3]

These attempts to make explanation in archaeology a matter of statistical, mathematical expression would seem to have great potential, but generally in sciences it is *theory* that determines how mathematics are to be used in that discipline, and in archaeology there is little that deserves the name "theory." And many archaeologists reject *a priori* a mathematical explanation of the dynamics of our past.

The emerging discipline of sociobiology, for example, includes among its proponents those who argue that much of what we are or have been culturally is deeply embedded in the chemistry of our chromosomes, that war, greed, competition, egotism, innovation, and a long list of other human behaviors can be partially suppressed by the weight of cultural institutions, but that they persist in our physiology as potent behavioral determinates.

There is a certain attractiveness to this idea. It reduces some cultural problems to the biochemical and genetic level—a level of analysis that modern science has shown to be very productive—and it also seems to make clear why human societies have inevitably been associated with war, conflict, and social inequity: because three million years of evolution have shaped us that way. And thus, except for some recent Frankensteinian-research with the mechanics of heredity, the sociobiological approach would seem to some extent to place fundamental changes in some human social patterns beyond the reach of human initiative.

The sociobiological perspective may well make highly significant contributions to the solution of anthropological problems, but it is unlikely that the causes and effects of the origins of culture, the evolution of agricultural economies, the rise of complex societies, or the other major problems of prehistory will be "resolved" by this approach. If there is a single major lesson of world prehistory, it would seem to be that cultures are marvelously adaptable: that, given compelling cultural reasons, people can use cultural forms to make cooperation or competition, murder or altruism, egalitarianism or privilege, or any other behaviors into vices or virtues, as the circumstances dictate.

Archaeologists are still searching for a general theory and model of culture, and currently archaeology is a bazaar of competing viewpoints. One of the most active of these is "Cognitive and Critical Archaeology," or the "Post Processualists."

The name post-processual is a bit propagandistic in that it implies that it is the kind of archaeology that will be done after positivistic processual archaeology is abandoned. Post-processualists believe that the search for a physics of artifacts is in essence a search for a unicorn; that archaeology can use a few scientific methods, such as radiocarbon dating, but that it will always be a humanistic enterprise.

> For the subjective idealism of scientistic archaeology we substitute a view of the discipline [of archaeology] as an hermeneutically informed dialectical science of past and present unremittingly embracing and attempting to understand the polyvalent qualities of the socially constructed world of the past and the world in which we live.[4]

The major flaw the post-processualists discern in the processualists is the assumption that archaeology can be a neutral, value-free, empirical science of artifacts. Young, for example, expresses this difference of opinion as, on the one hand those who think that "history is what *happened* in the past," as opposed to the post-processualist notion that history "is what a living society *does* with the past."[5]

Post-processualists emphasize what to most people may seem an obvious point—that an archaeologist's interpretations of the past are in part a function of his or her own sociocultural context. One might think that archaeologists would be less prone to this than, say, ethnologists, but archaeologists regularly report temples, social stratification, intensified storage, social class, warfare, states, and other "things" that do not, technically, exist as anything directly measurable. Our own culture determines to some extent what we make of the past. We can dig up people, sort through their feces, measure their bodies, sift their garbage for their food remains, translate their writings, measure their buildings, and do all kinds of scientific things to them, but in the end what we make of them will have a lot to do with our own lives and personalities.

The flavor of this form of analysis can be detected in the title of a recent study, "Social values, social constraints and material culture: the design of contemporary beer cans."

They assume that if one ignores the meaning that ancient peoples invested their materials with and concentrates only on the material, the meaning of the past can never be comprehended. As an illustration of this, Shanks and Tilley consider one of the most ordinary objects of British and Swedish life, beer cans and bottles, and they begin by noting that these two countries have different ideas about beer that can be presumed to be important in explaining their style and distribution:

> Whereas in Britain alcohol is not generally considered an item of key public or individual concern, in Sweden alcohol consumption is regarded as one of the most pressing of social issues, at least in governmental circles. If material culture-patterning is structured in relation to social processes in a systematic manner . . . then we might expect some considerable differences to exist between British and Swedish beer can design which can be meaningfully related to social strategies.[6]

In an extended quantitative analysis of beer bottles and cans, they found that compared to British beer cans, Swedish cans are more colorful and have more writing and design elements on them. Using multivariate statistics they distinguish groups within and between British and Swedish beers, and they relate these differences to differences in advertising, industrialization, attitudes toward women, social attitudes toward public drinking, and many other factors.

No doubt the conflict between those who see archaeology as a neutral science of the past and those who see it as fundamentally a political exercise will continue for many years to come. In summing up this situation, John Bintliff concluded that:

> 1) It is hardly possible to consider, despite the hopes of traditionalists, that Archaeology will reject such central emphases of the New Archaeology as: the testing of propositions, problem-orientation in fieldwork, quantitative methods wherever possible, the search for generalizations about our past. . . .
> 2) Cognitive Archaeology is here to stay, but nothing will be achieved by it unless it adopts the explicit, scientific procedures of the New Archaeology.[7]

To a large extent, however, the great majority of archaeological fieldwork has been devoted not to the larger theoretical questions about why cultures change, but rather to attempts to describe the lifeways of ancient peoples. If one accepts that this is possible and useful (some do not), then in these terms archaeology has made great progress. There is no need to review in detail the archaeological information we have considered in earlier chapters about early hominid taxonomy, Paleolithic technologies, the origins of domestication and agriculture, the rise of cultural complexity, and the many other prehistoric transformations that only a century ago were not even recognized, let alone understood. True, we have identified no "prime movers" (factors whose effects are so potent that they can be used to explain much of the cultural developments of past and future societies), but we have identified many patterns and parallels. Cross-cutting cultural developments all over the world are the same essential expressions of religion, warfare, population growth, emerging social stratification, and evolving economic productivity.

We have seen, for example, that human population growth has been linked to almost every cultural transformation, from initial tool use to the appearance of industrial empires, and it is indeed true that the increasing rate of technological innovation and general cultural change is closely correlated with increasing worldwide population numbers. This relationship may even be said to be causal: the rate of technological change might be linked simply to the number of minds available to solve problems and produce innovations. But this is clearly an incomplete explanation of our past. The archaeological record time and time again discloses that the relationship between population growth and cultural change is neither consistent nor direct, and it remains for archaeologists to demonstrate the causal connections between the world's increasing population size and the specific, crucial transformations of prehistory and history.

All the other "prime movers" have similar limitations. The recurrent nature of human interspecies violence, for example, from the casual cannibalism of early *Homo erectus* to World War II, clearly suggests that conflict is an evolutionary mechanism of considerable power; conflict must do something for societies, or it would be difficult to explain its depressing ubiquitousness in human affairs. But here too, the actual mechanisms whereby conflict interacted with other variables in most past societies to produce cultural change is unknown.

In sum, archaeologists have devoted many years to trying to establish simple correlations among cultural variables and/or environmental variables, but most now recognize that even if they are able to demonstrate a strong correlation in time and space between, say, population growth, warfare, and urbanism, they will still not have explained in a powerful way any of these factors or their relationships to each other. On the other hand, archaeology has made some progress in understanding both the past and what we can hope to know about it. Bill Nyne described Wagner's music as better than it sounds, and in a way archaeology is like this: its limitations are obvious but we understand human antiquity today to an extent unimaginable just a century ago. Currently there are many interesting—if undeveloped—ideas of how a more powerful discipline of archaeology might evolve.[8]

Lessons of Prehistory

To think of time—of all that retrospection,
To think of to-day, and the ages continued henceforward.

Have you guess'd you yourself would not continue?
Have you dreaded these earth-beetles?
Is to-day nothing? is the beginningless past nothing?
If the future is nothing they are just as surely nothing.
(Walt Whitman, from "To Think of Time.")

It would be gratifying if we could extract from our review of world prehistory some important predictions about the future of humankind but, as we have noted, archaeology will probably never be a predictive science. This is especially true when we consider the long-range future, particularly now that humanity has instruments with which to terminate all life on this planet. But even should we survive for millions of generations, the virtuosity of culture as an adaptive device makes extrapolations into the future an act of either ignorance or embarrassing temerity. What anthropologist, given the opportunity to stroll through Olduvai Gorge some Sunday morning two million years ago, could on the basis of that experience have predicted the gaudy technology and kaleidoscopic social and political forms of twentieth-century civilization? In a sense, evolutionary phenomena are by definition unpredictable.

Nonetheless, no student of world prehistory can overlook the persistent, powerful trends that tie the present, and perhaps the future, to the past. The evolution of technologies for energy capture, for example, seems to have been so basic to competitive success that we might expect that these technologies will continue to be important determinants of sociocultural and ideological forms. One obvious implication is that it might be naive to expect energy-dependent countries to refrain from using nuclear energy because of the slight attendant risk of nuclear disaster. The matter of "evolutionary potential" would also seem to have some applicability to the future, even though the world is much different now from what it was during most of the last nine thousand years, when agricultural potential was the major ingredient in evolutionary success. Throughout prehistory the evolutionary advantage frequently shifted to "marginal" groups, less complexly organized cultures located on the peripheries of the richer established cultures. In the contemporary world, where fossil fuels and raw materials are of great import, the advantage may already have shifted from the West to Russia, China, and Near Eastern and African nations.

Any number of future developments could of course change these expectations dramatically. Should a simple, inexpensive technology for using solar energy be developed, there may well be a general leveling of wealth and power among nations, particularly if what are now the world's poorer countries curb their rates of population growth.

Another apparent lesson of prehistory has to do with the uses of religion. I will let those with more confidence in their omniscience than myself pronounce on the ultimate nature of human religiosity; it is always sobering to recollect that neither the study of world prehistory nor any other art or science has much to offer in the way of a reason why we are out here, on a small planet in an incomprehensibly infinite universe, and to all evidence very much alone. Nonetheless, taken solely on the level of its effects on other aspects of culture, religion appears to operate principally as a highly adaptable, thermodynamically efficient mechanism of social control. We have in the archaeological and historical record an irrefutable demonstration that there is no act so repellent, be it sacrificing thousands of one's own countrymen or incinerating hundreds of thousands of one's "enemies," that it cannot be made not only acceptable, but entirely virtuous within the context of religious systems. The exploitation of the many by the few to build massive "worthless" pyramids and public buildings, the complete catalogue of sexual "perversions," the avoidance of this or that food—all have been incorporated into state religions with no more difficulty than Christian faith, hope, or charity.

Nor should the apparent decline of formal theistic religions in the modern world be interpreted as a sign of religion's demise as an important cultural mechanism. One need only look at contemporary Communist countries, where avowedly secular, atheistic cultures have replaced traditional religions with ideologies no different in their essentials from Catholicism or Buddhism. For the future, there appears no reason to suspect that belief systems will be any less important, although clearly their contents will vary. The competitive advantage, as always, will lie with cultures that evolve belief systems that motivate people in specific, "efficient" directions, as is well illustrated in centuries past by the link between capitalism and Calvinism, and irrigation agriculture and the Sumerian pantheon. Marx thought organized religion would disappear with the advent of democracy and social equality, and only the future will tell if he was right.

The lessons of the archaeological record concerning the origins and significance of social and economic inequalities are not as clear as they might seem at first glance. There is the inescapable fact that there has never been an economically differentiated, complex society, especially the supposedly "classless" societies, that was not also stratified into groups having differential access to wealth, power, and prestige; but it remains to be demonstrated that cultural complexity is inextricably linked to social stratification. It would appear that in the past the complexity of managing an economy based on agriculture and, later, agriculture in combination with fossil fuels, could work effectively only through administrative and social hierarchies and class-structured societies. But if in future centuries population densities are stabilized and perhaps reduced, if control of energy and food sources is decentralized and the production of material wealth made highly automated, it would seem at least possible that human societies will someday approximate the "social justice" of the late Pleistocene.

Perhaps the archaeological record's bleakest implication concerning our future has to do with that ambiguous concept of "the quality of life." We have already noted that today most of the world's people have a diet and standard of living in many ways inferior to that of Pleistocene hunters and gatherers, and although medical technology, solar power, and contraceptives may change this by bringing the Western industrial standard of living to all parts of the world, this is by no means a certainty. Even given sufficient energy and rapid industrialization, fundamental questions remain about the short-term prospects for the quality of life on this planet. In the wealthiest countries today the abundance of luxury goods seems to convey a sense of ease and fulfillment, but modern economies work only because the vast majority of the population are coerced or are willing to spend most of their lives at hard labor, often with competitive pressures that produce unbalanced, unfulfilled lives. And it is not only the poorer, laboring class whose quality of life is questionable. It has been observed that a citizen of Athens in the fifth century B.C. would consider today's professionals—physicians, politicians, professors, and football players alike—to be in the main incomplete, undeveloped people, whose "success" has required them to devote so much of their lives to their specialty that they are grotesquely incompetent at the oratorical, conversational, athletic, philosophical, agricultural, and aesthetic skills that the Greeks would insist on as necessary components of a "whole" person.

The past often takes on a rosy, romantic coloring from the perspective of the complexities and frustrations of the present, and we cannot ignore the illiteracy, warfare, and gross social exploitation that appear to be the very warp and woof of every ancient complex society, including ancient Athens; but many people feel that modern social and economic systems, whether they be capitalistic or socialistic, do not provide optimal environments for the balanced, liberal, personal development of the majority. Charles Darwin wrote, "Life in the crowded conditions of cities has many unattractive features, but in the long run these may be overcome, not so much by altering them, but simply by changing the human race into liking them." Gloomy ruminations of this sort have a way of showing up in succeeding centuries as quaint quotations demonstrating the curious myopia of past generations, and I devoutly hope it will be so in this case. But despite the many advantages of modern technology the long-term trend seems inexorably in the direction of greater specialization and compartmentalization of people, with consequent increasing alienation and frustration. How pleasant a society may be to live in has depressingly little to do with that society's evolutionary potential.

Social reformers in general might find the lessons of the archaeological record somewhat bleak. Complex societies are by definition complex systems, and the strong parallels in social forms that have evolved over the world in the past ten thousand years suggest, perhaps, that human social systems may not be easily changed for the better by altering one or two variables in them. The "war on drugs" and third-world poverty show how difficult managed change of societies can be.

If we take the story of the human past in its broadest context, we might even consider the *cosmic* implications of our evolutionary journey. Anyone who closely studies the past must eventually yearn for "another experimental case," the evolution of life on some other planet, which we could use as a standard of comparison. Astronomers tells us that carbon—the stuff of life—is widely distributed in the universe, and as far as we can probe other worlds the basic chemicals and physical forces seem just as they are on earth. Are we the only planet on which life has evolved? There are several answers, of course. As Gregg Easterbrook expressed them, they are (1) We have company; (2) we had company; we are alone in this galaxy; (3) we are alone, period; and (4) we are the first.[9]

The Future of Archaeology

The archaeological record is disappearing at such a rapid rate that the future of field archaeology on the problems of prehistory is very much in doubt. The worldwide wave of industrialism, the major destroyer of archaeological materials, seems destined to expand at ever-increasing speed.

Dismal as the loss of the archaeological record is, it may be that the major progress in archaeology in the next several generations will come not so much from the discovery of new bones and stones, but rather from a reconsideration of theoretical and methodological approaches and a reanalysis of existing data. We have recounted in this book the prospects of current analytical approaches, and these and other perspectives are likely to increase considerably our knowledge about our past. Does this mean that the general problem of, for example, the origins of cultural complexity will some day be as precisely solved as the formulae for making plastics? Probably not. These are very different kinds of questions with different criteria for solution. Yet, the history of science is replete with examples in which a seemingly impossible problem was not only solved but made routine: the general problems of prehistory we have discussed here may well fall into this category, to be solved in succeeding centuries in terms and with techniques which we now only dimly perceive. After all, people of only a few centuries ago no doubt would be astounded not only by our science and lush technology, but also by our knowledge of prehistory and the dynamics of culture.

Notes

1. Harris, *Cannibals and Kings*, p. x.
2. Salmon, *Philosophy and Archaeology*, pp. 26–29, *passim;* also see Schiffer, *Behavioral Archaeology*, p. 173. Watson, LeBlanc, and Redman, *Archeological Explanation.*
3. Schiffer, *Behavioral Archaeology*, p. 173.
4. Shanks and Tilley, *Re-Constructing Archaeology*, p. 243.
5. Young, "Since Herodotus, Has History Been a Valid Concept?" p. 7.
6. Shanks and Tilley, *Re-Constructing Archaeology*, p. 172.
7. Bintliff, *Archaeology at the Interface*, p. 28.
8. Trigger, "Archaeology at the Crossroads: What's New?" pp. 278–79.
9. Easterbrook, "Are We Alone?"

Bibliography

Bintliff, J. 1986. "Archaeology at the Interface: An Historical Perspective." In *Archaeology at the Interface*, eds. J. L. Bintliff and C. F. Gaffney. Oxford: B.A.R. International Series 300.

Easterbrook, G. 1988. "Are We Alone?" *The Atlantic* August 1988, pp. 25–38.

Harris, M. 1977. *Cannibals and Kings: The Origins of Cultures.* New York: Random House.

Salmon, M. H. 1982. *Philosophy and Archeology.* New York: Academic Press.

Schiffer, M. B. 1976. *Behavioral Archaeology.* New York: Academic Press.

Shanks, M. and C. Tilley. 1987. *Re-Constructing Archaeology.* Cambridge, England: Cambridge University Press.

Trigger, B. G. 1984, "Archaeology at the Crossroads: What's New?" *Annual Review of Anthropology* 13:275–300.

Watson, P. J., S. A. LeBlanc, and C. L. Redman. 1971. *Explanation in Archeology.* New York: Columbia University Press.

———. 1984. *Archeological Explanation.* New York: Columbia University Press.

Young, T. C., Jr. 1988. "Since Herodotus, Has History Been a Valid Concept?" *American Antiquity* 53(1):7–12.

CREDITS

1.1 Engraving by Theodor de Bry from *Americae*, 1592. Rare Book Division, The New York Public Library. **1.2** Frank Hole. **1.3** Sir William Hamilton, *Campi Phlgraei* (1765), Illus. by Peter Fabris from Vol. 1, American Museum of Natural History; photo courtesy of Kay Zakariasen and David Hanson. **1.4** American Museum of Natural History. **1.5** Hirmer Fotoarchiv, Munich. **1.6** American Museum of Natural History. **1.8** Radio Times Hulton Picture Library.

2.1 John Reader/Photo Researchers. **2.2** Danish National Museum. **2.3** Reproduced from *Cambridge Encyclopedia of Archaeology*, © Sceptre Books, a Division of Time Life International Ltd., London. **2.4** From J. W. Michaels, *Dating Methods in Archaeology* (Seminar Press, 1973), Fig. 1. © Academic Press, Inc. Reproduced by permission of J. W. Michaels and Academic Press, Inc. **2.5A** From M. D. Leakey, *Olduvai Gorge Vol. 3* (Cambridge Univ. Press, 1971), Fig. 8. Reproduced by permission of Cambridge Univ. Press. **2.5B** From B. Fagan, *People of the Earth*, 2nd ed. (Little, Brown, 1977), drawing p. 60. Reproduced by permission of Little, Brown & Co., Inc. **2.6** Hirmer Fotoarchiv, Munich. **2.7** Victor R. Boswell, Jr., © National Geographic Society. From *Ancient Egypt: Discovering Its Splendors*, 1978. **2.8** From J. D. Jennings, *Prehistory of North America* (McGraw-Hill, 1974), Fig. 1.4. © 1968, 1974 by McGraw-Hill, Inc., all rights reserved. Reproduced by permission of McGraw-Hill Book Co. **2.10** From J. Deetz, *Invitation to Archaeology* (Doubleday, 1967), Fig. 4. © 1967 by James Deetz. Reproduced by permission of Doubleday & Co., Inc. **2.12** Courtesy of D. D. Denton, Institute of Nautical Archaeology.

3.1 Reheinisches Landmuseum, Bonn. **3.2** American Museum of Natural History. **3.3** From *Cambridge Guide to Prehistoric Man*. Courtesy of Diagram Visual Information Limited. **3.4** San Diego Zoological Society. **3.5** After Jerison, 1976. **3.6** Reproduced from *Cambridge Encyclopedia of Archaeology*, © Sceptre Books, a Division of Time Life International Ltd., London. **3.7** After Howells, 1973. **3.8** Jay Matternes. **3.9** From J. B. Birdsell, *Human Evolution: An Introduction to the New Physical Anthropology*, 2nd ed. (Rand McNally, 1975), Fig. 6.7. Reproduced by permission of J. B. Birdsell. **3.10** Geza Teleki. **3.11** C. Owen Lovejoy. **3.12** From Skeleton, et al., CURRENT ANTHROPOLOGY, Vol. 27, #1 (1986): 33, fig. 5 and Mellars, et al., © 1986 by The Wenner-Gren Foundation for Anthropological Research. Courtesy of the University of Chicago Press. **3.13** From M. D. Leakey, *Olduvai Gorge Vol. 3* (Cambridge Univ. Press, 1971), Fig. 47. Reproduced by permission of Cambridge Univ. Press. **3.14** From W. Howells, *Evolution of Genus Homo* (Benjamin/Cummings Pub. Co., copyright © 1973), p. 47. Reproduced by permission of Benjamin Cummings Pub. Co.

4.2 From *Cambridge Guide to Prehistoric Man*. **4.3** A,B,C,D From *Cambridge Guide to Prehistoric Man*. Courtesy of Diagram Visual Information Limited. **4.4** From *Antiquity* 61, 1987. Courtesy of Antiquity and Robert Foley. **4.5** From *Cambridge Guide to Prehistoric Man*. Courtesy of Diagram Visual Information Limited. **4.6** From *Cambridge Guide to Prehistoric Man*. Courtesy of Diagram Visual Information Limited. **4.7** Henry de Lumley, Centre national de la recherche scientifique, Marseile. **4.8** Zdenek Burian. **4.9** From F. Bordes, *The Old*

Stone Age (Weidenfeld and Nicolson, 1968). Reproduced by permission of Weidenfeld and Nicolson. **4.10** From *Cambridge Guide to Prehistoric Man*. Courtesy of Diagram Visual Information Limited. **4.11** Courtesy of *Journal of Field Archaeology* and John J. Shea. **4.12** From *Newsweek*, January 11, 1988. Ian Tattersall, American Museum of Natural History; Richard Klein, University of Chicago; and Ib Ohllson. **4.13** From *Cambridge Guide to Prehistoric Man*. Courtesy of Diagram Visual Information Limited. **4.14** From *Cambridge Guide to Prehistoric Man*. Courtesy of Diagram Visual Information Limited. **4.15** From *Cambridge Guide to Prehistoric Man*. Courtesy of Diagram Visual Information Limited. **4.16** From *Cambridge Guide to Prehistoric Man*. Courtesy of Diagram Visual Information Limited. **4.17** From W. A. Fairservis, *The Threshold of Civilization*, drawings by Jan Fairservis (Charles Scribner's Sons, 1975), pp. 82–83. Reproduced by permission of Jan Fairservis. **4.18** Austrian Institute, New York. **4.19** French Cultural Service, New York. **4.20** From A. Leroi-Gourhan, "The Evolution of Paleolithic Cave Art," *Scientific American*, Feb. 1968, p. 66. Copyright © 1968 by Scientific American, Inc. All rights reserved. **4.21** Steven Brandt, University of Florida.

5.1 Philadelphia Museum of Art. **5.2** From Joe LeMonnier. With permission from *Natural History*, (11), (87); Copyright the American Museum of Natural History, 1987. **5.3** From *The Great Journey* by Brian Fagan. © 1987. Thames and Hudson, Inc. Reprinted by permission of the publisher. **5.4** From Joe LeMonnier. With permission from *Natural History*, (1), (87); Copyright the American Museum of Natural History, 1987. **5.5** From R. F. Flint, *Glacial and Quaternary Geology* (Wiley, 1971), Fig. 29.7 modified from Martin and Guilday in P. S. Martin and H. E. Wright, Jr., eds., *Pleistocene Extinctions: The Search for a Cause* (Yale Univ. Press, 1967, p. 1). Reproduced by permission of John Wiley & Sons, Inc., and Yale Univ. Press. **Table 5.1** From "The Archaeology of Radiocarbon Accelerator Dating," *Journal of World Prehistory*, I(2), 1987, © Plenum Publishing Corp. Reproduced by permission of the publisher. **5.6** From *Natural History*, 1/1987, © Natural History, American Museum of Natural History. Courtesy of C. G. Turner and American Museum of Natural History. **5.7 A,B** Tom Dillehay. **5.8** Tom Dillehay. **5.9** From J. J. Hester, *Introduction to Archaeology* (Holt, Rinehart, and Winston, 1976), Compiled from A. Krieger, "Early Man in the New World," in J. D. Jennings and E. Norbeck, eds., *Prehistoric Man in the New World* (University of Chicago Press, 1964). Reproduced by permission of the Univ. of Chicago Press.

6.1 Reproduced from *Cambridge Encyclopedia of Archaeology*, © Sceptre Books, a Division of Time Life International Ltd., London. **6.2** From K. P. Oakley, *Man the Tool Maker* (Univ. of Chicago Press, © 1949, 1961 by the Trustees of the British Museum. All rights resrved.), Fig. 39. Reproduced by permission of the Univ. of Chicago Press and the Trustees of the British Museum (Natural History). **6.6** From F. Hole et al., *Prehistory and Human Ecology of the Deh Luran Plain*, Mem. of the Mus. of Anthropology, No. 1 (Univ. of Michigan, 1969), Fig. 115. Reproduced by permission of Mus. of Anthropology Publications. **6.7** British School of Archaeology, Jerusalem. **6.9** State Antiquities Organization, Baghdad. **6.10** Peter Dorrell and Stuart Laidlaw. **6.11** R. S. Peabody Foundation for Archaeology.

6.12 From B. Fagan, *People of the Earth,* 2nd ed. (Little, Brown, 1977). Reproduced by permission of Little, Brown, & Co., Inc. **6.13** From *Cambridge Guide to Prehistoric Man.* Courtesy of Diagram Visual Information Limited. **6.14** From *Cambridge Guide to Prehistoric Man.* Courtesy of Diagram Visual Information Limited.

7.1 Photo, "Monou-teri men shooting at a wooden dummy," from *Yanomamo: The Fierce People* by Napoleon Chagnon, copyright © 1968 by Holt, Rinehart and Winston, Inc., reproduced by permission of the publisher. **7.2** Irven DeVore/Anthro-Photo. **7.3** From K. V. Flannery, ed., *The Early Mesoamerican Village*, Fig. 6.5, 1976, Academic Press. Reproduced by permission of Academic Press, Inc. **7.4** After Wright, 1977. **7.5** From J. Friedman and M. J. Rowlands, eds., *The Evolution of Social Systems*, Duckworth. **7.6** Compiled and drawn by M. Lehner.

8.3 British School of Archaeology in Jerusalem. **8.4** British School of Archaeology in Jerusalem. **8.6** From J. Mellaart, *Catal Huyuk* (McGraw-Hill, 1967), pp. 118, 120, 125. © 1967 Thames and Hudson Ltd. Reproduced by permission of J. Mellaart. **8.7** From J. Mellaart, *The Neolithic of the Near East* (Charles Scribner's Sons, 1975), Figs. 97, 91, 93, 94. © 1975 Thames and Hudson Ltd., London. Reproduced by permission of Charles Scribner's Sons and Thames and Hudson. **8.8** From J. Mellaart, *The Neolithic of the Near East* (Charles Scribner's Sons, 1975), Figs. 97, 91, 93, 94. © 1975 Thames and Hudson Ltd., London. Reproduced by permission of Charles Scribner's Sons and Thames and Hudson. **8.10** From G. A. Johnson, "Locational Analysis and the Investigation of Uruk Local Exchange Systems," in G. A. Sabloff, ed., *Ancient Civilization and Trade* (Univ. of New Mexico Press, 1975), Fig. 31. Reproduced by permission of the School of American Research, Sante Fe. **8.11** State Antiquities Organization, Baghdad. **8.12** From S. N. Kramer, "The Sumerians," *Scientific American*, Oct. 1957, p. 76. Copyright © 1957 by Scientific American, Inc. All rights reserved. **8.13 (top)** Hirmer Fotoarchiv, Munich. **8.13 (bottom)** University Museum, Philadelphia. **8.14** The Oriental Museum, Univ. of Chicago. **8.16** Georg Gerster/Comstock. **8.17** From Robert McC. Adams, *Heartland of Cities*, Fig. 17, 1981, University of Chicago Press. **8.18** From Charles Redman, *The Rise of Civilization*, Fig. 7.7, W. H. Freeman and Company.

9.3 Munich East Delta Expedition, D. Wildung. **9.4** Hirmer Fotoarchiv, Munich. **9.5** TWA. **9.6** From James E. Harris and Edward F. Wente, eds., *An X Ray Atlas of the Royal Mummies*, University of Chicago Press. **9.7** Mark Lehner. **9.8** From *Ancient Egypt: A Social History* by David O'Connor, © 1983 Cambridge University Press. Reproduced by permission of the publisher. **9.9** From the collection of Peter B. Rathbone. **9.10** From A. Wolinski adn C. M. Sheikholeslami, *The Culture of Ancient Egypt* (Univ. of Washington Continuing Education, 1978), Study Guide, p. 18. **9.11** and **9.12** Victor R. Boswell, Jr., © National Geographic Society. From *Ancient Egypt: Discovering Its Splendors*, 1978.

10.1 Chronology of SA Prehistory. **10.2** After Dales, 1966. **10.3** Josephine Powell. **10.4** Jan Fairservis. **10.5** James P. Blair, © National Geographic Society, from *Adventures of Archaeology.* Courtesy of the Department of Archaeology, Pakistan. **10.6** National Museum, New Delhi. **10.7** University Museum, Philadelphia.

11.2 From K. C. Chang, "In search of China's beginnings: New light on an old civilization," *American Scientist*, Vol. 69, pp. 151, 153. **11.3** From W. Watson, *Ancient China* (New York Graphic Society, 1974), Fig. 12. **11.5** From K. C. Chang, *The Archaeology of Ancient China*, 4/ed., Yale University Press. Courtesy of K. C. Chang and Yale University Press. © Yale University Press. **11.6** Judith M. Triestman, courtesy Doubleday & Co. **11.8** Institute of Archaeology, Academia Sinica, Peking. **11.9** William Thompson. **11.10** Courtesy of China Pictorial.

12.2 Reproduced from *Cambridge Encyclopedia of Archaeology,* © Sceptre Books, a Division of Time Life International Ltd., London. **12.3** After Evans. **12.4** From Sir Arthur Evans, *Place of Minos at Knossos*, Vol. III, 1921. Reproduced by permission of William and James, London. **12.5** Courtesy of Deutsches Archaeologisches Institut. **12.6** Reproduced from *Cambridge Encyclopedia of Archaeology,* © Sceptre Books, a Division of Time Life International Ltd., London. **12.7** Aerofilms, Ltd., London. **12.8** Hajo Hayen. **12.9** From M. Shinnie, *Ancient African Kingdom*, 1965, St. Martin's Press. Reproduced by permission of St. Martin's Press, Inc. **12.10** Reproduced from *Cambridge Encyclopedia of Archaeology,* © Sceptre Books, a Division of Time Life International Ltd., London.

13.1 Reproduced by permission of Equinox (Oxford) Ltd, Oxford, England and first published in *Atlas of Ancient America* (Facts on File Inc./New York and Phaidon Press/Oxford). **13.3** American Museum of Natural History. **13.5** Library of Congress. **13.6** Michael D. Coe. **13.7** Reproduced by permission of Equinox (Oxford) Ltd, Oxford, England and first published in *Atlas of Ancient America* (Facts on File Inc./New York and Phaidon Press/Oxford). **13.8** The Metropolitan Museum of Art, New York, Michael C. Rockefeller Mem. Coll. of Primitive Art. **13.9** Michael D. Coe. **13.10** Robert Drennan. **13.11** Jeffrey House. **13.12** Mexican National Tourist Council. **13.13** Jeffrey House. **13.14** and **13.15** From M. D. Coe *The Maya* (Praeger, 1966), Figs. 48 and 8. Reproduced by permission of M. D. Coe. **13.17** and **13.18** American Museum of Natural History. **13.19** Lee Boltin.

14.1 Reproduced by permission of Equinox (Oxford) Ltd, Oxford, England and first published in *Atlas of Ancient America* (Facts on File Inc./New York and Phaidon Press/Oxford). **14.2** *Cultures of Ancient Peru* (Smithsonian Institution Press, © 1974), Fig. 54. Reproduced by permission of the Smithsonian Institution Press. **14.3** and **14.4** Ferdinand Anton. **14.5** Marilyn Bridges. **14.6** From L. G. Lumbreras, *The Peoples and Cultures of Ancient Peru* (Smithsonian Institution Press, © 1974), Fig. 218. Reproduced by permission of the Smithsonian Institution Press. **14.7** National Museum of Natural History, Smithsonian Institution.

15.1 Reproduced by permission of Equinox (Oxford) Ltd, Oxford, England and first published in *Atlas of Ancient America* (Facts on File Inc./New York and Phaidon Press/Oxford). **15.2** From R. I. Ford, "Northeastern Archaeology: Past and Future Directions," *Annual Review of Anthropology*, 3, 1974, © 1974 by Annual Reviews, Inc. Reproduced by permission of Annual Reviews, Inc., and R. I. Ford. **15.3** Reproduced by permission of Equinox (Oxford) Ltd, Oxford, England and first published in *Atlas of Ancient America* (Facts on File

INDEX

Abri Vaufrey, 155
Achaemenids, 357
Acheulian tools, 142, 150, 153
Acosta, José de, 196
Adams, Robert McC., 126, 323–24, 338, 340, 351, 360–61
Adena culture, 565–66
Afontova, Gora, site at, 211
Agriculture, defined, 229
'Ain Ghazal, 326
Ain Mallaha, 246
Akhenaten, 389–91
Akkadians, 355–56
Alexander, Richard, 93
Alexander the Great, 392
Ali Kosh, 248, 334
Altamira Cave, 180–81
Alto Salaverry, 544
Ambrona, 149
Ammerman, A. J., 250, 309
Anasazi culture, 584–87
An-yang, 441
Arago skull, 154
Archaeometry, 56
Architecture, 243–44, 287–88, 336–37, 343–45, 384, 413–14, 441–42, 499–500
Ardrey, Robert, 105, 298
Arens, W., 147
Argissa-Maghula, 249
Arnold, Matthew, 24–25
Art, in Pleistocene Europe, 180–84
Artifact, defined, 39
Aryans, 420
Assyria, 356
Atahualpa, 550
Athens, J., 293–94
Atlantis, 197
Atlatl, 176, 231
Australopithecus africanus, 80
Australopithecus robustus, 113
Ayacucho, 537
Ayllu, 546
Aztec culture, 476–77, 512–19

Baboons, 103–4
Babylon, 318, 356
Babylonian Empire, 356
Ban Chiang, 439
Bandkeramik pottery, 462
Bands, cultural, defined, 281–82
Bantu, 467–68
Barker, Graeme, 250, 461–62
Barley, domestication of, 234–38, 268
Battle Ax People, 463–64
Beaker culture, 463–64
Beadle, George, 254
Beans, domestication of, 255
Belzoni, Giovanni, 3–4, 374
Bering Land Bridge, 200–204, 219
Beringia, 199–204, 215
Bible, 3, 196, 246, 257, 265, 318, 344, 354–55

Bilzingsleben, 155
Binford, Lewis: analysis of bones, 122, 144, 146–7, 154–55; Edge-zone hypothesis, 261–62; formation of cultural laws, 30–31; Mousterian tool kits, 161–62
Binford, Sally, 161–62
Bintliff, John, 599
Bipedalism, 99, 106–7, 112
Bir Kiseiba, 376
Birdsell, J. B., 121
Black, Henry, 80
Blanton, Richard, 495–97
Blumenschine, R., 119
Bogucki, Peter, 460
Book of Mormon, 196
Border Cave, 171
Bordes, François, 159–62
Boule, M., 157
Bow and arrow, first uses of, 176, 231
Brace, C. Loring, 157, 165
Braidwood, Robert, 260
Braun, David, 50, 563, 570
Brewer, Douglas, 57–58
Bricker, H., 172
Broad-spectrum revolution, 239
Brumfiel, Elizabeth, 286, 307–8, 501–2
Bryan, Alan, 216
Bufo marinus, 484
Buto, 380
Butzer, Karl, 31, 387

Cahokia, 571–72
Cajamarca, 544, 550
Calendrical systems, 183, 487–88, 508–9
Calpulli, 515
Cann, Rebecca, 166–68
Cannibalism, 105, 147, 164, 518
Canyon de Chelly, 587
Carbon-14 dating, 58–62, 209, 214
Carneiro, Robert, 298–301, 307, 472, 551
Casas Grandes, 584
Çatal Hüyük, 249, 326–30, 454
Caton-Thompson, Gertrude, 68, 377
Cattle, domestication of, 248–49
Cavalli-Sforza, L. L., 250, 309
Cerro Narrio cultures, 257
Cerro Sechin, 538
Chaco Canyon, 587
Chalcatzingo, 488
Champollion, J.-F., 397–98
Chan-chan, 545–46
Chang, K. C., 429–30, 433, 436, 444, 448–49
Chapin Mesa, 587
Chatelperronian industry, 173
Chavín de Huantar, 538–41
Chengchou, 439
Chiapa de Corzo, 504
Chica, 546
Chichen Itza, 512
Chichimec, 512
Chiefdoms, defined, 283–84

609

Chilca, 534–35
Chimpanzees, 104–6
Chimú culture, 545
Chinampas, 515
Cholula, 512
Chou culture, 444–47
Chronometric dating, 57–62
Chuño, 546
Chuquitanta. *See* El Paraiso
Clactonian, 153
Class conflict, and cultural complexity, 302–4, 352
Classification, 52–56
Close, Angela, 376
Clovis points, 200, 208, 211, 215–17
Cochise culture, 579
Codices, Mayan, 507
Coe, Michael D., 488
Cohen, Mark, 263–64
Columbus, Christopher, 196
Combe Grenal Valley, 159, 161–62
Commodity sealings, 341, 344
Condorcet, Marquis de, 19
Confucius, 445
Conkey, Margaret, 182, 184
Cortez, Hernan, 476–77, 519
Cowgill, George, 294, 501
Crelin, E., 162
Crete, early civilizations of, 456
Cro-Magnon man, 172
Cronin, J. E., 126
Crook, John H., 101
Cuicuilco, 491–92, 500
Cultural complexity, defined, 277–79, 289
Cultural processes, explanation of, 13–14, 29–34
Cultural reconstruction, 7–10
Cultural relativism, 280–81
Culture history, 10–13
Cuneiform, 346–47
Cuvier, Georges, 19
Cuzco, 546, 549

Dancey, William, 570
Dart, Raymond, 80
Darwin, Charles R., 21–26, 292, 603
Dating techniques, 56–64; dendrochronology, 57–58; potassium-argon, 62; radiocarbon, 58–62; relative, 63–64; seriation method, 63–64; stratification, 46–47; thermoluminescence dating, 62
Daughtery, Richard, 48
Day, Michael, 165–66
De Perthes, Boucher, 78
Decision making, and cultural evolution, 285
DeVore, Irven, 88, 99, 111–12
Diakonov, I., 302–3
Dikov, N., 211
Dillehay, Tom, 212
Djoser, 382
DK I, 117, 123
DNA, 84–86, 94, 166–68

Dog, domestication of, 240
Dolni Vestonice, 178
Domestication: defined, 226–27, 561; ecological background of, 234–38
Dominance hierarchy, 103–4
Dowson, T. A., 183
Drennan, Dick, 63, 284
Dry Creek, site at, 217
Dubois, Eugene, 79–80
Dunnell, Robert C., 309
Dyuktai Cave, 211
Dzibilchaltun, 502, 504

Earle, Timothy, 283, 286, 307–8, 490
Easterbrook, Gregg, 604
Edge-zone hypothesis, 261–62
Edwards, I. E. S., 384
Egypt: art and thought in, 392–97; early agriculture in, 376–78; ecology of, 373–74. *See also* Hieroglyphs; Mummification; Pyramids
El Niño, 532, 536
El Paraiso, 535–36
Elamites, 357
Eldredge, N., 126
Empire, defined, 285–86
Encephalization, ratio of, 92–93, 112
Engels, Friedrich, 301–2
Enlightenment, The, 17–19, 292
Environmental circumscription, 298–301, 472, 551
Erh-li-kang, 437–41
Erh-li-t'ou, 437
Eridu, 334–35, 344
Eskimos, 197–98, 202, 207
Estrus, 102–3, 105
Ethnoarchaeology, 66–67, 119–21, 281
Evolution, biological and cultural, 15, 19–26, 83–89
Excavation, 44–47
Eyzies, Les, 172

Factor analysis, 161
Faunal analysis, 52
Fayyum, 67–70, 377, 454
Feature, defined, 40
Feinman, G. M., 583
Feldman, M., 309
Fertile Crescent, 234–35, 319
Figgins, J. D., 200
Fisher, Elizabeth, 183
Fladmark, K., 203–4
Flannery, Kent V.: on domestication, 234, 252, 260, 262; on sedentarism, 242–45
Fleure, H., 260
FLK I, 117
Folsom points, 198–200, 217–18
Formative era, in Mesoamerica, 480–95
Fowler, Melvin, 563
Franklin, Benjamin, 18
Fremont culture, 587
Fried, Morton, 281
Friedman, J., 304–5
Frisch, Rose, 102

Fuhlrott, Johann Karl, 78
Funnel Beaker People, 460

Galileo, 18, 24
Geoarchaeology, 52
Giza, 383–85
Goats, domestication of, 240–42
Golding, William, 112
Gould, Stephen J., 25, 126, 254–55
Grayson, Donald K., 218–19
Great Bath (Mohenjo-daro), 414
Great Chain of Being, 16–17
Great Wall of China, 447
Greek world-view, 15–17, 603
Greenberg, Joseph, 207
Gregg, Susan, 31, 463
Grove, David, 488
Guinea pig, domestication of, 257
Guitarrero Cave, 257, 533

Halafian culture, 332–33
Hallstatt culture, 465–66
Halverson, John, 184
Hamilton, M. E., 88
Hammurabi, 356
Han culture, 446–47
Harappa, 411–12, 414
Harappan civilization, 407–24
Harlan, J., 260–61
Harris, Marvin, 264–66, 305–7, 503
Hasan Dag, volcano at, 327
Hassan, Fekri, 31, 69, 377, 380
Hayden, Brian, 48, 266–67
Hemudu, 434
Henry, Donald, 239
Hierakonpolis, 379–80
Hierarchy theory, 342–43
Hieroglyphs, 396–99
Higham, Charles, 470
Hittites, 457, 465
Hoabinihian tool assemblage, 180
Hoffman, Michael, 67, 379
Hohokam culture, 579–82
Hole, Frank, 260, 332, 337
Holloway, Ralph, 156
Homo erectus, 80, 124, 137–58
Homo sapiens neanderthalensis, 78–81, 157–75
Homo sapiens sapiens, 100, 136–37, 156–80, 206
Hooge, Paul, 570
Hopewell culture, 566–70
Houpeville, 161
Houston, Stephen, 505
Howell, F. Clark, 149
Huaca Prieta, 536
Huari, 544–45
Hyksos, 388–89

Ifrah, George, 349
Igbo-Ukwu, 468
Iltis, Hugh, 255
Imhotep, 382
Inanna, 350
Inca culture, 545–50

Indo-Europeans, 421
Indus River, cultural ecology of, 409
Infanticide, 121
Iron Age, of Europe, 465–66
Irrigation, and cultural complexity, 295–98, 358–59, 399
Isaac, G. L., 143
Isernia, 149
Isin, 356
Itzkoff, Seymour, 102
Izapa, 503

Jabrud Rock Shelter, 161
Japan, 180, 251, 469–70
Jarmo, 260, 329–30
Java, fossil finds from, 79–80
Jebel Sahaba, 375–76
Jefferson, Thomas, 197–98
Jelinek, Arthur, 171
Jericho, 246, 249, 325–26
Jerison, Henry, 92–93
Jett, Stephen, 583
Johanson, Donald, 98
Johnson, Gregory, 285, 336–42, 360, 422
Jomon culture, 469

Kalahari Bushmen, 119–21, 280–81
Kalibangan, 411–12, 414
Kaminaljuyú, 504
Kebara Cave, 171
Kebaran cultures, 238–39
Keene, Arthur, 31, 231
Khufu, 383
Kirch, Patrick, 470
Kiva, 585
Klasies River Mouth Cave, 171
Klein, Richard, 173
Knorosov, Yuri, 508
Knossos, 456
Kom el-Hisn, 386–87
Koobi Fora, 118, 142–43
Kostenki-Bershevo culture, 178–79
Kot Diji, 411
Kotosh, 537
Krapina, 164

La Chapelle-aux-Saints, 163
La Ferrassie, 163
La Ponga, 257
La Venta, 484–87
Laetoli, 40, 96, 98, 171
Laguna de los Cerros, 487
Lake Ushki, 211
Lamarck, Jean, 19
Lane, Mary Ellen, 68
Language, origins of, 162–63
Larsa, 356
Las Vegas, 257
Las Victorias, 488
Lascaux Cave, 181–82
Lattimore, Owen, 448
Leakey, Louis, 54, 117
Leakey, Mary, 96

Leakey, Richard, 115
LeBlanc, Stephen, 63
Lee, Richard, 280
Legge, A. J., 239
Lehner, Mark, 60–61, 383–85
Leroi-Gourhan, Andre, 183–84
Leroi-Gourhan, Arlette, 163
Lewis-Williams, J. D., 183
Libby, Willard, 58
Lieberman, P., 162
Lightfoot, K. G., 583
Llama, domestication of, 257
Loess, 430
Loss of fitness, and domestication, 227
Lothal, 412, 415
Lovejoy, Owen, 106–9
Lowther, Gordon, 119
Lucretius, 19
"Lucy," 98, 112–13
Lumley, Henri de, 7–8
Lungshan culture, 434–36
Lyell, Charles, 20

Maadi, 380
Ma'at, 394–96
McIntosh, S. K., and R. J., 467–68
MacNeish, Richard, 213, 216, 253
Maize, domestication of, 252–55, 257
Mal-ta, site at, 211
Malthus, Thomas, 20
Mangelsdorf, Paul, 252–53
Marx, Karl, 26–28, 301–2, 602
Marxism, 26–28
Marxist analyses of cultural evolution, 26–28,
   265, 301–5, 352, 362, 399–400
Materialist determinism, 26–28
Mayan calendar, 507–9
Mayan culture: decline of, 509–11; history,
   502–11; physical geography of, 502–4
Mayan writing, 507–8
Meadowcroft Rock Shelter, 209, 214–15
Mellaart, James, 326–29
Memphis, 388
Mendel, Gregor, 23
Menes, 381
Mesa Verde, 587
Mesoamerica, cultural ecology of, 477–80
Mesopotamia. *See* Southwest Asia
Middle Kingdom (in Egypt), 387–89
Mill, John Stuart, 20
Miller, Daniel, 410–13, 417, 423
Miller, Mary Ellen, 505
Millet, domestication of, 250–51
Minoan culture, 456
Minshat Abu Omar, 380
Mississippian culture, 570–77
Mochica culture, 542–43
Mogollon culture, 582–84
Mohenjo-daro, 412–15, 419
Monte Albán, 495–97
Monte Cicero, 163
Monte Verde, 209, 212–13, 220

Morgan, Lewis Henry, 26
Morris, Desmond, 105, 298
Mortuary evidence, 163, 288–89, 325, 333,
   352–54, 375, 442, 446, 488, 568
Moseley, Michael, 535
Moundville, 572
Mousterian tools, 159–62
Moyle, Peter, 583
Mugharet es-Shubbabiq Cave, 161
Mummification, 393
Murphy Site, 569
Mycenaean culture, 456–57

Nabta, 376
Naqada, 380
Narmer Palette, 378–81
Natchez, 574
Native Americans: arrival of, 198–215; physic-
   al features of, 206–8
Natufian culture, 238–39, 242–46
Natural-habitat hypothesis, 260
Nazca Valley, 542–43
Neander Valley, 78, 157
Neandertals. *See Homo sapiens neanderthalensis*
Neolithic, 431–34
Neve David, site at, 238
New Kingdom (in Egypt), 389–92
Nile River, cultural ecology of, 373–74
Niles, Susan, 549
Nissen, Hans, 338
Non Chai, 470

Oasis hypothesis, 259–60
Oaxaca, 255, 492–97
Oaxaca Barrio, 500
O'Brien, Michael, 561–62
Occupational specialization, 262
O'Connell, J., 120
Odell, George, 568
Old Crow, site at, 211–12
Old Kingdom (in Egypt), 380–87
Olduvai Gorge, 116–19, 142, 144
Olduwan tools, 118–19
Olmec culture, 482–90, 493–94
Olorgesailie, 142–44
Omo, 171
Optimal foraging theory, 230–31

Paleontology, 90–99
Paloma, 534, 536
P'an-keng, 441
Pan-p'o-ts'un, 432–33
Parsons, Jeffrey, 63, 497
Patayan culture, 587
Peake, H., 260
Pebble-tool complex, 209
Pedra Furada, 213
Peruvian cultures: ecological setting, 531–32;
   European conquest of, 529–31, 549–50; his-
   tory, 532–49
Petrie, Sir Flinders, 374–75, 378
Pig, domestication of, 249, 264–66

Pikimachay Cave, 213, 533
Pizarro, Francisco, 549–50
Pleistocene, 82–83, 203, 231
Pleistocene extinctions, 218–20
Plog, Fred, 584
Population growth: and agriculture, 262–64; and cultural complexity, 294, 298–301, 359–60, 399, 423, 551
Possehl, Gregory, 415, 420
Post-processualists, 598
Potato, domestication of, 257
Poverty Point, 563–64
Pozorski, Sheila, and Tom, 536
Pre-projectile point stage, 209
Preservation, 40–42
Price, Douglas, 460
Primates, evolution of, 99–116
Pueblo Bonito, 587
Puleston, Dennis, 503
Pulque, 518
Pyramids: Egyptian, 60–61, 383–85, 401, 498; Mesoamerican, 498

Qafzeh, 169–71
Qin Dynasty, 446
Quest for Fire, 115
Quilter, Jeff, 536
Quinoa, domestication of, 257
Quipu, 349, 546

Raikes, R., 419–20
Ramapithecus, 94
Ramenofsky, Ann, 576
Rameses III, 391–92
Rank-size rule, 342–43
Rathje, William, 40, 510
Redman, Charles, 330, 363–64
Reindeer, Pleistocene cultural uses of, 175–76
Remote sensing, 64–65
Renfrew, Colin, 184, 421, 462
Resource-stress model, 266–67
Rice, domestication of, 250–51
Rick, John, 532
Rig Veda, 420
Rindos, David, 227–29, 267, 562
Ritchie Site, 216–17
Rowlands, M. J., 304–5
Rowley-Conwy, P. A., 239

Saggs, H. W. F., 349, 399
Sahlins, Marshall D., 110–11, 265, 280
Saint-Cesare, 173
Salmon, Merilee, 596–97
Samarran culture, 332–33
Sampling, 55–56
San José Mogote, 493–94
San Lorenzo, 483
Sandalja site, 177
Sanders, William, 497
Saqqara, 383
Sargon of Akkad, 355
Scapulimancy, 436

Schacht, Robert, 301
Schaller, George, 119
Scheduling, mechanism of, 230, 237
Schiffer, Michael, 597
Schild, Romauld, 376
Schliemann, Heinrich, 4–5
Schmandt-Besserat, Denise, 344
Schwalbe, Gustave, 80
Scientific method, 17–18
Seals, stamp and cylinder, 341, 344, 416–18
Sedentary communities, defined, 229–30
Service, Elman, 280
Seshonk I, 392
Settlement patterns, 290–91, 415, 422
Sexual dimorphism, 100–103
Sexual selection, 101–3, 109–10
Shaffer, J., 422
Shang culture, 437–44
Shanidar Cave, 163
Shanks, M., 599
Shea, John, 171
Sheep, domestication of, 240–42
Shenan, Steven, 460
Shipman, Pat, 143
Siberia, 202, 210–11, 215
Silverman, Helaine, 542
Simulation modeling, 31
Singer, R., 171
Site, defined, 40
Skhul Cave, 170–71
Smith, Bruce, 561
Smith, Fred, 171
Smith, Michael, 515
Smith, William "Strata," 20
Social taxonomies, 281–86
Soffer, Olga, 177
Solecki, Ralph, 163
Solomon, 392
Sonviian tool assemblage, 180
Southwest Asia: archaeological record of, 324–58; physical geography of, 319–24
Soybean, domestication of, 251
Spencer, Herbert, 20–21
Squash, domestication of, 255
State, defined, 285
Stein, Julie, 41
Steinheim, 154–55, 166
Stellmoor site, 177
Steward, Julian, 29, 280–81, 292
Stocker, T., 536
Stonehenge, 464
Storey, Rebecca, 500
Streuve, V., 302
Stringer, Chris, 152–53
Suhm, P. F., 19
Sumerian civilization, 345–55, 358
Sumerian language, 346–49
Sunflower, domestication of, 561
Susa, 319, 356
Susiana Plain, 336–37, 340–42, 358, 360
Swanscombe, 153–55, 166
Swarts, ruins, 582

Swidden agriculture, 482–83
Symbols, 77
Symons, Donald, 109–10

Tabun Cave, 46, 170–71
Tahuantinsuyu, 546
Taima-taima, 216
Tainter, Joseph, 568
Tamaulipas, 255
Tambo Viejo, 542
Tanner, N., 105
Taung site, 80
Tehuacán, 257
Teleki, Geza, 99, 106
Tell, defined, 12
Tell Abu Hureya, 239–40
Tell as-Sawwan, 333–34
Tell el-Amarna, 391
Tell Mureybit, 246
Tene, La, 466
Tenochtitlan, 513–15
Teosinte, 253–55
Teotihuacan, 498–502, 504–5
Terra Amata, 150
Teshik-Tash, 163
Testart, Alan, 232–34
Thebes, 389
Thomsen, Christian, 19
Thutmosis III, 389
Tiahuanaco culture, 544–45
Tikal, 502, 504–5
Tilley, C., 599
Tlapacoya, 488
Tlatilco, 488, 491
Tolstoy, Leo, 25
Tolstoy, Paul, 498
Toltec culture, 510, 512
Tomaltepec, 494
Tooby, John, 88, 99, 111–12, 123
Tooth-size reduction, 165
Torralba, 149–50
Toth, Nicholas, 118–19
Townsend, Patricia, 121
Toynbee, Arnold, 320
Trail Creek Cave, 217
Tres Zapotes, 487
Tribes, defined, 282–83
Trigger, Bruce, 31
Trinkhaus, Eric, 152–53, 159, 164, 169, 173
Tula, 510, 512
Turner, C., 206–7
Tutankhamen, 391
Twain, Mark, 596
2001: A Space Odyssey, 75

Uaxactun, 504–5
Ubaid period, 337–38
'Ubeidiya, 144
Upper Paleolithic, 151–86
Ur death pit, 352–54
Urbanism, 400–401
Ur-Nammu, 355

Uruk, city of, 338–45
Uruk culture, 338–42

Valdivia culture, 257, 596–97
Valladas, Helene, 169
Valley of Mexico, 490–92
Vallois, H., 176
Vallonet Cave, 147–48
Varna, 463
Vaughn, Patrick, 48
Vegeculture, 251
"Venus" figurines, 179
Vertesszölös, 151
Virchow, Rudolf, 78

Wadi Kubbaniya, 376
Wadjah, 180
Warfare, and cultural evolution, 298–301, 306, 359–60, 399, 543, 551, 574
Wattle-and-daub construction, 481
Wave of Advance model, 250
Webster, David, 300
Weil, Simone, 595
Weisner, Polly, 48
Welles, Orson, 298
Wendorf, Fred, 375–77
Were-jaguar motif, 488
Whalen, Michael, 583
Wheat, domestication of, 235–38, 268
Wheeler, Sir Mortimer, 47, 421
White, Leslie A., 30, 77, 287
Whittle, Alasdair, 462
Wilson, Alan, 168
Wilson, David, 536
Wilson, John, 393
Winter, Marc, 494
Wittfogel, Karl, 295–97, 307
Wobst, H. Martin, 31
Wolpoff, M., 169
Woodland period, 565–70
Woolley, Sir Leonard, 47, 352–54
Worsae, Jens Jacob, 19
Wright, Henry, 285, 335–36, 340–42, 360, 410
Writing, origins of: in China, 437, 442; in Egypt, 397–99; in Greece, 348–49; in the Indus Valley, 416–17; in Mesoamerica, 507–8; in Mesopotamia, 341, 346–49
Wymer, J., 171

Xe, 504

Yang-shao culture, 431–35
Yerkes, Richard, 560, 565, 570
Yin phase, 441–44
Young, T. Cuyler, 32
Yuan K'ang, 19

Zagarell, Alan, 303, 362
Zawi Chemi Shanidar, 242
Zhoukoudian, 80–81, 145–47, 153, 179
Ziggurats, 344
Zinjanthropus, 117, 122